Arthur Hollick

The Later Extinct Floras of North America

Arthur Hollick

The Later Extinct Floras of North America

ISBN/EAN: 9783337271398

Printed in Europe, USA, Canada, Australia, Japan

Cover: Foto ©berggeist007 / pixelio.de

More available books at **www.hansebooks.com**

DEPARTMENT OF THE INTERIOR

MONOGRAPHS

OF THE

UNITED STATES GEOLOGICAL SURVEY

VOLUME XXXV

WASHINGTON
GOVERNMENT PRINTING OFFICE
1898

UNITED STATES GEOLOGICAL SURVEY

CHARLES D. WALCOTT, DIRECTOR

THE

LATER EXTINCT FLORAS OF NORTH AMERICA

BY

JOHN STRONG NEWBERRY

A POSTHUMOUS WORK

EDITED BY

ARTHUR HOLLICK

WASHINGTON
GOVERNMENT PRINTING OFFICE
1898

CONTENTS.

ILLUSTRATIONS.

EDITOR'S PREFACE.

This volume has been prepared, in its present form, under somewhat peculiar circumstances. An edition of twenty-five plates, without text, was printed about 1871. These were issued under the title, Illustrations of Cretaceous and Tertiary Plants of the Western Territories of the United States, in 1878. Subsequently a revised edition of these and forty-three additional plates was published, but was withheld from distribution, awaiting the completion of the text by Dr. Newberry. His sickness and subsequent death stopped further progress on the work, and much that he had accomplished went for naught. Upon two sets of the plates manuscript names for the figures were placed by him. One of these sets is in the possession of Dr. Lester F. Ward, and the other was in Dr. Newberry's library, which came into the possession of the Geological Department of Columbia University after his death. From these sources I was enabled to obtain a more or less complete list of the names which it was the evident intention of the author to give to the figures. Most of these were found to refer to descriptions previously published by Dr. Newberry,[1] or to species of other writers, notably Leo Lesquereux and Oswald Heer. A number, however, were apparently not referable to any published descriptions, and it became necessary to examine Dr. Newberry's manuscript, in order to connect these names, if possible, with his notes. A thorough search was also made

[1] Descriptions of fossil plants collected by Mr. George Gibbs, geologist to the United States Northwest Boundary Commission, under Mr. Archibald Campbell, United States Commissioner: Boston Journ. Nat. Hist., Vol. VII (1863), pp. 506–524.

Notes on the later extinct floras of North America, with descriptions of some new species of fossil plants from the Cretaceous and Tertiary strata: Ann. N. Y. Lyc. Nat. Hist., Vol. IX (April, 1868), pp. 1–76.

Brief descriptions of fossil plants, chiefly Tertiary, from western North America: Proc. U. S. Nat. Mus., Vol. V, 1882 [February and March, 1883], pp. 502–514.

for the type specimens, and all labels upon these were noted and compared with the names upon the plates and with the manuscript notes. By these means it has been possible to ascertain, in nearly every case, the name which Dr. Newberry intended to use.

Those who have had access to the plates upon which he placed his names have always endeavored to preserve these names by referring, whenever occasion demanded it, to "Newb. MSS. undistributed plates, U. S. Geol. Surv." This, however, could not be recognized as publication, and in the lapse of time some of the names were used for other species and under the rule of priority could no longer be retained for those of Dr. Newberry. It is also to be noted that names of such species as existed in manuscript only were liable to be superseded by published names of other authors, and under such circumstances Dr. Newberry's names would have to be dropped and the others substituted. One instance in this connection is *Sabal occidentalis*, Newb. MSS., which became *S. imperialis* Dn.

In arranging the text it has been thought desirable to quote Dr. Newberry's original published description in each instance, followed by his subsequent manuscript notes, whenever such could be obtained. In case a manuscript description was found for any unpublished species it has been included in full. In the event of no published or manuscript description having been found for any species, such name or memorandum as could be found in connection with the specimen was adopted and a note to that effect included over the editor's initials. In the case of but one figure could absolutely no clue be obtained as to its probable reference by Dr. Newberry.

In regard to the volume entitled Illustrations of Cretaceous and Tertiary Plants, etc., Dr. Newberry would never acknowledge any responsibility, the names accompanying the plates having been supplied by Lesquereux, at the request of Dr. F. V. Hayden, then director of the United States Geological Survey, without Dr. Newberry's sanction, and it was evidently his intention and desire to correct in the present volume several errors which appear in that one. In each instance, therefore, in which the same figure appears in both volumes the fact is noted, with any correction which was found necessary.

The work is confessedly incomplete in certain respects, due to loss of type specimens and absence or incompleteness of manuscript, and many of Dr Newberry's reasonings and conclusions would probably not be included if revised by him at the present time. These must, therefore, be accepted merely as reflecting his opinions at the time when they were written, the editor not feeling himself at liberty to alter them, and thus perhaps making Dr. Newberry appear to say what he might not have intended to say. It contains so much of value, however, and its contents are referred to so frequently, that the publication has become necessary both as a matter of scientific record and for purposes of research.

A. H.

THE LATER EXTINCT FLORAS OF NORTH AMERICA.

By John Strong Newberry.

DESCRIPTIONS OF SPECIES.

CRYPTOGAMIA.

PTERIDOPHYTA.

Order FILICINÆ.

Lygodium Kaulfussi Heer.

Pl. LXII, figs. 1–4.

Fl. Skopau; Beitr. naher Kent. Sachs.-Thuring. Braunkohl, Vol. XVIII (1861), p. 409, Pl. VIII, fig. 21; IX, fig. 1.

Lygodium neuropteroides Lesq. Hayden's Ann. Rept. 1870 [1872], p. 384; Tert. Fl. (1878), p. 61, Pl. V, fig. 4–7; VI, fig. 1.

Dr. C. A. White has collected from the Green River shales a splendid series of the fronds of a Lygodium which is apparently identical with that described by Lesquereux under the above name. These illustrate the growth of the plant far better than those he figures, and some of the more interesting and instructive ones are therefore now figured. Coming all from the same locality, indeed thickly impacted together and having the same nervation, they unquestionably represent a single species, and yet it will be seen that if diversity of form were accepted as affording specific distinctions half a dozen species might be made from them; hence we are taught by them that the fossil species of Lygodium already described are based on too insufficient material, and should have comparatively little weight until confirmed by further evidence. The number of figures now given, however, enable us to define this species in such a way that it is not liable to be mistaken.

As these fronds occur in the rock, the margins seem to be undulated and the lobes considerably curved and twisted. How much of this is due to contraction in drying before they were submerged and how much is natural it is now impossible to say; but specimens from Currant Creek, Oregon, exhibit the same peculiarity, the lobes being sometimes almost fan-shaped, the margins waved or involute, and recalling by their mode of growth the fronds of Marchantia, repeating what is so conspicuous in the Green River shales. We must therefore regard the characters enumerated as normal.

The nervation is in most specimens clearly defined and rather strong. It is crowded as compared with that of some other species, and is confluent along the middle of the lobes, precisely as in Neuropteris, without producing a midrib.

Professor Heer has described and figured in his great work on the plants of the Swiss Tertiaries (Fl. Tert. Helv., Vol. I, p. 42, Pl. XIII, fig. 3, and Vol. III, Pl. CXVII, fig. 25b) a species of Lygodium which evidently closely resembles this; so much so that unless some distinctive characters are furnished by the lobing of the fronds, they are likely to prove identical. Professor Heer names his species *L. acutangulum*, from the nervation, which is identical with that of the Green River specimens, but he describes the frond as three-lobed His specimens are, however, very imperfect, and two or three lobed specimens could be selected from the suite before me which would, taken by themselves, require a description corresponding precisely with that given by Heer.

Among the fronds collected by Dr. White at Green River is one which has much narrower lobes than the others, and it has apparently a finer nervation; but it is unfortunately much weathered, and the details of structure are rendered obscure. A figure is now given of it (Pl. LXII, fig. 2), but I am inclined to regard it as only one of the many forms of one protean species.

Since the above notes were written Messrs. Gardner and Ettingshausen have published their Monograph of the British Eocene Flora, Vol. I, Filices, and on Pl. VII have given a number of figures of *Lygodium Kaulfussi* Heer, with which they identify Lesquereux's species: a conclusion to which he also subscribes. It will be seen, however, by a comparison of Lesquereux's figures with those now given and with those published by Heer and Gardner that the American fern had larger pinnæ with broader and less undulate lobes, which are nearly of the same breadth from base to summit.

Among hundreds of specimens from Green River which I have examined there are very few which have the lobes of the pinnæ as narrow as are represented in the plates and descriptions of the fossil plant, and none which can be compared with the narrower and more undulate forms given by Gardner on Pl. VII, figs. 1 and 4, of Eocene Ferns. However, the nervation is essentially the same, and the fructification which has been recently found presents no obvious points of difference. I am therefore inclined to accept the view of Messrs. Gardner and Ettingshausen that all these so closely resembling fronds of Lygodium found in the later Cretaceous and older Tertiary rocks of Europe and America should be regarded as belonging to one species.

From the coal-bearing rocks of Fletts Creek and Carbonado, Washington, I have a few fronds and fragments of fronds of a species of Lygodium which offer no characters by which they can be distinguished from those found in the Green River group, and it seems to me probable that we have in all these specimens relics of one of those widespread and long-lived species which occur at different geological horizons among both animal and plant remains.

Formation and locality: Tertiary (Green River group). Green River, Wyoming.

ANEMIA PERPLEXA Hollick.[1]

Pl. XV, figs. 1, 1a; XVI, fig. 3; LXIII, figs. 1–4.

Sphenopteris (Asplenium) elongatum Newb. Boston Journ. Nat. Hist., Vol. VII (1863), p. 511.

Asplenium subcretaceum Sap.? Fl. Foss. Sez., Mem. Soc. Geol. France, Ser. II, Vol. VIII (1868), p. 315, Pl. XXIII, fig. 4.

Gymnogramma Haydenii Lesq.? Hayden's Ann. Rept. 1871 [1872], p. 295; Tert. Fl. (1878), p. 59, Pl. V, figs. 1–3.

Anemia subcretacea (Sap.). Gard. and Ett.? Monog. British Eocene Flora, Vol. I, Pt. II (1880), p. 45, Pls. VIII, IX.

"Frond bi- or tri-pinnate; pinnæ lanceolate, or linear, acute; lower ones broadly lanceolate, pinnatifid at base, margins deeply double-toothed,

[1] Under the rules of nomenclature as now accepted the original specific name given to this plant by Dr. Newberry can not be retained, as it is antedated by that of a living species—*Asplenium elongatum* Swartz (1806).

The relationships of the foreign, western, and eastern United States forms are further discussed by Dr. Newberry in his Flora of the Amboy Clays (Mon. U. S. Geol. Surv., Vol. XXVI, pp. 38–42), under the species of Asplenium and Anemia there described.

Dr. Newberry evidently intended to maintain the species now described and figured as distinct, and as the original name is not available I have been obliged to adopt an entirely new one.—A. H.

upper ones narrow lance linear, wedge-shaped at base, summit long-pointed, acute margins coarsely toothed; nervation strongly marked, acute-angled, medial nerve of pinnæ vanishing toward the summit, secondary nerves diverging from this at a very small angle, radiating to the margins, dichotomously forked."

A number of figures are now given of a fern, specimens of which have been collected at Point of Rocks, Wyoming; Golden and Erie, Colorado, and Bellingham Bay and Carbonado, Washington. In general character it so closely resembles *Gymnogramma Haydenii*, figured by Lesquereux (Tert. Fl., Pl. V, figs. 1-3), that it can hardly be considered distinct, but a few minor differences render it possible that we have here only two closely allied species. Lesquereux shows and describes the nervation of his fern as finer and simpler than that represented in our figures; but he states that the nervation is obscure in his specimens, and that it seems to have been buried in the parenchyma. The same is true of the specimens before us, and the distinctness of the nervation is exaggerated in the figures; but it can be plainly made out in some portions of the frond, and is more open and stronger than is shown in Lesquereux's plate. The reference of this plant to Gymnogramma is conjecture only; and the question of its botanical affinities can only be decided when fruiting fronds shall be found. The fossil is a marked one, however, and the figures and descriptions of it will serve a good purpose, whatever generic name may be hereafter given to it.

Previous to the description by Lesquereux (1871) Count Saporta had described (Fl. Foss. Sezanne (1868), p. 315, Pl. II, fig. 4) a very similar fern under the name of *Asplenium subcretaceum*. This was more fully illustrated by Gardner and Ettingshausen (Mon. British Eocene Flora, Vol. I, Pt. II (1880), p. 45, Pls VIII and IX), and called by them *Anemia subcretacea*. Lesquereux, Saporta, and the authors of the British Eocene Flora are agreed in considering the specimens from Wyoming, Sezanne, and Bournemouth as belonging to the same species. The large number of specimens of the fern which I have from Point of Rocks and Puget Sound show that while apparently identical with that figured by Lesquereux (Tert. Fl., p. 59, Pl. V, figs. 1-3), it differs so much from the foreign specimens that we must regard it as at least a strongly marked variety. Some fragments of pinnæ figured by Mr. Gardner—such as those given on Pl. VIII, fig. 1, Pl. IX, figs. 3 and 5—approach closely to the American plant, but we nowhere find here pinnæ with long, linear-notched pinnules which seem to form the most striking characteristic of the foreign fern. Among all my specimens I have nothing which resembles those figured on Pl. VIII, fig. 2, or Pl. IX, figs. 1, 2, 4, of Eocene Ferns.

Lesquereux's specimens were collected by Dr. Hayden on the divide between the headwaters of Snake River and Yellowstone Lake. Those now figured are from Bellingham Bay, Washington; Erie, Colorado, and Point of Rocks, Wyoming. The strata exposed in the last two localities are now generally conceded to be Cretaceous, although Lesquereux has claimed that they are Tertiary, and the discussion which these diverse views have excited has given special value to all new paleontological material from that region. If it should be agreed that all the ferns here associated together represent but a single species, that is no proof that the rocks which contain all of them are at one geological level. Nearly all the widespread species of fossil plants and animals have also considerable vertical range, and the American specimens are so much broader and stronger that they constitute a distinct variety, such as may have lived at a little earlier epoch than the European plants which are regarded as specifically identical with them. The proofs of the Cretaceous age of the Lower Laramie of Colorado and Wyoming, viz, numerous Dinosaurs and Cretaceous mollusks, with the absence of animal or plant remains that are elsewhere found in Tertiary rocks, may be regarded as decisive of this question. Hence we can only say that if the leaf beds of Sezanne be regarded as Tertiary, it does not at all follow that the Laramie group is so simply because it contains a species closely allied to, or a distinct variety of, a fern found in these beds abroad. According to Mr. Gardner, *Anemia subcretacea* occurs at Bournemouth, but we know that the Bournemouth beds are somewhat later than those of Gelinden and Sezanne, and that they are on the horizon of the Fort Union beds of the upper Missouri country.

Count Saporta does not approve Mr. Gardner's transfer of his *Asplenium subcretaceum* to Anemia, and his reasons are quoted by the latter in the memoir already referred to, page 46. It would seem, however, that this question can not be decided without the fructification, and that has not yet been found. This is somewhat remarkable, considering the fact that already thousands of specimens of *Anemia subcretacea* have been collected. If it were a species of Asplenium, it seems hardly possible that the fruit should be always absent, and this fact gives probability to the suggestion of Mr. Gardner that the fruit was borne upon independent fronds or stipes.

Mr. Gardner suggests that *Asplenium Foersteri* Deb. and Ett., described in the Urweltlichen Acrobryen des Kreidegebirges von Aachen und Maestricht, Pl. II, figs. 4, 7, 11, is also closely related to if not identical with *Anemia subcretacea;* but in a recent visit to Aachen I had an opportunity of examining some of Debey's original specimens, and it seemed to me they were very distinct from *A. subcretacea. A. Foersteri* is a thinner, more delicate fern, with few and slender nerves and with pinnæ irregularly lobed or undulate. I have identified this species among the plants from the Amboy clays, many of which also occur at Aachen. The Amboy clays are about on the horizon of the Dakota sandstones, and therefore very much older than the Laramie group.

Formation and locality: Cretaceous (Laramie group). Orcas Island, Bellingham Bay, Washington; Point of Rocks, Wyoming; Erie, Colorado.

ACROSTICHUM HESPERIUM Newb.

Pl. LXI, figs. 2–5.

Proc. U. S. Nat. Mus., Vol. V (March 21, 1883), p. 503.

"Frond large, pinnate; pinnæ linear, 1½ to 2 inches wide, 6 to 12 inches long, rounded at remote extremity, those in lower part of frond rounded or wedge-shaped at base, those above united by the entire base to the rachis and with each other; rachis of frond and midrib of pinnæ strong, smooth, somewhat sinuous; nervation reticulated, lateral nerves numerous, diverging from the midrib at an acute angle, anastomosing to form elongated six-angled areoles; fructification unknown."

This is a large and strong fern, represented in the collections by a number of specimens collected by Mr. C. A. White, which include portions from the lower and upper parts of the frond. In general aspect it much resembles *Acrostichum aureum* of Florida and the West Indies; but in that species the pinnæ are all separate and narrowed at the base, whereas in this plant near the summit of the frond they coalesce, forming a broadly palmated portion. Lesquereux, in his Tertiary Flora, p. 58, Pl. IV, fig. 2, describes a large fern with a somewhat reticulated nervation which he calls *Gymnogramma Gardneri.* The pinnæ must have been about as large and of similar form to those of the fern under consideration, and the nervation is also reticulated; but in Lesquereux's plant the midrib of the pinna is much stronger and is channeled, while the lateral nerves anastomose much less frequently, and it is evident that the specimens represent distinct species. Until the fructification of this fern shall be discovered, its generic relations can not be said to be established. However, the resemblance in nervation and proportions of the frond to Acrostichum is so strong that the reference to that genus seems justifiable.

Mr. J. Starkie Gardner, in his Monograph of the British Eocene Flora, Vol. I, p. 26, figures and describes a large Chrysodium found in the Bagshot beds of Bournemouth, England, which he calls *Chrysodium Lanzæanum*, and which closely resembles that now under consideration. I find hardly any points of difference, except that Mr. Gardner represents the Bournemouth species as having a strong pinnate frond which terminates in a single lanceolate pinna which is drawn down to an acute base; whereas in our species, as will be seen by reference to the figures now published, the frond terminates above in a palmate divergence of the terminal and upper lateral pinnæ, the bases of which all coalesce. It is interesting, however, to find a species so closely allied to this foreign one at nearly the same geological level in this country.

Formation and locality: Tertiary (Green River group). Green River, Wyoming.

PTERIS PENNÆFORMIS Heer. ?

Pl. XLVIII, fig. 5.

Fl. Tert. Helv., Vol. I (1855), p. 38, Pl. XII, figs. 1a–1d.
Pteris pseudopennæformis Lesq.? Tert. Fl. (1878), p. 52, Pl. IV, figs. 3, 4.

Formation and locality: Tertiary (Miocene?). Currant Creek, Oregon.

NOTE.—I have been unable to find any manuscript relating to the above, except brief memoranda on plate and specimen to the names and locality here quoted.—A. H.

PTERIS RUSSELLII Newb.

Pl. LXI, figs. 1, 1a.

Proc. U. S. Nat. Mus., Vol. V (March 21, 1883), p. 503.

"Frond large, pinnate; pinnæ crowded, linear in outline, narrow, long-pointed above, attached to rachis by entire base; decurrent; length,

16 to 20 centimeters; width, 10 millimeters; margins undulate below, irregularly and coarsely toothed above; nervation fine, but distinct; branches all forked, leaving midrib at an angle of about 45 degrees, all twice or three times forked."

Only the upper part of the frond of this fern appears on the specimens examined, but these show a species apparently distinct from any hitherto described. In general form the pinnæ resemble those of *Pteris pennæformis* Heer (Fl. Tert. Helv., Vol. I, p. 38, Pl. XII, figs. 1–1d), and *P. pseudopennæformis* Lesq. (Tert. Fl., p. 52, Pl. IV., figs. 3, 4), but it differs from the first by being a stronger plant, with wider and more coarsely toothed pinnæ, and less simple nervation; from the second, by the same characters and in having the nervation less crowded, the nerve branches issuing at a greater angle, and oftener forked.

Pteris erosa Lesq. (Tert. Fl., p. 53, Pl. IV, fig. 8) has broader pinnæ, of which the margins are set with finer and more numerous teeth.

The species is dedicated to Mr. I. C. Russell, who first collected it, in Vermejo Canyon, New Mexico. It has also been collected at Walsenburg, Florence, and Golden, Colorado.

Formation and locality: Cretaceous (Laramie group). Vermejo Canyon, New Mexico.

ONOCLEA SENSIBILIS FOSSILIS Newb.

Pl. XXIII, fig. 3; XXIV, figs. 1–5.

Onoclea sensibilis, L., Newberry in Ann. N. Y. Lyc. Nat. Hist., Vol. IX (April, 1868), p. 39; Ills. Cret. and Tert. Pl. (1878), Pl. VIII, fig. 1; IX, figs. 1–3.

"Frond pinnate, large; pinnæ, lanceolate in outline, with waved margins, more or less deeply lobed or pinnatifid, connate at their bases, forming a broad wing on the rachis of the frond; nervation strongly marked, more or less reticulated, the nerve of each lobe or pinnule springing from a common trunk, having a dendroid form, with waving branches, which often unite to form elongated lacunæ, of which the largest border the rachis of the pinnæ on either side, and are formed by the nerve branches of each lobe reaching over and touching, or closely approaching, the base of the nervation of the next superior lobe or pinnule."

The collection of fossil plants made at Fort Union by Dr Hayden contains a great number of examples of this beautiful fern, showing the upper and under surface of the frond, the variation of form of the pinnæ of different fronds, and different parts of the same frond.

The robust habit of this plant, the strong, waved, and reticulated nervation and broadly winged rachis, which seem to distinguish it at a glance from all known fossil species, suggested a comparison with some of the strong-growing tropical ferns, and it was only after a laborious examination of all the genera of exotic ferns contained in the herbaria to which I had access that I was led to turn my eyes nearer home, and found in Onoclea a striking and unexpected resemblance to it.

The common form of *Onoclea sensibilis* grows abundantly in all parts of our country, and is one of the first plants collected by the youthful botanist. In this we have the rachis of the frond more or less winged, and a nervation on the same general plan with that of the fern in question, but more distinctly reticulated than in some specimens of the fossil. (See Pl. XXIII, fig. 4.) By this I was at first misled, but in examining Dr. Torrey's var. *obtusilobatus* I found the exact counterpart of the exceptional forms in the lobation of the pinnæ and in the nervation. (See Pl. XXIII, figs. 5, 6.) The gradation of characters in this variety is very great. In some specimens we have a distinctly bipinnate frond; the pinnæ composed of numerous remote, even obovate, pinnules, and the nervation not reticulated, the nerves of the pinnules radiating and forked, but never joining. This is the extreme form, but even here the rachis of the frond is more or less winged. In an intermediate form we find the rachis winged, the pinnæ deeply lobed, and precisely the nervation of the fossil. Even in the common form the nervation is similar in plan, and the elongated spaces, destitute of nerve branches on either side of the rachis of the pinnæ, form a noticeable feature in both.

The general aspect of the frond and the nervation in some species of Woodwardia is not unlike that of the fossil now figured, and until we shall have found the fruit it will not be possible to prove that this is Onoclea and not Woodwardia. The resemblance of the fossil to Onoclea in the form of the frond, the lobation of the pinnules, and in the style of nervation is, however, stronger than to Woodwardia, as will be seen by a comparison of Pl. XXIII, fig. 4—a portion of the frond of the living Onoclea—with Pl. XXIV, figs. 4 and 5, corresponding portions of the fossil. Among the large number of specimens obtained of this fossil fern there are none which exhibit the fructification, an indication that this was borne on distinct fronds. If it were a species of Woodwardia it is almost certain that we should have found the fructification, since all the fronds of Woodwardia may be fruitful, and the fructification is generally observable in the fossil species of that genus.

Since the above notes were written I have obtained a number of specimens of Onoclea from the shores of Whatcom Lake, near Bellingham Bay, Washington. In this vicinity there is a great development of strata which are rich in fossil plants and are about the equivalents in time of the Laramie group; but, with few exceptions, the forms are distinct. This is one of the few which are common to the two localities.

Varying, as the living Onoclea does, in the size, outline, and nervation of the sterile frond—from 6 inches to 3 feet in height; from a finely reticulated to an open, dichotomous nervation; from a bipinnate frond with remote, obovate pinnules, to a pinnate form with wave-margined pinnæ and broadly alate rachis—it plainly includes all the characters of the fossils before us, and I therefore find it impossible to separate them.

This is apparently the plant described by Prof. E. Forbes (Quart. Journ. Geol. Soc. London, Vol. VII (1851), p. 103), under the name of *Filicites* (?) *hebridicus*, and obtained by the Duke of Argyle from the Island of Mull. It has also been met with by Professor Heer in collections of fossil plants from the Eocene beds of Atanekerdluk and other places in the arctic regions. (Fl. Foss. Arct., Vol. VII, p. 48, Pl. LXX, fig. 6.)

Formation and locality: Tertiary (Fort Union group). Fort Union, Dakota.

Lastrea (Goniopteris) Fischeri Heer?.

Pl. XLVIII, fig. 6.

Fl. Tert. Helv. Vol. I (1855), p. 34, Pl. IX, figs. 3a–3e.
Lastrea (*Goniopteris*) *Knightiana* Newb. Proc. U. S. Nat. Mus., Vol. V (March 21, 1883), p. 503.

"Frond large, tripinnate; pinnæ linear, 2 centimeters wide, 14 to 16 centimeters long; pinnules diverging at a large angle, united for two-thirds of their length, upper third free, pointed, and curved upward; venation clear and exact, midrib reaching the extremity of the pinnule; the lateral nerves about ten on either side, parallel, curved upward."

This beautiful fern may be readily recognized by the rigid exactness of its outline, the regularity and precision of its crowded nervation, and by the falcate curvature of the extremity of the acute pinnules. From the large angle made by the midrib of the pinnule with the rachis of the pinna the number of the pinnules on the frond seems crowded. In some of the pinnules the midrib has an elegant sigmoidal curve. This, with the parallel curvature of the lateral veins, gives a peculiar, exact, and elegant aspect to the plant.

The specimen figured was collected by Rev. Thomas Condon, at Currant Creek, Oregon, where it occurs matted together in masses. Lesquereux has also found what he considers to be the same species at John Day Valley, Oregon.

Of the described species, *Lastrea Fischeri* Heer (Fl. Tert. Helv. Vol. I, p. 34, Pl. IX, figs. 3a to 3e), resembles this most, but our plant is stronger, the pinnules are united for a greater portion of their length, are more acute, have a more crowded nervation and a distinctive upward curve. Yet these differences are rather of degree than kind, and hardly warrant the separation of the American and European plants.

From the species described by Lesquereux as *L. Goldiana* and *L. intermedia* (Tert. Fl., p. 56, Pl. IV, figs. 13 and 14), this may be distinguished by its acute, falcate, and more numerous pinnules.

Formation and locality: Tertiary (Miocene?). Currant Creek, Oregon.

Aspidium Kennerlyi Newb.

Pl. XVI, figs. 4, 5.

Boston Journ. Nat. Hist., Vol. VII (1863), p. 513.

"Frond pinnate; pinnæ deeply pinnatifid; pinnules oblong, obtuse, somewhat curved upward, united at their bases, margins acutely denticulate, sometimes entire; nervation strongly marked, secondary nerves mostly once-forked, basal nerve of each pinnule on the lower side often twice-forked."

This elegant species seems to have grown in the greatest abundance during the period of the deposition of the coal of Vancouvers Island, the shales over the Newcastle coal being so closely packed with its fronds as to show them crossing each other in every direction under every lamina that is raised. From their very abundance and consequent interference it is impossible to obtain the entire outline of a frond, or even of a pinna: the frond must, however, have been of considerable size, and the pinnæ 8 or 10 inches in length. These last are linear in outline, some of them somewhat curved, others quite straight, the difference being doubtless due to their different positions in the frond. The pinnules are usually arched upward, very broad at the base, rounded or obtusely pointed at the summit. Where well preserved, the margins of the larger ones are seen to be finely but distinctly denticulate. The nervation is quite strong, but the frond was evidently thick and firm, and though very prominent on the under side, on the upper the nerves are scarcely visible. The midrib is slightly sinuous, and vanishes toward the summit of the pinnule. The secondary nerves are generally once-forked, but the upper ones are simple, and the lower one on the lower side is often twice-forked, or rather two once-forked nerves spring from the same base.

Among fossil species this may be compared with *A. Filix antiqua*, Al. Br. (Heer, Fl. Tert. Helv. Vol. I, p. 35, Pl. XI, fig. 1), but though crenulated the pinnules in that species are not denticulate, and they are not curved. The nerves are also less strong and more simple than in our plant.

Formation and locality: Cretaceous (Puget Sound group). Nanaimo, Vancouver Island.

PECOPTERIS (CHEILANTHES) SEPULTA Newb.

Pl. LXII, figs. 5, 5a, 6.

Pecopteris (*Phegopteris*) *sepulta* Newb. Proc. U. S. Nat. Mus., Vol. V (March 21, 1883), p. 503.

"Frond small, delicate, pinnate; lower pinnæ straight, broadly linear in outline, rounded above, attached to rachis by the whole breadth of base; margins strongly lobed by the confluent pinnules; 1 centimeter wide by 5 centimeters long; upper pinnules crowded, conical in outline, gently curved upward, with waved or lobate margins; pinnules united by one-third of their length, oblong, obtuse; basal ones on lower side round, on the upper side flabellate, both attached by all their lower margin to the rachis of the frond; nervation strong and wavy, consisting of one many-branched nerve-stem in each pinnule, each branch once or twice forked; fructification unknown"

This elegant fern is apparently distinct from any species hitherto described. In general aspect it is not unlike *Pecopteris Torelli* Heer (Fl. Foss. Arct., Vol. I, p 88, Pl. I, figs. 15a, 15b), but in that species the pinnules are longer, more oblique, more acute, and the nervation more open. It also has some resemblance to *Cheilanthes Laharpii* Heer (Fl. Tert. Helv. Vol. I, p. 37, Pl. X, figs. 3a, 3b). That species is, however, more delicate, the pinnae more widely separated, the pinnules to a less degree united, the basilar pair similar to the higher ones, the nervation more open.

The upper portion of the frond of this fern, where the pinnae are not distinctly lobed, but simply undulate, bears a strong resemblance to that figured and described in Gardner and Ettingshausen's British Eocene Flora, Part II, p. 43, Pl. VI and Pl. X, figs. 2–4, under the name of *Gleichenia Hantonensis* (Wanklyn), but the secondary nerves are fewer and given off at a more acute angle.

The middle portion of the frond of our plant is, however, conspicuously different, since the pinnae are deeply lobed, forming distinct and peculiar pinnules at the base instead of being confluent as in *G. Hantonensis.* It seems to be probable, however, that both ferns belong to the same genus.

What this genus should be called must remain a matter of doubt until specimens shall be obtained in which the fructification is shown. Without better evidence than we yet possess, the reference of our plant to Gleichenia seems to be unwarranted.

The general form of the frond and the nervation are more like those of some species of Cheilanthes than of any other living ferns with which this has been compared; but it will be necessary to have the fructification before the identification with that genus can be regarded as established. It has been thought better, therefore, to place it in the convenient receptacle afforded by the fossil genus Pecopteris, with a suggestion of its probable affinities in the living flora of the world.

The figures given represent, 5, the middle portion of the frond; 6, the upper part, and 5a, the lower two pinnules at base of pinna on the under side enlarged. They were collected by Dr. C. A. White, from the Green River shales.

Formation and locality: Tertiary (Green River group). Green River, Wyoming.

Sphenopteris corrugata Newb.

Pl. I, fig. 6.

Ann. N. Y. Lyc. Nat. Hist., Vol. IX (April, 1868), p. 10; Ills. Cret. and Tert. Pl. (1878), Pl. II, fig. 6.

Hymenophyllum cretaceum Lesq. Hayden's Ann. Rept., 1872 [1873], p. 421; Cret. Fl. (1883), p. 45, Pl. XXIX, fig. 6 [excl. Pl. I, figs. 3, 4].

"Form of frond unknown; pinnules ovate or cuneiform, narrowed at the base, obtuse, lobed, often plicated longitudinally: nerves distinct, dichotomously branching from the base.

"The specimens of this fossil collected by Dr. Hayden are fragmentary and imperfect, but quite sufficient to show it to be different from any described species."

Since the above was written Lesquereux has published in his Cretaceous Flora descriptions of a fern from the Dakota sandstones, at Fort Harker, which he calls *Hymenophyllum cretaceum.* Of this he gives several figures on Pl. I, and another on Pl. XXIX. Of these the latter certainly represents our species, which is easily recognized by the wedge-shaped subdivisions and the plicate or corrugated surface: but the specimens figured on Pl. I belong to a different species, of which the frond was membranous and the rachis winged, and which approached much nearer to the living *Hymenophyllum.*

Formation and locality: Cretaceous (Dakota group). Blackbird Hill, Nebraska.

Order EQUISETACEÆ.

Equisetum Oregonense Newb.

Pl. LXV, fig. 7.

Proc. U. S. Nat. Mus., Vol. V (March 21, 1883), p. 503.

"Stem robust, 3 centimeters wide, longitudinal flutings numerous, about 24 in a half circumference; joints 5 centimeters distant; teeth triangular, short."

This species, collected by Rev. Thomas Condon, at Currant Creek, Oregon, is imperfectly represented in the collection, but there is enough of it to show it to be distinct from any other fossil yet found. It exceeds in magnitude any Tertiary species hitherto described in this country, and approaches more nearly to the larger forms of the Mesozoic rocks. It may be compared with *E. robustum* Newb., this volume, page 15, Pl. XVI, figs. 1, 2, but the stem is broader, the flutings double the number, and the teeth much shorter and blunter than in that species. *E. procerum* Heer (Fl. Tert. Helv. Vol. III, p. 158, Pl. CXLVI, fig. 1), from Locle, Switzerland, is larger, but differs widely from it by its coarser fluting, long and furrowed teeth.

Formation and locality: Tertiary (Miocene?). Currant Creek, Oregon.

Equisetum robustum Newb.

Pl. XVI, figs. 1, 2.

Boston Journ. Nat. Hist., Vol. VII (1863), p. 513.

"Stem robust, 8 lines wide, with about 24 strongly marked furrows; sheaths long; teeth long-pointed, acute, as many as the furrows; internodes a little longer than the diameter of the stem."

There is no living species of Equisetum which attains the size of the fossil before us, though it does not rival in this respect those found in the older Mesozoic rocks. Between the living and older extinct species it seems to form a connecting link, a stepping-stone by which the Calamites of the coal period and the gigantic Equiseta of the Trias have come down to the humble dimensions of their present representatives.

There is no described Tertiary species with which it will be likely to be confounded. *E. procerum* Heer (Fl. Tert. Helv. Vol. III, p. 158, Pl. CXLVI, fig. 1), is even larger, but will at once be distinguished from it by its smoother stem and far more numerous and less acute teeth.

Formation and locality: Cretaceous (Puget Sound group). Bellingham Bay, Washington.

Equisetum Wyomingense Lesq.

Pl. LXV, fig. 8.

Hayden's Ann. Rept., 1873 [1874], p. 409; Tert. Fl. (1878), p. 69, Pl. VI, figs. 8–11.

Formation and locality: Tertiary (Green River group). Green River, Wyoming.

Note.—So identified by Dr. Newberry, as indicated by memorandum on plate and label on specimen, but further information lacking.—A. H.

EQUISETUM sp.? Newb.

Pl. XXII, figs. 3, 4.

Fig. 3. "Radicle tubers of Equisetum (not described)." Ills. Cret. and Tert. Pl. (1878), Pl. VII, fig. 4.

Fig. 4. "Root of some ligneous plant (not described)." Ills. Cret. and Tert. Pl. (1878), Pl. VII, fig. 3.

NOTE.—The only manuscript by Professor Newberry which I have been able to find is a penciled memorandum on the plate referring these to Equisetum, viz:

Fig. 3. "Tuberous roots of Equisetum sp.?"

Fig. 4. "Aquatic rootlets of Equisetum sp.?"

Fig. 3 certainly represents *E. globulosum* Lesq., Proc. U. S. Nat. Mus., Vol. V (September 29, 1882), p. 444, Pl. VI, figs. 1, 2; Cret. and Tert. Fl. (1883), p. 222, Pl. XLVIII, fig. 3; but there is no indication that Dr. Newberry intended so to refer it.—A. H.

PHANEROGAMIA.

GYMNOSPERMÆ.

Order CYCADACEÆ.

NILSSONIA GIBBSII (Newb.) Hollick.

Pl. XV, figs. 2, 2a.

Tæniopteris Gibbsii Newb., Boston Journ. Nat. Hist., Vol. VII (1863), p. 512.

Nilssonia Johnstrupi Heer, Fl. Foss. Arct., Vol. VI, Abth. II (1882), p. 44, Pl. VI, figs. 1–6.

"Frond simple, petiolate, oblong, elliptical in outline, rounded at base and summit; margins entire, midrib strong, straight, smooth; lateral nerves leaving the midrib nearly at a right angle, simple, fine, parallel, numerous."

The above description was based on a single specimen collected by Mr. George Gibbs from the Cretaceous strata on Orcas Island, Washington, in 1858. From the character of the nervation and the entire margins it was supposed to be a fern, but Professor Heer has since obtained a number of specimens of the same plant from the Upper Cretaceous strata of Greenland, which seem to prove that it is the leaf of a cycad. (Fl. Foss. Arct., VI, Abth. II (1882), p. 44, Pl. VI, figs. 1–6.) He has named his plant *Nilssonia Johnstrupi*, but the specific name given by me has priority.

It is far more interesting to identify a plant from Orcas Island with one found in the Cretaceous strata of Greenland than to find it to be a new genus or species, as it helps us to establish a geological parallelism, and shows the wide diffusion of some species through the Cretaceous strata. By this plant and a few others the Vancouver and Orcas Island beds are connected with those of Atane, Greenland, and many common species correlate the Atane beds with the Amboy Clays of New Jersey.

Formation and locality: Cretaceous (Puget Sound group). Point Doughty, Orcas Island, Washington.

Order CONIFERÆ.

ARAUCARIA SPATULATA Newb.

Pl. I, Figs 5, 5a.

Ann. N. Y. Lyc. Nat. Hist., Vol. IX (April, 1868), p. 10; Ills. Cret. and Tert. Pl. (1878) Pl. II, figs. 5, 5a.

"The only specimen of this beautiful species contained in the collections of Dr. Hayden is a fragment of a branch, nearly half an inch in diameter. On this the leaves are thickly set, their bases slightly decurrent, being scarcely separated from each other. From these bases the leaves radiate in all directions, and are slightly recurved. They are half an inch in length, broadly spatulate, obtuse, and narrowed at the base. Along the medial line passes a distinct carina, which vanishes toward the apex."

From all living or fossil species, this seems very clearly distinguished by the form of the leaves. Two species of Araucarites have been described from the Cretaceous formation, of which descriptions are before me: *A. acutifolius* Endl. and *A. crassifolius* Endl. (Synops. Conif., pp. 301, 302), neither of which has spatulate leaves.

There is little doubt that this was a true Araucaria, and not very unlike, in its general aspects, some species now living.

It is also probable that these trees formed extensive forests on the land during the Cretaceous period, as I have found these strata in some localities in the West literally filled with large trunks of coniferous trees, many of which have rather the structure of Araucaria than of Pinus, Abies, or Juniperus, although all these genera were represented at that epoch.

Formation and locality: Cretaceous (Dakota group). Sage Creek, Nebraska.

ABIETITES CRETACEA Newb. n. sp.

Pl. XIV, fig. 5.

NOTE.—The only manuscript by Dr. Newberry in regard to this figure is on the label attached to the specimen.

The following description has been prepared from an examination of the specimen:

Branchlet slender; leaves one-half inch long, crowded, short petiolate, narrowly ovate-lanceolate, attenuate at both ends.—A. H.

Formation and locality: Cretaceous (Dakota group). Whetstone Creek, Santa Fe trail, northeastern New Mexico.

SEQUOIA CUNEATA Newb.[1]

Pl. XIV, figs. 3–4a.

Taxodium cuneatum Newb. Boston Journ. Nat. Hist., Vol. VII (1863), p. 517.

"Leaves numerous, short, broad, spatulate in form, rounder or subacute at summit, wedge-shaped below, narrowed into a very short petiole, or sessile upon the branchlets."

The specimens of this plant contained in the collection, though numerous, are too imperfect for satisfactory description. If found in strata of the same age, it might be considered but a variety of Taxodium; but if we can trust the accuracy of the very intelligent gentleman by whom it was collected, it is clearly of Cretaceous age, and therefore, in all probability, quite distinct from any described species.

The spatulate or cuneate form of the leaves, if this should be found to be a constant character, would serve to distinguish it at a glance from its Tertiary representatives.

Formation and locality: Cretaceous (Puget Sound group). Nanaimo, Vancouver Island.

---

[1] This species was transferred by Dr. Newberry from Taxodium to Sequoia in his manuscript.—A. H.

SEQUOIA GRACILLIMA (Lesq.) Newb.

Pl. XIV, fig. 6; XXVI, fig. 9. ?

Glyptostrobus gracillimus Lesq. Am. Journ. Sci., Vol. XLVI (July, 1868), p. 92; Cret. Fl. (1874), p. 52, Pl. I, figs. 8, 11–11f.
"Cone of *Sequoia* (not described)." Ills. Cret. and Tert. Pl. (1878), Pl. XI, fig. 9.

Lesquereux described (loc. cit.) a conifer which occurs frequently in the Dakota group in Nebraska, and also in the Cretaceous strata of New Jersey. It is characterized by a great number of slender, almost filiform, branches covered with acute lanceolate or ovate, sometimes subulate, leaves. Lesquereux speaks of their occurring in whorls of three, but in the large number of specimens before me I can find no evidence of a verticillate arrangement, and they seem to surround the stems spirally. They differ considerably in length, but the foliage can hardly be said to be dimorphous as in Glyptostrobus, Sequoia, and many other conifers, but usually on the older branches they are more closely appressed, more spreading above. Lesquereux compares this plant with Frenola of Australia, and suggests that it may be identical with Ettingshausen's *Frenelites Reichii*, from the chalk of Niederschœna. It has been my good fortune to obtain a number of cones of this plant, both from Nebraska and New Jersey, and I am able, therefore, to give a more complete description of it than has been heretofore possible. The cones are cylindrical, 2 to 2½ inches in length, one-half inch in diameter, and are formed of relatively large peltate scales, each with an umbilicus and central tubercle. [See Pl. XXVI, fig. 9.?] This is a totally different cone from that of Glyptostrobus, in which the divisions are squamiform with a fanlike, crenulated margin. The form of scale in the cones before us is similar to that of Sequoia and Taxodium, but the cones of the latter are usually globular, while those of Sequoia are often elongated, sometimes subcylindrical. The character of the foilage is near to that of some of the Sequoias, *S gigantea* and *S. Couttsiæ*, for example, while in Glyptostrobus the two forms of foliage are much more distinctly marked, the short appressed leaves closely investing the branches, resembling those before us, the open foliage quite different. The foliage of this plant is found in considerable abundance in the sandy layers of the Cretaceous on the Raritan River, and the cones were formerly numerous in the clay beds at Keyport, where they were associated with great quantities of lignite, very probably produced by the trees on which they were borne. In some cases the cones were replaced by pyrites, and these represent the original form and markings very perfectly, but require to be kept in alcohol or naphtha to prevent oxidation. They will be found in my memoir on the Flora of the Amboy Clays.

Formation and locality: Cretaceous (Dakota group). Whetstone Creek, New Mexico. (Excluding fig. 9.)

NOTE.—In the discussion of this species Dr. Newberry mentions having obtained cones from Nebraska and describes them, but does not refer to fig. 9, Pl. XXVI, which is therefore questioned by me.—A. H.

SEQUOIA HEERII Lesq.

Pl. XLVII, fig. 7.

Hayden's Ann. Rept., 1871 [1872], p. 290; Tert. Fl. (1878), p. 77, Pl. VII, figs. 11–13.

Formation and locality: Tertiary (Miocene). Bridge Creek, Oregon

NOTE.—The only reference by Dr. Newberry to this figure which I have been able to find is a pencil memorandum of the name, on the plate, and the specimen label giving the locality.—A. H.

SEQUOIA NORDENSKIOLDII Heer ?.

Pl. XXVI, fig. 4.

Fl. Foss. Arct., Vol. II (Miocene Fl. u. Fau. Spitzbergens, 1870), p. 36, Pl. II, fig. 13b; IV, figs. 1a, 1b, and 4–38.

***Taxites Langsdorfii* Brong. ? Prod. (1828), p. 108.**

***Sequoia Langsdorfii* (Brong.) Heer. Fl. Tert. Helv., Vol. I (1855), p. 54, Pl. XX, fig. 2; XXI, fig. 4.**

"*Sequoia Langsdorfii* ? Br." Newberry, Ann. N. Y. Lyc. Nat. Hist., Vol. IX (April, 1868), p. 46; Ills. Cret. and Tert. Pl. (1878), Pl. XI, fig. 4.

The leaves here figured are part of a large number of the same species collected by Dr. Hayden on the banks of the Yellowstone River. They are contained in fragments of a shaly argillaceous limestone, which have their surfaces covered by disconnected twigs with their leaves attached, that present the appearance of having been thrown down together, precisely as the deciduous branchlets of our cypress are detached by the frost. Among these are a few pieces of larger branches bearing short appressed leaves, which I have conjectured to be the permanent foliage of the tree.

These branches show at regular intervals the former points of attachment of deciduous (?) branchlets, but more of these are still in their places. They may have been dead twigs, some of which would naturally fall and accumulate with the leaves. The leaf-bearing branchlets are simple, and though lying together in great numbers and crossing at every angle, are distinct and disconnected. The probability would therefore seem to be that the foliage of the tree was deciduous, and although we have as yet no fruit to guide us, we may infer that it was not a Sequoia, but a Taxodium allied to our deciduous cypress. The leaves on the permanent branches are many-rowed, short, appressed, and awl-shaped. Those on the deciduous (?) branchlets are two-ranked, much longer, linear, acute or rounded, traversed by a strong median nerve, and decurrent at the base. The lower leaves on the branchlets are also generally shorter, sometimes much shorter, than those placed higher up.

In my notes on these specimens, given in The Later Extinct Floras, written before the publication of Professor Heer's series of works on the arctic flora, these specimens were doubtfully referred to *Sequoia Langsdorfii*, to which they bear a considerable resemblance, but the foliage seems to have been more open and the leaves more decidedly decurrent. In these characters they approach very closely to the foliage of *Sequoia Nordenskioldii*, of which the description is published in the Fl. Foss. Arct., Vol. II, Abth. III, Miocene Flora und Fauna Spitzbergens, p. 36, Pl. IV, figs. 4–38. The correspondence is so close that I have been led to regard them as probably identical. More material, including the fruit, will be necessary to discriminate between these closely resembling conifers, and this reference, which seems authorized by the character of the foliage, must be considered as provisional until confirmed by evidence which is more conclusive.

Formation and locality: Tertiary (Eocene ?). Yellowstone River, Montana.

SEQUOIA SPINOSA Newb.

Pl. LIII, figs. 4, 5.

Proc. U. S. Nat. Mus., Vol. V (March 21, 1883), p. 504.

"Branches slender; foliage open, rigid; leaves narrow, acute (acicular) arched upward, appressed or spreading; spirally divergent; staminate flowers in slender terminal aments, 2 inches long, 2 lines wide, anthers few, under peltate connective scales; cones ovate or subcylindrical, composed of rhomboidal or square peltate scales."

We have in the specimens before us, collected by Captain Howard, U. S. N., a new and strongly marked species of Sequoia, which is distinguishable at a glance from all of its known congeners by its remarkably sparse, rigid, slender, and acute leaves As usual among conifers of this group, there is some diversity in the character of the foliage, some of the leaves being closely appressed, others longer and more spreading. In general aspect the terminal branchlets resemble some of those belonging to *S. Couttsiæ* Heer (Phil. Trans., Vol. CLII, Pt. II; Foss. Fl. Bovey Tracey, Pl. LX, figs. 1, 2, 3, 6, 15, 44, 45; Fl. Foss. Arct., Vol. 1, Pl. XLV, fig 19), but the leaves are longer and more slender. None have been observed taking the squamose form exhibited by most of the foliage of *S. Couttsiæ* in the illustrations given by Professor Heer. The cones, too, are longer, being subcylindrical, while in *S. Couttsiæ* they are nearly globular. One of the cones is represented in fig. 5, Pl. LIII, unfortunately rather badly preserved. Quite a number are associated with the leaves in the specimens before us, but none more complete. The sterile aments are slender. the group of anthers much less crowded than usual. On some of the branchlets the foliage is more crowded and the leaves are broader than in the specimens figured on Pl. LIII, but this may be considered as a fair representation of its average character.

Formation and locality: Tertiary (Miocene). Cook Inlet, Alaska.

Taxodium distichum miocenum Heer.

Pl. XLVII, fig. 6; LI, fig. 3, in part; LII, figs. 2, 3 and 4 in part; LV, fig. 5, in part.

Miocene Baltische Flora (1869), p. 18, Pl. II; III, figs. 6, 7.

Formation and locality: Tertiary (Miocene). Birch Bay, Washington (Wilkes Exploring Expedition).

Note.—In the discussion of *T. occidentale* Dr. Newberry says that the specimens obtained at Birch Bay, Washington, by Professor Dana, and at Currant Creek, Oregon, by Rev. Thomas Condon, are hardly to be distinguished from the living *T. distichum*.—A. H.

TAXODIUM OCCIDENTALE Newb.

Pl. XXVI, figs. 1-3; LV, fig. 5, in part.?

Boston Journ. Nat. Hist., Vol. VII (1863), p. 576; Ann. N. Y. Lyc. Nat. Hist., Vol. IX (April, 1868), p. 45; Ills. Cret. and Tert. Pl. (1878), Pl. XI, figs. 1-3.

"Branchlets terete, leaves numerous, crowded, generally opposite, sessile, or very short petioled, one-nerved, flat, rounded at both ends."

Branchlets terete, leaves distichous, sessile on very short petioles; one-nerved, flat, rounded at both ends, the larger ones 4 centimeters wide by 20 centimeters long, the shorter ones elliptical, scarcely longer than wide.[1]

The characters and variations of the foliage of this plant are very well shown in the figures given of it. From these it will be seen that the leaves are unusually broad for their length, are distinctly rounded at both ends, are sessile or very short petioled, and are not at all decurrent. Some of them are also very short, the shortest almost circular, and they are borne on the secondary as well as tertiary branchlets.

In the notice of these leaves in The Later Extinct Floras they were compared with those of *Taxodium dubium* Heer, and it was stated that it differed from that species in having a larger number of leaves, less obliquely set on the branches, with rounded extremities, whereas in the foreign species the leaves are lanceolate in outline and acute at both ends. In his later works Professor Heer has expressed the opinion that *Taxodium dubium* is only a form of *T. distichum*, now living in our Southern States. This view has been generally accepted by fossil botanists, and the plants under consideration must be compared with the deciduous cypress. In looking over the large number of specimens which I have received from various localities I find that many of them can not be distinguished from the leaves of the living cypress. This is true of collections made by Professor Dana at Birch Bay, by Rev. Thomas Condon at Currant Creek, Oregon, and by Dr. Hayden in the lignite Tertiaries of the upper Missouri River. The specimens now figured, however, obtained by Dr. Hayden on the Yellowstone and Dr. Cooper in northern Montana, exhibit characters which would seem to be sufficient to separate them from the deciduous cypress, the leaves being relatively much broader and rounded at both ends.

[1] In addition to the original published description, as quoted, the above subsequent manuscript description is also included.—A. H.

Formation and locality: Tertiary (Eocene?). Yellowstone River, Montana and northern Montana. (Excluding Pl. LV, fig. 5, in part,.)

GLYPTOSTROBUS EUROPÆUS (Brong.) Heer.

Pl. XXVI, figs. 6–8a; LV, figs. 3, 4.[1]

Fl. Tert. Helv., Vol. I (1855), p. 51, Pl. XIX; XX, fig. 1.
Taxodium Europæum Brong. Ann. Sci. Nat., Vol. XXX (1833), p. 168.
"*Glyptostrobus Europæus* (Brong.)." Newberry, Ann. N. Y. Lyc. Nat. Hist., Vol. IX (April, 1868), p. 43; Ills. Cret. and Tert. Pl. (1878), Pl. XI, figs. 6–8a.

"Branches slender, bearing many branchlets; leaves of two forms, one short, thick, and appressed, the other longer (one-half inch), slender, divergent, acute, the shorter form carinated, the longer less distinctly, if ever so; male catkins small, terminal, globular, composed of a few shield-shaped scales; fertile cones larger, ovoid in form, scales narrow, wedge-shaped at base, at summit expanded, semicircular, with waved or crenate margins, the dorsum of each more or less distinctly marked with 10 to 12 acute, radiating carinæ."

One of the most interesting plants of the European Tertiary is the Glyptostrobus, first discovered by Brongniart, and subsequently fully illustrated in the magnificent work of Prof. O. Heer, Flora Tertiaria Helvetiæ, Vol. I, p. 52, Pl. XVIII; XXI, fig. 1; Vol. III, p. 159. The genus is now only represented on the earth's surface by *G. heterophyllus* and *G. pendulus* of China, but during the middle Tertiary epoch was widely spread over both hemispheres. Most of the exposures of our older Tertiary strata have furnished specimens of some one of the various phases of what is regarded by Professor Heer as a single species, but which has been described under the three names of *G. Europæus*, *G. Ungeri*, and *G. Œningensis*

What are probably but varieties of this same plant were collected by the United States Exploring Expedition under Captain Wilkes, at Birch Bay, Washington, by George Gibbs, esq., geologist to the Northwestern Boundary Commission (see Boston Journ. Nat. Hist., Vol. VII, No 4 (1863), p. 517), and are represented by numerous specimens in the collection of fossil plants made by Dr. Hayden on the Yellowstone and Upper Missouri.

[1] Dr. Newberry's only manuscript for Pl. LV, figs. 3, 4, is a pencil memorandum referring them to "*Glyptostrobus Ungeri* Heer."—A. H.

In this country, as in Europe, the foliage of Glyptostrobus exhibits two forms wherever the plant is found; the short appressed, and the longer divergent leaves. In addition to this the specimens from the northwest coast have common character by which they may be distinguished at once from those collected by Dr. Hayden. The Western plant is more slender, the appressed leaves sharper and more delicate, the divergent leaves much longer, corresponding more nearly to the European form described as *G. Ungeri*, while those from the Upper Missouri resemble more the variety known as *G. Europæus*. The cones, however, found with the Missouri specimens are more like those of *G. Ungeri* than *G. Europæus*, the dorsum of the scale being marked by short, radiating carinæ, as in the former, the margin being waved, but not regularly scalloped, as in the latter.

From the extreme West we have as yet no cones which can be certainly referred to this plant, so that the most important element in the comparison is wanting, but it would seem that here, as in Europe, the different phases of the plant belonging to the genus Glyptostrobus are so linked together that they should be regarded as forming but a single species. At least we have not yet obtained sufficient material to justify us in attempting to define the limits of other species.

The two living species of Glyptostrobus which Fortune found growing in China seem to resemble the fossil forms as much as they do each other, and it is perhaps doubtful whether they should not all be united under the same name. The living and fossil plants are associated with fan-palms, and belong to the flora of the southern temperate zone, or that of a latitude ten degrees south of the localities where the fossils occur.

Formation and locality: Tertiary (Fort Union group). Fort Union, Dakota, and Birch Bay, Washington (Wilkes Exploring Expedition).

THUJA INTERRUPTA Newb.

Pl. XXVI, figs. 5–5d.

Ann. N. Y. Lyc. Nat. Hist., Vol. IX (April, 1868), p. 42; Ills. Cret. and Tert. Pl. (1878), Pl. XI, figs. 5, 5a.

"Branchlets flat, narrow, linear, pinnate, opposite, except at the summit of the branch, somewhat remote, connected only by the slender woody axis on which the leaves of the branchlets are not decurrent; leaves in four rows appressed, those of the upper and lower ranks orbicular or obovate, shortly mucronate, lateral ones longer, subulate, terminating in awnlike points; larger branches naked or bearing closely appressed linear scalelike leaves."

This is a very distinct and beautiful species collected by Dr. Hayden, near Fort Union, Dakota, presenting marked differences from any known living or fossil members of the genus.

Its most remarkable character is its slender and graceful habit, and the separation of the pairs of leafy branchlets along the naked and slender branch. The leaves, too, are less crowded than in most other species, and the lateral ranks are prolonged into acute awnlike points, all of which must have given it an aspect considerably unlike that of any species hitherto described.

At the time this species was described no true Thuja had been recognized in the fossil state. *Thuites salicornoides* (Ung. Chlor. Prot. Pl. II, fig. 1; XX, fig. 8) is regarded by Endlicher and Heer as a Libocedrus, to which it certainly seems, judging from the figures and descriptions given of it, to be more closely allied than to Thuja. Since that time, however, a number of fossil plants have been referred to the genus Thuja, principally derived from the amber. One species, *T. saviana*, Gaud., Neue Denkschr. Schweitz. Gesell., Vol. XVII (1860) Fl. Foss. Ital., 3d Memoir, p. 12, Pl. I, figs. 4–20; II, figs. 6, 7, has been established upon the fruits as well as the foliage, so that there can be no question in regard to its botanical position. Another species, *T. mengeanus*, Goepp. and Ber. Monogr. Foss. Conif. (1850), p. 181, Pl. XVIII, figs. 10, 11, resembles so closely our *T. occidentalis* that it has been referred by Goeppert to that species. Besides this, half a dozen additional species obtained from the amber have been described by Goeppert from meager material and consequently somewhat vaguely. It may be considered established, however, that during the Tertiary age the genus Thuja was in existence and well represented in the coniferous flora. The species now under consideration is represented by a large number of specimens, though usually of small size, in the collections made at Fort Union by Dr. Hayden, and has also been met with by Mr. George M. Dawson in the Tertiary lignite strata of Canada. No fruit has been found that can be certainly connected with the leaves, but there is in the collection one imperfect cone derived from the same locality with the branches of Thuja which resembles closely in structure the cone of *T. occidentalis*.

Formation and locality: Tertiary (Fort Union group). Fort Union, Dakota.

ANGIOSPERMÆ.

MONOCOTYLEDONEÆ.

Order GRAMINEÆ.

PHRAGMITES sp.? Newb.

Pl. XXII, figs. 5, 5a.

Ann. N. Y. Lyc. Nat. Hist., Vol. IX (April, 1868), p. 38; Ills. Cret. and Tert. Pl. (1878), Pl. VII, figs. 5, 5a.

"Among the plants collected by Dr. Hayden from the Miocene beds near Fort Union are numerous fragments of what seems to be a species of Phragmites. These consist of portions of broad, unkeeled, flaglike leaves, marked by numerous longitudinal nerves, of which there are eight or nine more strongly marked, and between these about seven much finer, connected by alternate cross bars. No keel is shown in any of these fragments. In general structure these leaves closely resemble those of *P. Oeningensis* Heer (Fl. Tert. Helv., Vol. I, p. 64, Pl. XXIV); but the material is not sufficient to determine whether our species is identical with that.

"*Formation and locality:* Fort Union, Dakota (Dr. Hayden)."

Order PALMÆ.

SABAL CAMPBELLI Newb.

Pl. XXI, figs. 1, 2.

Boston Journ. Nat. Hist., Vol. VII (1863), p. 515.

"Leaf large, 8 feet in diameter, with fifty to seventy folds; petiole long, 16 lines or more in width, flat above, without a central keel and unarmed; nerves numerous and fine, about fifty in each fold—six principal nerves on each side of the midrib, with three intermediate nerves between each pair, the middle one being strongest."

In its general character this palm bears a strong resemblance to *Sabal major*, Ung. sp. (Chlor Prot., p. 42, Pl. XIV, fig. 2; Fl Tert. Helv., Vol. I,

p. 88, Pl. XXXV; XXXVI, figs. 1, 2), the size of the leaf, the number of folds, and the character of the nervation being nearly the same; but in our plant the petiole is flat or slightly arched, without the central keel of *S. major* Unfortunately we have as yet obtained no specimen showing the under side of the leaf, and therefore want the important diagnostic character of the length of the point of the petiole.

From *Sabal Lamanonis* this species may be distinguished by its greater size, more numerous leaf-folds, finer and more crowded nervation, and by its flat unkeeled petiole.

Fan-palms are not now found on the Pacific coast above Cape St. Lucas (lat. 23° north), though the average temperature would permit them to grow perhaps as far north as San Francisco (lat. 38°). In the valley of the Mississippi and on the Atlantic coast they extend northward to the parallel of 35°.

Formation and locality: Cretaceous (Puget Sound group). Bellingham Bay, Washington.

SABAL GRANDIFOLIA Newb. n. sp.

Pl. XXV; LXIII, fig. 5; LXIV, figs. 2, 2a.

Sabal Campbelli Newb. (in part). Boston Journ. Nat. Hist., Vol. VII (1863), p. 515.
"*Sabal Campbelli* Newb." Ills. Cret. and Tert. Pl. (1878), Pl. X.

Leaves very large, 8 to 10 feet in diameter, with eighty to ninety folds; petiole 1½ to 3 inches wide, flat or slightly arched above without a keel above or below; margins smooth, terminating in an arch, often unsymmetrical, on the upper side, from which the folds radiate; on the under side prolonged into a spine, 6 inches or more in length.

This species was first made known by specimens brought by Dr. Hayden from the valley of the Yellowstone. These represent both the under and upper surfaces of the leaf, and among them are fragments from the central and marginal portions. Some of these specimens are the originals of the figures given on Pls. XXV and LXIV. A portion of a leaf supposed to belong to this species is represented in Pl. LXIII, fig. 5. This was from Fischers Peak, New Mexico.

In the great number of the remains of palms found in the Tertiary and Cretaceous rocks of the west—trunks, leaves, and fruit—it has been very difficult to define distinct species, and it is probable that many years will elapse before perfect order can be brought out of the present confusion. The species now under consideration may, however, be identified by the large size of its leaf, its plain unkeeled petiole drawn out into a long acute spine on the under side, the very numerous folds, and the crowded, subequal nervation.

The only species that rivals it in size and is liable to be confounded with it is *Sabalites Grayanus* Lesq. (Tert. Fl., p. 112, Pl. XII, fig. 2), reported as found at "Golden, Colorado; Point of Rocks, Wyoming; Vancouver Island, and in Mississippi." Only fragments have, however, been found in some of these localities, and it is scarcely probable that their identification with the specimens from Golden will be confirmed by future observation. In the figure given by Lesquereux of the type of his species, the point of the petiole is not more than half as long as in some of the leaves of *Sabal grandifolia;* and if the strongly keeled petiole, of which a portion is represented on the plate cited above, can be accepted as normal for *S. Grayanus*, this would in itself be sufficient to distinguish the species. The petiole of the leaf of *S. grandifolia* is smooth and gently arched above and below, never keeled.

I formerly supposed this species to be identical with that found at Bellingham Bay, Washington (*S. Campbelli*, Newb.), and figured on Pl. XXI of this monograph, but that species has somewhat smaller leaves, with a less number of folds and less crowded nervation.

The best specimens yet obtained of *Sabal grandifolia* are those collected by Dr. Hayden in the Yellowstone Valley; but others, which indicate an almost equal size and exhibit essentially the same characters, were obtained by Mr. I. C. Russell from the green sandstones of the Laramie group on Fischers Peak, Colorado, and I have specimens representing this species from Walsenburg, Florence, Coal Basin, and other places where there are outcrops of the Laramie. Fan-palms occur in the Cretaceous rocks of Orcas Island and in the coal series of Fletts Creek, near Tacoma, Washington, but they are smaller and with fewer folds. Fragments of palm leaves were obtained by Dr. Evans on Vancouvers Island, and these have been referred to *Sabalites Grayanus* by Lesquereux, but they were very imperfect and of little value in the comparison of species.

Formation and locality: Cretaceous (Laramie group). Fischers Peak, Colorado, and Tertiary (Eocene ?), Yellowstone River, Montana.

Sabal imperialis Dn.

Pl. XVI, figs. 6, 6a.

Trans. Roy. Soc. Canada, Vol. I, Sec. IV, 1882 [1883], p. 26, Pl. VI; Vol. XI, Sec. IV, 1893 [1894], p. 57, Pl. XIV, fig. 61.

Sabal sp. Newb. Boston Journ. Nat. Hist., Vol. VII (1863), p. 515.

"Fragments only of a fan palm are contained in the collections made at Nanaimo; if, as now appears probable, the beds containing it are Cretaceous, it will doubtless prove to be a new species.

"The only tangible characters exhibited in the specimens yet obtained are in the nervation.

"The nerves are very fine, nearly sixty in each fold—six stronger ones on each side of the midrib, and between each two of these three finer ones, of which the middle is strongest."

Formation and locality: Cretaceous (Puget Sound group). Nanaimo, Vancouvers Island.

Sabal Powellii Newb.

Pl. LXIII, fig. 6; LXIV, figs. 1, 1a.

Proc. U. S. Nat. Mus., Vol. V (March 21, 1883), p. 504.

"Leaves of medium size, 4 or 5 feet in diameter, petiole smooth, unarmed, terminating above in a rounded or angular area from which the folds diverge, beneath concavely narrowing to form a spike 3 to 4 inches in length; rays about fifty, radiating from the end of the petiole, perhaps sixty in the entire leaf compressed to acute wedges where they issue from the petiole; strongly angled and attaining a maximum width of about 1 inch; nerves fine, about twelve stronger ones on each side of the keel, with finer intermediate ones too obscure for enumeration."

These leaves, as will be seen by the figures given, bear considerable resemblance to those described by Lesquereux under the name of *Flabellaria Eocenica* (Tert. Fl., p. 111, Pl. XIII, figs. 1–3), but a large number of specimens in the collections made at Green River, agreeing among themselves in all essential particulars, enable us to clearly define the species and show its distinctness from any yet found on this continent. From *Flabellaria Eocenica* it differs in having a larger number of folds and a longer point of support on the under side of the leaf. From *Sabal Campbelli* Newb.

it may be distinguished by its smaller size, less number of folds, and somewhat shorter spike of the petiole. *Sabal Grayanus* Lesq., is larger, with nearly double the amount of rays and a keeled petiole. *Sabal grandifolia* Newb. is much larger and like *S. Grayanus* has twice as many folds. These large species may be distinguished from each other by the concavely pointed and keeled petiole of *S. Grayanus*.

In the figures given, that on Pl. LXIII, fig. 6, represents the under side of the leaf at its base, showing pointed spike formed by the prolongation of the petiole. Pl. LXIV, fig. 1, represents the summit of the petiole and base of the leaf on the upper side. Here the rays are inserted on either side of a nearly symmetrical angle of the petiole, but other specimens show that the line of insertion of the rays is sometimes obliquely arched, precisely as in the figure of the base of the leaf of *Sabal grandifolia*, shown in fig. 2, Pl. LXIV. Fig. 1a of the same plate represents two folds of the leaf of *Sabal Powelli*, given of the natural size, to show the nervation.

Formation and locality: Tertiary (Green River group). Green River Station, Wyoming.

MANICARIA HAYDENII Newb.

Pl. LXIV, fig. 3.

Proc. U. S. Nat. Mus., Vol. V (March 21, 1883), p. 504.

"Frond large, leaves pinnately plicated, folds 1½ centimeters in width above, slightly narrowed below; flat or gently arched, smooth, springing from the midrib at an angle of 25 degrees above, 30 degrees below (in the specimens figured); folds attached to the midrib obliquely by the entire width, and to each other by their entire length (?); nervation fine, uniform (?), parallel."

The specimen figured is only a small portion of an entire leaf, and is inadequate to supply material for a satisfactory description. It is, however, evidently the central portion of a palm leaf of which the general form was elongated and the length probably many times the breadth. It was composed of a large number of pinnate, united, flattened folds, divergent from the midrib at an acute angle. These folds were not keeled like those of Flabellaria and Sabal, but either plain or gently arched: whether they were united throughout their entire length or were free toward the margin of the leaf is not certainly known, as we have nowhere seen the entire breadth of the leaf; but it is probable that they were joined to the margin. Until more complete specimens of this plant shall be obtained nothing positive can be said of its relations to living palms; but it is evidently allied to Heer's *Manicaria formosa* (Fl. Tert. Helv. I, p. 92, Pl. XXXVIII), and to the living Manicaria of South America. It certainly also belongs to the same genus with Lesquereux's palm leaves which he has grouped under the new generic name of Geonomites, but it has seemed to the writer more closely allied to Mannicaria than Geonoma. Its specific relations are also somewhat doubtful. It most resembles *Geonomites tenuirachis* Lesq. (Tert. Fl., p. 117, Pl. XI, fig. 1), but in the figured specimen of that plant the folds of the leaf spring from the midrib at a much more acute angle than in the specimen before us. This difference could be reconciled if it were certain that Lesquereux's specimens came from near the summit of the leaf, where the folds generally approach the direction of the midrib. Dr. Hayden reports the specimen to which the name of *Geonomites tenuirachis* was given as coming from the Raton Mountains and from strata which are older than that which furnishes the specimen now described. So far as now known there are no species common to the Raton Mountain beds and the Green River Tertiary. There is a strong probability, therefore, that the differences indicated have specific value.

Formation and locality: Tertiary (Green River group). Green River Station, Wyoming.

Order SMILACEÆ.

SMILAX CYCLOPHYLLA Newb.

Pl. LIV, fig. 3, in part.

Boston Journ. Nat. Hist., Vol. VII (1863), p. 520.

"Leaves circular or round, ovate, cordate or slightly peltate at base, five-nerved, central and interior pair of lateral nerves strongly marked, basilar pair delicate and scarcely reaching the middle of the leaf; secondary nervation forming a polygonal network more or less rectangular."

Unfortunately, the only specimen of this plant which I have—that collected by Professor Dana and figured in his Geology of the United States Exploring Expedition, Atlas, Pl. XXI, fig. 10—is imperfect, the upper part of the leaf being wanting. So far as its outline is indicated by the part which remains, it would seem to have been nearly orbicular. If such was the case, it resembled in general aspect the leaves of *S. orbicularis* Heer (Fl. Tert. Helv., Vol. III, p. 167, Pl. CXLVII, figs. 18, 19), and perhaps as much those of the living *S. rotundifolia*.

From *S. orbicularis* it differs, however, in the shortness of the exterior pair of lateral nerves and in the polygonal reticulation of the secondary nervation.

Formation and locality: Tertiary (Eocene?). Birch Bay, Washington.

Order IRIDACEÆ.

IRIS sp.? Newb.

Pl. XXII, fig. 6.

Ills. Cret. and Tert. Pl. (1878), Pl. VII, fig. 6.

NOTE.—The only manuscript relating to this specimen which I have been able to find is the above designation, in pencil, on the margin of the plate. Locality not known.—A. H.

MONOCOTYLEDON OF UNCERTAIN AFFINITIES.

MONOCOTYLEDON gen. et sp.? Hollick.

Pl. XLVI, fig. 9.

NOTE.—This figure apparently represents the lower portion of a leaf of some monocotyledon, but neither the specimen nor any manuscript referring to it was found except a memorandum of the locality on the plate margin, and there is no indication of Dr. Newberry's ideas concerning its affinities.—A. H.

Formation and locality: Tertiary (Miocene). Bridge Creek, Oregon.

DICOTYLEDONEÆ.

Order JUGLANDACEÆ.

JUGLANS NIGELLA Heer.

Pl. LI, figs. 2 (in part), 4.

Fl. Foss. Arct., Vol. II, Abth., II (1869), p. 38, Pl. IX, figs. 2-4.

NOTE.—So identified by Dr. Newberry, as indicated by memorandum on margin of plate.—A. H.

Formation and locality: Tertiary (Miocene). Admiralty Inlet, Alaska.

JUGLANS OCCIDENTALIS Newb.

Pl. LXV, fig. 1; LXVI, figs. 1–4c.

Proc. U. S. Nat. Mus., Vol. V (March 21, 1883), p. 507.

"Leaves somewhat variable in form and size, from 3 to 8 inches in length and 1 to 2 inches in width, but generally 6 inches long by 1½ inches wide, broad-lanceolate in outline, widest in the middle, summit acute, base rounded, often unsymmetrical; margins entire; nervation delicate; midrib straight; lateral nerves, about twenty on each side, gently curved upward, the lower ones branched and anastomosing near their extremities, the upper simple and terminating in the margins; tertiary nervation very delicate, or obscure from being buried in the parenchyma of the leaf, forming an open and irregular network. Fruit small, elongated, somewhat prismatic; divisions of the envelope lenticular in outline, narrow, thin."

The figures given of this species, collected by Dr. C. A. White, illustrate very well the average size and form of the leaves. The number contained in the collection is large, and they seem to have been extremely abundant in the locality where they were obtained. In a few instances they are found attached to the stems that bore them, but are generally separated and more or less torn and broken. The tree was evidently a strong-growing and luxuriant one, for some of the leaves are not less than 8 inches in length; the nervation is fine and often not discernible, probably from the thickness of the leaf; in some specimens, however, it is more distinct and has all the characters of that of the genus to which the leaves have been referred. The fruit, of which fortunately one specimen was found in immediate contact with the leaves, is small, marked with raised lines, elongate in form, and resembles more the fruit of *Carya oliveformis* than any other of our living species. It might be inferred from the small size of the nut and its elongated form that it was immature, but near it lies a segment of the envelope which has apparently exfoliated at maturity. As only one specimen of the fruit has been discovered, it is possible that it does not represent the average size and form. This fruit is distinctly that of a Carya and not of a Juglans, as now defined, but the leaf is more like that of the latter than the former genus. It distinctly falls within the old genus Juglans, but can hardly be reduced to either of its subdivisions which have now been given generic value.

A species of Juglans collected near the same locality as this has been described by Lesquereux under the name of *J. Schimperi*, but his description and figures indicate a plant different from this one. He describes the leaves of his species as being broadest near the base, long and narrow, having a nervation that differs from that of the leaves before us; the lateral nerves being camptodrome—that is, uniting in festoons along the borders and the tertiary nervation forming rectangular areoles—while in our species a large part of the lateral nerves terminate in the margins and the tertiary nervation is more open and irregular.

Formation and locality: Tertiary (Green River group). Green River, Wyoming.

CARYA ANTIQUORUM Newb.

Pl. XXXI, figs. 1–4.

Ann. N. Y. Lyc. Nat. Hist., Vol. IX (April, 1868), p. 72; Ills. Cret. and Tert. Pl. (1878), Pl. XXIII, figs. 1–4.

"Leaves pinnate, large, leaflets lanceolate, long-pointed, acute, sessile, finely serrate, middle leaflet broadly lanceolate, widest above the middle, narrowed to the base, which is somewhat unequal; lateral leaflets narrow, lanceolate, unsymmetrical throughout, somewhat falcate; nervation sharply defined, conspicuously parallel, medial nerve straight in the terminal leaflets, more or less curved in the lateral ones; secondary nerves springing from the midrib at a large angle, numerous, subparallel, all arched upward, their extremities prolonged parallel with the margins of the leaf; the upper ones strongly arched, but terminating more directly in the margins; tertiary nerves distinct, mostly simple, straight, and parallel among themselves, connecting adjacent secondary nerves nearly at right angles."

The form, serration, and nervation of these leaves are entirely those of Carya, and while without the fruit it may not be possible to fix their place in the series more definitely than to say that they represent the genus Juglans as formerly constituted, including Carya, we may at least refer them with confidence to a place within the limits of that genus. The leaves of the species of Carya and Juglans are very similar, so much so that some of the Caryas, such as *C. oliveformis*, have leaves that could in the fossil state hardly be distinguished from those of Juglans.

The specimens before us, however, seem to me to be more widely separated from those of the known species of Juglans than are those of the Pecan, and there seems little doubt that the tree, if now living, would fall within the limits of Carya.

In some specimens the lateral nerves are remarkably straight and numerous, giving to the leaf very much the aspect of those of Æsculus; but, from a comparison of the many leaves of this plant in the collection of Dr. Hayden, I infer that they were not palmately grouped, but pinnate, the form of the bases of the leaves indicating this.

The tertiary nervation is also quite different from that of Æsculus. In the latter genus it usually forms an exceedingly fine network filling the interspaces between the secondary nerves, in which the straight transverse latticelike bars so characteristic of the fossils before us are wanting. At least this is the case with our American "Buckeyes." In *Æ. Hippocastanum* of the Old World something of the kind is visible, but in prevalence and regularity very unlike that in the fossil.

In has been questioned whether these leaves should be referred to Juglans or Carya, and after somewhat extensive comparisons I was led to include them in the latter genus. In looking over the descriptions that have been given of various fossil species of Juglans we find that quite a large number of them should be rather reckoned as pertaining to Carya, taking the fruit as a criterion. For example, in the *J. corrugata* of Ludwig (Palæontogr., Vol. VIII, p. 178, Pl. LXX) the form and the nervation of the leaf is very much like this before us, only the nervation is a little less regular and the marginal serration is coarser. The fruit associated with these leaves is more nearly allied to that of our *J. nigra* than it is to the fruit of the common species of Carya, whereas in the illustrations of *J. levigata*, Brong., given by Ludwig (Palæontogr., Vol. VIII, p. 134, Pl. LIV, figs. 1–6), we have leaves which correspond in a general way with these, as far as form and marginal serration are concerned; nervation exceedingly regular, but more camptodrome, and the fruit distinctly that of Carya. It will be necessary to wait the discovery of the fruits which were connected with these strongly marked leaves, an event which will be likely to occur at no distant date, before deciding to which subdivision of the old genus Juglans it belongs.

Formation and locality: Tertiary (Eocene?). Mouth of Yellowstone River, Montana.

Order MYRICACEÆ.

MYRICA (?) TRIFOLIATA Newb. n. sp.

Pl. XIV, fig. 2.

Leaves in threes, lance-linear in outline, acute at summit and base; margins remotely and coarsely marked with appressed teeth; nervation delicate.

These are leaves which are manifestly distinct from any others from the Dakota sandstones yet described, and are referred to Myrica with doubt, as nothing but the general resemblance of form and marginal serration can be cited as proof of affinity. In due time, however, more material illustrating the species will be discovered, and, we may hope, also the fruit. At present it stands simply as a positive addition to the list of arborescent plants hitherto found in the Dakota group, but one of which the botanical relations must be determined by future observations.

Formation and locality: Cretaceous (Dakota group). Whetstone Creek, northeastern New Mexico.

Order SALICACEÆ.

POPULUS ACERIFOLIA Newb.

Pl. XXVIII, figs. 5–8.

Ann. N.Y. Lyc. Nat. Hist., Vol. IX (April, 1868), p. 65; Ills. Cret. and Tert. Pl. (1878), Pl. XIII, figs. 5–8.

"Leaves long-petioled, broad-ovate in outline, often somewhat three-lobed, obtuse, slightly cordate at base, margins coarsely and unequally crenate; nervation radiate, strong; medial nerve straight, giving off one pair of lateral nerves near the center of the leaf, and above these about three smaller ones on each side. From the base of the midrib spring two pairs of lateral nerves on each side. Of these the lower and smaller pair diverge at an angle of 60 degrees to 70 degrees with the midrib, are nearly straight, give off numerous short branches on the lower side, and terminate in the lateral margin below the middle. The second and larger pair of laterals diverge from the midrib at an angle of about 35 degrees to 45 degrees, are straight or slightly curved upward, terminating in the margins above the middle, or in the lobes, when lobes are developed; from these spring three or four branches on the outside, which, simple or branching, terminate in the scallops of the border. The tertiary nervation, shown very distinctly in some of the specimens, forms a network similar to that of the leaves of living species of Populus, of which the areolæ exhibit considerable diversity of form and size, being polygonal, with a roundish outline, or quadrangular."

Collected by Dr. F. V. Hayden.

The general aspect of these leaves is much like that of some of the living maples, but they are less distinctly trilobate. The crenation of the margin is coarse, irregular, and obtuse or rounded, as is usually the case with the leaves of a group of poplars, the leaves of which in other respects most resemble these. The surface is, in many specimens, somewhat roughened, as though in the living leaf it was canescent; also a common character among poplars, but rare or unknown among maples. The leaves of the maples are generally thin, and the network of the tertiary nerves is remarkably fine and uniform, affording a reliable generic character. This is visible in the leaves of all the recent maples, and is beautifully shown in the impressions of the leaves of *A. pseudoplatanus*, given in Ettingshausen and Pokorny's Physiotypia Plant. Austria, Pl. XVII, fig. 10.

Among fossil species this perhaps resembles most *P. leucophylla* (Foss. Flor. v. Gleichenberg, Denkschrift, k. k. Acad. Wien., Vol. VII (1854), p. 177, Pl. IV, figs. 6–9), but is much more distinctly crenate-toothed on the margin. The teeth of *P. leucophylla* are either obsolete or remote and acute, making a sinuate-dentate margin.

Formation and locality: Tertiary (Fort Union group). Fort Union, Dakota.

Populus cordata Newb.

Pl. XXIX, fig. 6.

Ann. N. Y. Lyc. Nat. Hist., Vol. IX (April, 1868), p. 60: Ills. Cret. and Tert. Pl. (1878), Pl. XIV, fig. 6.

"Leaves orbicular or round heart-shaped, deeply cordate at the base; margins strongly toothed, except the inner border of the lobes of the base;

nervation radiate; medial nerve straight, simple below, branched near the summit; lateral nerves, three pairs diverging at nearly equal angles, from a common point of origin; lower lateral nerves small, simple, arched upward at their summits, terminating in the margins; second pair of lateral nerves springing from the basal point of radiation nearly at right angles with the midrib, arching upward as they approach the lateral margins, and supporting each about three branches on the inner side; third pair of lateral nerves diverging from the midrib at its base at an angle of about 45 degrees, bearing one or two lateral branches, and terminating in the margin above the middle of the leaf."

Of this neat species there are no complete specimens in the collection made by Dr. Hayden, none of them showing the summit of the leaf. Enough is, however, discernible in them to show that they represent a species of Populus different from any other in the collection and from any before described. Of the species at present growing on the North American continent the leaves of *P. heterophylla* approach most nearly to these, but the nervation of the leaves of that tree is never so distinctly radiate.

In the character of its marginal dentations this species resembles *P. mutabilis crenata* Heer, but is clearly distinguished from that by its cordate base and corresponding radiate venation.

Populus Zaddachi Heer (Fl. Tert. Helv., Vol. III, p. 307) has a still closer resemblance to this than either of the species mentioned, and it has been regarded by Lesquereux as identical with it, but in all the figures of that species published the dentation of the margin is less strong and acute and the nervation is less radiate.

In *P. cordata* the basilar pair of lateral nerves reaches the margins below the middle of the leaf, and the second pair of lateral nerves spring from nearly the same point, while in *P. Zaddachi* the basilar pair reach the margin above the middle and the second pair leave the midrib considerably above the origin of the basilars.

The leaf figured by Professor Heer (Fl. Foss. Alaskana; Fl. Foss. Arct., Vol. II, Abth. II, Pl. II, fig. 5), has the character of the fossil before us and would seem to represent the same species. Yet notwithstanding the differences already pointed out, this is referred by Professor Heer to *P. Zaddachi*. The nervation is, however, so different from that of the typical forms of that species that I am compelled to regard them as distinct till proof is furnished to the contrary.

Formation and locality: Tertiary (Eocene?). Banks of Yellowstone River, Montana.

POPULUS (?) CORDIFOLIA Newb.

Pl. III, fig. 7; V, fig. 5.

Ann. N.Y. Lyc. Nat. Hist., Vol. IX (April, 1868), p. 18; Ills. Cret. and Tert. Pl. (1878), Pl. V, fig. 5.

"Leaves heart-shaped, slightly decurrent on the petiole; margins entire; nerves fine but distinctly defined; medial nerve straight or slightly curved, running to the margin; lateral nerves, six on each side, given off at an angle of about 50 degrees, nearly parallel among themselves, straight near the base of the leaf, slightly curved toward the summit; lower lateral nerves giving off on the lower side about four simple or once-forked, slightly curved branches, which terminate in the basilar margin; second pair of lateral nerves giving off about three similar branches on the lower side, which run to the lateral margins; third pair supporting about two, and fourth pair one branch on the lower side near the summit; tertiary nerves springing from the secondary nearly at right angles, slightly arched and running across nearly parallel to connect the adjacent secondary nerves."

Collected by Dr. F. V. Hayden.

In its general aspect this species closely resembles the preceding, but several specimens which I have before me agree in being less rounded and more heart-shaped, and the lateral nerves are more numerous and given off at a larger angle.

In these leaves the basilar nerves reach the lateral margins below the middle, and their second branches, as a consequence, have more the aspect of some of the leaves of the Cupuliferæ, such as Corylus. The latticelike arrangement of the tertiary veins in this, as in the other species of the group, is very characteristic of the Cupuliferæ, though not strictly limited to them. If we could imagine a Corylus with rounded or broadly cordate leaves, of which the margins were entire, we should have a very near approach to these plants.

Formation and locality: Cretaceous (Dakota group). Blackbird Hill, Nebraska.

POPULUS CUNEATA Newb.

Pl. XXVIII, figs. 2-4; XXIX, fig. 7.

Ann. N.Y. Lyc. Nat. Hist., Vol. IX (April, 1868), p. 64; Ills. Cret. and Tert. Pl. (1878), Pl. XIII, figs. 2-4, under *P. nervosa* var.; and Pl. XIV, fig. 7, under *P. Nebrascencis.*

"Leaves small, obovate, somewhat wedge-shaped at the base, obtusely pointed at the summit, coarsely, obtusely, and irregularly dentate on the margins, three-veined, basilar nerves given off at an acute angle, terminating above the middle of the margin; secondary nerves few-forked, and often inosculating."

This species is represented by numerous specimens in the collection made by Dr. Hayden. It will be seen to be distinctly separable from any of the species published with it, and the same may be said in regard to those published elsewhere. In general form it bears some resemblance to *P. attenuata*, Al. Braun (Heer, Fl. Tert. Helv., Vol. II, p. 15, Pl. LVII, figs. 8-12, and Pl. LVIII, figs. 1-4); also to some forms of *P. mutabilis*? Heer; but the nervation is less crowded than in those species, and both are acuminate-pointed. An elongated form is shown on Pl. XXIX, fig. 7.

Formation and locality: Tertiary (Eocene ?). Banks of Yellowstone River, Montana.

POPULUS CYCLOPHYLLA Heer.

Pl. III, figs. 3, 4; IV, fig. 1.

Proc. Phila. Acad. Nat. Sci., 1858, p. 266. Lesq., Ills. Cret. and Tert. Pl. (1878), Pl. III, figs. 3, 4; Pl. IV, fig. 1, under *P. litigiosa* Heer.

Populites cyclophylla (*Populus*) Heer. Lesq., Am. Journ. Sci., Vol. XLVI (July, 1868), p. 93.

Populites cyclophylla (Heer)? Lesq., Cret. Fl. (1874), p. 59, Pl. IV, fig. 5; Pl. XXIV, fig. 4.

The specimens upon which Heer founded his species are given on Pl. III, and are before me as I write. The smaller specimen represented by fig. 3 is characteristic and normal, except that it is not more than half the average size of the leaves of this species. Fig. 4 is but a fragment, and it is very doubtful whether it should be considered as belonging to *P. cyclophylla.* The leaf figured on Pl. IV is about of the average size, and though incomplete, may be accepted as a fair representative of the species. Such leaves are not uncommon in the Dakota group at Fort Harker, and a precisely similar one is figured by Lesquereux on Pl. IV of his Cretaceous Flora. It is more than doubtful whether any of these leaves belong to a true Populus; the nervation is more distinctly and regularly pinnate than in any living species of the genus, and the probability is that we have here the relics of a genus of trees now extinct, but closely related to the poplars.

Formation and locality: Cretaceous (Dakota group). Blackbird Hill, Nebraska, and Fort Harker, Kansas.

POPULUS (?) DEBEYANA Heer.

Pl. IV, fig. 3; V, fig. 7.

Nouv. Mem. Soc. Helv. Sci. Nat., Vol. XXII (1866), p. 14; Pl. I, fig. 1.

Juglans Debeyana (*Populus?*) Heer, Lesq. Am. Journ. Sci., Vol. XLVI (July, 1868), p. 101.

Juglans (?) *Debeyana* Heer, Lesq. Cret. Fl. (1874), p. 110, Pl. XXIII, figs. 1–5; Ills. Cret. and Tert. Pl. (1878), Pl. IV, fig. 3; V, fig. 7.

A number of leaves in the collection made by Dr. Hayden are clearly identical with that referred with doubt by Professor Heer to Populus from the generalities of its nervation, and impressions of what would seem to have been glands at the base on either side of the point of insertion of the petiole. In our specimens, however, there are no glandular impressions, and the departure from the normal type of nervation in Populus noticed by Professor Heer is still more conspicuous.

The strong pair of basilar nerves so characteristic of the poplars is entirely wanting, the inferior lateral nerves being small, and the stronger ones, which succeed them above, are not opposite. In view of the marked departure which these leaves exhibit from the nervation and form of the typical poplars, Professor Heer suggests that they may represent an extinct genus of the order Salicineæ, but it seems to me their affinities are closer with the Magnoliaceæ, and that it is even probable that they represent a species of the genus Magnolia.

Lesquereux has suggested that this leaf should be referred to Juglans, comparing it with *J. latifolia* Heer, from the Tertiary of Switzerland; but a considerable number of specimens before me fail to convince me of the justice of this reference, and yet they hardly suggest any other botanical relations. The leaves were evidently very thick and leathery, and the nervation is crowded and strong. It will be necessary that some other parts of the plant shall be obtained before this question can be satisfactorily settled.

Formation and locality: Cretaceous (Dakota group). Blackbird Hill, Nebraska.

Populus elliptica Newb.

Pl. III, figs. 1, 2.

Ann. N. Y. Lyc. Nat. Hist., Vol. IX (April, 1868), p. 16.

Ficus ? rhomboideus Lesq. Am. Journ. Sci., Vol. XLVI (July, 1868), p. 96; Ills. Cret. and Tert. Pl. (1878), Pl. III, figs. 1, 2.

Phyllites rhomboideus, Lesq. Cret. Fl. (1874), p. 112, Pl. VI, fig. 8.

"Leaves long-petioled, suborbicular or transversely elliptical, slightly cuneate at the base, and apiculate at summit; lower half of leaf entire; superior half, or more, very regularly and rather finely obtusely serrate, or crenate, the points of the teeth inclining upward: primary nerves usually five, sometimes three, radiating from the base at equal angles: from these the secondary nerves spring at acute angles."

This is an exceedingly neat and well-defined species, very fully represented in Dr. Hayden's collections. It is symmetrical in form, broader than high, forming a transverse ellipse, from the opposite sides of which rise the corresponding and equal projections of the apiculate summit and slightly decurrent base. The crenation of the upper portion of the leaf is very regular and neat, the teeth of small size, and turned upward. The general aspect of the leaf is not very different from that of some specimens of the living *P. tremuloides*, but the entire margins at the lower half of the leaf, the more elliptical outline, shorter point, and larger and more regular teeth, mark its specific differences with sufficient distinctness, while the correspondence which the leaves of the two species present in the general characters of form, nervation, and crenation, affords satisfactory evidence of generic identity.

In the Tertiary plants collected by Dr. Hayden on the upper Missouri a species of Populus occurs (*P. rotundifolia*), which exhibits a striking resemblance in general form to that now under consideration. In that species, however, the crenation of the superior margin is uniformly coarser and less acute, and the nervation is more delicate.

Formation and locality: Cretaceous (Dakota group). Blackbird Hill, Nebraska.

POPULUS FLABELLUM Newb

Pl. XX, fig. 4.

Boston Journ. Nat. Hist., Vol. VII. (1863), p. 524.

"Leaves flabellate, orbicular or reniform, obtuse, wedge-shaped at base, slightly decurrent onto the petiole. Margins entire or waved; principal nerves three, two lateral ones reaching nearly to the summit; secondary nerves fine, flexuous, forked."

There is no living species of Populus of which the normal form of the leaves approaches very closely to that of those under consideration, though one, three-nerved like these, may be occasionally found among the round-leavèd poplars. Among the Tertiary plants collected by Dr. Hayden on the Yellowstone is a species, yet unpublished, very much like this, both in the form and nervation of the leaves, and among the Cretaceous plants collected by him in Nebraska is another nearly equally like it; but in both these the upper margins of the leaves are more or less crenulated.

Formation and locality: Cretaceous (Puget Sound group). Chuckanutz, near Bellingham Bay, Washington.

POPULUS GENETRIX Newb.

Pl. XXVII, fig. 1.

Ann. N. Y. Lyc. Nat. Hist., Vol. IX (April, 1868), p. 64; Ills. Cret. and Tert. Pl. (1878), Pl. XII, fig. 1.

"Leaves large, cordate in form, acuminate; margins serrate, with rather small appressed teeth; three-nerved; nervation sparse but strong; midrib straight, with few small branches; basilar nerves very strong, given off at an acute angle, much branched at the summit, reaching nearly to the margin far above the middle; from each of the basilar-lateral nerves spring five to six exterior branches, the lower ones very strong and branched, the upper slender and simple."

In general aspect this leaf is very similar to that of the living *P. balsamifera*, and apparently differs from it only in its nervation. It is more decidedly three-nerved than those of any of the living group which it may be supposed to represent—*P. balsamifera*, *P. candicans*, *P. monilifera*, etc.; yet one may occasionally find a leaf of either of these species which in this respect approaches the fossil before us. The dentation of the margin is essentially that of *P. balsamifera*, and it can hardly be doubted that we have here the progenitor of one or more of the group of poplars with which I have compared it, and which now grow in the region where these fossil plants were collected.

The different species of Populus among the Tertiary plants collected by Dr. Hayden are far more generally three-nerved than are the living species which now inhabit this country. In this respect they resemble more the foreign *P. alba;* and it may be said that the majority of species described in this memoir are more closely allied to the section Coriaceæ than to the Balsamitæ.

Professor Schimper, in his *Paléontologie Végétale*, Volume II, page 690, refers this species to *Populus balsamoides* Goepp., basing this conclusion upon manuscript information received from Professor Heer. It is impossible, however, to harmonize the discrepancies which exist between the specimens before us and the figures and descriptions of Professor Heer. (Fl. Tert. Helv., Vol. II, p. 18, Pl. LIX; LX, figs. 1–3; LXIII, figs. 5, 6; Vol. III, p. 173.) In all the figures and descriptions given of *P. balsamoides* the medial nerve is far stronger than the lateral nerves. These form many pairs, of which the lower reach the margins below the middle of the leaf. On comparing the figure now given it will be seen that the differences are very marked, for the leaf of *P. genetrix* is practically three-nerved; at least the midribs and the two chief lateral nerves are nearly of equal strength. The lower pair of lateral nerves may be considered as mere branches of the second pair. From these differences I am compelled to regard *P. genetrix* and *P. balsamoides* as distinct species.

Formation and locality: Tertiary (Eocene ?). Banks of Yellowstone River, Montana.

POPULUS LITIGIOSA Heer.

Pl. III, fig. 6.

Nouv. Mem. Soc. Helv. Sci. Nat., Vol. XXII (1866), p. 13, Pl. I, fig. 2; Ills. Cret. and Tert. Pl. (1878), Pl. III, fig. 6.

The specimen of which the figure is cited above is that of which a tracing was sent by Mr. Meek to Professor Heer, and on which he based his description. This specimen is too imperfect to furnish a full diagnosis of the species or to afford comparison with the other fossil plants with which it is associated.

It is evident, however, that the general form of the leaf and the character of the nervation are similar to those of *P. cyclophylla* Heer, but it would seem that the margin is somewhat waved, and the nervation is rather more open than in the larger specimens of the species with which I have compared it. The basal pair of nerves also form a slightly greater angle with the midrib, and branches given off from them below are longer, supplying a broader expanse of the leaf. Like several of the other less common leaves of the Dakota group, these must remain as somewhat doubtful material until further collections shall add to our knowledge of them.

Formation and locality: Cretaceous (Dakota group). Blackbird Hill, Nebraska.

Populus microphylla Newb.

Pl. III, fig. 5.

Ann. N. Y. Lyc. Nat. Hist., Vol. IX (April, 1868), p. 17; Ills. Cret. and Tert. Pl. (1878), Pl. III, fig. 5.

"Leaves very small, scarcely an inch in length, roundish in outline, somewhat wedge-shaped at base, where they are entire; the upper part of the leaf rounded and deeply toothed, teeth conical, acute or slightly rounded at the summits; nerves radiating from the base, branching above, the branches terminating in the dentations of the margin."

This very neat species, from the collection made by Dr Hayden, might be supposed to be only a form of *P. elliptica,* with which it is associated, but a number of specimens of each show no shading into each other, and it is scarcely possible that so wide a variation of marginal dentation should exist in the same species. Although the leaves of *P. elliptica* are two or three times as large as those of the species under consideration, the teeth of the margins are less than half the size and are of a different type, being inclined upward, the sides of each tooth of unequal length, while the dentations of *P. microphylla* are conical in outline, with nearly equal sides.

Formation and locality: Cretaceous (Dakota group). Blackbird Hill, Nebraska.

POPULUS NEBRASCENSIS Newb.

Pl. XXVII. figs. 4, 5.

Ann. N. Y. Lyc. Nat. Hist., Vol. IX (April, 1868), p. 62; Ills. Cret. and Tert. Pl. (1878). Pl. XII. figs. 4, 5.

"Leaves long-petioled, 2 to 3 inches long, ovate, pointed, regularly rounded at the base, coarsely and irregularly toothed except near the base where the margins are entire; nervation strong, radiating from the base of the leaf; medial nerve straight, simple (or supporting very small nerves), except near the summit, where two or three larger branches rise from it; lateral nerves, two pairs on each side, springing from a common point of origin; lower pair arched upward, nearly parallel with the margin of the leaf, to which they send off one or more simple branches; second pair of laterals diverging from these at an angle of 30 degrees, arching upward, and running parallel with the midrib, terminating in the margin near the summit, each giving off about three exterior branches, which curve upward and terminate in the dentations of the border."

This species, by its general form and nervation, approaches closely to *P. smilacifolia*, but the base is rounded (sometimes slightly wedge-shaped), never distinctly cordate; the superior lateral nerves are not quite so much drawn together toward the summit, and the margins are differently and much more coarsely dentate.

A large number of specimens of this species present constant and distinctive characters. They exhibit considerable variation in size, being from 1 to 3 inches in length, but in form, nervation, and marginal dentation they are alike.

These specimens, from the collections made by Dr. F. V. Hayden, are derived from different localities, and without doubt represent a distinct species which was spread over the Tertiary continent.

By the character of the impressions left on the stone, as well as by the coarse and unequal dentation of the margins, we may infer an affinity between this and the downy-leafed poplars of the present epoch, such as *P. alba* of Europe, etc., while in the smooth surface and finely denticulate or entire margin of *P. smilacifolia* we have evidence of resemblance to *P. tremuloides*.

There is no fossil species for which this can well be mistaken. Some of the forms of *P. crenata* Unger (Foss. Fl. Sotzka, p. 166 [36], Pl. XXXVI [XV], figs. 2–5) resemble these leaves, but they are not so distinctly radiate nerved. Unger represents the teeth of the margin as more acute, and more like those of *P. tremula*, with which he compares his fossil species.

Some varieties of *Populus Zaddachi* Heer (Fl. Tert. Helv., Vol. III, p. 307; Fl. Foss. Arct., Vol. I, p. 98, Pl. VI, figs. 1–4; XV, fig. 1b) are somewhat like this species, and it has been suggested by Mr Lesquereux that they are identical; but in all the figures of that species published the margins are serrate-dentate, whereas in the leaves before us they are much more closely crenate-dentate; also most of the leaves are cordate at the base, and this is a feature given by Heer in his description, but among quite a large number of the leaves of *P. Nebrascensis* which have served as a basis for the specific description, the form is ovate, the base rounded, sometimes a little produced, but never cordate or even emarginate.

Formation and locality: Tertiary (Eocene?). Banks of the Yellowstone River, Montana.

Populus nervosa Newb.

Pl. XXVII, figs. 2, 3.

Ann. N. Y. Lyc. Nat. Hist., Vol. IX (April, 1868), p. 61; Ills. Cret. and Tert. Pl. (1878), Pl. XII, figs. 2, 3.

"Leaves rounded in outline, margins nearly entire, or slightly serrate at the base, sharply but not deeply toothed on the sides, on the summit strongly doubly serrate, with a tendency to become three-lobed; nervation strongly marked and crowded; basal nerves springing from the midrib above the margin, given off at an angle of 30 degrees or more, reaching the margin above the middle, where they terminate in the most prominent teeth or lobes; from these basilar nerves are given off five or six strong lateral nerves, which arch upward and, more or less forked, terminate in the marginal teeth; above the basilar nerves three or four pairs of strong lateral nerves are given off from the midrib, which run parallel with the basilar pair, and terminate, like them, in the compound teeth of the upper margin. The lateral nerves are connected by numerous strong secondary nerves, which are generally simple and slightly arched, sometimes broken, and anastomosing with each other. This latter character gives a lattice-like appearance to the leaf, to a degree unusual in the genus."

Collected by Dr. F. V. Hayden.

The strong nervation of this species is one of its most marked characters, and has suggested the name given to it. By this and the double dentation of the superior margin, as well as by their acerine form, these leaves are easily distinguishable from any of those with which they are associated and any hitherto described.

Formation and locality: Tertiary (Eocene?). Banks of Yellowstone River, Montana.

Populus nervosa elongata Newb.

Pl. XXVIII, fig. 1.

Populus nervosa var. B. *elongata* Newb. Ann. N. Y. Lyc. Nat. Hist., Vol. IX (April, 1868), p. 62; Ills. Cret. and Tert. Pl. (1878), Pl. XIII, fig. 1.

"Leaves ovoid or oblong in outline, wedge-shaped at base, abruptly pointed at summit, basal margins entire, sides rather finely toothed, superior margin, coarsely, somewhat doubly dentate; nervation strongly marked, less crowded than in var. A.; basal nerves springing from the midrib above the basal margin nearly straight, reaching the sides above the middle and terminating in the first large dentations of the upper margin; exterior lateral nerves of the basal pair, three or four in number, remote, nearly simple, curved upward, and terminating in the lateral teeth; secondary nerves above basal pair, three on each side of the midrib, parallel with the basal pair, and connected with them, each other, and the midrib, by numerous strong, generally simple, lattice nerves."

Collected by Dr. F. V. Hayden.

The nervation of these leaves is essentially the same as that of those last described, and which, notwithstanding the difference of form that they represent, I am inclined to consider as belonging to the same species. This diversity of form is not greater than may be seen in the leaves of any poplar tree, and the differences of dentation are not greater than those observed in different leaves of many living and fossil species. The origin of the large basilar nerves *above* the base of the leaves, the strong and latticed nervation, and the dentation of the same general character, with the fact that all the specimens are from the same locality, all combine to lead me to consider the two forms as specifically identical.

Formation and locality: Tertiary (Eocene?). Yellowstone River, Montana.

Populus polymorpha, Newb.

Pl. XLVI, figs. 3, 4; XLVII, fig. 4; XLIX, figs. 4, 7, 8, 9 [misprinted 1]; LVIII, fig. 4.

Proc. U. S. Nat. Mus., Vol. V (March 21, 1883), p. 506.

"Leaves petioled, ovate, rounded or slightly wedge-shaped at the base, acute or blunt-pointed at the summit; margins coarsely and irregularly crenate, dentate, or crenate-dentate; nervation strongly marked, pinnate; in the more elongated forms, about eight branches on each side of the midrib given off at an acute angle; in the broader forms the lower nerves issue at nearly a right angle; the upper ones at an angle larger than in the preceding form."

The leaves of this tree are the most numerous of all represented in the collection from Oregon made by Rev. Thomas Condon, several hundred in greater or less completeness being included in the specimens which have been passed in review. They show a marked diversity of form, some being long ovoid or elliptical, rather pointed at base and summit; others ovoid or roundish with a rounded base; some are light and delicate, others have strong nerves, and evidently were thick and leathery in texture. More generally a base similar to that on Pl. XLIX, fig. 9, accompanies a summit coarsely dentate or crenate.

It is with some hesitation that this leaf has been referred to Populus, but it presents greater affinities in nervation and marginal markings with this group than any with which they have been compared. The general aspect of the leaf represented on Pl. XLVI, fig. 4, is quite that of some of the poplars, particularly of the group represented by the abele (*P. alba*, L.), while the specimens figured on Pl. XLIX, figs. 4 and 7, and Pl. XLVII, fig. 4, are so different from the prevailing style of poplar leaves that the propriety of referring them to this genus seems questionable. There are, however, connecting links between all these different forms, and the general resemblance of the group to the leaves of the poplars is strong enough to warrant their provisional association.

Among the fossil leaves which have been described as species of Populus some of the many forms of *P. mutabilis* Heer show a considerable resemblance to these before us, and one phase of *Populus leucophylla* Ung. (Fl. Gleichenberg, p. 177 [21], Pl IV, figs. 6–9), especially that represented in fig. 9 of the plate cited, could hardly be distinguished from some of the Bridge Creek leaves

Formation and locality: Tertiary (Miocene). Bridge Creek, Oregon.

POPULUS RHOMBOIDEA Lesq.

Pl. XX, figs. 1, 2.

Am. Journ. Sci , Vol. XXVII (1859), p. 360.

In the collection of the Northwest Boundary Commission are numerous specimens which I have referred with some doubt to species of Populus described by Lesquereux. My specimens are, however, too imperfect to permit me to decide with certainty the question of their identity. Associated as they are with Inoceramus, there can be no reasonable doubt of their Cretaceous age.

Among the fossil leaves brought from Orcas Island, there are some which bear considerable resemblance to these, but they are too imperfect to render the comparison satisfactory.

Formation and locality: Cretaceous (Puget Sound group). Nanaimo, Vancouvers Island.

POPULUS ROTUNDIFOLIA Newb.

Pl. XXIX, figs. 1–4.

Proc. U. S. Nat. Mus., Vol. V (March 21, 1883), p. 506; Ills. Cret. and Tert. Pl. (1878), Pl. XIV, figs. 1–4, under *P. cuneata*.

"Leaves of small size, rarely more than an inch in diameter, approximately circular in outline, either quite round or transversely or longitudinally elliptical; slightly wedge-shaped at the base, and decurrent on the long petiole; basal margin entire; upper half of leaf coarsely crenate, dentate, and usually short pointed at the summit; nervation flabellate, consisting of a median and two principal lateral nerves, which give off numerous branches"

When the leaf is more than usually elongated, as in fig. 3, the basilar nerves spring from the midrib a little below the junction of the main lateral branches. The normal form is well represented in fig. 1, but it is not unusual to see those which are slightly flabelliform, like fig. 4. The tissue of the leaf would seem to have been thick and leathery, since the surfaces are unusually smooth, and the nerves sunk in the parenchyma are often scarcely perceptible.

The leaves described above present some anomalies in form and structure as compared with most of our poplars, since they are frequently flabelliform, and were apparently of much thicker and denser tissue than those of any living species. They present, however, a marked resemblance to those described and figured in this report under the names of *P. elliptica* and *P. flabellum*, one from the Dakota group of Kansas, the other from the Upper Cretaceous of Orcas Island on the northwest coast, and *P. cuneata* from the Tongue River Tertiary; and all the group, in form, nervation, and serration, have sufficient likeness to some of the living poplars, particularly to *P. tremuloides* of America and *P. pruinosa* of Songaria, to warrant their being included in the same genus.

There are some tropical trees of which the leaves present considerable resemblance to our fossils, especially one of the Proteaceæ (*Adenanthos cuneatus* of Australia), the leaves of which are small, cuneate at base, rounded at summit, where they are coarsely crenate, having almost precisely the form of one of the specimens of the fossil in question. This is, however, apparently an abnormal form, and the similarity which I have noticed is perhaps accidental and certainly of little value. The nervation of these fossil leaves is considerably different from that of Adenanthos, and a mere resemblance in form, however close, would hardly warrant us in supposing that the fossil plant could have any very near affinity with one so far removed geographically and botanically from the flora with which it is associated.

Probably all the specimens represented by figs. 1, 2, 3, and 4 belong to one species, though that from which fig. 3 was taken was obtained in a different locality from any of the others and has a somewhat different aspect. Taken by itself this might readily be supposed to belong to a rosaceous plant, perhaps a Rubus, Pyrus, or Cratægus; but it would be difficult to find its exact counterpart in any living species of these genera. It is perhaps safer to consider it only an unusual form of fig. 1 and refer it provisionally to the same species. Its geological value will be secured by the truthful figure given of it.

Formation and locality: Tertiary (Eocene?). Yellowstone River, Montana; Fort Union, Dakota; Carbon Station, Wyoming.

Populus smilacifolia Newb.

Pl. XXIX, fig. 5.

Ann. N. Y. Lyc. Nat. Hist., Vol. IX (April, 1868), p. 66; Ills. Cret. and Tert. Pl. (1878), Pl. XIV, fig. 5.

"Leaves ovate, pointed, slightly cordate at the base; margins finely and obtusely crenulated; nervation radiate, delicate and sparse; medial nerve straight, giving off only fine and scarcely perceptible lateral nerves below, and two or three longer branches near the summit; two pairs of lateral nerves radiate with the medial nerve from the same point at the base of the leaf; of these the lower two are small, nearly simple, and arched evenly upward; the other two, nearly as strong as the midrib, spring from the base at an angle of about 25 degrees, and after diverging to the middle of the leaf, curve upward toward the summit, near which they terminate in the margins. These lateral nerves support four or five simple or once-forked branches, each given off exteriorly, which curve upward, and terminate in the lateral margins. The tertiary nerves are given off nearly at right angles from the secondaries and form a delicate polygonal or quadrangular network over the surface of the leaf."

Collected by Dr. F. V. Hayden.

The lower pair of lateral nerves should properly be considered as branches of the larger ones, so that the leaf is more distinctly three-veined than that of any living species of Populus. This character, with the smooth surface and nearly entire margins, gives these leaves the general aspect of those of Smilax and suggested the name given them. Their nervation, however, is sufficiently distinct from that of Smilax, and is clearly that of Populus, though in a somewhat exaggerated form. In Smilax three or five nerves radiate from the base of the leaf and terminate together at the summit, which those of the leaves of Populus never do. In Smilax, too, the principal nerves give off no large branches, but all the interspaces are filled with a labyrinth of anastomosing veins, forming a very different network from that of Populus.

The marginal serration of the present species would seem to have been much like that of the leaves of the living *P. tremuloides*, but still finer, while the size of the leaf was considerably larger.

Formation and locality: Tertiary (Fort Union group). Fort Union, Dakota.

POPULITES ELEGANS Lesq.?

Pl. VIII, fig. 3.

Am. Journ. Sci., Vol. XLVI (July, 1868), p. 94.

NOTE.—So identified by Dr. Newberry, as indicated by memorandum on the margin of the plate.—A. H.

Formation and locality: Cretaceous (Dakota group). Fort Harker, Kansas.

SALIX ANGUSTA Al. Br.?

Pl. LXV, fig. 2.

In Bruckm. "Fl. Oening. Foss." Würtemb. Naturwiss. Jahresh. (1850), p. 220.
S. angustifolia Al. Br., in Buckland, Geol. and Mineral., p. 512 (1837).[1]

A very narrow-leaved willow; is exceedingly common in the Green River beds, some slabs of the rock being quite covered with the leaves. These are narrow, lanceolate, tapering gradually to a long and strong petiole and to a long, narrow, and acute point above. The margins are entire and sharply defined, the midrib strong, the lateral nerves numerous and fine.

In general form these leaves agree very well with the excellent figures of *Salix angusta*, given by Heer in his Fl. Tert. Helv., Vol. II, p. 30, Pl. LXIX, figs. 1–11, but the base is in our specimens narrower, so much so that the blade seems to be decurrent on the petiole. The leaves from Green River apparently represent the same species as that figured by Lesquereux, (Tert. Fl., p. 168, Pl. XXII, figs. 4, 5) but perhaps not that shown in fig. 5, as in all the many specimens now before me the base is narrower and more

[1] The oldest published name for this species is *S. angustifolia* Al. Br., 1837, but this name was preoccupied by the living species. This fact was apparently recognized by Braun, as he subsequently changed it to *S. angusta*, which is here adopted.—A. H.

wedge-shaped than the latter. Unfortunately the specimen represented in fig. 4 has the base and summit broken away, and the identification is therefore not absolutely certain, but as it was obtained in the same region where Dr. White collected the narrow-leaved willows before us there is every probability that they are the same.

Whether the narrow-leaved willow of the Green River beds is identical with that found in the so-called Miocene or Oeningen is, however, an open question. That both are willows there can be no reasonable doubt, but the leaves of so many species of willow are narrow lanceolate with tapering bases and summits that it is quite impossible to be sure of an identification based on a mere general resemblance. All we can say, therefore, is that during the deposition of the Green River Tertiary beds willow trees grew on the banks of the rivers and lakes of that region, having long, narrow leaves with simple margins and undistinguishable by any well-marked character from those obtained from the Tertiary of Oeningen.[1]

Formation and locality: Tertiary (Green River group). Green River, Wyoming.

SALIX CUNEATA Newb.

Pl. II, figs. 1, 2.

Ann. N. Y. Lyc. Nat. Hist., Vol. IX (April, 1868), p. 21; Ills. Cret. and Tert. Pl. (1878), Pl. I, figs. 1, 2 [fig. 1 under *Salix Meekii*].

"Leaves of medium size, sessile or short-petioled, entire, elongate, narrow, acute at both ends, broadest toward the apex, gradually narrowed below to the base; medial nerve distinct; secondary nerves delicate, springing from the midrib at an angle of about 20 degrees near the middle of the leaf, 15 to 20 degrees below, straight and parallel near the bases, gently arched above and inosculating near the margins."

Collected by Dr. F V. Hayden.

This species presents some marked characters by which it may be distinguished from those before described. It is true that the variations of form among the leaves of our recent species of willow are almost infinite, and even in the same species and from the same tree leaves may be obtained

[1] A comparison of our figure with those of Heer and Lesquereux leads me not only to doubt their identity, but to think that ours is more likely to be a Eucalyptus. The marginal nervation is certainly more characteristic of the latter genus than of Salix.—A. H.

of such different aspect that, taken separately, they might readily be mistaken for those of different species. Since the difficulty in the determination of recent willows is so great that it has become proverbial, specific distinctions derived from the leaves only, especially in those obtained from the same locality, may justly be looked upon with suspicion. Here, as elsewhere, however, it is probable that recent botany will derive some aid from the careful study of fossil plants, and the nervation will probably be found to afford constant characters where the outlines of the leaves can hardly be relied on.

It may be seen by reference to the foregoing descriptions of Salices that a number of characters combine to distinguish what, for geological convenience, I have chosen to regard as distinct species.

Salix Meekii is lanceolate, tapering nearly equally to both ends, which are alike acute; this leaf is petioled and the nervation regular and delicate.

S. flexuosa is sessile, linear, and rather abruptly narrowed to point and base; nervation obscure, apparently very delicate and uniform.

S. cuneata is comparatively thick and leathery, the form symmetrical, lanceolate, pointed, but scarcely acute at both ends; the midrib strong, prolonged into a short, robust petiole; secondary nerves unequal, given off at a large angle, thick at base, slender, tortuous, and irregularly confluent near the margins.

In *S. membranacea* the leaves are large and thin, broadest near the base, which is rounded, summit long-pointed and acute; nervation distinct and regular, but delicate.

Formation and locality: Cretaceous (Dakota group). Mouth of Big Sioux River, Nebraska.

Salix flexuosa Newb.

Pl. II, fig. 4; XIII, figs 3, 4; XIV, fig. 1.

Ann. N. Y. Lyc. Nat. Hist., Vol. IX (April, 1868), p. 21; Ills. Cret. and Tert. Pl. (1878), Pl. I, fig. 4.

"Leaves narrow, linear, pointed at each end, sessile or very short-petioled: medial nerve strong, generally somewhat flexuous; secondary nerves pinnate, leaving the principal nerve at an angle of about 40 degrees, somewhat branched and flexuous, but arching so as to inosculate near the margins."

This is perhaps only a variety of *S. Meekii*, which it resembles in its nervation, as far as can be observed in specimens fossilized in sandstone, but, although much narrower in its general form, it is less acuminate at either extremity, and is apparently sessile. As in some of our living narrow-leaved willows, these leaves are generally somewhat flexuous, and as they are seen lying in their natural curves on the surfaces of the rock they have as familiar and perfectly willowlike a look as leaves of *Salix angustifolia* would if artificially fossilized in the manner followed by Goeppert.

Since the above description was written I have collected this species from a number of widely separated localities and found it to hold its character with great constancy.

Formation and locality: Cretaceous (Dakota group). Big Sioux River, Blackbird Hill, Cedar Spring, etc., Nebraska, and Whetstone Creek, New Mexico.

SALIX FOLIOSA Newb. n. sp.

Pl. XIII, figs. 5, 6.

Leaves long-petioled, broadly linear; 8 to 9 inches long by 1 inch wide; suddenly narrowed to the base; acute at the summit; margins entire, sometimes undulate; nervation delicate.

Leaves of this species occur in great abundance on the banks of Whetstone Creek in northeastern New Mexico, and characteristic figures are given of specimens collected by myself in that locality. The leaves are larger than those of any other known Cretaceous Salix, unless it be *S. membranacea;* but it differs from that in its leaves being wedge-shaped instead of rounded at the base.

From the locality referred to, where the fossils are contained in a fine-grained, light-colored sandstone, in which the most delicate tissues would be preserved, we may expect the fruit of these and other fossil plants to be discovered, with a decided illumination of the botanical affinities of the plants of the Dakota group.

Formation and locality: Cretaceous (Dakota group). Whetstone Creek, New Mexico.

SALIX MEEKII Newb.

Pl. II, fig. 3.

Ann. N. Y. Lyc. Nat. Hist., Vol. IX (April, 1868), p. 19; Ills. Cret. and Tert. Pl. (1878), Pl. I, fig. 3 [under *S. cuneata*].

"Leaves petioled, thin and delicate, lanceolate, acute at both ends, nervation delicate, midrib slender, secondary nerves fine, springing from the medial nerve at an angle of 35 degrees, gently arched and anastomosing near the margins; network of tertiary veins somewhat lax, but composed of nervules of such tenuity as to be rarely visible."

This is the plant of which an outline sketch was sent Professor Heer by Mr. Meek. In that sketch the general form was alone given, the details of nervation, as well as the texture of the leaf, not being deducible from it. Professor Heer considered it a Laurus, and as probably identical with *Laurus primigenia* Ung., a common species in the Tertiary of Europe. Aside from the *a priori* improbability of this plant, found in the Middle Cretaceous rocks, being identical with one which in the Old World dates back no further than the Miocene, there are characters in the fossil itself which seem to separate it from even the genus Laurus. The nervation has a different aspect from that of any of the Lauraceæ with which I am acquainted, being both more lax and delicate, the secondary nerves less accurately arched, and their summits more wavy; the patterns formed by their anastomosis less regular and determinate. In these respects, as well as in its comparatively thin and delicate texture, it resembles much more the willows than the laurels.

It seems hardly worth while to compare the plant before us with any of the living willows, for everything indicates that all the species of the Cretaceous, both vegetable and animal, long since perished Among the great number of fossil species found in the Tertiary strata there are several which have a general resemblance to it and from which it might be unwise to regard it as distinct if they were from the same formation. *Salix elongata* Web. (Palæontogr. (1852), Pl. XIX, fig. 10) has nearly the same form, but the secondary nerves are given off at a larger angle and are much more arched.

From its associate species in the Cretaceous strata it seems not difficult to distinguish it. *Salicites Hartigi* Dunker (Palæontogr. (1856), p. 181,

Pl. XXXIV, fig. 2) is apparently much more strongly nerved. The general form was perhaps similar, although Dunker's specimen wants both point and base.

Formation and locality: Cretaceous (Dakota group). Blackbird Hill, Nebraska.

SALIX MEMBRANACEA Newb.

Pl. II, figs. 5–8,[1] 8a.

Ann. N. Y. Lyc. Nat. Hist., Vol. IX (April, 1868), p. 19; Ills. Cret. and Tert. Pl. (1878), Pl. I, figs. 5–8a [fig. 8a not named on plate].

"Leaves petioled, large, smooth, and thin, lanceolate, long-pointed, rounded or abruptly narrowed at the base, near which they are broadest; margins entire; medial nerve slender, often curved, secondary nerves remote, very regularly and uniformly arched from their bases, terminating in or produced along the margins till they anastomose; tertiary nerves given off nearly at right angles, forming a very uniform network of which the areoles are polygonal and often quadrate."

This is a strongly marked species, collected by Prof. George H. Cook, of which I have specimens fossilized in fine clay and exhibiting with great distinctness all the details of nervation. It was evidently thin and membranous in texture, though attaining a large size. Like most of the willows, it is frequently unsymmetrical, one side being most developed and the midrib curved.

The leaf is broadest near the base, and is thence narrowed into a long and acute point.

Formation and locality: Cretaceous (Raritan). Amboy Clays, Raritan River, New Jersey.

Order BETULACEÆ.

CARPINUS GRANDIS Ung.

Pl. LIV, fig. 3, in part; LV, fig. 6.

Synop. Foss. Pl. (1845), p. 220.

Leaves which seemed to represent this very widespread species of Carpinus were collected by Professor Dana at Birch Bay, near the mouth

[1] This specimen may also be found figured in Flora of the Amboy Clays, Pl. XXIX, fig. 12. (Mon. U. S. Geol. Surv., Vol. XXVI.)—A. H.

of Frazer River, and appear in Pl. XXI, fig. 10, of the Atlas which accompanies the Geology of the Wilkes Exploring Expedition. Upon the same slab are seen the branches of *Glyptostrobus Ungeri* (?), the branchlets and cone of *Taxodium distichum miocenum*, leaves of *Rhamnus Gaudini* (?), and *Smilax cyclophylla* Newb. Some of these are reproduced on Pl. LIV, fig. 3 (Carpinus and Smilax); Pl. LV, figs. 3 to 6 (Glyptostrobus, Taxodium, Carpinus). Very few fossil plants were brought from this locality, but they seem to represent a horizon somewhat different from that which has supplied any other specimens in the collection. Leaves of various kinds appear to be exceedingly abundant and beautifully preserved there, and it is to be hoped that the locality may be visited by some other collectors, who shall bring us a fuller representation of its riches.

Formation and locality: Tertiary (Eocene ?). Birch Bay, Washington.

CORYLUS AMERICANA FOSSILIS Newb.

Pl. XXIX, figs. 8–10.

Corylus Americana Walt. Newb. Ann. N. Y. Lyc. Nat. Hist., Vol. IX (April, 1868), p. 59; Ills. Cret. and Tert. Pl. (1878), Pl. XIV, figs. 8–10.

Among the variety of specimens of the leaves of *C. Americana* with which I have compared these fossils, there are some which, if fossilized, would form impressions absolutely undistinguishable from them, and I have therefore found it impossible to fix upon any characters by which they can be separated. As compared with the fossils which I have referred to *C. rostrata*, these leaves are a little more rounded in outline, the nervation somewhat more open and delicate, the marginal teeth more nearly equal in size, and more obtuse.

Of all the species of Corylus, living or fossil, which have been described, there is none of which the leaves so much resemble the ones under consideration as do those of *C. Americana*.

Collected by Dr. F. V. Hayden.

Formation and locality: Tertiary (Fort Union group). Fort Union, Dakota.

CORYLUS MACQUARRII (Forbes) Heer.

Pl. XXXII, fig. 5; XLVIII, fig. 4.

Alnites ? MacQuarrii Forbes. Quart. Journ. Geol. Soc. London, Vol. VII (1851), p. 103, Pl. IV, fig. 3.
Corylus MacQuarrii Heer. Urwelt. d. Schw. (1865), p. 321.
Corylus grandifolia Newb. Ann. N. Y. Lyc. Nat. Hist., Vol. IX (April, 1868), p. 59; Ills. Cret. and Tert. Pl. (1878), Pl. XV, fig. 5.

"Leaves large (5 to 6 inches long), short-petioled, unequally cordate at the base, pointed above, coarsely and unequally dentate; nervation strong; midrib straight or curved, not sinuous; lateral nerves, six to seven pairs; lower pair diverging at a larger angle than the upper ones, and supporting a number of short, generally simple, branches, on the lower side, which terminate in the basal margin; second pair diverging at an angle of 45 degrees, reaching the margin about the middle, supporting about four branches on the outside; upper pair simple or branched once, rarely twice."

Collected by Dr. F. V. Hayden.

This was evidently a large, thick, roughish leaf, having more the aspect and texture of the leaves of the mulberry than of the hazel. The nervation is, however, much nearer that of the latter genus. Indeed, in all essential characters it is the same as that of the three species of Corylus with which it is associated. The dentation of the margin, also, is acute, unequal, partially double, much more like that of the leaves of Corylus than of any of those with which I have compared it.

As is remarked in the description of *C. orbiculata*, a large amount of material has been collected and described since the description of *C. grandifolia* was written, and it has been shown that numerous leaves of Corylus of large size occur in the Tertiary beds of many parts of North America and extend to the European continent. Comparing our specimens with these figures and descriptions, we are led to believe that our *C. grandifolia* is only a large and strong form of *C. MacQuarrii*.

Formation and locality: Tertiary (Fort Union group). Fort Union, Dakota.

Corylus orbiculata Newb.

Pl. XXXII, fig. 4.

Ann. N. Y. Lyc. Nat. Hist., Vol. IX (April, 1868), p. 58; Ills. Cret. and Tert. Pl. (1878), Pl. XV, fig. 4.

"Leaves small, orbicular, or nearly so, slightly and unequally cordate at base, blunt-pointed above; margins set with fine and nearly equal teeth; nervation strong; midrib curved and slightly sinuous; lateral nerves about seven pairs, mostly straight and nearly parallel among themselves, lower pair sending off each seven to eight short, simple or forked branches which terminate in the teeth of the edge; second pair supporting each about three branches of similar character; upper lateral nerves simple, or having each two to three branches near the summit; tertiary nerves parallel, distinct."

Collected by Dr. F. V. Hayden.

This is another hazel-like leaf, of which, without the fruit, the classification must be somewhat doubtful. The general form is more like that of the leaves of Tilia (*T. Americana* and *T. Europæa*), being much rounder than those of any species of Corylus with which I am familiar. The nervation is, however, different from that of Tilia and is, in fact, altogether that of Corylus. In Tilia the leaves are usually broadly cordate; the nervation of the base and lateral portions of the leaf being supplied from the first or basal pair of lateral nerves, which are largely developed, much branched, and reach considerably above the middle point of the lateral margin. In Corylus, on the contrary, the basal nerves are short and supply only the basal margins; the second pair of lateral nerves is relatively more developed than in Tilia, Morus, etc., and in the number and parallelism of the lateral nerves their leaves approach more nearly to the strictly feather-veined leaves of Fagus, Alnus, etc.

Since the above description was written Professor Heer has published his splendid series of volumes on the arctic flora, and has in a number of places made reference to or given figures and descriptions of *Corylus Mac-Quarrii*, which shows that this was a very variable species, and perhaps the leaf under consideration, to which from its circular form I gave the name *C. orbiculata*, is but one of the numerous varieties of this plant, which seems to have been widely spread over all the North American continent during Tertiary times. Further collections made in the country bordering the upper Missouri will doubtless supply a larger amount of material illustrating this species, and may prove it to be worthy of recognition as distinct from all others. Taken by itself it presents such striking differences from the other species of Corylus known that it has seemed to me best to give it a distinct name.

Formation and locality: Tertiary (Fort Union group). Fort Union, Dakota.

CORYLUS ROSTRATA FOSSILIS Newb.

Pl. XXXII, figs. 1–3.

Corylus rostrata Ait. Newb. Ann. N. Y. Lyc. Nat. Hist., Vol. IX (April, 1868), p. 60; Ills. Cret. and Tert. Pl. (1878), Pl. XV, figs. 1–3.

"These leaves offer no characters by which they can be distinguished from those of the living 'beaked hazel-nut.' They are clearly those of a hazel, and show such a perfect correspondence with those of one of the species living in the region where these fossils occur that, until the fruit shall be found and the question definitely settled, I have thought it best to consider them as identical."

Collected by Dr. F. V. Hayden.

Corylus insignis Heer (Fl. Tert. Helv., Vol. II, p. 43. Pl. LXXIII, figs. 11–17; Fl. Foss. Arct., Vol. II, Abth. IV, p. 469, Pl. XLIX, fig. 5) is closely allied to the plant under consideration, and should perhaps be united with it.

Formation and locality: Tertiary (Fort Union group). Fort Union, Dakota.

BETULA ANGUSTIFOLIA Newb.

Pl. XLVI, fig. 5; XLVII, fig. 5.

Proc. U. S. Nat. Mus., Vol. V (March 21, 1883), p. 508.

"Leaves petioled, oblong-lanceolate, 3 inches long by 1 inch wide; wedge-shaped or slightly rounded at the base, acuminate at summit; margins finely serrate below, coarsely and doubly serrate above: nerves slender, about eight branches on each side of the midrib."

These leaves, of which there are a number in the collection made by Rev. Thomas Condon, are distinguished from the other species of Betula with which they are associated by their narrower and more elongated form and the coarse, double-crowded dentation of the upper portion of the leaf. They are also separated by these characters from the numerous other species of the genus mentioned by Professor Heer as found in the Tertiary of the northern part of this continent, *B. macrophylla* (Fl. Foss. Arct., Vol. I, p. 146, Pl. XXV, figs. 11–19), *B. prisca* Ett. (Heer, Fl. Foss. Arct., Vol. I, p. 148, Pl. XXV, figs. 20–25; Fl. Foss. Arct., Vol. II, Abth. II, p. 28, Pl. V, figs. 3–7). They bear a closer resemblance to the leaves of *B. ostryæfolia* Sap. (Fl. Foss. Sezanne, p. 345 [57], Pl. XXV [IV], fig. 8), and *B. Sezannensis* Wat. (Pl. Foss. Bass. Paris, p. 130, Pl. XXXIV, fig. 6); but both these species are crenato-dentate, while in the leaves before us the teeth are acute. Among living species this may be compared with *B. lenta* Willd., but is narrower and the marginal dentation is less uniform.

The leaf figured on Pl. XLVII, fig. 5, differs somewhat from those which have been here associated with it, in its more rounded base, coarser dentation below, more open and opposite nervation, and it may represent another species. Part of these differences, however, is probably due to difference in the preservation of the marginal dentation.

Formation and locality: Tertiary (Miocene). Bridge Creek, Oregon.

Betula heterodonta Newb.

Pl. XLIV, figs. 1–4; XLV, figs. 1, 6.

Proc. U. S. Nat. Mus., Vol. V (March 21, 1883), p. 508.

"Leaf 2 to 4 inches in length, long petioled, ovate, acuminate, rounded at the base; margins coarsely and irregularly serrate, the principal denticles receiving the terminations of the nerve branches; the sinuses between these sometimes plain, sometimes set with a few small teeth; nervation delicate, about eight branches given off from each side of the midrib."

The collection from Oregon, made by Rev. Thomas Condon, contains a large number of leaves belonging to this species. These present considerable variety in size, as will be seen in the figures. There is also some diversity in the degree of denticulation of the margin. The examples which show this best among those figured are Pl. XLIV, fig. 2; Pl. XLV, fig. 1. Here we see the lateral nerve branches running into prominent teeth of the margin as in many other species of Betula, such as *B. nigra* L., *B. Blancheti* Heer, but the sinuses between these large teeth are sometimes entire, sometimes bear a few small teeth. The marginal markings are hardly shown in the largest leaf now figured, and it is represented simply to give the form and dimensions, but in the other figures it will be seen that the variation in the dentation is considerable.

In form and general aspect the leaf represented in fig. approaches closely to *B. grandifolia* Ett., as shown by Heer in Fl. Foss. Arct., Vol. II, Abth. II, Pl. V, fig. 8, but the marginal dentation is different. The smaller leaves may in a like manner be compared with Heer's figure of *B. prisca* (*loc. cit.*, fig. 3), but here again the dentation is unlike that of our specimens. It is, however, possible that further observations will lead to the combination of the two species referred to, *B. prisca* and *B. grandifolia*, which are not very unlike with those which occur in such abundance in the locality from which our specimens were derived. The differences, however, are so clearly perceptible that without further information to the contrary the union of these species is not warranted. On Pl. XLV, fig. 6, is represented a samara.

Formation and locality: Tertiary (Miocene). Bridge Creek, Oregon.

BETULA sp.? Newb.

Pl. LVII, fig. 4.

NOTE.—Marked as above on the margin of the plate by Dr. Newberry. Further information lacking, but locality probably Bridge Creek, Oregon.—A. H.

ALNUS ALASKANA Newb.

Pl. XLVIII, fig. 8.

Proc. U. S. Nat. Mus., Vol. V (March 21, 1883), p. 509.

"Leaf large, oblong ovoid, acuminate, rounded, or slightly heart-shaped at base; nervation crowded, sixteen to eighteen branches on each side of the midrib; margins set with very numerous, small, uniform, acute teeth."

We have here a strongly marked species of Alnus, apparently distinct from any hitherto described. Its conspicuous characteristics are its very crowded nervation, the broad, oblong ovoid outline, and the minute and regular serration of the margin. In this latter character it resembles *A. ciliata*, living in eastern North America, but differs in the form of the leaf and in the greater number of lateral nerve branches. From *A serrata*, figured on Pl. XXXIII of this monograph, it is at once distinguished by the very much finer marginal dentation, as well as by the greater number of nerve branches. The remarkably fine denticulation of the margin is a character which distinguishes it from *A. Kiefersteinii* and *A. nostratum* the species most commonly preserved in the Tertiary rocks.

Formation and locality: Tertiary (Miocene). Kootznahoo Archipelago, latitude 57° 35′, longitude 134° 19′, Alaska. Collected by United States steamer *Saginaw*, February 18, 1869.

Alnus serrata Newb.

Pl. XXXIII, fig. 11.

Ann. N. Y. Lyc. Nat. Hist., Vol. IX (April, 1868), p. 55; Ills. Cret. and Tert. Pl. (1878), Pl. XVI, fig. 11.

"Leaves oval or elliptical, slightly cordate at the base, rounded or subacute at summit; margins serrate throughout, serrations fine, sharp, and appressed below, coarse and double above; nervation pinnate, strongly marked; basilar pair of lateral nerves short and simple, upper ones branched near the extremities."

Collected by Dr. F. V. Hayden.

These leaves have nearly the form of *Alnus Kieffersteinii* Ung. (Chlor. Prot., p. 115, Pl. XXXIII, figs. 1–4), and a nervation similar in kind, but more crowded. The marginal serration is also coarser.

Formation and locality · Tertiary (Eocene ?). Banks of Yellowstone River, Montana.

Alnus serrulata fossilis Newb. n. sp.

Pl. XLVI, fig. 6.

Among the leaves from Bridge Creek occurs one very beautifully preserved, which is represented in fig. 6, Pl. XLVI. It will be seen at a glance that it closely resembles the leaves of *A. serrulata*, and I have been unable to find any characters upon which to base a distinction. More material will of course be needed before the fact may be considered established that our most common alder was growing in the Tertiary. There would be nothing surprising, however, in such a discovery; indeed, it was to be expected that this species, so widespread as it now is, should have some representative in the Tertiary flora. We know that our living flora of North America is the progeny by direct descent of the Tertiary flora, and the result of investigation will undoubtedly be to increase the number of species considered identical in the two floras.

Formation and locality: Tertiary (Miocene). Bridge Creek, Oregon.

ALNUS sp.? Newb.

Pl. XLVI, fig. 7.

NOTE.—Accompanying this figure, on the margin of the plate, and on the specimen label, are memoranda by Dr. Newberry referring it to this genus and giving the locality. Further information in relation to it is lacking.—A. H.

Formation and locality: Tertiary (Miocene). Bridge Creek, Oregon.

ALNITES GRANDIFOLIA Newb.

Pl. IV, fig. 2.

Ann. N. Y. Lyc. Nat. Hist., Vol. IX (April, 1868), p. 9 (name only); Ills. Cret. and Tert. Pl. (1878), Pl. IV, fig. 2.

Leaf orbicular, with coarsely and obtusely dentate margins; nervation strong, consisting of a straight midrib with six to seven lateral branches, which are nearly opposite and diverge at less than a right angle with each other. Branchlets spring from these on the outer side; several from the lower pair, two from the second pair, which, like the upper of the two given off from each of the third pair, terminate in the dentations of the border. The tertiary nerves which connect the secondary branches are imperfectly parallel, somewhat closely approximate and continuous, forming a more regular lattice work than is formed in any of the associated leaves except those of Protophyllum.

No complete specimens of this strongly marked leaf have as yet been obtained, and nothing but an approximation can be made to its botanical affinities. It is, however, so distinctly marked that it deserves notice if for nothing else than that the attention of collectors may be drawn to it. It has been provisionally placed in the ill-defined genus Alnites, because it bears considerable resemblance to some of the leaves of Alnus, but perhaps quite as much to those of Hamamelis. The existence of closely related species of the latter genus in the floras of America and Japan gives reason to suppose that this was an element in the old flora which spread from America into Asia and Europe, and therefore gives a probability of its being found in the Tertiary and even Cretaceous flora.

Formation and locality: Cretaceous (Dakota group). Blackbird Hill, Nebraska.

Order FAGACEÆ.

FAGUS CRETACEA Newb.

Pl. I, fig. 3.

Ann. N. Y. Lyc. Nat. Hist., Vol. IX (April, 1868), p. 23 (named, but not specifically described); Ills. Cret. and Tert. Pl. (1878), Pl. II, fig. 3.

Leaves 2 to 3 inches in length, ovate in outline, pointed above and below, petioled, nervation sharply defined, regular, lateral nerves parallel, straight below, gently arched above, terminating in the margins, which are sometimes gently undulate, the nerves terminating in the prominences of the margins; in other leaves the margins are quite entire and nothing of this last-mentioned character is seen.

Collected by Dr. F. V. Hayden.

This pretty species is represented in the collection by but a single specimen. This is, however, remarkably well preserved, giving the general form and the details of nervation with great distinctness. From the character of the nervation I have little hesitation in referring it to the genus Fagus. Some of the Rhamnaceæ, particularly species of Rhamnus and Frangula, have leaves which would be very like the one before us if fossilized; but in the fossil plant the lateral nerves are sharply defined, numerous, almost perfectly parallel among themselves, and run quite to the margins, which are seen to be slightly waved, the termini of the nerves being most prominent and the intervals between them forming shallow sinuses. In Rhamnus, however, even in *R. frangula*, of which the leaves so much resemble this, the margins are not waved and the lateral nerves do not terminate as distinctly in them as they do in Fagus and in our fossil.

A striking similarity may be noticed between some of the leaves of the living *Fagus sylvatica*, and this, though there is no probability of that species having begun its life so early in the history of the globe as the first part of the Cretaceous period. The resemblance is noted only as giving good grounds for the reference of the fossil to the genus Fagus. It will be necessary, however, to find the fruit before the fact can be accepted as fully proven of the existence of beeches during the Cretaceous.

A large number of fossil species of Fagus have been described from the Tertiaries of Europe by Unger, Dunker, Heer, etc., but the genus has never before been obtained from the Cretaceous formation.

Formation and locality: Cretaceous (Dakota group). Smoky Hill, Kansas.

QUERCUS ANTIQUA Newb.

Pl. XIII, fig. 2.

Ann. N. Y. Lyc. Nat. Hist., Vol. IX (April, 1868), p. 26.

"Leaves of medium size, lanceolate in outline, acute, often somewhat flexuous; margins serrate-dentate, with strong, obtuse teeth, which are appressed or turned toward the summit; midrib strong and reaching the apex; lateral nerves numerous, of unequal strength, gently arched upward, terminating in the marginal teeth."

The specimens upon which this description is based are fossilized in a somewhat coarse ferruginous sandstone, which has not preserved the minor details of the nervation; but the generalities of form and structure, which are clearly enough shown, seem to indicate that it represented in the Cretaceous flora the chestnut oaks of the present epoch. Several Tertiary species bear considerable resemblance to it, as *Q. Mediterranea* Ung., and *Q Haidingera* Ett.; but in both these species the marginal dentations are less uniform in size, and, when having a similar outline, are smaller.

Formation and locality: Cretaceous (Dakota group). Banks of Rio Dolores, Utah.

QUERCUS BANKSLÆFOLIA Newb.

Pl. XVIII, figs. 2–5.

Bost. Journ. Nat. Hist., Vol. VII (1863), p. 522.

"Leaves very long, linear, lanceolate, long-pointed and acute at either end; margins set with numerous nearly uniform, acute, appressed teeth turned toward the superior extremity: midrib strong, running the entire length of the leaf; lateral veins numerous, simple, strongly marked, parallel, arched upward, terminating in the teeth of the margin; reticulated nervation buried in the thick parenchyma of the leaf, and generally invisible in the fossil state."

This beautiful leaf resembles, in the style and strength of its nervation, those of the living chestnut oak, but is more slender than any other species, living or fossil, which has come under my observation.

Among described fossil species *Q. Drymeja* Ung. (Chlor. Prot., p. 113, Pl. XXXII, figs. 1–4), *Q. lonchitis* Ung. (Fl. Sotzka, Pl. IX, figs. 3–8), and *Q Saffordi* Lesq. (Geol. Survey of Arkansas, p. 319, Tab. VI, fig. 3) seem to approach it most closely, the former two, indeed, being very nearly allied to it; but in these species the leaves are broader and the lateral nerves are more remote. In *Q. Saffordi* the leaf is, perhaps, equally slender, but the teeth are coarser and less depressed, and the nervation much less strong and regular, resembling in this respect that of the willow oaks (*Q. Phellos*, etc.). The living species with which our plant may be compared are *Q. Xalapensis* and, judging from Professor Heer's description of it, *Q. Sartorii* Liebman. Both of these are from Mexico.

Formation and locality: Cretaceous (Puget Sound group). Chuckanutz, near Bellingham Bay, Washington.

QUERCUS CASTANOIDES Newb.

Pl. LXV, fig. 6.

Proc. U. S. Nat. Mus., Vol. V (March 21, 1883), p. 506.

"Leaf linear-lanceolate, acute, 6 inches long by 1 inch broad; margins remotely and somewhat irregularly set with coarse, in some cases spinous, teeth; nervation strong; midrib straight, sharply defined; lateral branches unequally spaced, simple, forked near the extremity, terminating in the marginal denticles."

Only imperfect fragments of this leaf are contained in the collection made by Dr. C. A. White, but these are quite sufficient to show the species to be distinct from any other known. The irregularity in the dentation of the margin and in the spacing of the main nerve branches separate this from the chestnuts and bring it within the genus Quercus, and it would seem to be allied to the living and fossil chestnut oaks. More leaves and the fructification will be needed before a complete description can be written, but it is hoped that the figure now given will serve for its recognition, and that since it is plainly different from any of its associates it will be in the future identified and its structure and relations be more fully made out.

Formation and locality: Tertiary (Green River group). Green River, Wyoming.

QUERCUS CASTANOPSIS Newb.

Pl. LVI, Fig. 4.

Proc. U. S. Nat. Mus., Vol. V (March 21, 1883), p. 505.

"Leaves oblong-elliptical, rounded at the base; nervation regular; midrib straight, branches parallel, simple, terminating in the principal teeth of the margin; margins doubly dentate, the larger teeth receiving the extremities of the nerve branches, and each carrying a minor denticle; upper surface smooth; texture of the leaf coriaceous."

Collected by Mr. S. M. Rothhammer.

But a single specimen of this leaf is before us, yet this is so peculiar and strongly marked that it seems to deserve description. In general aspect it closely approaches the leaves of Castanea and Fagus, but the margins are doubly dentate, a feature I have not found in any of the beeches or chestnuts. It is present, however, in some of the chestnut oaks, as in *Q. Olafseni* Heer (Fl. Foss Arct., Vol. I, p. 109, Pl. X, fig. 5; XI, figs. 7–11; XLVI, fig. 10). It seems safer, therefore, to refer the leaf to Quercus rather than to the other genera mentioned.

Formation and locality: Tertiary (Eocene?). Yellowstone River, Montana.

QUERCUS CONSIMILIS Newb.

Pl. XLIII, figs. 2–5, 7–10.

Proc. U. S. Nat. Mus., Vol. V (March 21, 1883), p. 505.

"Leaves petioled, lanceolate, acuminate, wedge-shaped or rounded at base, where they are often unequal; margins usually dentate, occasionally only undulate, sometimes entire below, denticulate above; teeth acute, often spinous, sometimes short and closely appressed; nervation fine and regular; lateral nerves slender, parallel, generally arched upward; below, where margin is entire, camptodrome; above, craspedodrome, the branches terminating in the marginal teeth; tertiary nervation consisting of minute branches connecting the lateral nerves either directly or anastomosing, with fine quadrangular network filling the intervals. Fruit ovoid; when mature 2 centimeters in length by 15 millimeters in breadth; cupule scaly, covering nearly half of the glans."

Collected by Rev. Thomas Condon.

Of the leaves of this species the collection contains many hundreds which show a considerable diversity of size and form; some are only 2 inches in length, others 6; some have the margin acutely toothed throughout, in others the margin of the lower part of the leaf is entire, the upper denticulate; while in others still the margins are entire or gently undulate to near the summit where they are always more or less denticulate. These leaves closely resemble those that have been described under the name of *Q. Drymeja* Ung. (Chlor. Prot., p. 113, Pl. XXXII, figs. 1–4; Foss. Fl. Sotzka, p. 163 [33], Pl. XXIX [VIII], figs. 1, 2; Heer, Fl. Tert. Helv., Vol. II, p. 50, Pl. LXXV, figs. 18–20), and also some of them, those in which the margins are closely and sharply denticulate, are not unlike *Q. lonchitis* Ung. (Foss. Fl. Sotzka, p. 33, Pl. IX [XXX], figs. 3–8), but the prevailing character is such as apparently distinguishes them from either of these species or any other described, namely, first, the base broader than in *Q. Drymeja*, frequently entire for one-third or one-half of the length of the leaf; second, the margins generally denticulate, but sometimes merely undulate or entire except near the summit—a range of variation which does not seem to prevail in the species named.

In the figures given on Pl. XLIII, fig. 2 represents the more common or average form and size, figs. 3, 4, and 5 the more denticulate variety. In order to make the series complete it would have been necessary to occupy the entire plate with representations of the different forms observable in the collection. In many of the specimens the preservation is complete, the outlines being sharply defined, the minutest subdivisions of the nerves being retained. The number of acorns and cupules contained in the collection is also large, and while it is possible that not all belong to this species, as it is by far the most abundant we are compelled to connect the abundant acorns with the numerous leaves. In fig. 7 an immature acorn is shown still attached to the stem; in fig. 8, a detached cupule; in fig. 9, the base of a large acorn; in fig. 10, a large cupule seen from above.

Formation and locality: Tertiary (Miocene). Bridge Creek, Oregon.

QUERCUS CORIACEA Newb.

Pl. XIX, figs. 1–3; XX, fig. 5.

Bost. Journ. Nat. Hist., Vol. VII (1863), p. 521.

"Leaves lanceolate, long-pointed, acute, wedge-shaped at base, decurrent on the petiole; margins entire, or rarely bearing a few acute teeth toward the summit; nervation strongly marked; midrib strong; lateral nerves numerous, subparallel, branching and inosculating at the summit."

This is one of the willow oaks represented among recent species by *Q. imbricaria*, etc.

The figures given illustrate the variations of form exhibited in the collection. From these it will be seen that, with the general character of *Q. chlorophylla* Ung. and *Q. elæna* Ung., it is distinct from both, the first being rounded above and with finer nerves, the second larger and narrower, with a nervation finer and closer and the summits of the lateral nerves more distinctly and regularly united.

Formation and locality: Cretaceous (Puget Sound group). Chuckanutz, near Bellingham Bay, Washington.

QUERCUS DUBIA Newb.

Pl. XXXVII, fig. 5.

Proc. U. S. Nat. Mus., Vol. V (March 21, 1883), p. 506; Ills. Cret. and Tert. Pl., (1878), Pl. XX, fig. 5, under *Phyllites cupanioides*.

"Leaf ovoid in outline, unsymmetrical; margins strongly and remotely toothed; teeth subacute or obtuse; nervation delicate; midrib flexuous; lateral branches, about six on a side, somewhat waved, branched, and interlocking, and terminating in the marginal denticles; surface smooth, consistence probably somewhat coriaceous."

Collected by Dr. F. V. Hayden.

This is a strongly marked leaf which most resembles that of some of the live oaks. The texture was evidently leathery, the surface smooth; the nervation is that of Quercus or Ilex, as well as the marginal dentation. The species with which it may be compared are *Q. aspera* Ung. (Chlor. Prot., p. 108, Pl. XXX, figs. 1–3), *Q. Buchii* Web. (Paleontogr. (1852), p. 171 [57], Pl XIX [II], fig. 4), and *Q ilicoides* Heer (Fl. Tert. Helv., Vol. II,

p. 55, Pl. LXXVII, fig. 16); but from all these it may be distinguished by its coarse, generally obtuse, marginal denticles.

Formation and locality: Tertiary (Eocene?). Tongue River, Montana.

Quercus elliptica Newb.

Pl. XVIII, fig. 1; XX, fig. 3.

Bost. Journ. Nat. Hist., Vol. VII (1863), p. 523.

"Leaves elliptical or ovate, rounded or somewhat wedge-shaped at base, pointed above; margins entire. Surface smooth, consistence thick and leathery; nervation strong; lateral nerves numerous, diverging at a large angle, slightly arched upward, often sinuous, forked and anastomosing above."

In its nervation this species resembles several of the laurel-leaved oaks already described from the Tertiary rocks of Europe, such as *Q. nereifolia*, *Q. Heerii*, *Q. elaena*, etc., but is distinguishable from all these and other otherwise similar species by its broad elliptical or ovate outline. The margins in the specimens before us are apparently entire, but they are probably sometimes toothed, as in most allied species.

Formation and locality: Cretaceous (Puget Sound group). Chuckanutz, near Bellingham Bay, Washington.

Quercus flexuosa Newb.

Pl. XIX, figs. 4–6.

Bost. Journ. Nat. Hist., Vol. VII (1863), p. 521.

"Leaves 4 to 6 inches long, lanceolate, often more or less curved, pointed, acute, narrowed at the base to the petiole; margins somewhat irregularly sinuate-dentate; nervation strongly marked, lateral nerves forked and anastomosing at the summit."

This is apparently one of the chesnut oaks, but has not the regularity of nervation which characterizes most of that group, of which *Q. castanea* may be taken as a type.

Among fossil species there are many to which it bears considerable resemblance, such as *Q. Gaudini* Lesq., *Q. Gmelini* Ung. *Q. urophylla* Ung., etc., but from these and all others described it seems to be sufficiently distinct. In *Q. Gaudini* the secondary nerves are curved and run along the margins. In the other species mentioned they are less numerous and more curved and the marginal teeth are coarser.

Formation and locality: Cretaceous (Puget Sound group). Chuckanutz, near Bellingham Bay, Washington.

QUERCUS GRACILIS Newb.

Pl. LXVII, fig. 4.

Proc. U. S. Nat. Mus., Vol. V (March 21, 1883), p. 504.

"Leaves narrow, lanceolate, long-pointed, acute, wedge-shaped at the base; margins set with remote, low, acute teeth; nervation regular and fine; nerve branches fifteen to twenty on each side, curved gently upward, and terminating in the marginal teeth."

Collected by Dr. J. S. Newberry.

This is another of the lanceolate, serrate-leaved oaks of which *Q. Drymeja* Ung. (Chlor. Protog., p. 113, Pl. XXXII, figs. 1–4) may be considered as a type. It differs from that species, however, in its more crowded nervation, smaller teeth, and shallower sinuses.

In the figure given the nervation is represented as too strong, and the marginal teeth are not sufficiently acute. Several very beautifully preserved specimens are before us, which give a very exact and complete view of it, and its resemblance to *Q. Drymeja* is so strong that if it had occurred in the same horizon and locality there would have been no propriety in separating them; but in addition to the differences that have been mentioned, the geological horizons are so different that the probability of finding any identity of species is extremely small. For the present, therefore, it has been thought best to regard this as distinct from the great number of leaves that have been in North America and Europe referred to *Q. Drymeja*.

Formation and locality: Cretaceous (Montana group). Point of Rocks, Wyoming.

QUERCUS GRÖNLANDICA Heer.

Pl. LI, fig. 3, in part; LIV, figs. 1, 2.

Fl. Foss. Arct., Vol. I (1868), p. 108, Pl. VIII, fig. 8; X, figs. 3, 4; XI, fig. 4; XLVII, fig. 1.

NOTE.—So identified by Dr. Newberry, as indicated by memorandum on the margin of the plate and on specimen label.—A. H.

Formation and locality: Tertiary (Miocene). Cook Inlet, Alaska.

QUERCUS LAURIFOLIA Newb

Pl. LIX, fig. 4; LX, fig. 3.

Proc. U. S. Nat. Mus., Vol. V (March 21, 1883), p. 505.

"Leaves petioled, lanceolate, 6 inches in length by 1½ inches in width, equally narrowed to the point and petiole; margins entire, or faintly toothed, or undulate; nervation regular; midrib strong, straight, lateral branches, about ten pairs, arching gently upward, terminating in the margins."

Collected by S. M. Rothhammer, on the expedition of Gen. Alfred Sully.

Although reluctant to add one more to the large number of ill-defined species of oak which have been established upon the fossil leaves brought from the far west, this seems to be inevitable, inasmuch as the leaves before us are in all probability those of Quercus and distinct from any hitherto described. The most striking feature in these leaves is their elegant lanceolate and symmetrical form, broadest in the middle and narrowing regularly to the pointed base and summit. The craspedodrome nervation and the undulate or faintly toothed margins seem to separate these leaves from Laurus and connect them with the oaks. The figures given but imperfectly represent the leaves in question, but it is hoped that the description will permit their identification when found.

Formation and locality: Tertiary (Eocene ?). Burned shales over lignite beds, Fort Berthold, Dakota.

QUERCUS PAUCIDENTATA Newb.

Pl. XLIII, fig. 1.

Proc. U. S. Nat. Mus., Vol. V (March 21, 1883), p. 505.

"Leaves oblanceolate, 6 inches in length by 1½ in breadth, narrowed to the base, sometimes unsymmetrical, long-pointed, and acute at the summit; margins entire below, coarsely toothed above; nervation strong and regular, about ten branches on each side of the midrib, which curve upward, festooned below, terminating in the teeth above."

Collected by Rev. Thomas Condon.

No complete specimens of these leaves are contained in the collection, the one figured being the best. The texture was evidently thick and leathery. The apex is erroneously represented in the figure, as subsequent development of the specimen shows that it terminates in a long-drawn acute point. Among described species this may be compared with *Q. Nimrodis* Ung. (Foss. Fl. Sotzka, p. 163 [33], Pl. XXXI [X], figs. 1–3), and *Q. Meriani* Heer (Fl Tert. Helv., Vol. II, p. 53, Pl. LXXVI, fig. 12), but in those species the marginal teeth are stronger and are not, as in this, confined to the summit. The substance of the leaf of the specimens before us was evidently very thick and leathery.

Formation and locality: Tertiary (Miocene). Bridge Creek, Oregon.

Quercus salicifolia Newb.

Pl. I, fig. 1.

Ann. N. Y. Lyc. Nat. Hist., Vol. IX (April, 1868), p. 24; Ills. Cret. and Tert. Pl. (1878), Pl. II, fig. 1.

"Leaves petiolate, smooth, thick, entire, lanceolate, abruptly pointed at both ends; medial nerve strong, straight, or more or less curved; secondary nerves of unequal size, strong near their points of origin, becoming fine, flexuous, and branching as they approach the margins of the leaf, where some of them inosculate by irregular curves, while others terminate in the margins."

Collected by Dr. F. V. Hayden.

This species differs considerably in its general aspect from the willow-like leaves with which it is associated, and must have been much thicker and smoother. The midrib is very strong, terminating below in a thick, but short, petiole. The lateral nerves are much less uniform and regular than those of the leaves to which I have referred. They are at first strong, but soon diminish, and many of them extend but halfway to the margin, the others being unequally curved and branching irregularly or anastomosing with each other. The finer details of nervation are not given in the specimens before me, and perhaps more ample material will show that our fossil should not be regarded as a Quercus, but, as far as its characters are given, they agree best with those of that genus. The texture of the leaf was evidently thick and its surface glossy, more so than in any Salix now living; the nervation, too, is more of the oaks than willows; the alternation of larger with smaller secondary nerves, all diminishing rapidly and irregularly branched and flexuous above, are characters common to the leaves of all the willow-oaks. Some leaves of the living *Q. imbricaria* would closely resemble these if fossilized in the same manner. In the Lauraceæ with lanceolate leaves the nervation is generally much more exact and regular than in the specimen before us, the side nerves being generally curved gracefully and more or less uniformly upward, their extremities anastomosing, or, more rarely, reaching the margin. If the fine reticulation of the tertiary nerves was distinctly visible there would perhaps be little difficulty in determining with a good degree of certainty the generic relations of this fossil. In the oaks this reticulation is very fine, the areolæ of rather uniform size and quadrangular or polygonal, about as broad as long. In the willows the meshes are larger, more irregular, and more or less elongated.

Formation and locality: Cretaceous (Dakota group). Blackbird Hill, Nebraska.

Quercus simplex Newb.

Pl. XLIII, fig. 6.

Proc. U. S. Nat. Mus., Vol. V (March 21, 1883), p. 505.

"Leaves lanceolate, long-pointed, narrowed, and slightly rounded at the base; margins entire; nervation fine and regular."

In collections made by Rev. Thomas Condon at Bridge Creek, Oregon, are numerous leaves similar to that described above. Some are larger, but all present the same characters. The form of the leaf is similar to that of *Q. consimilis*, with which it is associated and from which it differs only by its entire margin. Since in that species the margins are sometimes nearly entire, it is possible that in the leaves before us that character may be intensified, giving an entire variety. Of this, however, proof can only be obtained by further collections.

Formation and locality: Tertiary (Miocene). Bridge Creek, Oregon.

Quercus sinuata Newb.

Pl. XIII, fig. 1.

Ann. N. Y. Lyc. Nat. Hist., Vol. IX (April, 1868), p. 27.

"Leaves small, obovate in general outline, narrowed to the petiole, or slightly decurrent; margins deeply lobed, lobes rounded, broader than the sinuses that separate them, three nearly equal on either side, summit broadly rounded or obscurely lobed, often oblique; nervation strong and simple, midrib straight or slightly flexed, giving off lateral branches, which run to the margins of each lateral lobe."

The general form of this leaf is much like that of our living *Q. obtusiloba*, though it is smaller and more symmetrical. Among the many fossil species which have been described there is none which approaches this very closely, most of them bearing either simple, entire leaves, or toothed, rather than lobed ones.

Formation and locality: Cretaceous (Dakota group). Banks of Rio Dolores, Utah.

QUERCUS SULLYI Newb.

Pl. LX, fig. 2.

Proc. U. S. Nat. Mus., Vol. V (March 21, 1883), p. 506.

"Leaves ovate, pointed, wedge-shaped, or rounded at the base; margins set remotely or closely, with acute, spiny-pointed teeth; nervation strong, somewhat flexuous; lower pair of lateral nerves giving off numerous branches; middle and upper pairs simple below, forked at the summit."

Collected by S. M. Rothhammer, on the expedition of Gen. Alfred Sully.

The characteristics of these leaves are but imperfectly shown in the figure, but the general form, margin, and nervation can be very well made out from the numerous fragments contained in the collection made by the Sully Expedition. It is evident that we have here one of the Ilex-like oaks, and indeed it may be a question whether it is not rather a holly than an oak. The leaf was generally unsymmetrical, the nervation strong but flexuous, the surface roughened by the tertiary nerve branches. In a general way these leaves resemble those of the common evergreen oak, *Quercus agrifolia* of California, but the spines of the margin are smaller and more numerous, the leaves more elongate and pointed. It is evident, however, that the tree which bore them belonged to the same group of oaks.

Formation and locality: Tertiary (Eocene ?). Burned shales over lignite beds, Fort Berthold, Dakota.

Order ULMACEÆ.

Ulmus speciosa Newb.

Pl. XLV, figs. 2–5, 7, 8.

Proc. U. S. Nat. Mus., Vol. V (March 21, 1883), p. 507.
Ulmus pseudo-Americana Lesq., Cret. and Tert. Fl. (1883), p. 249, Pl. LIV, fig. 10.

"Leaves 4 to 6 inches in length by 2 inches in width, petioled, long-ovoid, or elliptical in outline, pointed at summit; margins coarsely and doubly serrate; nervation strong, regular, fifteen to twenty parallel branches one either side of midrib. Fruit large, 27 centimeters in diameter, subcircular, emarginate."

This large and fine species of elm is represented by hundreds of specimens in the collection made by Rev. Thomas Condon, and while most are imperfectly preserved, there are some which show all the details of form and structure. The general aspect of the leaves is not unlike that of *U. Bronnii* Ung. (Chlor. Prot., p. 100, Pl. XXVI, figs. 1–3), but is fully twice as large and coarsely and doubly serrate.

The leaf represented by fig. 8 is one of many which occur in the collection, all presenting nearly the same character; that is, they are smaller than those just described, with much finer marginal dentation. That dentation is, however, double and like that of the larger leaves, though less pronounced, and there are no characters presented by these leaves which would justify us in regarding them as representing a distinct species. For the present, therefore, it has been thought better to leave these as small forms of *U. speciosa*.

Among living species *U. fulva* approaches closer to those now under consideration than any other, and the differences between the fossil and living forms are not so great but that we may very well regard one as the progenitor of the other. In *U. fulva* the leaves are smaller and relatively broader, being ovoid in outline, but the character of the marginal dentation and of the nervation is essentially the same.

The samara, represented by fig. 7, is supposed to be the fruit of the large elm described above. It is very similar in size and character to the fruit credited to *U. Bronnii* by Ung., but is somewhat broader. It has not yet been distinctly connected with the leaves we have called *U. speciosa*, but there seems to have been no other tree growing in the locality where these specimens are found of which this could well be the fruit.

The fruit of *U. fulva* has nearly the same form as this, but is only about half as large, while the fruit of *U. Americana* is still smaller and is obovoid and cilliated.

Formation and locality: Tertiary (Miocene). Bridge Creek, Oregon.

Planera crenata Newb.

Pl. LVII, fig. 3.

Proc. U. S. Nat. Mus., Vol. V (March 21, 1883), p. 508.

"Leaves oblong, ovate; short petioled; 5 centimeters long by 25 millimeters wide; base rounded; summit blunt-pointed; margins coarsely crenate; nervation simple, delicate, six simple branches on each side of the midrib terminating in the crenations of the margin."

Collected by Dr. F. V. Hayden.

In general aspect these leaves resemble some of the varieties of *P. Ungeri*, but differ from them in the crenate margins, the lobes being fewer and all rounded. In these respects it differs also from the species described in this volume, *P. longifolia* Lesq. (Pl. LVIII, fig. 3), *P. variabilis* Newb. (Pl. LXVI, figs. 5, 6, 7), and *P. nervosa* Newb. (Pl. LXVII, figs. 2, 3).

Formation and locality: Tertiary (Eocene ?). Tongue River, Montana.

Planera longifolia Lesq.

Pl. LVIII, fig. 3.

Hayden's Ann. Rept., 1872 [1873], p. 371; Tert. Fl. (1878), p. 189, Pl. XXVII, figs. 4–6.

Note.—So identified and located by Dr. Newberry, as indicated by memorandum on margin of plate. Further information lacking.—A. H.

Formation and locality: Tertiary (Miocene). Florissant, Colorado.

Planera microphylla Newb

Pl. XXXIII, figs. 3, 4.

Ann. N. Y. Lyc. Nat. Hist., Vol. IX (April, 1868), p. 55; Ills. Cret. and Tert. Pl. (1878), Pl. XVI, figs. 3, 4.

"Leaves very small, ovate-lanceolate, generally unsymmetrical, curved or falcate, cordate at base, pointed but rarely acute, coarsely and bluntly toothed; nervation strong; lateral nerves diverging at an angle of about 50 degrees in five to six pairs branching toward the summit, and inosculating along the margins; tertiary nerves strong, leaving the secondaries nearly at right angles, much branched and anastomosing to form a coarse and irregular network."

Collected by Dr. F. V. Hayden.

In its general form this leaf has a striking resemblance to *Planera Ungeri* Ett. (Abhandl. k. k. geolog. Reichsanstalt. Wien, Vol. II (1851), Foss. Fl. Wien, p. 14, Pl. II, figs. 5–18), *Ulmus Zelkovæfolia* Ung. (Chlor. Prot., p. 94, Pl. XXIV, figs. 7–13; XXVI, figs. 7, 8), but it is apparently considerably smaller, narrower, and more coarsely toothed.

Formation and locality: Tertiary (Fort Union group). Fort Union, Dakota.

Planera nervosa Newb.

Pl. LXVII, figs. 2, 3.

Proc. U. S. Nat. Mus., Vol. V (March 21, 1883), p. 508.

"Leaves ovate or lanceolate, pointed, wedge-shaped, or rounded at the base, petioled; margins set with coarse, appressed teeth; nervation strong, crowded, regular; lateral nerves simple, parallel, terminating in the teeth of the margins."

Collected by Dr. C. A. White.

The most striking feature in these leaves is their strong, crowded, regular nervation, from thirteen to nineteen nearly equidistant simple nerve branches issuing from either side of the midrib. The nervation is equally regular in *P. longifolia*, Lesq., Tert. Fl., p. 189, Pl. XXVII, figs. 4–6; this volume, p. 81, Pl. LVIII, fig. 3, but is lighter, and the marginal dentation is coarser, the teeth more obtuse.

Fig. 4, on Lesquereux's plate cited above, resembles more the leaves before us and apparently belongs to a species distinct from the other two leaves with which it is there associated, possibly to this one. The leaves of *P. longifolia* are found in great abundance at Florissant, Colorado, and they are so much alike that there is no difficulty in separating them from other described species; while in the localities where the leaves of *P. nervosa* occur there are none which have the few long, horizontally cut teeth of *P. longifolia*. Hence while there is considerable resemblance in the general aspect of these leaves, there can be little question that they are specifically distinct.

Formation and locality: Tertiary (Green River group). Green River, Wyoming.

PLANERA VARIABILIS Newb.

Pl. LXVI, fig. 5–7.

Proc. U. S. Nat. Mus., Vol. V (March 21, 1883), p. 508.

"Leaves lanceolate, to broad ovate; usually unsymmetrical, petioled; summit acute, sometimes long-pointed; base rounded or wedge-shaped; margins coarsely crenulate-dentate or serrate, with remote, appressed teeth; midrib straight, strong; lateral nerves delicate, frequently alternating stronger and finer, gently arched upward, terminating in the teeth of the border; the finer intermediate ones sometimes fading out before reaching the margin."

Collected by Dr. C. A. White.

Some of the various forms of leaves ascribed to *Planera Ungeri* fairly represent those before us, and their generic resemblance is apparent; but in our plant the leaf is more pointed, the serratures are coarser, generally more obtuse, and, when acute, more appressed.

Planera longifolia Lesq., has larger, more symmetrical, and less acute leaves, with coarser triangular teeth. (See Pl. LVIII, fig. 3.)

From the other species described in this volume this may be distinguished by its greater size, more ovate form, coarser serrations, and relatively smaller crenations. *Planera emarginata* Heer (Fl. Tert. Helv., Vol. II, p. 61, Pl. LXXIX, fig. 24) has much more acute teeth and more bristling aspect.

Several figures have been given of this species, in order to show the diversity of form it assumes, and it could be easily imagined that they were specifically different; but coming as they do from one locality, and in the large collections made from this, we have an unbroken series, all pervaded by a similarity of aspect, we must conclude that they are all from one kind of tree. Possibly future collections will prove that the narrower, more rigid form, with the deeply cut and acute serrations, and parallel, nearly straight lateral veins, shown in fig. 7, belongs to a different species; but in the very large number of Planera leaves before me it is impossible to make any division without making several. They are, therefore, all grouped together for the present.

Formation and locality: Tertiary (Green River group). Green River Station, Wyoming.

CELTIS PARVIFOLIA Newb.

Pl. LIII, fig. 6.

Proc. U. S. Nat. Mus., Vol. V (March 21, 1883), p. 510.

"Leaves small; oblong-ovate in outline; rounded and unsymmetrical at the base, pointed at the summit; margins, except at the base, coarsely dentate; nervation sparse; two principal branches on each side of midrib, one pair springing from the base and throwing off branchlets, another strong pair issuing from the midrib at the middle of the leaf, other delicate branches given off near the summit."

Collected by Dr. F. V. Hayden.

In its general aspect, as well as its details of structure, this leaf very closely resembles *C. Australis*, differing from our living *C. occidentalis*, as well as from the fossil species that are found in the Tertiary beds of this country by its simpler nervation, its smaller size, and the relatively coarser serration of the margin.

Formation and locality: Tertiary (Eocene?). Tongue River, Montana.

Order MORACEÆ.

FICUS (?) ALASKANA Newb.

Pl. LI, fig. 1; LII, fig. 1; LV, figs. 1, 2.

Proc. U. S. Nat. Mus., Vol. V (March 21, 1883), p. 512.

"Leaves large, reaching 8 to 10 inches in length and breadth; trilobed, generally unsymmetrical; lobes pointed, usually obtuse; margins entire or locally undulate; nervation strong, conspicuously reticulate; principal nerves, three, giving off branches, which divide near the margins, sometimes connecting in festoons, sometimes craspedodrome; tertiary nervation forming a coarse network of usually oblong meshes filled with a fine polygonal reticulation; upper surface of the leaf smooth and polished, lower roughened by the reticulation of the nerves."

Collected by Captain Howard, U. S. N.

These beautiful leaves have been referred with much doubt to Ficus. They present considerable resemblance to some of the leaves of *Ficus tiliæfolia* Heer, particularly the lobed form shown in Fl. Tert. Helv., Vol. III, p. 183, Pl. CLII, fig. 14, and the nervation is sometimes similar, though generally less distinctly camptodrome. The differences, however, between our leaves and the usually simple unsymmetrical obliquely based leaves of *T. tiliæfolia* show specific and perhaps generic distinctness. The localities which furnished the specimens now figured show by the great abundance of leaf impressions brought from there that they were at one time the home of rich and luxuriant vegetation, the slabs which carry these leaves being crowded with those of many different genera and species closely impacted together. Among these are the great oak leaves, 1 foot to 15 inches in length and 6 inches in width (*Q. Grönlandica*), *Taxodium distichum miocenum*, *Juglans nigella*, *Prunus variabilis*, large leaves of *Platanus* and *Pterospermites*, *Corylus MacQuarrii*, etc. This Ficus (?) seems to have been as abundant as any other, and collectors who shall visit the locality hereafter, by taking proper pains, will be able to find abundant and satisfactory representatives of all these and many other plants, and will undoubtedly obtain conclusive evidence of their botanical relations.

Formation and locality: Tertiary (Miocene). Cook Inlet and Admiralty Inlet, Alaska.

FICUS ASARIFOLIA MINOR Lesq.

Pl. LXVII, figs. 5, 6.

Hayden's Ann. Rept., 1874 [1876], p. 303; Tert. Fl. (1878), p. 208. Not *F. asarifolia* Ett., Fl. Bilin., p. 156, Pl. XXV, figs. 2, 3, 6.

NOTE.—These specimens unquestionably represent the variety of the species referred by Lesquereux to *F. asarifolia* Ett. in Hayden's Annual Report, 1874 [1876], p. 303; but this species has serrated margins, while in ours the margins are entire or slightly undulate. This distinction was recognized by Dr. Newberry in a memorandum on the plate, but he failed to state what name he intended to give to the American leaves.—A. H.

Formation and locality: Cretaceous (Montana group). Point of Rocks, Wyoming.

FICUS (?) CONDONI Newb.

Pl. LVI, fig. 1; LVII, fig. 1; LVIII, fig. 1.

Proc. U. S. Nat. Mus., Vol. V (March 21, 1883), p. 512.

"Leaves large, sometimes nearly 2 feet in length, three to five-lobed, slightly decurrent, and the petiole sometimes stipulate; margins entire, or gently undulate; nervation very strongly marked and closely reticulate, roughening the surface, camptodrome, but nerve branches sometimes terminating in the margins of the middle lobe."

Collected by Rev. Thomas Condon, to whom the species is dedicated as a recognition of the important contribution he has made to paleontology in the discovery and exploitation of these interesting plant beds.

The remains of this remarkable plant occur in great abundance in the Bridge Creek Tertiary beds, and it is represented in the collections made there by a large number of specimens. Some of these indicate leaves 18 inches to 2 feet in length and nearly as much in breadth. The most striking feature which they exhibit after their great size is the marked reticulation of the surface, which has given a peculiar lacelike roughening to the rock in the leaf impression. This character, as well as the general form and nerve structure, is fairly well given in the figures, and no one having seen them will have difficulty in recognizing the fossil.

The reference to the genus Ficus wants the confirmation of the fruit before it can be accepted as established, but among all the leaves with which these have been compared there are none to which they bear so great resemblance as to those of the Moraceæ, and especially with those of the leaves of Ficus and Artocarpus. The nervation is strikingly like that of a number of species of Ficus, such as *F. scabriuscula*, *F. oppositifolia*, *F. Roxburghiana*, *F. sycomorus*, and perhaps to none more than to that of the common fig, *F. Carica*. Hence, with regret in adding to the already large number of ill-defined fossil species of Ficus, it has seemed best to provisionally refer these leaves to that genus, giving them a place to which, without the evidence of the fruit, they are apparently as much entitled as any others. Sometime the fructification will be found, and then all doubt will be set at rest. There is good evidence that the genus Ficus was well represented in the luxuriant, warm temperate or subtropical flora which prevailed over so much of North America during the Tertiary age, as it is now in the forests of tropical and subtropical America. At the same time it is necessary to say that of the large number of species of Ficus more than 20, which have been described as occurring in our Tertiary rocks, the identification has been in many instances based upon evidence that must be regarded as unsatisfactory.

One of the most striking characters of these leaves is formed by the reflexed stipule-like lobe at the base of the leaf. This is a feature that it has in common with some species of Platanus, especially *Platanus basilobata* Ward (Synopsis, Flora Laramie Group, 6th Ann. Rept. U. S. Geol. Surv. for 1884–85 [1886], Pls. XLII, XLIII), and something of the kind is frequently found in strong growing shoots of the living *Platanus occidentalis*. As I have said in my description of *Platanus nobilis*, there are some characters in the leaves of that tree which raise the question whether it was a true Platanus, and yet my reference of it to that genus has been confirmed by Sir William Dawson and Dr. Lester F. Ward. The former has found leaves which he considers those of *P. nobilis* having this basilar lobe, and he has suggested that Dr. Ward's *P. basilobata* should be named *P. nobilis* var. *basilobata*. I should not be surprised if, in the light of more material, *P. nobilis* and the species now under consideration should be united in a new genus; but without additional material such a step would be hardly wise.

Formation and locality: Tertiary (Miocene). Bridge Creek, Oregon.

FICUS MEMBRANACEA Newb.

Pl. LIX, fig. 2.

Proc. U. S. Nat. Mus., Vol. V (March 21, 1883), p. 512.

"Leaves sessile, 4 to 6 inches in length, by 2½ to 3½ in width: ovate, abruptly and usually blunt-pointed, narrowed to the base, generally unsymmetrical, margin entire, nervation delicate, open, camptodrome; ten or more branches given off on either side of the midrib, curving upward, and forming a festoon near the margin."

Of these leaves there are quite a number in the collection from Alaska, made by Captain Howard, and such as exhibit considerable diversity of form, as will be seen by the figures. That shown on Pl. LIX is imperfect and imperfectly represented; it is smaller than the average and more pointed, and the base should be prolonged and narrowed. The reference of these leaves to Ficus is provisional and can not be insisted upon. The nervation is that of this genus, and a considerable resemblance is shown to those described by Lesquereux (Tert. Fl., p. 194, Pl. XXVIII, figs. 9–12) under the name of *F. oblanceolata*, but they are larger, broader in form, and the nervation is much more open. The texture of these leaves was evidently thin and membranous, a character plainly shown by the contrast in appearance which they present to oaks, poplars, prunes, etc., with which they are associated; this is also indicated by the delicate, open, and flexuous nervation.

Formation and locality: Tertiary (Miocene). Cook Inlet, Alaska.

FICUS PLANICOSTATA Lesq.?

Pl. XLVI, fig. 1.

Hayden's Ann. Rept., 1872 [1873], p. 393; Tert. Fl. (1878), p. 201, Pl. XXXI, figs. 1-8, 10-12.

The leaf here figured is hardly sufficient for satisfactory determination; it is imperfect at the summit and throughout part of the margin; however, the insertion of the petiole and the nervation give it characters which are separated widely from any other leaves with which it is associated in the collection. The petiole is broad, and is inserted obliquely in the base of the leaf. The nervation is beautifully camptodrome, the branches of the basal pair of lateral nerves, as well as the summits of the lateral nerves above, forming a most beautiful and regular festoon. This is essentially the nervation of *F. planicostata*, and although the specimen is much smaller and narrower than the average of the leaves ascribed to that species, I have thought best to refer it doubtfully to this place until further material will permit the definitive settlement of the question.

Formation and locality: Tertiary (Miocene). Bridge Creek, Oregon.

FICUS RETICULATA (Lesq.) Hollick.

Pl. XII, figs. 2, 3.

Laurophyllum reticulatum Lesq. Hayden's Ann. Rept. 1872 [1873], p. 425; Cret. Fl. (1874), p. 76, Pl. XV, figs. 4, 5.
Ficus laurophyllum Lesq. Hayden's Ann. Rept. 1874 [1876], p. 342, Pl. V, fig. 7.
Ficus laurophylla Lesq. Cret. and Tert. Fl. (1883), p. 49, Pl. 1, figs. 12, 13.

Quite a number of specimens of these very distinctly marked leaves are contained in the collection received from Mr. Sternberg from Fort Harker, and still larger and finer ones since obtained through other channels show that the leaves sometimes attained a size considerably greater than that represented in fig. 2, but it was as wide and much longer. All these are alike in showing a smooth and polished surface, a thick, leathery texture, a remarkably strong, straight, smooth midrib; pinnate, delicate, irregularly spaced, branched camptrodome lateral nerves. Except that they are more lanceolate and pointed, these leaves would hardly be distinguishable from those of *Ficus elastica* if fossilized. In form, exactness of outline, and strength of midrib, they resemble the leaves of Nerium, but the nervation is quite different.

Formation and locality: Cretaceous (Dakota group). Fort Harker, Kansas, and Blackbird Hill, Nebraska.

Protoficus inæqualis Newb.

Pl. LVIII, fig. 2; LX, fig. 1.

Proc. U. S. Nat. Mus., Vol. V (March 21, 1883), p. 512.

"Leaves 4 to 5 inches long, by 3 inches wide; oval, pointed at the summit, narrowed and rounded at the unsymmetrical base; margins entire or in part undulate; nervation strongly defined but open; about seven branches on each side of the midrib, the lower two or three giving off branches below, the upper simple, arched upward, terminating in the margin, the intervals between the branches spanned by numerous, generally simple tertiary nerves."

Collected by Dr. F. V. Hayden.

The general aspect of these beautiful leaves is not well given in the figure. They seem to have been thick and polished above, roughened below by the strongly marked nervation. They resemble in many respects the leaves of Protoficus, described by Count Saporta, from the travertines of Sezanne, as will be seen by comparing his figure of *Protoficus crenulata* (Fl. Foss. Sezanne, p 67, Pl. VI, fig. 5). Our leaves differ from that, however, in this, that the base is unsymmetrical, the margin is entire or undulate, and the nervation is craspedodrome. This latter character is not common, but is not unprecedented among the figs, the leaves of several species of which bear considerable resemblance to these, e. g., *F. sycomorus.*

It will also be noticed that the leaves under consideration are not unlike those described by Lesquereux under the name of *Ficus planicostata* var. *Goldiana* (Tert. Fl., p. 202, Pl. XXXIII, figs. 1–3), but differ from them in the inequality of the base and the details of nervation. It seems highly probable, however, that they should form species of the same genus.

Formation and locality: Tertiary (Eocene ?). Tongue River, Montana.

Order ARISTOLOCHIACEÆ.

Aristolochia cordifolia Newb.

Pl. XXXIX; XL, fig. 7; LX, fig. 4.

Ann. N. Y. Lyc. Nat. Hist., Vol. IX (April, 1868), p. 74; Ills. Cret. and Tert. Pl. (1878), Pl. XXII, under *Catalpa crassifolia;* XXV, fig. 7.
Catalpa crassifolia Newb. Op. cit., p. 56.

"Leaves large, fleshy, ovate, heart-shaped at base, pointed above, sometimes unsymmetrical; margins entire; nervation strongly developed; midrib straight or flexuous; lateral nerves about seven pairs; lower pair strongest, not reaching the middle of the leaf, giving off each about four branches on the lower side, of which the lower ones spring from the base of the laterals and are much branched; upper laterals branched at their summits, branches uniting to form a festoon somewhat remote from the margin; tertiary nervation invisible."

Collected by Dr. F. V. Hayden.

These leaves are referred in the Annual Report of the New York Lyceum of Natural History with hesitation to Catalpa, which they considerably resemble in form and nervation; but a large number of specimens submitted to inspection since the description was written exhibit characters which lead me to suspect that they represent a species of Aristolochia. This additional material shows the leaves to have been sometimes very large, more than 1 foot in diameter, broadly cordate in outline, often unsymmetrical. Fig. 4, given on Pl. LX, exhibits the broader and more rounded form and the open festooned nervation; but this is scarcely more than one-third of the linear dimensions of the largest. The texture of the leaf seems to have been very thin, the nervation is sparse and open, though the principal nerves must have been somewhat fleshy. There are also associated with these leaves slender tortuous stems that seem to be portions of a vine. Taking these facts into consideration, I have been led to refer these leaves to Aristolochia and to compare them with the large, broadly cordate leaves of *A. sipho.* Future collections will undoubtedly furnish material which will render it possible to speak with confidence in regard to the generic relations of the plant.

Formation and locality: Tertiary (Eocene?). Banks of Amil Creek, Dakota.

Order NYMPHÆACEÆ.

CABOMBA (?) GRACILIS, Newb.

Pl. XXII, fig. 1; XXIII, fig. 1.

Cabomba gracilis Newb. Proc. U. S. Nat. Mus., Vol. V (March 21, 1883), p. 514. Ills. Cret. and Tert. Pl. (1878), Pl. VII, fig. 1, under "aquatic rootlets of Equisetum"; VIII, fig. 2, under "Equisetum."

"Stem slender, smooth; submerged leaves set at intervals of half an inch to an inch apart on the stem, opposite dichotomously and frequently branched, segments narrowly linear, or filiform, flattened, smooth, truncated, scarcely distinguishable from the stem and leaves of *C. Caroliniana*."

A large number of intertwining, smooth, narrow stems, with opposite, many-forked, linear leaves, are contained in some of the collections made from the Tertiary beds of the upper Missouri by Dr. F. V. Hayden. They were at first regarded as aquatic rootlets, but an examination of a multitude of well-preserved specimens shows that they are leaves and not roots, and comparing them with living plants they are found to imitate with a most perfect exactness the stems of leaves of Cabomba. The smaller specimens, like that figured, accurately represent the filiform variety of *Cabomba Caroliniana* of our Southern States. Mingled with these stems and leaves are obscure fragments of what may have been the peltate leaves, since some of them show a sort of umbilicus as though the point of attachment of the stem. Had there been but one or two of these specimens corresponding to the above description, their nature would have been left in so much doubt as to render it unwise to call attention to them; but occurring as they do in connection with other aquatic plants in very large numbers, and having a definite and invariable character, the stems smooth and lacking all the characteristics of creeping root stalks or aquatic roots, the leaves expanded, each pair in its own plane, and the pairs alternating, show that we have here to do with the stem and foliage of an aquatic plant of a marked and peculiar character. To this character no living plant seems to approach so nearly as Cabomba, and here the resemblance is so close that the probabilities become very strong that the reference to that genus will be confirmed hereafter by the discovery of the floating leaves and flowers.

Formation and locality: Tertiary (Fort Union group). Fort Union, Dakota

CABOMBA INERMIS (Newb.) Hollick.

Pl. XXII. fig. 2; XXIII, fig. 2.

Psilotum inerme, Newb. Ann. N. Y. Lyc. Nat. Hist., Vol. IX (April, 1868), p. 38; Ills. Cret. and Tert. Pl. (1878), Pl. VII, fig. 2, under "aquatic rootlets of Equisetum"; VIII, fig. 3, under *Psilotum inerme*.

Associated with the last-described species are a large number of dichotomously forked, flattened leaves, which are imperfectly represented in the figure given. These have all the general character of the smaller ones, but are many times larger—5 to 6 inches in length—so large, indeed, that it seems impossible that they should have appertained to the same species. A distinct and significant name has therefore been given to them.[1] These leaves are flattened and smooth, and have precisely the aspect of the broader leaves of the living Cabomba. Groups of these springing from a common base were formerly likened to Psilotum, and described in the Later Extinct Floras as *P. inerme;* but the study of additional material has led to the conviction that the probabilities are very much stronger that we have here a representation of a species of Cabomba. The isolated position of Cabomba in our modern flora is an indication that it is a relic of the past, and it was to be expected that in the sediments of the ancient fresh-water lakes of the far West, where the leaves of palms are preserved, affording evidence of a warm climate like that of our Southern States, traces of the former existence of Cabomba should be found. With some of the groups of leaves of the plant now under consideration are imperfect traces of fructification which in their general character confirm the reference to Cabomba, and yet are not sufficiently well preserved to thoroughly establish the botanical affinities of these plants. It is to be hoped that attention being called to this peculiar group of fossils, they will be specially sought and found in the Fort Union beds in such condition as to establish beyond question their botanical affinities.

In fig. 2, Pl. XXII, a pair of leaves is represented which are intermediate in size between the two forms described, and these are erroneously shaded in such a way as to give the impression of rounded stems; in fact, these leaves are perfectly flat and correspond in form and structure to the

[1] Dr. Newberry's manuscript name for this species is *C. grandis*, but under the accepted rules of nomenclature the original published specific name *inerme* must stand.—A. H.

others, but the plant was evidently somewhat decayed and mutilated before fossilization.

Taking the series of figures now given and referred to Cabomba, they might be supposed to represent three species or different phases of one, but the very large number of the smallest form contained in the collection, and the close correspondence in size and form exhibited by them, seems clearly to justify the conclusion that they represent but a single species, while the larger form also generally exhibits the same characteristics. The intermediate size represented in fig. 2, Pl. XXII, has few representatives in the collection, and hardly affords material for the creation of a distinct species. It has been thought better, therefore, to refer this to the larger one, to which it is most nearly allied in size.

Formation and locality: Tertiary (Fort Union group). Fort Union, Dakota.

BRASENIA (?) ANTIQUA Newb.

Pl. LXVIII, fig. 7.

Brasenia antiqua Newb. Proc. U. S. Nat. Mus., Vol. V (March 21, 1883), p. 514 (not *B. antiqua* Daws., Trans. Roy. Soc. Canada, III, sec. 4, p. 15, 1885 [1886]).

"Stems long, flexuous, cylindrical (now flattened), smooth, many times branched toward summit, bearing pedunculate spheroidal capitula consisting of numerous club-shaped pods."

We have here the remains of an aquatic plant, having the general structure of Brasenia as regards stem and fruit, but the specimens are too imperfect to enable us to decide with confidence on its botanical relations. No leaves or flowers have yet been found, and the seeds are scarcely sufficient for its classification. Our common water shield, *Brasenia peltata*, is a very widely disseminated plant, as it is found on both sides of our continent and in Japan and the East Indies. This indicates that it has long been an inhabitant of the earth's surface, and whether the specimen before us can be accepted as evidence of its existence in North America during the Tertiary, the probabilities are strong that Brasenia was an inhabitant of the old lakes of the West and that its remains will be met with.

Formation and locality: Tertiary (Green River group). Green River, Wyoming

Order MAGNOLIACEÆ.

MAGNOLIA ALTERNANS Heer?

Pl. V, fig. 6.

Nouv. Mem. Soc. Helv. Sci. Nat., Vol. XXII (1866), p. 20, Pl. III, figs. 2-4; IV, figs. 1, 2.

NOTE.—So identified, provisionally, by Dr. Newberry, as indicated by memorandum on margin of plate. Locality probably Blackbird Hill, Nebraska.—A. H.

MAGNOLIA ELLIPTICA Newb. n sp.

Pl. XII, fig. 1.

Leaf 6 inches long by 3½ inches broad, elliptical in outline, rounded at the base, acute at the summit; midrib strong and straight; lateral nerves numerous, strong, nearly simple, arched upward, parallel, inosculating near margin (camptodrome).

Collected by Dr. F. V. Hayden.

Among described species, this approaches nearest to *M. Hilgardiana* Lesq. of the Tertiary of the Mississippi, but is shorter, broader, more rounded at the base, and more abruptly pointed at the summit.

There is some doubt in regard to the age of the strata from which this plant was derived, and it is possible that it is tertiary and is but a phase or variety of the species with which it has been compared.

Formation and locality: Tertiary (Eocene ?). Tongue River, Montana.

MAGNOLIA OBOVATA Newb.

Ann. N. Y. Lyc. Nat. Hist., Vol. IX (April, 1868), p. 15.

"Leaves large, obovate, entire, thick and smooth; pointed and slightly decurrent on the petiole; nervation strong; midrib straight and extending to the summit; lateral nerves pinnate, set at somewhat unequal distances, straight and parallel below, forked and inosculating above, forming a festoon parallel with the margin; tertiary nerves forming an irregular network of polygonal and relatively large areoles."

NOTE.—As may be seen by comparing the descriptions, this species is manifestly identical with the one described by Dr. Newberry under the name *Nyssa vetusta* (see p. 125 of this monograph), and inasmuch as the latter name has priority of place in the publication where they both originally appeared, the name *Magnolia obovata* becomes a *nomen nudum*. How this could have escaped Dr. Newberry's attention or the attention of subsequent workers and reviewers is strange.—A. H.

MAGNOLIA ROTUNDIFOLIA Newb.

Pl. LIX, fig. 1.

Proc. U. S. Nat. Mus., Vol. V (March 31, 1883), p. 513.

"Leaves petioled, large (8 inches in length by 6 inches in width), round-ovate in outline, rounded or blunt-pointed above and slightly wedge-shaped below; margins entire; nervation open and delicate: four to six lateral branches given off from the midrib at remote and irregular distances, curving gently upward, and forming festoons near the margin."

Collected by Dr. F. V. Hayden.

In general form this fine species would seem to be somewhat like *M. regalis* Heer (Fl. Foss. Arct., Vol. IV, Abth. I, p. 81, Pl. XX; XXI, figs. 1, 2) and *M. Nordenskiöldia* Heer (*op. cit.*, p. 82, Pl. XXI, fig. 3; XXX, fig. 1), but with a much more slender and less crowded nervation than the first and a more rounded form than the second. A number of specimens in the collection show some diversity of form, and it is possible that the leaf figured is more rounded and less pointed than the average, but unless there should be very great departure from this standard there is little probability of this species being united with any other. The nervation is almost precisely that of the living *M. acuminata*, and there can not be any reasonable doubt that it is a representative of the same genus.

Formation and locality: Cretaceous (Laramie group). Fischers Peak, Colorado.

LIRIODENDRON MEEKII Heer.

Pl. VI, figs. 5, 6.

Proc. Phil. Acad. Nat. Sci. 1858, p. 265; Nouv. Mem. Soc. Helv. Sci. Nat., Vol. XXII (1866), p. 21, Pl. IV, figs. 3, 4; Ills. Cret. and Tert. Pl. (1878), Pl. VI, figs. 5, 6 [fig. 6 under *L. primaevum*].

NOTE.—So identified by Dr. Newberry, as indicated by memoranda on margin of plate and on specimen label.—A. H.

Formation and locality: Cretaceous (Dakota group). Blackbird Hill, Nebraska

LIRIODENDRON PRIMÆVUM Newb.

Pl. VI, fig. 7.

Ann. N. Y. Lyc. Nat. Hist., Vol. IX (April, 1868), p. 12; Ills. Cret. and Tert. Pl. (1878), Pl. VI, fig. 7. [Not named on plate.]

"Leaves three-lobed, upper lobe emarginate, all the lobes rounded; nervation delicate, principal nerve straight or slightly curved, terminating in the sinus of the superior lobe; secondary nerves gently arching upward, simple or forked near the extremities, a few more delicate ones alternating with the stronger."

Collected by Dr. F. V. Hayden.

This leaf is considerably larger than that of *L. Meekii* Heer, less deeply lobed, and the lobes more broadly rounded. In its general aspect this species approaches much nearer the living tulip tree and the Tertiary species of Europe (*L. Procaccinii* Ung.) than that described by Professor Heer from the collections of Dr. Hayden (*L. Meekii*). The leaves of the former species are, however, generally more deeply lobed and the lobes are acute, but I have collected leaves of *L. tulipifera* of small size with all the lobes rounded and in all respects remarkably like that under consideration. On the whole this is so like the leaf of our tulip tree that there can be little doubt that it represents a species of the same genus which grew on our continent at the commencement of the Cretaceous epoch. This is one of the most important facts deduced from the collections of Dr. Hayden, for the genus *Liriodendron* is now represented by but a single known species, which is confined to North America. During the Miocene Tertiary epoch, however, it formed part of the flora of Europe, as well preserved leaves of a species very closely allied to, if not identical with, the living one grew in Italy, Switzerland, and Iceland.

Thus this comes into the interesting category of Magnolia, Liquidambar, Sassafras, etc., genera which flourished both in Europe and America during the Miocene epoch, but which have long since ceased to exist on the European continent.

These specimens also teach us the still more interesting truth that Liriodendron, Sassafras, Magnolia, Quercus, Salix, Platanus, Populus, and many others of our living genera date back on this continent to a period long anterior to the dawn of the Tertiary age, and having survived all the changes of the incalculable interval now form the most conspicuous elements in our existing forests.

Formation and locality: Cretaceous (Dakota group). Blackbird Hill, Nebraska.

Order BERBERIDACEÆ.

BERBERIS SIMPLEX Newb.

Pl. LVI, fig. 2.

Proc. U. S. Nat. Mus., Vol. V (March 21, 1883), p. 514.

"Leaves pinnate, with three or more pairs of leaflets; leaflets ovoid, rounded or emarginate at base, acute, with two to four large spiny teeth on each side."

Collected by Rev. Thomas Condon.

This, so far as known, is the first example of the occurrence of a Berberis in the fossil state in America, and of this we have only a single specimen, though that is unmistakable in its character. It is evidently allied to *B. aquifolium*, which grows so abundantly in the region where the fossil was found, but differs from it in the small number and large size of the teeth on the margins of the leaflets in the fossil. It is true that occasionally the smaller variety of *B. aquifolium* (*B. repens* Lind.) has leaflets very much like these, and I have before me as I write a specimen which I collected at Lake City, Colorado, in which some of the leaflets are almost precisely like these, differing from the fossil only in the less prolonged acute apex, and the narrower, somewhat wedge-shaped base. The surface of the fossil is quite smooth, showing almost nothing of the details of nervation; and this in a rock where the finer nerve markings are often most beautifully shown, as in the leaf represented on the same plate and which was obtained from the same beds. Hence we may conclude that in texture the leaf was thicker and its surface smoother than in *B. aquifolium*, in which the strong reticulated nervation is distinctly shown on both sides. In some specimens of *B. Nepaulensis* from the Himalayas we find a closer resemblance to the fossil plant than is offered by any of our native species, viz, sessile and slightly cordate leaflets with a simpler nervation, showing on the under side only the midrib and a basal pair of branches; teeth three to five on each side, the point produced as in the fossil.

Formation and locality: Tertiary (Miocene). Bridge Creek, Oregon.

Order LAURACEÆ.

SASSAFRAS CRETACEUM Newb.

Pl. VI, figs. 1–4; VII, figs. 1–3, VIII, figs. 1, 2.

Ann. N. Y. Lyc. Nat. Hist., Vol. IX (April, 1868), p. 14; Ills. Cret. and Tert. Pl. (1878), Pl. VI, figs. 1–4.

S. Mudgii Lesq. Am. Journ. Sci., Vol. XLVI (July, 1868), p. 99; *S. Mudgei* Lesq. Cret. Fl. (1874), p. 78, Pl. XIV, figs. 3, 4; XXX, fig. 7.

S. subintegrifolius Lesq. Am. Journ. Sci., Vol. XLVI (July, 1868), p. 99; *S.* (?) *subintegrifolium* Lesq. Cret. Fl. (1874), p. 82, Pl. III, fig. 3 (misquoted fig. 5.)

S. Harkeriana Lesq. Hayden's Ann. Rept., 1872 [1873], p. 425; *S. Harkerianum* Lesq. Cret. Fl. (1874), p. 81, Pl. XIII, figs. 3, 4; XXVII, fig. 2.

S. obtusus Lesq. Hayden's Ann. Rept., 1871 [1872], p. 303; *S. obtusum* Lesq. Cret. Fl. (1874), p. 81, Pl. XIII, figs. 2–4.

Populites salisburiæfolia Lesq.? Am. Journ. Sci., Vol. XLVI (July, 1868), p. 94.

S. (*Araliopsis*) *cretaceum* Newb. var. *dentatum* Lesq. Hayden's Ann. Rept., 1874 [1876], p. 344; *S. cretaceum* Newb. Lesq. in Cret. Fl. (1874), p. 80, Pl. XI, figs. 1, 2.

S. acutilobum Lesq. Cret. Fl. (1874), p. 79, Pl. XIV, figs. 1, 2.

S. (*Araliopsis*) *cretaceum* Newb. var. *obtusum* Lesq. Cret. Fl. (1874), p. 80, Pl. XII, fig. 3; XIII, fig. 1.

"Leaves petiolate, decurrent at base, very smooth above, strongly nerved below; three-lobed; lobes entire and acute. The nervation is all strongly defined; the central nerve straight or nearly so; the lateral primary nerve springing from it at an angle of 30 degrees; secondary nerves regularly arched till they approach the margin of the lobes, when they are abruptly curved and run together. From these the tertiary nerves are given off at a right angle, and from these the quaternary nerves spring at a similar angle, together forming a network of which the areoles are subquadrate."

Collected by Dr. F. V. Hayden.

It is perhaps not certain that the relationship between this beautiful fossil and the living Sassafras is as intimate as I have suggested, for Dr. Hayden obtained no fruits with the leaves, though from the abundance of the latter it is to be hoped that they may yet be found in the same locality. Until the fructification shall be procured, the suggestion that a species of our modern genus Sassafras flourished as far back as the epoch of the deposition of the Middle Cretaceous strata, may be accepted with a certain degree of mental reservation. It is true, however, that there is a most marked correspondence, both in external form and nervation, between the living and the fossil plants, the differences being no greater than we might expect to find between species of the same genus. The nervation of the fossils is stronger and more regular, and the whole aspect of the leaf rather neater and more symmetrical.

With the material already before us we may at least infer that there was living in the American forests of the Cretaceous period a Lauraceous tree, bearing trilobate leaves, having the general aspect and nervation of those of our Sassafras.

The large collections made from the Dakota group at Fort Harker and elsewhere since the above note was written have included a great number of trilobate leaves, which are not separable by any constant and well-marked character from those which formed the basis of the above description, viz, figs. 1 to 4, Pl. VI. On these, however, Lesquereux has established a number of species of Sassafras, namely, *S. acutilobum* (the form figured on Pl. VII, fig. 1), *S. Harkerianum* (shown in our fig. 2, Pl. VIII), *S. Mudgei*, (Pl. VII, fig. 2) *S. obtusum* (Pl. VIII, fig. 1), *S. subintegrifolius* (Pl. VII, fig. 3), etc.

A very large number of beautifully preserved specimens collected by Mr. Sternberg at Fort Harker, and which have been submitted to me for examination, show so many connecting links between these different forms that I am quite unable to separate them into distinct species.

Formation and locality: Cretaceous (Dakota group). Blackbird Hill, Nebraska; Fort Harker and Smoky Hill Fork, Kansas.

SASSAFRAS CRETACEUM RECURVATUM (Lesq.) Newb.

Pl. IX, fig. 2.

Sassafras recurvatus Lesq. Hayden's Ann. Rept., 1872 [1873] p. 424.
Platanus recurvata Lesq. Cret. Fl. (1874), p. 71, Pl. X, figs. 3–5.

NOTE.—Dr. Newberry considered this leaf to be a variety of his *S. cretaceum*, as indicated by a memorandum on the margin of the plate.—A. H.

Formation and locality: Cretaceous (Dakota group). Fort Harker, Kausas

Cinnamomum Heerii Lesq.

Pl. XVII, figs. 1–3.

Am. Journ. Sci., Vol. XXVII (1859), p. 361; Trans. Am. Phil. Soc., Vol. XIII (1869), p. 431, Pl. XXIII, fig. 12; Cret. Fl. (1874), p. 84, Pl. XXVIII, fig. 11.

Guided only by the brief description given by Lesquereux, I can not be positive that the species of Cinnamomum before us is identical with that procured by Dr. Evans from Vancouvers Island. In Lesquereux's specimens the summit of the leaf was wanting, but he conjectures that the lateral nerves extended to the point. Among my specimens are several in which the upper extremity of the leaf is preserved.

From these it appears that the lateral nerves terminate in the margin before reaching the point. This would separate it from *C. Buchi*, and would bring it nearer to *C. Scheuchzeri* or *C. lanceolatum*. My specimens, however, indicate a larger and thicker leaf than that of either of these species.

It would be a matter of no little interest to determine the relations of the specimens of Cinnamomum contained in the Northwest Boundary Collection with those brought from Vancouver Island and Bellingham Bay by Dr. Evans, as that would probably permit us to decide whether the plant beds of Orcas Island should be grouped with those of the mainland or with those of Nanaimo.

Formation and locality: Cretaceous (Puget Sound group). Orcas Island, Washington.

Order HAMAMELIDACEÆ.

Liquidambar Europæum Al. Br.

Pl. XLVII, figs. 1–3.

In Buckl. Geol. and Mineral., p. 513 (1837).

In the collection of fossil plants made by Rev. Thomas Condon at Bridge Creek, Oregon, occur a number of fragments of the leaves of a Liquidambar which I am unable to distinguish from some of the forms of the species known as *L. Europæum* Al. Br. The leaves are large, five to seven lobed, the lobes ovoid, long-pointed, and finely serrate. A fragment of a leaf apparently precisely like this is figured by Heer in his Flora of Alaska (Fl. Foss. Arct., Vol. II, Abth. II, p. 25, Pl. II, fig. 7), and is referred by him to *L. Europæum*. The fruit associated with the leaves at Bridge Creek, as represented in fig. 3, is smaller than that of the living

Liquidambar of the Atlantic coast of North America, and the capsules are smaller. The leaves of Liquidambar are found generally distributed through the Middle Tertiary of Europe and have been described from many localities. They exhibit a great diversity in size and form, as is true of the living species above referred to, and it is the opinion of Heer and Schimper that this is the descendant of the fossil one.

Lesquereux has described a species of Liquidambar from the Pliocene deposits of Chalk Bluff, California, which he regards as distinct from *L. Europæum*. The largest specimen which he figures has almost exactly the form of those before us, but he says that they are usually small, and three-lobed. Probably this also is to be regarded as only a variety of *L. Europæum*, and all forms as hardly distinguishable from the living *L. styraciflua*. This species is quite variable. In northern Mexico the tree and leaves are small and the latter are all three-lobed. In Louisiana the Sweet Gum often forms the greater part of the forest growth; the trunk attains the height of 60 to 80 feet, with a diameter of 2 to 3 feet. The tree grows along the coast as far north as Massachusetts, and has leaves 6 to 7 inches in diameter. They are generally five-lobed, but I have found on the same tree leaves that were three-, five-, and seven-lobed.

Formation and locality: Tertiary (Miocene). Bridge Creek, Oregon.

LIQUIDAMBAR OBTUSILOBATUS (Heer) Hollick.

Pl. V, fig. 4; XII, fig. 4.

Phyllites obtusilobatus Heer. Proc. Acad. Nat. Sci. Phila. (1858), p. 266.

Acerites pristinus Newb. Ann. N. Y. Lyc. Nat. Hist., Vol. IX (April, 1868), p. 15.

Liquidambar integrifolius Lesq. Am. Journ. Sci., Vol. XLVI (July, 1868), p. 93; Cret. Fl. (1874), p. 56, Pl. II, figs. 1–3; XXIV, fig. 2; XXIX, fig. 8; Ills. Cret. and Tert. Pl. (1878), Pl. V, fig. 4, under *Acerites pristinus*.

This is the leaf first described by Professor Heer, from an outline sketch, in the Proceedings of the Academy of Natural Sciences, Philadelphia, 1858, page 266, under the name of *Phyllites obtusilobatus*. When, in 1868, the Later Extinct Floras of North America was published, an imperfect specimen was described by the writer as *Acerites pristinus*. Subsequently several much better specimens were obtained by Lesquereux which led him to refer it to the genus Liquidambar. His description is given in American Journal of Science, Vol. XLVI (July, 1868), page 93, and in his Cretaceous Flora, page 56, where it is illustrated by numerous figures. Nearly all of these represent somewhat deeply five-lobed leaves, of which the lobes are pointed and sometimes acute. The figure given on Pl. XII of this monograph shows that the lobes may sometimes become broadly rounded.

Since this note was written I have found in the Amboy Clays of New Jersey—a formation about on a level geologically with the Dakota group—leaves which I can not distinguish from those figured by Mr. Lesquereux.[1] All these five-lobed entire margined leaves contrast somewhat strongly with those of the living species, and I am disposed to doubt the propriety of referring them to the same genus. The leaves of *L. styraciflua* are quite variable in size and form, but always have pointed lobes and serrated margins. In Northern Mexico all the "sweet gum" trees have three-lobed leaves, rarely more than 3 inches in diameter, while in New Jersey the leaves are from five to seven lobed and generally from 5 to 6 inches in diameter.

In the Puget Sound group a small three-lobed leaf occurs which could hardly be distinguished from these of the Mexican variety of the common species. These, like those of *L. Europæus*, as figured by Unger and Heer, can not be doubted to be Liquidambar, but the leaves now under consideration seem to me more likely to belong to the group of three- to five-lobed Aralias that are so common in the Dakota and Amboy groups.

Formation and locality: Cretaceous (Dakota group). Blackbird Hill, Nebraska, and Fort Harker, Kansas.

Order PLATANACEÆ.

PLATANUS ASPERA Newb.

Pl. XLII, figs. 1–3; XLIV, fig. 5; LIX, fig. 3.

Proc. U. S. Nat. Mus., Vol. V (March 21, 1883), p. 509.

"Leaves attaining a diameter of 1 foot or more; petioled; rounded at the base, more or less three-lobed, sometimes nearly ovoid; nervation strong, about nine branches on each side of the midrib; margins deeply, and often compoundly toothed."

Collected by Rev. Thomas Condon.

---

[1] Dr. Newberry probably has reference to *Aralia rotundiloba* Newb. Flora of the Amboy Clays, p. 118, Pl. XXVIII, fig. 5; XXXVI, fig. 9 (Mon. U. S. Geol. Surv., Vol. XXVI).—A. H.

We have here in the specimens which are figured and others similar, representatives of a fine species of Platanus which is apparently distinct from any hitherto described. In general form it most resembles *P. Haydenii* Newb., and may prove to be only a variety of this species; but the leaves of *P. Haydenii* obtained in Wyoming have only an undulate or bluntly toothed margin; it is well known, however, that this is a character which is exceedingly variable, and specific distinctions can hardly be based upon it. However, the marginal teeth shown in figs. 1 and 2, the base and summit of the leaf, are so peculiar in their size and their compound character that without connecting links we should not be justified in uniting these leaves with any others. In fig. 3 of the plate cited it will be noticed that the dentation at the base of the middle lobe is smaller and more like that in *P. Haydenii*, but the margins in this specimen are so incomplete that they afford information of but little value. Its chief importance is its demonstration of the large size and distinctly trilobate outline of some of the leaves of this tree.

The leaf figured in Pl. XLIV, fig. 5, presents a marked difference of form from those represented on Pl. XLII, but the character of the margins is the same, and it seems probable that this is only the ovoid form which the young and some of the mature leaves are prone to assume. Until further light shall be thrown on the subject it is safest to consider all the leaves mentioned in this note as belonging to the same species.

Formation and locality: Tertiary (Miocene). Bridge Creek, Oregon.

PLATANUS HAYDENII Newb.

Pl. XXXVI; XXXVIII; LVI, fig. 3.

Ann. N. Y. Lyc. Nat. Hist., Vol. IX (April, 1868), p. 70; Ills. Cret. and Tert. Pl. (1878), Pl. XIX; XXI.

"Leaves large, long-petioled, when mature three, perhaps rarely five lobed; lobes nearly equal, long-pointed, acute; on either side of the middle lobe five to eight obtuse teeth; margins of the lateral lobes sinuately toothed to near the base; younger leaves ovate, acuminate, coarsely toothed throughout, except near the base, which is slightly decurrent; nervation strong, radiate from the base, primary nerves three, which are nearly straight, and terminate in the three lobes of the border. From the midrib spring seven or eight pairs of lateral nerves above the basilar pair; these diverge at an angle of about 35 degrees, are slightly flexed at the base, straight or nearly so above, where they are somewhat truncated, their branches terminating in the marginal teeth. The basilar nerves diverge from the midrib at an angle of about 35 degrees and run nearly straight to the extremities of the lateral lobes. They each give off on the lower side seven or eight branches, of which the second or third is strongest. These are more or less curved and branched, the branches terminating in the teeth of the margin. Fruit two to three lines long, prismatic, clavate."

Collected by Dr. F. V. Hayden.

This fine species, which is well represented in the collection, is closely related to *Platanus aceroides*, so common in the Miocene strata of Europe. There are, however, noticeable differences, which seem to me to have a specific value. The leaves of *P aceroides*, though exhibiting a great variety of form, are, I believe, always acutely toothed, while in the specimens before us the teeth are never acute, except those which in the young leaves represent the lateral lobes of the mature form. In *P. aceroides* also, according to Heer (Fl. Tert. Helv., Vol. II, p. 71, Pl. LXXXVII and LXXXVIII, figs. 5–15), the nervation is more sparse, the angle of divergence of all the nerves greater, the number of lateral branches of the midrib less, and the number of marginal teeth considerably greater. Professor Heer says (*loc. cit.*) that in *P. aceroides* the middle lobe of the leaf has two to three dentations on either side, while in *P. Haydenii* the mature leaf has eight to ten teeth on each side of the middle lobe. The difference before specified in the form of the marginal teeth is very marked and strikes the eye at a glance. In *P. aceroides* they are few, long, and acute, sometimes even uncinate, while in *P. Haydenii* they are more numerous, less prominent, and always obtuse, sometimes merely giving a wavy outline to the margin of the leaf.

Detached seeds are all that we have of the fruit, and these, though plainly derived from a Platanus, in their condition of fossilization afford no good characters with which to compare this species with the two now living on this continent, or with the living and fossil species of the Old World.

P. aceroides, according to Heer, had fruit in racemes like the Mexican plane tree, while the fruit of *P. occidentalis* is single. In general aspect the species now before us is more like the eastern than the western of our

American sycamores, to the former of which it has considerable likeness and may very well have been its progenitor.

The fine leaf figured on Pl. XXXVIII, from La Bontes Creek, is probably a young or abnormal state of this species, as it occurs with the ordinary trilobate form.

Formation and locality: Tertiary (Eocene ?). Banks of the Yellowstone River, Montana.

Platanus latiloba Newb.

Pl. I, fig. 4.

Ann. N. Y. Lyc. Nat. Hist., Vol. IX (April, 1868), p. 23; Ills. Cret. and Tert. Pl. (1878), Pl. II, fig. 4.

Platanus obtusiloba Lesq. Am. Journ. Sci., Vol. XLVI (July, 1868), p. 97.

Sassafras (*Araliopsis*) *mirabile* Lesq. ? Cret. Fl. (1874), p. 80, Pl. XII, fig. 1.

"Leaves petiolate, three-lobed, decurrent at the base, lobes broad, obtuse, or abruptly acuminate; principal nerves three, secondary nerves issuing from these at an acute angle, tertiary nerves leaving the secondary at a right angle, forming a network over the surface of the leaf, of which the areolæ are subquadrate."

Collected by Dr. F. V. Hayden.

Judging from the imperfect specimens which we have of this species, it is quite distinct from any described. Having the general form and nervation of the leaves of *P. occidentalis*, the margins are much less deeply sinuate, the lobes less acuminate, and the entire outline of the leaf more simple. The same is true of its relations with *P. orientalis* of the Old World. The fossil species, of which several have been described by Unger and Goeppert, are quite distinct from this. The species described by Unger (*P. Sirii* and *P. grandifolia*) are much more deeply lobed, while that figured by Heer, Goeppert, and Ettingshausen (*P. aceroides*) is less deeply lobed, but more strongly toothed. All fossil species heretofore known are from the Tertiary strata, this being the first instance where the genus has been found in rocks of the Cretaceous epoch.

A large number of nearly complete specimens of the leaf described above have recently been obtained from the Dakota sandstones near Fort Harker, Kansas. Some of these have come into the possession of Lesquereux, who has included them in the genus Sassafras, and has figured and described some of them in his Cretaceous Flora under the name of *Sassafras (Araliopsis) mirabile*.

Count Saporta has raised the question whether any of the trilobate leaves referred by Lesquereux and myself to Sassafras really belong to this genus, and has suggested that their affinities are more likely to be with Aralia. This question can only be definitely settled by the discovery of the fruits of the tree which bore these leaves; these will undoubtedly be found when they are carefully looked for by collectors. Waiting such time, however, we may say that some of the many trilobate leaves found in the Dakota group by their form and nervation are much more like the leaves of Sasafras than those of any other living genus. In these the form is elegantly trilobate, the margins entire, the lobes rounded or obtusely pointed; the nervation is camptodrome. Possibly these leaves will be found to shade into those now under consideration, but judging from the material now before us the difference is considerable. For example, these leaves are larger, have a waved and sometimes even denticulate margin above, while the nerves are stronger and straighter, terminating in the denticles of the border. In all these respects they are more like the leaves of Platanus than those of Sassafras, and they are therefore for the present retained in the genus to which they were referred in the first published description.

Formation and locality: Cretaceous (Dakota group). Blackbird Hill, Nebraska.

PLATANUS NOBILIS Newb.

Pl. XXXIV; XXXVII, fig. 1; L, fig. 1.

Ann. N. Y. Lyc. Nat. Hist., Vol. IX (April, 1868), p. 67; Ills. Cret. and Tert. Pl. (1878), Pl. XVII; XX, fig. 1, under *P. Haydenii*.

"Leaves large, 1½ feet in length and breadth, petioled, three-lobed or subfive-lobed, lobes acute, margins of lobes and base entire, or near the summits of the lobes delicately sinuate-toothed; nervation strongly marked, generally parallel; medial nerve straight, two basilar nerves of nearly equal length and strength diverge from it at an angle of 30 to 35 degrees, are straight throughout, and terminate in the apices of the principal lateral lobes. Above the basilar nerves about 16 pairs of lateral nerves are given off from the midrib at about the same angle; these are nearly straight and parallel, terminating in the teeth of the margin. From each of the basilar nerves diverge about the same number of pairs of branches as from the midrib, and these are also nearly straight and parallel, and terminate directly in the margin. Of these the second or third exterior one on each side is often much the stronger of the series, and is then prolonged into a small but distinct lateral, triangular, acute lobe, giving the leaf a somewhat pentagonal form. From this basilar branch of the lateral nerves, twelve or more short, generally simple, branchlets spring on the lower side, and four to five on the upper side near the summit, all of which terminate in the margins. The tertiary nerves connect the adjacent secondary nerves nearly at right angles; sometimes they are straight and parallel, but oftener more or less broken and branching where they meet, near the middle of the interspaces. Where the systems of nervation of the lateral and middle lobes come in contact, the tertiary nerves are stronger and form a somewhat irregular network, of which the areolæ are large and subquadrate."

Collected by Dr. F. V. Hayden.

In general aspect these magnificent leaves are considerably unlike those of any known species of Platanus, and I have felt some hesitation in referring them to that genus. The texture was evidently thicker and the surfaces smoother than in the leaves of most Sycamores, and, on the whole, they recall the leaves of Cecropia or some other of the broad, leathery, polished leaves borne by the trees of the tropics. On close examination, however, they are found to present the radical structure of the leaves of Platanus, and, aside from their association with so many genera plainly belonging to the flora of the temperate zone, their form and nervation seem to me to afford at least presumptive evidence that they were borne by a tree of that genus. They will, perhaps, suggest to the fossil botanist the leaves described by Unger under the names of *Platanus Hercules*, *P. Jatrophæfolia*, etc. (Chlor. Prot., p. 137, Pl. XLV, figs. 6, 7, etc.), and which he subsequently removed from that genus. But those palmate, many-lobed leaves were very unlike these now before us, and resemble much more the leaves of Jatropha or Sterculia than those of Platanus.

The crowded, somewhat heavy and regular nervation of these leaves, their thick texture and polished surface, must have given the tree on which they grew an aspect quite different from that of *P. occidentalis;* but *P. orientalis*, and sometimes *P. racemosa*, have thick and polished leaves, and the deviation from the common form is not so great in these fossils as in the living species I have named, or the fossil species named by Unger, *P grandifolia* and *P. Sirii* (Chlor. Prot., p. 136, Pl. XLV, figs. 1–5, and Foss. Fl. Sotzka, p. 36 [166], Pl. XV [XXXVI]), fig. 1.

In size these leaves exceed those of any known species of Sycamore, and if we are correct in referring them to Platanus, they may be considered the only relics we have of by far the noblest species of the genus. Some of the leaves are a foot and a half in length and of about equal breadth, and yet they do not so far exceed the ordinary size of the leaves of the Sycamores as do the leaves of *Acer macrophyllum* those of other species of maple.

Since the above notes were written, Lesquereux has described (Tert. Fl., p. 237, Pl. XXXIX, figs. 2–4) some trilobate, sometimes five-lobed leaves, which he compares with *Platanus nobilis*, and is inclined to regard them as identical; but it will only be necessary to refer to the figures now given, especially that on Plate L, to show that the differences are such as to distinctly separate them. In *Aralia notata* Lesq., the general plan is not unlike that of the leaves in question (which is true also of most trilobate leaves), but here the resemblance ceases, for in *A. notata* the margins are entire and the lateral nerves connect in festoons along the margin (camptodrome), whereas in *P. nobilis* the lateral branches terminate in the teeth with which the margins of the lobes are set (craspedodrome).

In the Report of Progress of the Geological and Natural History Survey of Canada for 1879–80, Appendix N, Prof. J. W. Dawson gives notes on a number of species of plants collected on the Souris River, and among others he mentions *Platanus nobilis*, of which good specimens were procured by Dr. Selwyn and Dr. G. M. Dawson, and he confirms, by observations on these specimens, my reference to the genus Platanus. He also mentions a feature which does not appear in any of the specimens I have seen, namely, two short basal lobes extending backward on the petiole. This is not, however, unprecedented in the leaves of Platanus, as I have seen something of the kind in the large leaves borne by young and vigorous plants of *P. occidentalis*. The figure given on Pl. L is of the natural size, and attests the magnitude claimed for some of the leaves of this magnificent tree. When it is realized that the main nerves of the middle and lateral lobes must unite at a point some inches below the part represented at the bottom of the figure, and that the central lobe was at least 6 inches and the lateral lobes 3 or 4 inches larger than represented, it will be seen that the leaf could not have been much less than a foot and a half in length and breadth. These dimensions are rivaled by no living species of Platanus, but I have fragments of the leaves of *P. Raynoldsii* which could have been little less in size.

The leaf figured on Pl. XXXVII, fig. 1, is an immature form of this species. This is established by its occurrence with the larger and more deeply lobed leaves, with which it is connected by intermediate forms.

Formation and locality: Tertiary (Eocene ?). Near Fort Clark, Dakota.

PLATANUS RAYNOLDSII Newb.

Pl. XXXV.

Ann. N. Y. Lyc. Nat. Hist., Vol. IX (April, 1868), p. 69; Ills. Cret. and Tert. Pl. (1878), Pl. XVIII.

"Leaves of large size, sub-orbicular or rudely triangular in outline, more or less rounded below, three-pointed above, often decurrent on to the petiole, margins at base entire, on the sides and above, coarsely and obtusely double-serrate, the lobes of the upper margin short and broad, less produced than in most other species; nervation strong but open, having the general character of *P. occidentalis* and of the fossil species *P. aceroides.*"

Collected by Dr. F. V. Hayden.

The younger leaves are rounded in outline and decurrent on the petiole. Those more fully developed (which are sometimes 15 inches in length and breadth), more triangular in form, not always decurrent, and having lobes more produced, offer considerable resemblance to those of *P. aceroides*, an extinct species from the Miocene of Europe, the nervation being similar in kind and not greatly different in degree. The leaf is, however, always less angular than in *P. aceroides* and *P. Haydenii*, and the character of the marginal serration is essentially different from that of any known species. In *P. aceroides* the margins are set with long, acute, curved, simple teeth, as in the living *P. occidentalis;* in *P. Haydenii* the margins are for the most part only sinuate; and in *P. nobilis* the middle lobes only are toothed, and those but slightly; while in the species before us, with the exception of the basal margin, the whole outline is marked by a broad, strong, double dentation.

The figure given on Pl. XXXV is that of a complete leaf about half the size, linear, of the largest contained in the collection.

In texture the leaf was apparently similar to that of *P. occidentalis*, rather thin and more or less roughened.

Formation and locality: Tertiary (Eocene ?). Banks of Yellowstone River, Montana.

Order ROSACEÆ.

PYRUS CRETACEA Newb.

Pl. I, fig. 7.

Ann. N. Y. Lyc. Nat. Hist., Vol. IX (April, 1868), p. 12; Ills. Cret. and Tert. Pl. (1878), Pl. II, fig. 7.

"Leaves petioled, small, roundish-oval or elliptical, often slightly emarginate, entire or finely serrate; medial nerve strong below, rapidly diminishing toward the summit; lateral nerves four or five pairs with intermediate smaller ones, diverging from the midrib at unequal angles, curved toward the summits, where they anastomose in a series of arches parallel with the margin; tertiary nerves forming a network of which the areolæ are somewhat elongated."

Collected by Dr. F. V. Hayden.

There are a number of leaves in the collection, of which the characters, as far as they are discernible, agree more closely with those of the species of Pyrus than with any other with which I have compared them. All the traces of their original structure which remain, however, are quite insufficient to permit their generic limitation to be determined with any degree of certainty. The leaves of many of the allied genera of the Rosaceæ have so much in common that even with the leaves of the living plants it would be difficult, if not impossible, to separate them. The fossils before us are, however, very characteristic of the formation which contains them, and for that reason require notice, and, as far as practicable, description.

There are several other leaves in the collection which seem to me to have belonged to Rosaceous trees, and there is perhaps no a priori improbability that Pyrus began its existence on this continent with its congeners and companions in our forests of the present day.

Formation and locality: Cretaceous (Dakota group). Smoky Hill, Kansas.

Amelanchier similis Newb.

Pl. XL, fig. 6.

Ann. N. Y. Lyc. Nat. Hist., Vol. IX (April, 1868), p. 48; Ills. Cret. and Tert. Pl. (1878), Pl. XXV, fig. 6.

"Leaves petioled, ovate, obtuse or acuminate, rounded or slightly cordate at the base; margin coarsely toothed, except near the petiole, where it is entire; nervation pinnate, delicate; medial nerve straight, six to seven pairs of lateral nerves diverging from the midrib at an angle of about 40 degrees, slightly curved upward, especially near the summit, the upper ones nearly simple, but giving off a perceptible branch near the summit on the lower side, which runs into the next tooth below. The lower pair spring from the extreme base of the leaf, are strong and simple, and strike the margin where the dentation commences. The second pair of lateral nerves each send off two or three slender nerves from near the summit to the teeth of the adjacent margin; tertiary nerves very fine, leaving the secondaries at right angles, and forming a fine network of which the areolæ are nearly quadrate."

Collected by Dr. F. V. Hayden.

The number of specimens of this species in the collection is small and all but one are imperfect. This one is the impression of a thin, delicate leaf, of which all the details of nervation are preserved as perfectly as they could have appeared in the living plant. The other specimens indicate that the leaves were usually pointed, often acute.

From the nervation and character of dentation of these leaves, I think we may at least say that the plant which bore them was Rosaceous, and among the Rosaceous genera with which I have compared them they approach most nearly to *Amelanchier*, some of the leaves of *A. Canadensis* being entirely undistinguishable from them in form or nervation.

A. Canadensis now grows over all the temperate parts of the continent and would seem from its wide range to be an old resident of the continent and as likely to be represented in the Tertiary as any other of our plants.

Formation and locality: Tertiary (Eocene?). Banks of Yellowstone River, Montana.

CRATÆGUS FLAVESCENS Newb.

Pl. XLVIII, fig. 1.

Proc. U. S. Nat. Mus., Vol. V (March 21, 1883), p. 507.

"Leaves small, about 1 inch in length and breadth; lobed; lobes rounded and bearing a few teeth or crenulations; the summit of the leaf trilobed, with two lateral lobes below on either side."

Several small, lobed leaves are contained in the collection made by Rev. Thomas Condon, which bear such resemblance to those of some species of Cratægus that we seem to be justified in referring them to this genus. Of these the one figured is the most complete in outline; this in its general proportions and markings approaches closely to the leaves of *C. flava* Ait., but in that species the leaves are usually somewhat larger and the lobes are set with several acute teeth.

Eighteen fossil species of Cratægus have been described, and of these three from the Tertiary deposits of North America, namely, *C. antiqua* Heer (Fl. Foss. Arct., Vol. I, p. 125, Pl. L, figs. 1, 2), *C. Warthana* Heer, and *C. æquidentata* Lesq. (Tert. Fl., p. 297, Pl. LVIII, figs. 4, 4a); but these are much larger and have rhomboidal and undivided leaves; indeed, it is not certain that they all belong to the genus Cratægus.

Of foreign species there is none with which this is likely to be confounded. *C. dyssenterica* Mass. (Fl. Foss. Senigall, p. 414, Pl. XIX, fig. 1), is similarly lobed, but the leaves are larger and much more deeply cut.

The resemblance of the leaves before us to those of the living *C. flava* is so close that it is quite possible that the present is the derivative from the ancient species, a possibility suggested in the specific name chosen.

Formation and locality: Tertiary (Miocene). Bridge Creek, Oregon.

PRUNUS VARIABILIS Newb.

Pl. LII, figs. 3 and 4 (in part), 5.

Proc. U. S. Nat. Mus., Vol. V (March 21, 1883), p. 509.

"Leaves short-petioled, very variable in form; lanceolate or broadly lance-ovate, 2 to 3 inches long by 1 to 2 inches wide; acuminate at the summit, wedge-shaped at base; margins thickly set with minute, acute, appressed teeth."

Numerous leaves, which evidently belong to the genus Prunus, occur in the collections from Alaska made by Captain Howard, and sometimes several on the same slab that exhibit no differences except the marked variation in form shown in the figures and alluded to in the name given. Compared with the living species, these leaves have much the aspect of some of the forms of *P. Virginiana*, the marginal serration being very much the same, though the leaves of the living plant are usually obovate.

A species of Prunus is described by Professor Heer from the Tertiary strata of Greenland under the name of *P. Scottii* (Fl. Foss. Arct., Vol. I, p. 126, Pl. VIII, fig. 7), but the only leaves he describes and figures are much larger and longer and more coarsely toothed than these.

Numerous species of Prunus have been described from the Tertiary of the Old World, but so far as we can judge there are none that have the somewhat peculiar lanceolate leaf, broader in the middle than elsewhere and narrowed at both ends, terminating in a long point, like the one under consideration. It has been thought necessary, therefore, to distinguish this by a special specific name.

Formation and locality: Tertiary (Miocene). Cook Inlet, Alaska.

Order LEGUMINOSÆ.

Cassia sp.? Newb.

Pl. XLVI, fig. 10.

Note.—The only information which I have been able to obtain in regard to this figure is the manuscript note, "Cassia fruit," by Dr. Newberry, on the margin of the plate, and the locality given on the specimen label.—A. H.

Formation and locality: Tertiary (Miocene). Bridge Creek, Oregon.

Leguminosites Marcouanus Heer.

Pl. V, fig. 3.

Proc. Acad. Nat. Sci. Phila. (1858), p. 265; Ills. Cret. and Tert. Pl. (1878), Pl. V, fig. 3, under *Phyllites obcordatus*.

The original tracing of this leaf, on which Professor Heer has written the name given it, enables me to identify it with certainty and to correct an error which has been committed in reference to it, namely, that its name has been given to another larger, broader, obovate leaf found with it, and described by Professor Heer with the name of *Phyllites obcordatus*.

The general form of these leaves is not unlike, but the one now under consideration is narrower, slightly unequal at the base, and has a remarkably sparse nervation, as will be seen by referring to the figures.

Formation and locality: Cretaceous (Dakota group). Blackbird Hill, Nebraska.

Order ANACARDIACEÆ

RHUS (?) NERVOSA Newb.

Pl. XXXIII, figs. 5, 6.

Rhus nervosa Newb. Ann. N. Y. Lyc. Nat. Hist., Vol. IX (April, 1868), p. 53; Ills. Cret. and Tert. Pl. (1878), Pl. XVI, figs. 5, 6.

"Leaves pinnate, leaflets oblong or linear in outline, rounded or cordate at the base, pointed above; margins coarsely and acutely serrate; nervation pinnate, strong; lateral nerves numerous, leaving the midrib at an acute angle, simple or somewhat branched, parallel, gently arched upward, and terminating in the teeth of the border."

Collected by Dr. F. V. Hayden.

The specimens of this plant scarcely afford material for satisfactory classification. They bear a strong resemblance to the pinnate leaflets of some of our shrubby species of Rhus, especially of *R. copallina* and *R. typhina*. The nervation and marginal serration are essentially the same, and the texture of the leaf would appear to have been similar, but the nerves are stronger and the dentation coarser than in most specimens of these species with which I have compared it. With the trifoliate and oak-leaved species it has little in common, and will not be likely to be confounded with any of the fossil species which have been described.

The general form of the leaf is not unlike *R. Meriani* Heer (Fl. Tert. Helv., Vol. III, Pl. CXXVI, figs. 5–11), but the margins of the leaves of that species are not as deeply toothed.

Formation and locality: Tertiary (Fort Union group). Fort Union, Dakota.

Order ACERACEÆ.

ACER sp.? Newb.

Pl. XLVI, fig. 8.

NOTE.—The only information which I have been able to obtain in regard to this figure is the manuscript note, "Acer fruit," by Dr. Newberry, on the margin of the plate, and the locality as given on the specimen label.—A. H.

Formation and locality: Tertiary (Miocene). Bridge Creek, Oregon.

NEGUNDO TRILOBA Newb.

Pl. XXXI, fig. 5.

Ann. N. Y. Lyc. Nat. Hist., Vol. IX (April, 1868), p. 57; Ills. Cret. and Tert. Pl. (1878), Pl. XXIII, fig. 5.

"Leaves thin and delicate, but distinctly nerved, pinnate in one or more pairs, leaflets lanceolate or lance-ovate, long-pointed, rounded or slightly cordate at base, short-petioled; margins coarsely, remotely, and irregularly toothed; terminal leaflet trilobate, the margins toothed or serrated; nervation of lateral leaflets pinnate, nine or ten pairs of lateral nerves diverging from the midrib at an angle of about 50 degrees, arching upward, more or less branched toward the summit. Of these the basal pair are shortest and simple, following the course of the adjacent margin; the second pair are strongest, and throw off each three or four curved branches on the lower side."

Collected by Dr. F. V. Hayden.

The general aspect, including texture, form, dentation, and nervation of the lateral leaflets is strikingly like that of the corresponding parts of the leaf of the living *Negundo aceroides*. The genus Negundo is represented among living plants by but a single species, and this is so like Acer in all but its leaves that Professor Gray intimates that it should hardly be considered distinct from that genus. A fossil species has been discovered in the Tertiaries of Europe, *N. Europæum* Heer (Fl. Tert. Helv., Vol. III, p. 60, Pl. CXVIII, figs. 20–22), but it would seem to have been a smaller species than the living one, and had obovate wedge-based leaves quite different from those before us.

If, in the light of more and better material, it should prove that a species of Negundo lived on the American continent during the Tertiary age, it would be a fact of no little interest, and would strengthen the claims of *Negundo aceroides* to a distinct generic place in the botanical series. In that case, however, its trilobate terminal leaflet would still further indicate its acerine affinities.

Formation and locality: Tertiary (Fort Union group). Fort Union, Dakota.

Order SAPINDACEÆ.

Sapindus affinis Newb.

Pl. XXX, fig. 1; XL, fig. 2.

Ann. N. Y. Lyc. Nat. Hist., Vol. IX (April, 1868), p. 51; Ills. Cret. and Tert. Pl. (1878), Pl. XXIV, fig. 1; XXV, fig. 2.

"Leaves pinnate in many pairs of leaflets, with a single lanceolate terminal one; leaflets smooth, thick, lanceolate, long-pointed, acute, sessile or short-petioled, unsymmetrical, rounded or wedge-shaped at base; nerves fine and obscure, ten or more branches diverging from the midrib on either side at somewhat unequal distances, and of unequal size. These arch upward, giving off several lateral branches at right angles, or nearly so, and die out near the margins, or are carried around in a curve parallel with it, and thus connect."

Collected by Dr. F. V. Hayden.

These leaves are most strikingly like those of Sapindus, and taken by themselves would afford perhaps sufficient ground for uniting them with that genus. They are also very like a series of leaves found in the Tertiaries of Europe, figured by Professor Heer, in Fl. Tert. Helv., Vol. III, p. 61, Pls. CXIX, CXX, CXXI, under the names of *Sapindus falcifolius*, *S. densifolius*, and *S. dubius*. The nervation is also the same; so there can hardly be a doubt that our plant and those of Professor Heer are generically identical, and, if the proofs before him of the identity of his fossils with the living genus Sapindus are sufficient, we must conclude that the specimens before us are also the representatives of that genus. In our specimens, however, the leaves are constantly shorter and broader than in the species I have mentioned, and are often rounded at the base, so that I have been compelled to regard them as specifically distinct.

Formation and locality: Tertiary (Eocene?). Mouth of Yellowstone River, Montana.

SAPINDUS (?) MEMBRANACEUS Newb.

Pl. XXX, figs. 2, 3.

Sapindus membranaceus Newb. Ann. N. Y. Lyc. Nat. Hist., Vol. IX (April, 1868), p. 52; Ills. Cret. and Tert. Pl. (1878), Pl. XXIV, figs. 2, 3.

"Leaves pinnate in many pairs of leaflets, and terminating in a large ovate, often unsymmetrical one; lateral leaflets lanceolate, acute, wedge-shaped at base, unsymmetrical, thin and membranous, with entire margins; nervation fine and sparse, many pairs of lateral nerves being given off by the midrib (from which also spring many small lateral branchlets), and these arching upward inosculate near the margin or die out."

Collected by Dr. F. V. Hayden.

This is similar in nervation and in the general form of the lateral leaflets to the preceding species (*S. affinis*), but the whole plant is more delicate, the leaf thinner, the nervation finer, the terminal leaflet several times as large and of a different form.

Formation and locality: Tertiary (Fort Union group). Fort Union, Dakota.

Order RHAMNACEÆ.

RHAMNUS ELEGANS Newb.

Pl. L, fig. 2.

Ann. N. Y. Lyc. Nat. Hist., Vol. IX (April, 1868), p. 49.

"Leaves lanceolate, entire, rounded or abruptly narrowed at the base, long-pointed and acute above, broadest part one-third the distance from the base to apex; nervation regular and sharp, but delicate; midrib strongly marked, lateral nerves twelve to fifteen, nearly equidistant on either side, gently arched upward, and terminating in the margins; tertiary nerves numerous, fine, spanning the distance between the branch nerves, and dividing this space into narrow, sub-rectangular areoles."

Collected by Miss Kate Haymaker.

This is a remarkably neat and symmetrical leaf, both as regards its outline and nervation. Its lines are all graceful, with little of the rigidity that characterizes the leaves of most of the Rhamnaceæ, and more of the aspect of the leaf of a Lauraceous tree; but the numerous parallel side-nerves, terminating all in the margins, form a character which the Laurels never have.

Of described species it most resembles Weber's *R. Decheni* (Palæontogr. Vol. II, p. 204 [90], Pl. XXIII [VI], fig. 2), but differs from it in having an ovate, lanceolate form, and the nervation is a little more crowded.

Formation and locality: Cretaceous (Laramie group). Belmont, Colorado.

Rhamnus Eridani Ung.

Pl. XLVIII, fig. 7

Gen. et Sp., Pl. Foss. (1850), p. 465.

The leaf represented in fig. 7 is unique in the collection made at Bridge Creek, Oregon, but though imperfect it is very distinctly marked, and apparently belongs to the genus Rhamnus, and so closely resembles some of the figures of *Rhamnus Eridani* Ung., especially that described in Fl. Foss. Arct, Vol. I, p. 123, Pl. XLIX, fig. 10, that I have not felt justified in regarding them as distinct.

Formation and locality: Tertiary (Miocene). Bridge Creek, Oregon.

Rhamnites concinnus Newb.

Pl. XXXIII, figs. 7 (8?).[1]

Ann. N. Y. Nat. Hist., Vol. IX (April, 1868), p. 50; Ills. Cret. and Tert. Pl. (1878), Pl. XVI, figs. 7, 9 (fig. 9 under *Viburnum asperum*).

"Leaves petioled, long ovate, acute, rounded at the base, coarsely and nearly equally mucronate-dentate; nervation pinnate, remarkably precise and parallel throughout; medial nerve straight; lateral nerves, nine to ten pairs, diverging at an angle of about 20 degrees, slightly arched upward, parallel among themselves, basilar pair reaching to margin below the middle of the leaf, sending off each about eight short, simple, slightly curved, parallel branches to the dentations of the baso-lateral margin; superior lateral nerves simple, or once-forked at the summit; tertiary nerves very numerous, simple, parallel, connecting the lateral secondary nerves and the branches of the basilar nerves nearly at right angles."

Collected by Dr. F. V. Hayden.

---

[1] The description applies without doubt to fig. 7, but does not agree with fig. 8. This latter specimen, however, is plainly labeled in Dr. Newberry's handwriting as belonging to this species, although it would appear to be more logical if allied with fig. 9, same plate (*Viburnum asperum* Newb.)—A. H.

These beautiful leaves are so definite in form and structure and so perfectly preserved that we should have no difficulty in referring them to their appropriate genus if we could find among living trees their precise generic counterpart, but up to the present time I have not been able to satisfy myself that they are generically related to any living plants. The nervation is in some respects very like that of Berchemia, e. g., *B. volubilis*, the "Supple Jack" of our Southern States. Nowhere else do I remember to have seen the same parallelism of the secondary and Tertiary nerves, but the serration of the margin is coarser than in any of the Rhamnaceæ with which I am acquainted, and the development of the basilar pair of lateral nerves is much greater than in Berchemia. This latter character is not without example in Rhamnus, as it is even more conspicuous in some species of the genus, as, for example, in *R. celtifolia* of the Cape of Good Hope. A cross between that species and our Berchemia, with a greater development of the marginal dentation than either exhibits, would give us the fossil before us.

Considering it to exhibit more of the character of the Rhamnaceæ than of any other family, I have placed it doubtfully there.

Formation and locality: Tertiary (Fort Union group). Fort Union, Dakota.

ZIZYPHUS LONGIFOLIA Newb.

Pl. LXV, figs. 3–5.

Proc. U. S. Nat. Mus., Vol. V (March 21, 1883), p. 513.

"Leaves 4 to 7 inches long by 6 to 12 lines wide; lanceolate, long-pointed, wedge-shaped at base, and long petioled; margins waved, or more or less distinctly toothed: midrib well defined from base to summit; basal pair of lateral nerves approaching closely to the margin near the middle of the leaf, then curving gently inward and anastomosing with the higher lateral nerves, of which there are three or more set alternately and curving upward, forming a festoon near the margin; tertiary nerves very finely reticulated."

Of this species a large number of specimens occur in the Green River Shales in certain layers where they are associated with the ferns Lygodium and Acrostichum They may be at once distinguished from those of any other described species of Zizyphus by their elongated and lanceolate form. In the same slabs which contain these leaves are a few which, though imperfect, apparently represent Lesquereux's *Z. cinnamomoides.* These are ovate or ovate-lanceolate in outline, and yet may be only a variety of the species described above. They differ, however, widely from the description of *Z. cinnamomoides* of Lesquereux.

Formation and locality: Tertiary (Green River group). Green River, Wyoming.

Order VITACEÆ.

Vitis rotundifolia Newb.

Pl. LI, fig. 2, in part; LIII, fig. 3.

Proc. U. S. Nat. Mus., Vol. V (March 21, 1883), p. 513.

"Leaf broadly rounded or sub-triangular in outline, cordate at the base, and with an acute point at the summit, and at the extremity of each of the angles; intermediate portions of the margin coarsely and bluntly toothed; strongly three-nerved; tertiary nervation distinct and flexuous."

Collected by Captain Howard.

The general aspect of this leaf is but imperfectly given in the drawings, inasmuch as the strength of the nervation has been somewhat exaggerated, but the leaf was apparently thicker and with stronger nervation than in most of the vines.

Among living species it bears the strongest resemblance to *V. labrusca,* but is less distinctly angled and more strongly dentate on the margin. Professor Heer has described three species of Vitis that occur in the arctic regions, *V. Olriki* (Fl. Foss. Arct., Vol. I, p. 120, Pl. XLVIII, fig 1), *V. arctica* (*op. cit.,* Pl. XLVIII, fig. 2), and *V. Islandica* (*op. cit.,* p. 150, Pl. XXVI, figs. 1e, 1f, 7a), but all these had leaves which were more elongated triangles in form and of lighter structure.

Formation and locality: Tertiary (Miocene). Admiralty Inlet, Alaska.

Order TILIACEÆ.

Grewia crenata (Ung.) Heer.

Pl. XLVI, fig. 2; XLVIII, figs. 2, 3.

Fl. Tert. Helv., Vol. III (1859), p. 42, Pl. CIX, figs. 12–21; CX, figs. 1–11.
Dombeyopsis crenata Ung., Gen. et Sp. Pl. Foss. (1850), p. 448.

Formation and locality: Tertiary (Miocene). Bridge Creek, Oregon.

Order ARALIACEÆ.

ARALIA MACROPHYLLA Newb.

Pl. LXVII, fig. 1; LXVIII, fig. 1.

Proc. U. S. Nat. Mus., Vol. V (March 21, 1883), p. 513.

"Leaves large, long-petioled, palmately five-parted from the middle upward, divisions conical in outline, sometimes entire, often remotely, occasionally coarsely toothed; nervation strong and regular; the midribs of the divisions strong and straight, those from the second lateral lobes springing from near the bases of the first lateral lobes; secondary nerves numerous, distinct, curved gently upward; where the margins are entire, partially camptodrome; where dentate, terminating in the teeth; tertiary nerves anastomosing to form quadrangular and very numerous areoles."

Collected by Dr. C. A. White.

In general form and nervation these leaves are very similar to the typical fossil species of the genus, viz: *A. Whitneyi* Lesq., *A. angustiloba* Lesq., of the Pliocene of California, and *A. Hercules* (Ung.) Sap. (Ann. Sci. Nat. Bot., 5me Ser., Vol. IV, p. 295 [151], Pl. IX, fig. 2), of the Miocene of Radoboj, Croatia (*Platanus Hercules* Ung., Chlor. Prot., p. 138, Pl. XLVI), and especially *A. Saportanea* Lesq. of the Dakota Cretaceous. From all these, however, it differs specifically in several characters. Unger's species agrees in having the midribs of the lobes radiating from the base, while in the species described by Lesquereux, enumerated above, the lower pair spring from the first laterals some distance above their bases, as though the primary form was a tripartite leaf, the lateral lobes contracted where they join, thus acquiring a spatulate outline; and his *A. grandifolia* has more coarsely toothed, *A. Jatrophæfolia*, seven-parted leaves. In the localities where they are found the leaves of *A. macrophylla* are exceedingly abundant, sometimes matted together so as to obscure their outlines. These show that they vary in size, in the number of lobes, and in the character of the margins, occasionally one occurring which is only three-lobed, while almost all are five, and the margins are sometimes nearly entire, while in other leaves they are all strongly, even spinously dentate. The leaves vary from 3 to 12 inches in length, and the lobes are sometimes long and narrow, in others much broader. This variability indicates that the leaves having narrow entire lobes found in the Dakota group and named *A. quinquepartita*, *A. tripartita*, and *A. cuneata*, by Mr. Lesquereux, are but forms of one species. *Aralia Whitneyi* Lesq. has seven-parted leaves, these less deeply lobed, and with entire margins; *A. angustiloba* more deeply cut leaves with narrower and entire lobes (Mem. Mus. Comp. Zoöl., Vol. VI, No. 2 (1878), p. 22, Pl. V, figs. 4, 5).

Perhaps of all described species of Aralias *A. Saportanea* Lesq., from the Dakota group of Kansas (U. S. Geol. and Geog. Surv. of Colorado, Hayden (1874), p. 350, Pl. I), approaches nearest to those under consideration, but are distinguished by minor characters, smaller size, less deeply dentate margins, etc. This species is found, however, in our Middle Cretaceous strata, forming part of the most ancient angiosperm flora, and while the species are unquestionably distinct, their great resemblance may be fairly taken as an indication that one is the progenitor of the other. The group of leaves now before us has been, perhaps without sufficient proof, referred to the genus Aralia, and it is highly desirable that this question should be decided by the discovery of fruit or flowers. But whether Aralia or not, they constitute a marked feature in the older angiosperm floras in this country and in Europe, and their geological interest and value is to a certain degree independent of their botanical relations. It has been suggested by Count Saporta that not only the trilobed leaves from the Dakota Cretaceous, which I have described as Sassafras, but also the great leaves of *Platanus nobilis*, figured in this volume, should be referred to Aralia, as the platanoid leaves described by Unger as *P. Hercules*, etc., have been; but there is little resemblance between the quinquepartite, narrow-lobed, toothed leaves of *A. Saportanea* Lesq. and its associates with three lobes, broadly rounded, sometimes almost obsolete and entire, in *Sassafras cretaceum*, and it only requires a glance at the figure of the huge leaf of *Platanus nobilis*, given on Pl. L of this monograph, to be satisfied that its affinities are with Platanus rather than Aralia.

Formation and locality: Tertiary (Green River group). Green River, Wyoming.

ARALIA (?) QUINQUEPARTITA Lesq.

Pl. IX, fig. 1.

Hayden's Ann. Rept., 1871 [1872], p. 302; Cret. Fl. (1874), p. 90, Pl. XV, fig. 6.

The possession of a better specimen than that on which Lesquereux based the description of the species, one, in fact, that is nearly entire, prompts the publication of the figure now given.

Since the appearance of the Cretaceous Flora, Lesquereux has figured and described a number of species of Aralia (Report of Dr. F. V. Hayden, 1874, pp. 348, 349), of which his *Aralia concreta* and *A. tripartita* are perhaps only forms of the species under consideration.

Formation and locality: Cretaceous (Dakota group). Fort Harker, Kansas.

ARALIA TRILOBA Newb.

Pl. XL, figs. 4, 5.

Ann. N. Y. Lyc. Nat. Hist., ol. IX (April, 1868), p. 58; Ills. Cret. and Tert. Pl. (1878), Pl. XXV, figs. 4, 5.

"Leaves pinnate or ternate; lateral leaflets long-oval, rounded, or slightly heart-shaped, and unequal at base, pointed at summit, sharply serrate throughout; nervation pinnate; texture thin; surfaces smooth.

"Trilobate leaf similar in surface, texture, nervation, and marginal serration, but unequally three-lobed; lobes acute, long-pointed."

Collected by Dr. F. V. Hayden.

The character of these leaves is very well shown in the specimens before me. They seem to indicate a species of Aralia, and have a marked resemblance to some of the leaves of our two most common species, *A. racemosa* and *A. nudicaulis*. The trilobate leaf is not commonly found in our Aralias, but there is always a tendency to the production of such a form, and I have frequently remarked it in *A. racemosa*, as it grows at the West. That is, however, a larger and stronger plant than this was.

Formation and locality: Tertiary (Eocene?). Fort Clarke, Dakota.

Order CORNACEÆ.

CORNUS NEWBERRYI Hollick.[1]

Pl. XXXVII, figs. 2–4.

Cornus acuminata Newb. Ann. N. Y. Lyc. Nat. Hist., Vol. IX (April, 1868), p. 71. (not *C. acuminata* Weber, Palæontogr., Vol. II (1852), p. 192); Ills. Cret. and Tert. Pl. (1878), Pl. XX, figs. 2–4, under *C. acuminata*.

"Leaves ovate or ovate-lanceolate, long-pointed, acute, entire, narrowed at the base, and slightly decurrent; midrib distinct, straight or curved toward the summit, following the course of the frequently deflexed point; lateral nerves numerous, regular, and nearly parallel, simple, lower ones straight with a slightly curved summit, upper ones becoming progressively more arched upwards, when near the apex of the leaf curved in so as nearly to join the extremity of the midrib; tertiary nervation so fine as to be hardly perceptible in the fossil state."

The specimens of these leaves contained in the collection of Dr. Hayden are quite numerous and pretty well preserved. Although there is no fruit of Cornus associated with them, there can be little doubt that they are properly referred to that genus. The aspect of the leaves of Cornus is peculiar, and such as is usually readily recognizable at a glance. This facies is given by the outline as well as the nervation. The outline is usually more or less accurately oval, the margin entire, the base rounded or slightly wedge-shaped, the summit pointed and laterally flexed. The nervation is very clearly defined, the midrib strong at the base, tapering gradually till it reaches the extreme point of the apex; the lateral nerves pinnate, approximated below, more remote above; all simple, arched upward, those near the summit being drawn in to join the midrib.

This latter characteristic is visible in all the species of Cornus known and is particularly noticeable in the common herbaceous species of *C. Canadensis*. It is also very marked in *C. Florida*, *C. sericea*, *C. alternifolia*, etc.

The tertiary nervation is generally delicate and sparse, the tertiary branchlets running across obliquely, but with nearly a straight course, between the adjacent lateral nerves. In all these characters, as far as they

[1] Dr. Newberry's original published name, *C. acuminata* (1868), was antedated by Weber's, *C. acuminata* (1852), given to another species. It therefore became necessary to change the name.—A. H.

are retained in the fossils before us, we find an entire correspondence with the living genus Cornus, and refer these leaves to that place in the botanical series with as much confidence as the foliary appendages alone can give.

Lesquereux suggests that this plant is identical with his *Juglans rhamnoides* (Tert. Fl., p. 284), but after a careful comparison of specimens I am compelled to consider them as distinct. The nervation of these leaves is that of Cornus and not of Juglans, and no species of the latter genus has the long, strong petiole on which the blade is decurrent, as in the specimens before us.

Formation and locality: Tertiary (Eocene?) Fine laminated sandstone, with *Platanus Haydenii* and *Populus Nebrascensis.* Yellowstone River, Montana.

NYSSA (?) CUNEATA Newb.

Pl. XVII, figs. 4–6.

Ficus ? cuneatus Newb. Bost. Journ. Nat. Hist., Vol. VII (1863), p. 524.

"Leaves obovate or elliptical, shortly acuminate at summit, wedge-shaped at base, decurrent onto the petiole; nervation distinct, flexuous, reticulated; midrib strong; lateral nerves eight or nine pairs gently arched upward, the lower ones curved at the extremities, anastomosing near the margin, the upper ones forked above the branches, meeting and forming a coarse network."

The specimens of this plant are too few and two obscurely preserved to permit any accurate determination; for the present it may be left in the genus Nyssa, to some species of which it certainly bears a close resemblance, both in outline and nervation.

Formation and locality: Cretaceous (Puget Sound group). Orcas Island, Washington.

NYSSA VETUSTA Newb.

Pl. I, fig. 2; IV, fig. 4.

Ann. N. Y. Lyc. Nat. Hist., Vol. IX (April, 1868), p. 11; Ills. Cret. and Tert. Pl. (1878), Pl. II, fig. 2, under *Magnolia obovata.*

Magnolia obovata Newb. Ann. N. Y. Lyc. Nat. Hist., Vol. IX (April, 1868), p. 15; Ills. Cret. and Tert. Pl. (1878), Pl. IV, fig. 4.

"Leaves large, obovate, entire, thick, and smooth, pointed and slightly decurrent on the petiole; nervation strong; midrib straight and extending to the summit; lateral nerves pinnate, set at somewhat unequal distances, straight and parallel below, forked and inosculating above, forming a festoon parallel with the margin; tertiary nerves forming an irregular network of polygonal and relatively large areoles."

Collected by Dr. F. V. Hayden.

Of this species there are numerous specimens in the collections made by Dr. Hayden in as good preservation as the material in which they are fossilized will permit. The nervation is strongly marked, and all its more prominent characters as appreciable in the fossil as they were in the fresh leaves. In nervation, consistence, and outline these leaves are almost undistinguishable from those of the "Pepperidge" (*Nyssa multiflora*). The primary and secondary nervation of some species of Magnolia also exhibit a strong resemblance to that of these fossils, but a less complete correspondence than Nyssa presents. Without the fruit, or at least leaves preserved in a fine argillaceous sediment in which the finer details of nervation are given, the affinity suggested must be to some extent conjectural.

Formation and locality: Cretaceous (Dakota group). Blackbird Hill, Nebraska.

Order SAPOTACEÆ.

SAPOTACITES HAYDENII Heer.

Pl. V, fig. 1.

Proc. Acad. Nat. Sci. Phil. (1858), p. 265; Ill. Cret. and Tert. Pl. (1878), Pl. V, fig. 1.

Professor Heer compares this leaf with one described by him in his Flora Tertiaria Helvetiæ under the name of *S. mimusops*. He further described it as "diminishing toward the base, rounded toward the apex, rather deeply emarginate. From the midrib, which gradually becomes slender and dies out, proceed at acute angles very numerous secondary nerves, which have the peculiarity of ramifying very much."

This is one of the leaves described by Professor Heer from tracings sent him by Mr. Meek, and the specimen now figured is that from which the tracing was made. As it has not before been figured, and is frequently referred to in the earlier discussions of the flora of the Dakota group, it has seemed desirable that a figure should be given of it so that it may be identified. The original tracing of Mr. Meek, on which Professor Heer wrote the name given to the leaf, as well as the original, are before me as I write, so there can be no mistake about the identification of the species. I have seen no other specimens than this one, and have nothing to add to the description given by Professor Heer, except that the emargination of the summit is in part at least the result of fracture and may not be a constant character. The peculiar crowded nervation will serve to distinguish this leaf from the others described by Professor Heer and noticed elsewhere (*Leguminosites Marcouanus* and *Phyllites obcordatus*), both of which have similar obovate outlines and emarginate summits.

Formation and locality: Cretaceous (Dakota group). Blackbird Hill, Nebraska.

Order OLEACEÆ.

Fraxinus affinis Newb.

Pl. XLIX, fig. 5.

Proc. U. S. Nat. Mus., Vol. V (March 21, 1883), p. 510.

"Leaves petioled, lanceolate, long-pointed, attenuate at base; margins coarsely and irregularly toothed at and above the middle."

Collected by Rev. Thomas Condon.

This leaf has almost precisely the form, serration, and nervation of some folioles of *F. Americana* now living, but it is narrower and has a more crowded nervation than the average leaflets of that species.

Among fossil ashes this approaches closely to *F. excelsifolia* Webb. (Palæontogr. IV, p. 150, Pl. XXVII, fig. 3), but the dentation in that species is much coarser and the nervation more remote.

Professor Heer has described two species of Fraxinus (*F. predicta* and *F. denticulata*), both of which Lesquereux thinks he has identified among the Tertiary leaf impressions obtained from the West. The fragments he figures, however, are too imperfect for the identification of the species. They are both described by Professor Heer as sessile, while the leaf before us is distinctly petioled.

Formation and locality: Tertiary (Miocene). Bridge Creek, Oregon.

FRAXINUS DENTICULATA Heer?

Pl. XLIX, fig. 6.

Fl. Foss. Arct., Vol. I (1868), p. 118, Pl. XVI, fig. 4.

NOTE.—The only manuscript which I have found relating to this figure is a marginal note on the plate referring it to "*Fraxinus dentata* Heer?," evidently meaning *F. denticulata*, and the specimen label giving the locality.—A. H.

Formation and locality: Tertiary (Miocene). Bridge Creek, Oregon.

FRAXINUS INTEGRIFOLIA Newb.

Pl. XLIX, figs. 1–3.

Proc. U. S. Nat. Mus., Vol. V (March 21, 1883), p. 509.

"Leaves short-petioled or sessile; lanceolate; broadest near the base, which is abruptly narrowed and wedge-shaped; summit narrowed, extremity rounded; margins entire; nervation reticulate, camptodrome; lateral branches connected in elegant festoons near the margins; intervals filled with a network of roundish, polygonal meshes."

Collected by Rev. Thomas Condon.

These leaves have been referred with some doubt to Fraxinus, but the nervation is almost exactly like that of *F. prædicta* Heer (Fl. Tert. Helv. III, p. 22, Pl. CIV, figs. 12 to 13g), and the general form is similar, except that in that species the folioles are unsymmetrical and are generally more or less dentate.

Formation and locality: Tertiary (Miocene). Bridge Creek, Oregon.

Order CAPRIFOLIACEÆ.

VIBURNUM ANTIQUUM (Newb.) Hollick.[1]

Pl. XXXIII, figs. 1, 2.

Tilia antiqua Newb. Ann. N. Y. Lyc. Nat. Hist., Vol. IX (April, 1868), p. 52; Ills. Cret. and Tert. Pl. (1878), Pl. XVI, figs. 1, 2, under *Tilia antiqua*.

Viburnum tilioides Ward. Bull. U. S. Geol. Surv. No. 37 (1887), p. 107, Pl. L, figs. 1–3; LI, figs. 1–8; LII, figs. 1, 2.

"Leaves 4 to 5 inches long, nearly as wide, often somewhat unsymmetrical, cordate at base, abruptly acuminate at summit, coarsely and

[1] This species was referred to the genus Tilia, by Dr. Newberry, in his original description, but Dr. Lester F. Ward has clearly shown that it belongs in the genus Viburnum.—A. H.

nearly equally toothed; nervation strong, medial nerve straight, bearing eight or nine pairs of lateral nerves, which diverge at an angle of about 45 degrees. The basilar pair of lateral nerves each sending off five or six branches on the lower side, which are again branched and terminate in the teeth of the margin. The second pair of lateral nerves have each four similar branches, the third pair three, the fourth pair two, the fifth pair one, though there are frequent departures from this rule. The tertiary nerves are strongly marked, leaving the secondary nerves nearly at right angles, crossing directly between the adjacent ones, or anastomosing with some irregularity in the middle of the interspaces."

Collected by Dr. F. V. Hayden.

There are many fragments of these leaves in the collection before me, embedded in a very fine and hard argillaceous limestone, and very beautifully preserved. They exhibit considerable resemblance to the leaves of Morus, especially *M. rubra*, but in that plant the basilar nerves of the leaves are more developed and reach the margins higher up. The marginal dentation is also generally more acute in the leaves of the mulberry and the leaves more pointed. The nervation of these fossil leaves is almost precisely that of our common species of Tilia, but in that the marginal dentation is much sharper. In a Southern species, however, *T. heterophylla*, I have found leaves which seem to be the exact counterpart of these; leaves with a roughish surface, strong and regular nervation, just after this pattern, and with a coarse, obtuse, and regular dentation. I am, therefore, inclined to refer these fossils to Tilia, and to regard them as the relics of a species closely allied to, if not identical with, *T. heterophylla*.

Formation and locality: Tertiary (Eocene?). Near Fort Clarke, Dakota.

VIBURNUM ASPERUM Newb.

Pl. XXXIII, fig. 9.

Ann. N. Y. Lyc. Nat. Hist., Vol. IX (April, 1868), p. 54; Ills. Cret. and Tert. Pl. (1878), Pl. XVI, fig. 8.

"Leaves ovate in outline, rounded or slightly cordate at base, acute and long-pointed above, margins all cut by relatively large acute teeth; nervation strong, crowded; midrib straight; lateral nerves alternate, about nine on each side, the lowest and strongest bearing each five to six simple branches on the lower side; the lateral nerves of the middle of the leaf carrying one to two branches at the summits, the upper ones simple, all terminating in the marginal teeth; tertiary nerves numerous, connecting the secondaries nearly at right angles, and generally parallel."

Collected by Dr. F. V Hayden.

The nervation of these leaves is strong, regular, and crowded. The marginal serration is simple, coarse, and sharp, much like that of the leaves of many species of Viburnum.

Formation and locality: Tertiary (Fort Union group). Fort Union, Dakota.

Viburnum cuneatum Newb.

Pl. LVII, fig. 2.

Proc. U. S. Nat. Mus., Vol. V (March 21, 1883), p. 511.

"Leaves petioled, long-obovate, 10 centimeters or more in length by 4 centimeters in width; margins entire below the middle; above, set with coarse sub-acute or acute teeth; nervation strong, simple; midrib straight, giving off at an acute angle seven or eight simple, strong nerve branches on either side, which terminate in the teeth of the margin."

Collected by Dr. F. V. Hayden.

The general aspect of this peculiar leaf is as much like that of Cornus as Viburnum, and if the basal portion alone were shown, few botanists would doubt the propriety of referring it to Cornus. But the upper part of the leaf is very strongly dentate, the simple strong nerve branches terminating in these teeth, a character unknown in the species of Cornus, living or fossil. Some species of Viburnum exhibit a somewhat similar nervation and the dentate margin is much more in character here than in Cornus. It has been thought best, therefore, to refer it provisionally to Viburnum, a genus which seems to have been quite prevalent in late Cretaceous and Tertiary times on this continent, running into a great number of distinct species.

It is true, however, that the lateral nerves in the leaves of Viburnum are always branched, though in some specimens of *Viburnum dentatum* perhaps only one or two of the branches in a leaf give off branchlets. The dentation is quite that of *V. dentatum.* Further collections, which will undoubtedly be made in the region where this leaf was found, will doubtless determine to which of these genera these belong, the counterbalancing characters of nervation and margin leaving it a question which it is now impossible to decide.

Formation and locality: Tertiary (Eocene ?). Tongue River, Montana.

VIBURNUM LANCEOLATUM Newb.

Pl. XXXIII, fig. 10.

Ann. N. Y. Lyc. Nat. Hist., Vol. IX (April, 1868), p. 54; Ills. Cret. and Tert. Pl. (1878), Pl. XVI, fig. 10.

"Leaves small, narrow, ovate or ovate-lanceolate, rounded or slightly wedge-shaped at the base, pointed above, coarsely and sharply serrate-dentate throughout; nervation strong; midrib straight; lateral nerves about five pairs, diverging from the midrib at an angle varying from 15 to 20 degrees, all slightly and uniformly arched upward, the basilar pair each throwing out at an acute angle about six simple branches, which terminate in the teeth of the margin, the upper branches supporting each one or two similar branches near the summits; tertiary nervation fine, and undistinguishable in the fossil state."

Collected by Dr. F. V. Hayden.

In the regularity and precision of the nervation these leaves resemble those of Carpinus, but in most species of that genus the serration of the margins is double, while here it is single, and, except in one or two Old World forms, the nervation of the leaves of the living species of that genus is considerably different, the basilar pair of lateral nerves being much shorter and simple or less branched.

The style of nervation observable in these fossils occurs in one or two species of Rhamnus, but is there very exceptional, and the marginal serration of Rhamnus is rarely, if ever, so coarse as in the plant before us.

In Zizyphus we have a similar nervation, and not a dissimilar style in Celtis, but in neither of these have we such marginal teeth. In Viburnum, however, we have some examples of leaves exhibiting a closer resemblance to the fossils than any I have cited above, as in *Viburnum erosum* Thurnbg., from Korea, and *V. odoratissimum* of Japan. In both these plants we find leaves with a great development of the basilar pair of nerves, and a coarse, acute, and regular dentation of the margin.

Formation and locality: Tertiary (Fort Union group). Fort Union, Dakota.

DICOTYLEDONEÆ OF UNCERTAIN AFFINITIES.

PROTOPHYLLUM MINUS Lesq.

Pl. IX, fig. 3.

Cret. Fl. (1874), p. 104, Pl. XIX, fig. 2; XXVII, fig. 1.

NOTE.—So identified by Dr. Newberry, as indicated by memorandum on margin of plate.—A. H.

Formation and locality: Cretaceous (Dakota group). Fort Harker, Kansas.

PROTOPHYLLUM MULTINERVE Lesq.

Pl. VII, fig. 4.

Cret. Fl. (1874), p. 105, Pl. XVIII, fig. 1.
Pterospermites multinervis Lesq. Hayden's Ann. Rept. 1871 [1872], p. 302.

The figure now given shows the basal portion of a leaf which may have been 6 inches in diameter. It is intended to exhibit its peculiar sub-peltate character by which it may be at once recognized. More or less complete leaves of this species are quite common in the Cretaceous rocks of Kansas, and a large number are in my possession. None of these are absolutely perfect, but some are so nearly so as to permit me to add something to the description given by Lesquereux.

The leaf when in normal form was nearly orbicular, being slightly pointed above, uniformly rounded at the base, and evidently somewhat cupped by the interior insertion of the petiole. The margin was entire or slightly undulate, the nerves strong, regular, approximately parallel, camptodrome, the branches terminating in the prominences of the margin where it is undulate.

The resemblance of these leaves to those obtained from the Tertiary of Greenland and described by Heer under the name of *Pterospermites* (*P. dentatus*, *P. integrifolius*, *P. spectabilis*, and *P. alternans*) is very striking and gives presumptive evidence of botanical affinity.

The large leaves brought by Dr. W. H. Dall from Alaska and figured on Pls. LIII and LIV evidently belong in the same category and may not be specifically different from Heer's *P. spectabilis*. No satisfactory conclusion, however, can be reached in regard to the relations of this group of leaves until the fruits belonging to the same tree shall be found.

Formation and locality: Cretaceous (Dakota group). Fort Harker, Kansas.

PROTOPHYLLUM STERNBERGII Lesq.

Pl. X; XI.

Cret. Fl. (1874), p. 101, Pl. XVI; XVIII, fig. 2.
Pterospermites Sternbergii Lesq. Hayden's Ann. Rept. 1872 [1873], p. 425.

The specimens figured on Pls. X and XI represent but parts of some of these magnificent angiospermous leaves found in the Dakota group of Kansas. They apparently represent Lesquereux's *P. Sternbergii*, but are perhaps not distinct from those described by him first as *Credneria Lecontiana*, and subsequently *Protophyllum Lecontianum*.

The leaf figured on Pl. X seems to have been nearly round and at least 12 inches in diameter; that represented on Pl. XI was more ovate and was still larger. Both were included in the collections made at Fort Harker by Mr. Charles H. Sternberg, and Lesquereux has done only justice to him by attaching his name to the finest species contained in the large collection of fossil plants which he made there.

As previously remarked, no satisfactory relationship has been established between Protophyllum and living genera of plants, but I would suggest that some of the species of Coccoloba, such as *C. pubescens*, present many points of similarity of structure.

Formation and locality: Cretaceous (Dakota group). Fort Harker, Kansas.

PTEROSPERMITES DENTATUS Heer.

Pl. LIII, figs. 1, 2; LIV, fig. 4.

Fl. Foss. Arct., Vol. I (1868), p. 138, Pl. XXI, fig. 15b; XXIII, figs. 6, 7.

The leaves here represented are probably not distinct from those described by Professor Heer under the above name, although the fragment which he had did not permit him to give a full characterization or satisfactory figures. His description consists of three words: "*Foliis, sub-peltatis, dentatis,*" all of which is true of the much more complete specimens before us, but they also show that the base of the leaf is entire, or nearly so, the upper margin variably dentate or nearly entire. These specimens also show that the leaves of *P. dentatus*—if we accept that name for the species—are variable in size, in the strength of the nervation, and in their degree of perfoliation. Hence it is highly probable that the three species described by Professor Heer from the arctic regions, namely, that cited above, and his *P. spectabilis* and *P. alternans* (Fl. Foss. Arct., Vol. II, Abth. IV, p. 480, Pl. XLIII, fig. 15b; LIII, figs. 1–4, and LIV, fig 3), will ultimately be combined in one.

The specimens before us were brought by Mr. W. H. Dall from the Yukon River, in Alaska. They show that the plant which bore them was of strong, luxuriant growth, probably a tree of large size. No other species is immediately associated with this in the collection made by Mr. Dall, but the formation in which it occurs is undoubtedly of the same age with that at Cooks and Admiralty inlets—the so-called Arctic Miocene—and this tree formed a part of the luxuriant vegetation which included the gigantic *Quercus Grönlandica*, *Ficus Alaskana*, etc., and covered Alaska in Tertiary times.

Formation and locality: Tertiary (Miocene). Yukon River, Alaska.

PHYLLITES CARNEOSUS Newb.

Pl. XLI, figs. 1, 2.

Ann. N. Y. Lyc. Nat. Hist., Vol. IX (April, 1868), p. 75; Ills. Cret. and Tert. Pl. (1878), Pl. XXVI, figs. 1, 2.

"Leaves large, fleshy, and strongly nerved, orbicular in outline, cordate or rounded, often unsymmetrical at the base, obtuse at summit, margins wavy or coarsely and deeply scalloped; nervation strongly marked throughout; medial nerve straight, or nearly so, frequently produced into a long and strong petiole; lateral nerves in six to eight pairs, all more or less forked; lower pair short and curving downward soon after leaving the midrib; second pair also curved outward near the base, and reaching the baso-lateral margin by a course nearly at right angles to the line of the midrib; third pair strongest, much branched on the lower side above the middle: upper pairs once or twice forked near the summit; tertiary nerves parallel, simple, straight or gently arched, given off at right angles from the secondary, which they connect."

Collected by Dr. F. V. Hayden.

Up to the present time I have failed to identify these leaves with those of any genus known, living or fossil. In general form they resemble those of Coccoloba, and must have belonged to some plant having much the habit of *C. uvifera;* but the leaves of that plant are entire, and the nervation is quite different. One of the other species of Coccoloba, which grows in the West Indies, *C. diversifolia*, has leaves with a marginal serration, and a nervation more like that of the leaves before us, but both margins and nerves are unlike.

The leaves which I have designated by the name of *Phyllites cupanioides*, as it seems to me, should be generically united with these.

Formation and locality: Tertiary (Fort Union group). Fort Union, Dakota.

PHYLLITES CUPANIOIDES Newb.

Pl. XLI, figs. 3, 4.

Ann. N. Y. Lyc. Nat. Hist., Vol. IX (April, 1868), p. 74; Ills. Cret. and Tert. Pl. (1878), Pl. XXVI, figs. 3, 4, under *P. venosus.*

"Leaves large, fleshy, ovate, elliptical in outline, rounded at base, sub-acute at summit, margins coarsely and obtusely toothed above, simple or waved below; nervation pinnate, strong; midrib straight or flexuous, lateral nerves, about six on each side, crowded below, more remote above, basilar pair short and simple, uniting above with the tertiary branches of the second pair to form a marginal festoon, middle secondaries each bearing one or two branches near the summits, upper one simple; tertiary nervation distinct, forming lattice-like bars connecting the secondary nerves at right angles."

These fine leaves exhibit a resemblance in their texture and crenate margins to those to which I have given the name of *Phyllites carneosus.* They are, however, of different form, and have more simple and rectilinear nervation. The collection of Dr Hayden contains a great number of fragments of this species, but up to the present time I have failed to find among living plants any which afford a satisfactory comparison with them. A general similarity in form and nervation to Cupania, and especially to *C. Americana*, has suggested the name adopted, but it can not be said that the correspondence is very close.

Formation and locality: Tertiary (Fort Union group). Fort Union, Dakota.

PHYLLITES OBCORDATUS Heer.

Pl. V, fig. 2.

Proc. Phil. Acad. Nat. Sci., 1858, p. 266; Ills. Cret. and Tert. Pl. (1878), Pl. V. fig. 2, under *Leguminosites Marcouanus.*

This is the leaf described by Professor Heer from a tracing by Mr. Meek and figured in Dana's Manual of Geology with the name *Leguminosites Marcouanus*, and described and figured by Lesquereux in his Cretaceous Flora, page 90, Pl. XXVIII, fig. 2, under the name of *Bumelia Marcouana.* The original tracing now before me, bearing Professor Heer's name written with his own hand, renders the identification easy and certain, and shows, as remarked elsewhere, that the names of this and the associated obovate emarginate leaf have been interchanged. Lesquereux, supposing that Professor Heer had applied the name Leguminosites to this leaf, which he has shown to be long-petioled, and therefore almost certainly not belonging to a leguminous plant, changed the name to Bumelia, but as mentioned elsewhere, the name Leguminosites was applied to another leaf, and this must stand as Phyllites until some good reason can be given for transferring it to another genus, and in that case it would be necessary to retain the specific name *obcordatus.*

Formation and locality: Cretaceous (Dakota group). Blackbird Hill, Nebraska.

PHYLLITES VANONÆ Heer.

Pl. III, fig. 8.

Nouv. Mem. Soc. Helv. Sci. Nat., Vol. XXII (1866), p. 22, Pl. I, fig. 8; Ills. Cret. and Tert. Pl. (1878), Pl. III, fig. 8, under *Diospyros primæva.*

NOTE.—So identified by Dr. Newberry, as indicated by memorandum on specimen and margin of plate.—A. H.

Formation and locality: Cretaceous (Dakota group). Blackbird Hill, Nebraska.

PHYLLITES VENOSUS Newb.

Pl. XXX, fig. 4.

Ann. N. Y. Lyc. Nat. Hist., Vol. IX (April, 1868), p. 75; Ills. Cret. and Tert. Pl. (1878), Pl. XXIV, fig. 4.

"Leaves thick and fleshy, irregularly oval in outline, rounded or slightly heart-shaped at base, blunt-pointed above, unsymmetrical throughout, margins entire or serrate, nervation strong, pinnate, midrib flexuous, lateral nerves arched upward, branching at summit."

Collected by Dr. F. V. Hayden.

I have been able to detect no relationship between these leaves and those of any living plants, and publish the figures and description given in hopes that others may be more successful. They have the general aspect of those of a Lauraceous tree, but I suspect they are related to those now described under the names of *P. carneosus* and *P. cupanioides*.

Formation and locality: Tertiary (Fort Union group). Fort Union, Dakota.

NORDENSKIOLDIA BOREALIS Heer.

Pl. LXVIII, figs. 4–6.

Fl. Foss. Arct., Vol. II, Abth. III (1870), p. 65, Pl. VII, figs. 1–13.

Professor Heer describes a capsulary dry fruit which he has called by the name given above. It occurs in groups, is spheroidal, dehiscent, with ten to twelve carpels of which the section is wedge-shaped, the smaller angle turned inward to a central vertical axis. Professor Heer compares this fruit with that of *Cistus ladaniferus*, to which it has a general resemblance. It was collected at Cape Staratschin (Spitzbergen) with *Nymphæa arctica* and fragments of Phragmites and of Sparganium; also at Atanekerdluck (Greenland). From its associates in Spitzbergen it would seem to be the fruit of an aquatic plant. In the Green River Shales Dr. White has collected numerous specimens which are apparently identical with those described by Heer. Some of these are grouped in such a way that it is evident that the fruit was compound; that is, a number were aggregated in a spike or crowded panicle, while the scattered capsules represented in our figs. 5 and 6 are distinctly pedunculated and apparently terminated in a rostrum, the prolongation of a central axis.

After a somewhat extended comparison with the fruits of various plants, I am compelled to question the conclusion that these have any botanical affinity with Cistus, and it seems to me the plant here represented was more likely allied to Allisma. By the examination of the fruit of our *Allisma plantago* it will be seen to be a rounded head, flattened or excavated above, consisting of a number of triangular capsules combined precisely as in the Nordenskioldia. This resemblance, taken in connection with the apparent aquatic habit of the plant, justifies at least a conjecture that we have in these fruits relics of an allismoid plant larger and stronger than our living *Alisma plantago*, but further collections will be needed to justify or disprove this inference.

Formation and locality: Tertiary (Green River group). Green River, Wyoming.

CARPOLITHES SPINOSUS Newb.

Pl. LXVIII, figs. 2, 3.

Proc. U. S. Nat. Mus., Vol. V (March 31, 1883), p. 514.

"Fruit enclosed in an exocarp composed of three elliptical or lentiform segments, furrowed along the middle line of the dorsum and bristling with erect, acute spines 6 to 8 millimeters long; peduncle cylindrical, strong, 1 inch or more in length."

Collected by Prof. I. C. Russell.

A figure is given of this fruit because of its remarkable character rather than with the hope of establishing its botanical relations. Its occurrence associated with many palm leaves and its tripartite division afford presumptive evidence that it belongs to the palms, but no living palm fruit suggests itself as an analogue. Apparently all that we see here is a husk or envelope which probably inclosed an elliptical nut that was partially protected by the bristling spines of the outer surface.

Formation and locality: Cretaceous (Laramie group). North Branch of Purgatory River, Colorado.

CARPOLITHES LINEATUS Newb.

Pl. XL, fig. 1.

Ann. N. Y. Lyc. Nat. Hist., Vol. IX (April, 1868), p. 31 (name only); Ills. Cret. and Tert. Pl. (1878), Pl. XXV, fig. 1.

NOTE.—The only manuscript which was found relating to this figure is a memorandum of the name and locality on the plate margin. The following description was prepared from an examination of the figure: Fruit rounded, elliptical in outline, five-eighths inch long by one-half inch wide, beaked, finely striate in direction of greater dimension.—A. H.

Formation and locality: Tertiary (Fort Union group). Fort Union, Dakota.

CALYCITES POLYSEPALA Newb.

Pl. XL, fig. 3.

Ann. N. Y. Lyc. Nat. Hist., Vol. IX (April, 1868), p. 31 (name only); Ills. Cret. and Tert. Pl. (1878), Pl. XXV, fig. 3.

NOTE.—The only manuscript which was found relating to this species is a memorandum of the name and locality, on the plate margin, in Dr. Newberry's handwriting.

The following description was prepared from an examination of the figure: Organism calyx-like, sub-circular in outline, about 1¼ inches in diameter, consisting of six divisions (sepals ?), each of which is about three-eighths inch long by three-sixteenths inch wide at base, tapering to an acute point.—A. H.

Formation and locality: Tertiary (Fort Union group). Fort Union, Dakota.

TABLE OF DIS

List of species, showing locali

	Page of this work.	Species.	Raritan River, New Jersey, Raritan Formation.	Smoky Hill, Kansas, Dakota group.	Fort Harker, Kansas, Dakota group.	Blackbird Hill, Nebraska, Dakota group.	Big Sioux River, Nebraska, Dakota group.	Cedar Spring, Nebraska, Dakota group.	Decatur, Nebraska, Dakota group.	Rio Dolores, Utah, Dakota group.	Whetstone Creek, New Mexico, Dakota group.	Sage Creek, South Dakota, Dakota group (?).	Keyport, New Jersey, Mattewan Formation.	Nanaimo, Vancouvers Island, Puget Sound group.	Chuckanuts, Washington, Puget Sound group.	Bellingham Bay, Washington, Puget Sound group.	Point of Rocks, Wyoming, Montana Formation.	Vermejo Canyon, New Mexico, Laramie group.	Fischers Peak, Colorado, Laramie group.	Raton Mountains, Colorado, Laramie group.
			1	2	3	4	5	6	7	8	9	10	11	12	13	14	15	16	17	18
1	1	Lygodium Kaulfussi Heer																		
2	3	Anemia perplexa Hollick														+	+			
3	6	Acrostichum hesperium Newb																		
4	7	Pteris pennæformis Heer ?																		
5	7	Pteris Russellii Newb																+		
6	8	Onoclea sensibilis fossilis Newb																		
7	10	Lastrea (Goniopteris) Fischeri Heer?																		
8	11	Aspidium Kennerlyi Newb												+						
9	12	Pecopteris (Cheilanthes) sepulta Newb																		
10	14	Sphenopteris corrugata Newb				+														
11	14	Equisetum Oregonense Newb																		
12	15	Equisetum robustum Newb														+				
13	15	Equisetum Wyomingense Lesq																		
14	16	Equisetum sp. ? Newb																		
15	16	Nilssonia Gibbsii (Newb.) Hollick														+				
16	17	Araucaria spatulata Newb										+								
17	18	Abietites cretacea Newb									+									
18	18	Sequoia cuneata Newb												+			+			
19	19	Sequoia gracillima (Lesq.) Newb	+								+		+							
20	20	Sequoia Heerii Lesq																		
21	20	Sequoia Nordenskioldii Heer?																		
22	21	Sequoia spinosa Newb																		
23	22	Taxodium distichum miocenum Heer																		
24	23	Taxodium occidentale Newb																		
25	24	Glyptostrobus Europæus (Brong.) Heer																		
26	25	Thuja interrupta Newb																		
27	27	Phragmites sp.? Newb																		
28	27	Sabal Campbelli Newb														+				

TRIBUTION.

ties mentioned in the text.

Purgatory River, Colorado, Laramie group.	Walsenburg, Colorado, Laramie group.	Florence, Colorado, Laramie group.	Erie, Colorado, Laramie group.	Coal Basin, Colorado, Laramie group.	Belmont, Colorado, Laramie group.	Marshalls, Colorado, Laramie group.	Black Butte, Wyoming, Laramie group.	Hams Fork, Wyoming, Laramie group.	Carbon, Wyoming, Laramie group (?).	Spring Canyon, Montana, Livingston Formation.	Golden, Colorado, Denver Formation, Laramie group.	Carbonado, Washington, Eocene.	Fletts Creek, Washington, Eocene.	Henrys Fork, Utah, Eocene (?).	Amil Creek, North Dakota (?), Eocene (?).	La Bontes Creek, Nebraska, Eocene (?).	Fort Berthold, North Dakota, Fort Union group.	Fort Clark, North Dakota, Fort Union group.	Fort Union, North Dakota, Fort Union group.	Yellowstone River, Montana, Fort Union group.	Tongue River, Montana, Fort Union group (?).	Green River, Wyoming, Green River group.	Dalles of the Columbia, Oregon, Miocene (?).	Florissant, Colorado, Miocene.	McBees Canyon, Oregon, Miocene.	Currant Creek, Oregon, Miocene.	Bridge Creek, Oregon, Miocene.	Cooks Inlet, Alaska, Miocene.	Admiralty Inlet, Alaska, Miocene.	Yukon River, Alaska, Miocene.	Kootznahoo, Alaska, Miocene.	Birch Bay, Washington, Miocene.	Locality not known.	
19	20	21	22	23	24	25	26	27	28	29	30	31	32	33	34	35	36	37	38	39	40	41	42	43	44	45	46	47	48	49	50	51	52	
												+	+									+				+								1
			+					+			+L	+																						2
																						+												3
														+												+								4
	+	+									+L																							5
																			+															6
																										+								7
																																		8
																						+												9
																																		10
																									+	+								11
																																		12
																						+												13
																																	+	14
																																		15
																																		16
																																		17
																																		18
																																		19
																											+							20
																				+														21
																												+						22
																												+				+		23
																				+														24
																			+													+		25
																			+															26
																			+															27
																																		28

List of species, showing localities

	Page of this work.	Species.	Raritan River, New Jersey, Raritan Formation.	Smoky Hill, Kansas, Dakota group.	Fort Harker, Kansas, Dakota group.	Blackbird Hill, Nebraska, Dakota group.	Big Sioux River, Nebraska, Dakota group.	Cedar Spring, Nebraska, Dakota group.	Decatur, Nebraska, Dakota group.	Rio Dolores, Utah, Dakota group.	Whetstone Creek, New Mexico, Dakota group	Sage Creek, South Dakota, Dakota group.(?)	Keyport, New Jersey, Mattowan Formation.	Nanaimo, Vancouvers Island, Puget Sound group.	Chuckanuts, Washington, Puget Sound group.	Bellingham Bay, Washington, Puget Sound group.	Point of Rocks, Wyoming, Montana Formation.	Vermejo Canyon, New Mexico, Laramie group.	Fischers Peak, Colorado, Laramie group.	Raton Mountains, Colorado, Laramie group.
			1	2	3	4	5	6	7	8	9	10	11	12	13	14	15	16	17	18
29	28	Sabal grandifolia Newb																	+	
30	30	Sabal imperialis Dn												+						
31	30	Sabal Powellii Newb																		
32	31	Manicaria Haydenii Newb																		
33	32	Smilax cyclophylla Newb																		
34	33	Iris sp.? Newb																		
35	33	Monocotyledon gen. et sp. ? Hollick																		
36	33	Juglans nigella Heer																		
37	34	Juglans occidentalis Newb																		
38	35	Carya antiquorum Newb																		
39	37	Myrica trifoliata Newb									+									
40	37	Populus acerifolia Newb																		
41	38	Populus cordata Newb																		
42	40	Populus (?) cordifolia Newb				+														
43	41	Populus cuneata Newb																		
44	41	Populus cyclophylla Heer			+	+														
45	42	Populus (?) Debeyana Heer				+														
46	43	Populus elliptica Newb				+														
47	44	Populus flabellum Newb													+					
48	44	Populus genetrix Newb																		
49	45	Populus litigiosa Heer			+	+														
50	46	Populus microphylla Newb				+														
51	47	Populus Nebrascencis Newb																		
52	48	Populus nervosa Newb																		
53	49	Populus nervosa elongata Newb																		
54	50	Populus polymorpha Newb																		
55	51	Populus rhomboidea Lesq												+						
56	51	Populus rotundifolia Newb																		
57	53	Populus smilacifolia Newb																		

mentioned in the text—Continued.

Purgatory River, Colorado, Laramie group.	Walsenburg, Colorado, Laramie group.	Florence, Colorado, Laramie group.	Erie, Colorado, Laramie group.	Coal Basin, Colorado, Laramie group.	Belmont, Colorado, Laramie group.	Marshalls, Colorado, Laramie group.	Black Butte, Wyoming, Laramie group.	Hams Fork, Wyoming, Laramie group.	Carbon, Wyoming, Laramie group (?).	Spring Canyon, Montana, Livingston Formation.	Golden, Colorado, Denver Formation, Laramie group.	Carbonado, Washington, Eocene.	Fletts Creek, Washington, Eocene.	Henrys Fork, Utah, Eocene (?).	Amil Creek, North Dakota (?), Eocene (?).	La Boutes Creek, Nebraska, Eocene (?).	Fort Berthold, North Dakota, Fort Union group.	Fort Clark, North Dakota, Fort Union group.	Fort Union, North Dakota, Fort Union group.	Yellowstone River, Montana, Fort Union group.	Tongue River, Montana, Fort Union group (?).	Green River, Wyoming, Green River group.	Dalles of the Columbia, Oregon, Miocene (?).	Florissant, Colorado, Miocene.	McBees Canyon, Oregon, Miocene.	Currant Creek, Oregon, Miocene.	Bridge Creek, Oregon, Miocene.	Cooks Inlet, Alaska, Miocene.	Admiralty Inlet, Alaska, Miocene.	Yukon River, Alaska, Miocene.	Kootznahoo, Alaska, Miocene.	Birch Bay, Washington, Miocene.	Locality not known.	
19	20	21	22	23	24	25	26	27	28	29	30	31	32	33	34	35	36	37	38	39	40	41	42	43	44	45	46	47	48	49	50	51	52	
+	+		+																	+														29
																																		30
																						+												31
																						+												32
																																+		33
																																	+	34
																											+							35
																													+					36
																						+												37
																				+														38
																																		39
																			+															40
																				+														41
																																		42
																				+	+													43
																																		44
																																		45
																																		46
																																		47
																				+	+													48
																																		49
																																		50
																				+														51
																			+	+														52
																				+														53
																											+							54
																																		55
									+										+	+														56
																			+															57

List of species, showing localities

	Page of this work.	Species.	1 Raritan River, New Jersey, Raritan Formation.	2 Smoky Hill, Kansas, Dakota group.	3 Fort Harker, Kansas, Dakota group.	4 Blackbird Hill, Nebraska, Dakota group.	5 Big Sioux River, Nebraska, Dakota group.	6 Cedar Spring, Nebraska, Dakota group.	7 Decatur, Nebraska, Dakota group.	8 Rio Dolores, Utah, Dakota group.	9 Whetstone Creek, New Mexico, Dakota group.	10 Sage Creek, South Dakota, Dakota group (?).	11 Keyport, New Jersey, Mattowan Formation.	12 Nanaimo, Vancouvers Island, Puget Sound group.	13 Chuckanuts, Washington, Puget Sound group.	14 Bellingham Bay, Washington, Puget Sound group.	15 Point of Rocks, Wyoming, Montana Formation.	16 Vermejo Canyon, New Mexico, Laramie group.	17 Fischers Peak, Colorado, Laramie group.	18 Raton Mountains, Colorado, Laramie group.
58	54	Populites elegans Lesq.?			+															
59	54	Salix angusta Al. Br.?																		
60	55	Salix cuneata Newb					+													
61	56	Salix flexuosa Newb				+	+	+			+									
62	57	Salix foliosa Newb									+									
63	58	Salix Meekii Newb				+			+											
64	59	Salix membranacea Newb	+																	
65	59	Carpinus grandis Ung																		
66	60	Corylus Americana fossilis Newb																		
67	61	Corylus MacQuarrii (Forbes) Heer																		
68	62	Corylus orbiculata Newb																		
69	63	Corylus rostrata fossilis Newb																		
70	63	Betula angustifolia Newb																		
71	64	Betula heterodonta Newb																		
72	65	Betula sp.? Newb																		
73	65	Alnus Alaskana Newb																		
74	66	Alnus serrata Newb																		
75	66	Alnus serrulata fossilis Newb																		
76	67	Alnus sp.? Newb																		
77	67	Alnites grandifolia Newb				+														
78	68	Fagus cretacea Newb		+																
79	69	Quercus antiqua Newb								+										
80	69	Quercus banksiæfolia Newb													+					
81	70	Quercus castanoides Newb																		
82	71	Quercus castanopsis Newb																		
83	71	Quercus consimilis Newb																		
84	73	Quercus coriacea Newb													+					
85	73	Quercus dubia Newb																		
86	74	Quercus elliptica Newb													+					

mentioned in the text—Continued.

19 Purgatory River, Colorado, Laramie group.	20 Walsenburg, Colorado, Laramie group.	21 Florence, Colorado, Laramie group.	22 Erie, Colorado, Laramie group.	23 Coal Basin, Colorado, Laramie group.	24 Belmont, Colorado, Laramie group.	25 Marshalls, Colorado, Laramie group.	26 Black Butte, Wyoming, Laramie group.	27 Hams Fork, Wyoming, Laramie group.	28 Carbon, Wyoming, Laramie group (?).	29 Spring Canyon, Montana, Livingston Formation.	30 Golden, Colorado, Denver Formation, Laramie group.	31 Carbonado, Washington, Eocene.	32 Fletts Creek, Washington, Eocene.	33 Henrys Fork, Utah, Eocene (?).	34 Amil Creek, North Dakota (?), Eocene (?).	35 La Bontes Creek, Nebraska, Eocene (?).	36 Fort Berthold, North Dakota, Fort Union group.	37 Fort Clark, North Dakota, Fort Union group.	38 Fort Union, North Dakota, Fort Union group.	39 Yellowstone River, Montana, Fort Union group.	40 Tongue River, Montana, Fort Union group (?).	41 Green River, Wyoming, Green River group.	42 Dalles of the Columbia, Oregon, Miocene (?).	43 Florissant, Colorado, Miocene.	44 McBees Canyon, Oregon, Miocene.	45 Currant Creek, Oregon, Miocene	46 Bridge Creek, Oregon, Miocene.	47 Cooks Inlet, Alaska, Miocene.	48 Admiralty Inlet, Alaska, Miocene.	49 Yukon River, Alaska, Miocene.	50 Kootznahoo, Alaska, Miocene.	51 Birch Bay, Washington, Miocene.	52 Locality not known.	
																																		58
																						+												59
																																		60
																																		61
																																		62
																																		63
																																		64
																																+		65
																			+															66
																			+															67
																			+															68
																			+															69
																											+							70
																											+							71
																											?							72
																															+			73
																				+	+													74
																											+							75
																											+							76
																																		77
																																		78
																																		79
																																		80
																						+												81
																				+														82
																											+							83
																																		84
																					+													85
																																		86

List of species, showing localities

	Page of this work.	Species.	Raritan River, New Jersey, Raritan Formation.	Smoky Hill, Kansas, Dakota group.	Fort Harker, Kansas, Dakota group.	Blackbird Hill, Nebraska, Dakota group.	Big Sioux River, Nebraska, Dakota group.	Cedar Spring, Nebraska, Dakota group.	Decatur, Nebraska, Dakota group.	Rio Dolores, Utah, Dakota group.	Whetstone Creek, New Mexico, Dakota group.	Sago Creek, South Dakota, Dakota group (?).	Keyport, New Jersey, Mattewan Formation.	Nanaimo, Vancouvers Island, Puget Sound group.	Chuckanuts, Washington, Puget Sound group.	Bellingham Bay, Washington, Puget Sound group.	Point of Rocks, Wyoming, Montana Formation.	Vermejo Canyon, New Mexico, Laramie group.	Fischers Peak, Colorado, Laramie group.	Raton Mountains, Colorado, Laramie group.
			1	2	3	4	5	6	7	8	9	10	11	12	13	14	15	16	17	18
87	74	Quercus flexuosa Newb.													+					
88	75	Quercus gracilis Newb.															+			
89	75	Quercus Grönlandica Heer																		
90	76	Quercus laurifolia Newb.																		
91	76	Quercus paucidentata Newb.																		
92	77	Quercus salicifolia Newb.				+														
93	78	Quercus simplex Newb.																		
94	78	Quercus sinuata Newb.								+										
95	79	Quercus Sullyi Newb.																		
96	80	Ulmus speciosa Newb.																		
97	81	Planera crenata Newb.																		
98	81	Planera longifolia Lesq.																		
99	81	Planera microphylla Newb.																		
100	82	Planera nervosa Newb.																		
101	83	Planera variabilis Newb.																		
102	84	Celtis parvifolia Newb.																		
103	84	Ficus (?) Alaskana Newb.																		
104	85	Ficus asarifolia minor Lesq.															+			
105	85	Ficus (?) Condoni Newb.																		
106	87	Ficus membranacea Newb.																		
107	88	Ficus planicostata Lesq.																		
108	88	Ficus reticulata (Lesq.) Hollick			+	+														
109	89	Protoficus inæqualis Newb.																		
110	90	Aristolochia cordifolia Newb.																		
111	91	Cabomba (?) gracilis Newb.																		
112	92	Cabomba inermis (Newb.) Hollick																		
113	93	Brasenia (?) antiqua Newb.																		
114	94	Magnolia alternans Heer?				?														
115	94	Magnolia elliptica Newb.																		

mentioned in the text—Continued.

Purgatory River, Colorado, Laramie group.	Walsenburg, Colorado, Laramie group.	Florence, Colorado, Laramie group.	Erie, Colorado, Laramie group.	Coal Basin, Colorado, Laramie group.	Belmont, Colorado, Laramie group.	Marshalls, Colorado, Laramie group.	Black Butte, Wyoming, Laramie group.	Hams Fork, Wyoming, Laramie group.	Carbon, Wyoming, Laramie group (?).	Spring Canyon, Montana, Livingston Formation.	Golden, Colorado, Denver Formation, Laramie group.	Carbonado, Washington, Eocene.	Fletts Creek, Washington, Eocene.	Henrys Fork, Utah, Eocene (?).	Amil Creek, North Dakota (?), Eocene (?).	La Bontes Creek, Nebraska, Eocene (?).	Fort Berthold, North Dakota, Fort Union group.	Fort Clark, North Dakota, Fort Union group.	Fort Union, North Dakota, Fort Union group.	Yellowstone River, Montana, Fort Union group.	Tongue River, Montana, Fort Union group (?).	Green River, Wyoming, Green River group.	Dalles of the Columbia, Oregon, Miocene (?).	Florissant, Colorado, Miocene.	McBees Canyon, Oregon, Miocene.	Currant Creek, Oregon, Miocene.	Bridge Creek, Oregon, Miocene.	Cooks Inlet, Alaska, Miocene.	Admiralty Inlet, Alaska, Miocene.	Yukon River, Alaska, Miocene.	Kootznahoo, Alaska, Miocene.	Birch Bay, Washington, Miocene.	Locality not known.	
19	20	21	22	23	24	25	26	27	28	29	30	31	32	33	34	35	36	37	38	39	40	41	42	43	44	45	46	47	48	49	50	51	52	
																																		87
																																		88
																												+						89
																	+																	90
																											+							91
																																		92
																											+							93
																																		94
																	+																	95
																											+							96
																					+													97
		+																																98
																			+															99
																						+												100
																						+												101
																					+													102
																												+	+					103
																																		104
																											+							105
																												+						106
	+	+					+				+D																+							107
																																		108
																					+													109
															+						+													110
																			+															111
																			+															112
																						+												113
																																		114
																					+													115

List of species, showing localities

	Page of this work.	Species.	Raritan River, New Jersey, Raritan Formation.	Smoky Hill, Kansas, Dakota group.	Fort Harker, Kansas, Dakota group.	Blackbird Hill, Nebraska, Dakota group.	Big Sioux River, Nebraska, Dakota group.	Cedar Spring, Nebraska, Dakota group.	Decatur, Nebraska, Dakota group.	Rio Dolores, Utah, Dakota group.	Whetstone Creek, New Mexico, Dakota group.	Sage Creek, South Dakota, Dakota group (?).	Keyport, New Jersey, Mattewan Formation.	Nanaimo, Vancouvers Island, Puget Sound group.	Chuckanuts, Washington, Puget Sound group.	Bellingham Bay, Washington, Puget Sound group.	Point of Rocks, Wyoming, Montana Formation.	Vermejo Canyon, New Mexico, Laramie group.	Fischers Peak, Colorado, Laramie group.	Raton Mountains, Colorado, Laramie group.
			1	2	3	4	5	6	7	8	9	10	11	12	13	14	15	16	17	18
116	94	Magnolia obovata Newb				+														
117	95	Magnolia rotundifolia Newb																	+	
118	96	Liriodendron Meekii Heer				+														
119	96	Liriodendron primævum Newb				+														
120	97	Berberis simplex Newb																		
121	98	Sassafras cretaceum Newb		+	+	+														
122	99	Sassafras cretaceum recurvatum (Lesq.) Newb			+															
123	100	Cinnamomum Heerii Lesq														+				
124	100	Liquidambar Europæum Al. Br																		
125	101	Liquidambar obtusilobatus (Heer) Hollick			+	+														
126	102	Platanus aspera Newb																		
127	103	Platanus Haydenii Newb																		
128	105	Platanus latiloba Newb				+														
129	106	Platanus nobilis Newb																		
130	109	Platanus Raynoldsii Newb																		
131	110	Pyrus cretacea		+																
132	111	Amelanchier similis Newb																		
133	112	Cratægus flavescens Newb																		
134	112	Prunus variabilis Newb																		
135	113	Cassia sp.? Newb																		
136	113	Leguminosites Marcouanus Heer				+														
137	114	Rhus (?) nervosa Newb																		
138	115	Acer sp.? Newb																		
139	115	Negundo triloba Newb																		
140	116	Sapindus affinis Newb																		
141	117	Sapindus (?) membranaceus Newb																		
142	117	Rhamnus elegans Newb																		
143	118	Rhamnus Eridani Ung																		
144	118	Rhamnites concinnus Newb																		

mentioned in the text—Continued.

Purgatory River, Colorado, Laramie group.	Walsenburg, Colorado, Laramie group.	Florence, Colorado, Laramie group.	Erie, Colorado, Laramie group.	Coal Basin, Colorado, Laramie group.	Belmont, Colorado, Laramie group.	Marshalls, Colorado, Laramie group.	Black Butte, Wyoming, Laramie group.	Hams Fork, Wyoming, Laramie group.	Carbon, Wyoming, Laramie group (?).	Spring Canyon, Montana, Livingston Formation.	Golden, Colorado, Denver Formation. Laramie group.	Carbonado, Washington, Eocene.	Flotts Creek, Washington, Eocene.	Henrys Fork, Utah, Eocene (?).	Antil Creek, North Dakota (?), Eocene (?).	La Bontes Creek, Nebraska, Eocene (?).	Fort Berthold, North Dakota, Fort Union group.	Fort Clark, North Dakota, Fort Union group.	Fort Union, North Dakota, Fort Union group.	Yellowstone River, Montana, Fort Union group.	Tongue River, Montana, Fort Union group (?).	Green River, Wyoming, Green River group.	Dalles of the Columbia, Oregon, Miocene (?).	Florissant, Colorado, Miocene.	McBees Canyon, Oregon, Miocene.	Currant Creek, Oregon, Miocene.	Bridge Creek, Oregon, Miocene.	Cooks Inlet, Alaska, Miocene.	Admiralty Inlet, Alaska, Miocene.	Yukon River, Alaska, Miocene.	Kootznahoo, Alaska, Miocene.	Birch Bay, Washington, Miocene.	Locality not known.	
19	20	21	22	23	24	25	26	27	28	29	30	31	32	33	34	35	36	37	38	39	40	41	42	43	44	45	46	47	48	49	50	51	52	
																																		116
																																		117
																																		118
																																		119
																											+							120
																																		121
																																		122
																																		123
																											+							124
																																		125
																											+							126
							+?				+D					+				+														127
																																		128
																		+																129
							+?				+D									+														130
																																		131
																				+														132
																											+							133
																												+						134
																											+							135
																																		136
																			+															137
																											+							138
																			+															139
																				+														140
																			+															141
					+	+																												142
																											+							143
																			+															144

List of species, showing localities

	Page of this work.	Species.	Raritan River, New Jersey, Raritan Formation. 1	Smoky Hill, Kansas, Dakota group. 2	Fort Harker, Kansas, Dakota group. 3	Blackbird Hill, Nebraska, Dakota group. 4	Big Sioux River, Nebraska, Dakota group. 5	Cedar Spring, Nebraska, Dakota group. 6	Decatur, Nebraska, Dakota group. 7	Rio Dolores, Utah, Dakota group. 8	Whetstone Creek, New Mexico, Dakota group. 9	Sage Creek, South Dakota, Dakota group (?). 10	Keyport, New Jersey, Mattewan Formation. 11	Nanaimo, Vancouvers Island, Puget Sound group. 12	Chuckanuts, Washington, Puget Sound group. 13	Bellingham Bay, Washington, Puget Sound group. 14	Point of Rocks, Wyoming, Montana Formation. 15	Vermejo Canyon, New Mexico, Laramie group. 16	Fischers Peak, Colorado, Laramie group. 17	Raton Mountains, Colorado, Laramie group. 18
145	119	Zizyphus longifolia Newb																		
146	120	Vitis rotundifolia Newb																		
147	120	Grewia crenata (Ung.) Heer																		
148	121	Aralia macrophylla Newb																		
149	122	Aralia (?) quinquepartita Lesq			+															
150	123	Aralia triloba Newb																		
151	124	Cornus Newberryi Hollick																		
152	125	Nyssa (?) cuneata Newb														+				
153	125	Nyssa vetusta Newb				+														
154	126	Sapotacites Haydenii Heer				+														
155	127	Fraxinus affinis Newb																		
156	128	Fraxinus denticulata Heer?																		
157	128	Fraxinus integrifolia Newb																		
158	129	Viburnum antiquum (Newb.) Hollick																		
159	129	Viburnum asperum Newb																		
160	130	Viburnum cuneatum Newb																		
161	131	Viburnum lanceolatum Newb																		
162	132	Protophyllum minus Lesq			+															
163	132	Protophyllum multinerve Lesq			+															
164	133	Protophyllum Sternbergii Lesq			+															
165	133	Pterospermites dentatus Heer																		
166	134	Phyllites carneosus Newb																		
167	135	Phyllites cupanioides Newb																		
168	136	Phyllites obcordatus Heer				+														
169	136	Phyllites Vanonæ Heer			+	+														
170	136	Phyllites venosus Newb																		
171	137	Nordenskioldia borealis Heer																		
172	138	Carpolithes spinosus Newb																		
173	138	Carpolithes lineatus Newb																		
174	139	Calycites polysepala Newb																		

mentioned in the text—Continued.

19. Purgatory River, Colorado, Laramie group.	20. Walsenburg, Colorado, Laramie group.	21. Florence, Colorado, Laramie group.	22. Erie, Colorado, Laramie group.	23. Coal Basin, Colorado, Laramie group.	24. Belmont, Colorado, Laramie group.	25. Marshalls, Colorado, Laramie group.	26. Black Butte, Wyoming, Laramie group.	27. Hams Fork, Wyoming, Laramie group.	28. Carbon, Wyoming, Laramie group (?).	29. Spring Canyon, Montana, Livingston Formation.	30. Golden, Colorado, Denver Formation, Laramie group.	31. Carbonado, Washington, Eocene.	32. Fletts Creek, Washington, Eocene.	33. Henrys Fork, Utah, Eocene (?).	34. Amil Creek, North Dakota (?), Eocene (?).	35. La Bontes Creek, Nebraska, Eocene (?).	36. Fort Berthold, North Dakota, Fort Union group.	37. Fort Clark, North Dakota, Fort Union group.	38. Fort Union, North Dakota, Fort Union group.	39. Yellowstone River, Montana, Fort Union group.	40. Tongue River, Montana, Fort Union group (?).	41. Green River, Wyoming, Green River group.	42. Dalles of the Columbia, Oregon, Miocene (?).	43. Florissant, Colorado, Miocene.	44. McBees Canyon, Oregon, Miocene.	45. Currant Creek, Oregon, Miocene.	46. Bridge Creek, Oregon, Miocene.	47. Cooks Inlet, Alaska, Miocene.	48. Admiralty Inlet, Alaska, Miocene.	49. Yukon River, Alaska, Miocene.	50. Kootznahoo, Alaska, Miocene.	51. Birch Bay, Washington, Miocene.	52. Locality not known.	
																						+												145
																													+					146
																											+							147
																						+												148
																																		149
																		+	+															150
																				+														151
																																		152
																																		153
																																		154
																											+							155
																											+							156
																											+							157
																		+	+															158
																			+															159
																					+													160
																			+															161
																																		162
																																		163
																																		164
																														+				165
																			+															166
																			+															167
																																		168
																																		169
																			+															170
																						+												171
+																																		172
																			+															173
																			+															174

PLATES.

PLATE I.

PLATE I.

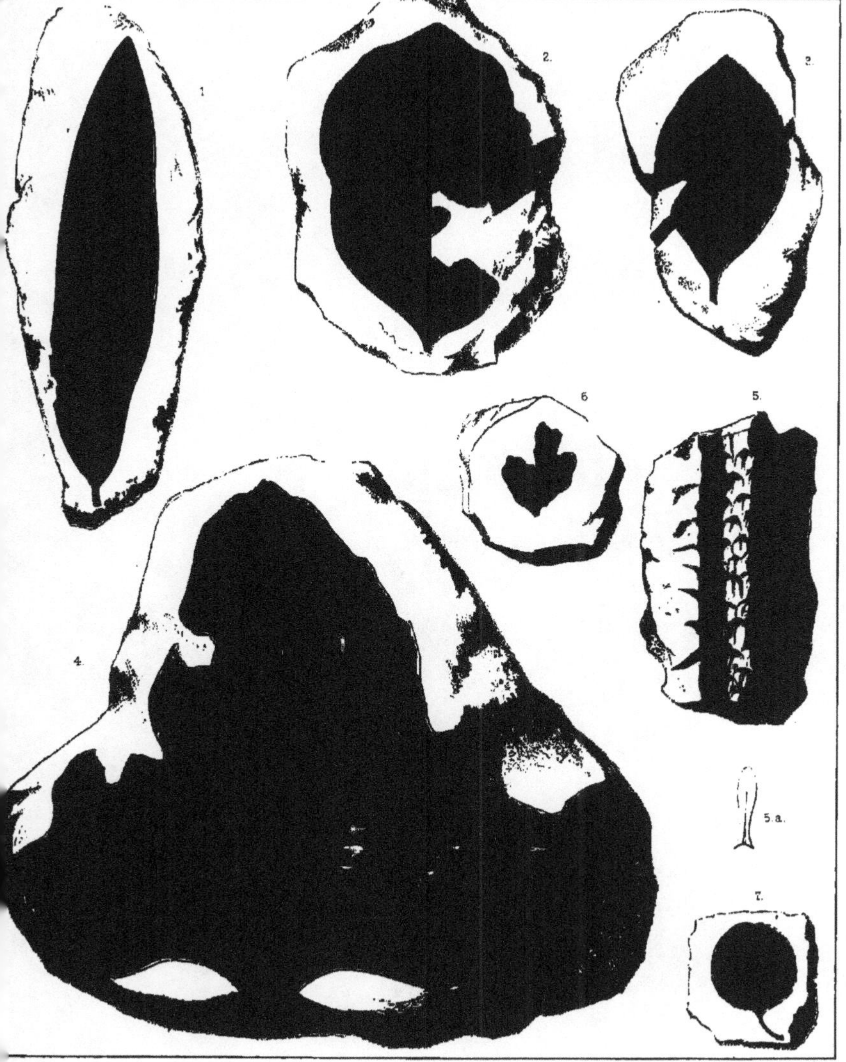

... & Son, Lith.

PLATE II.

PLATE II.

	Page.
Figs. 1, 2. Salix cuneata Newb	55
3. Salix Meekii Newb	58
4. Salix flexuosa Newb	56
5–8a. Salix membranacea Newb	59

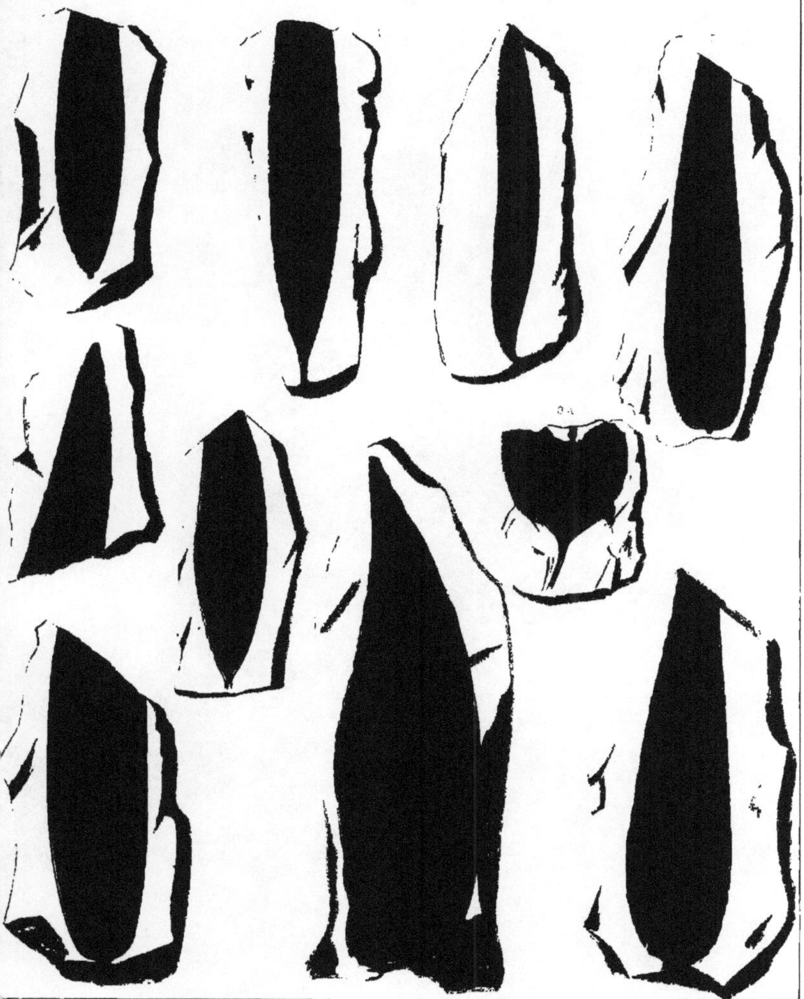

PLATE III.

PLATE III.

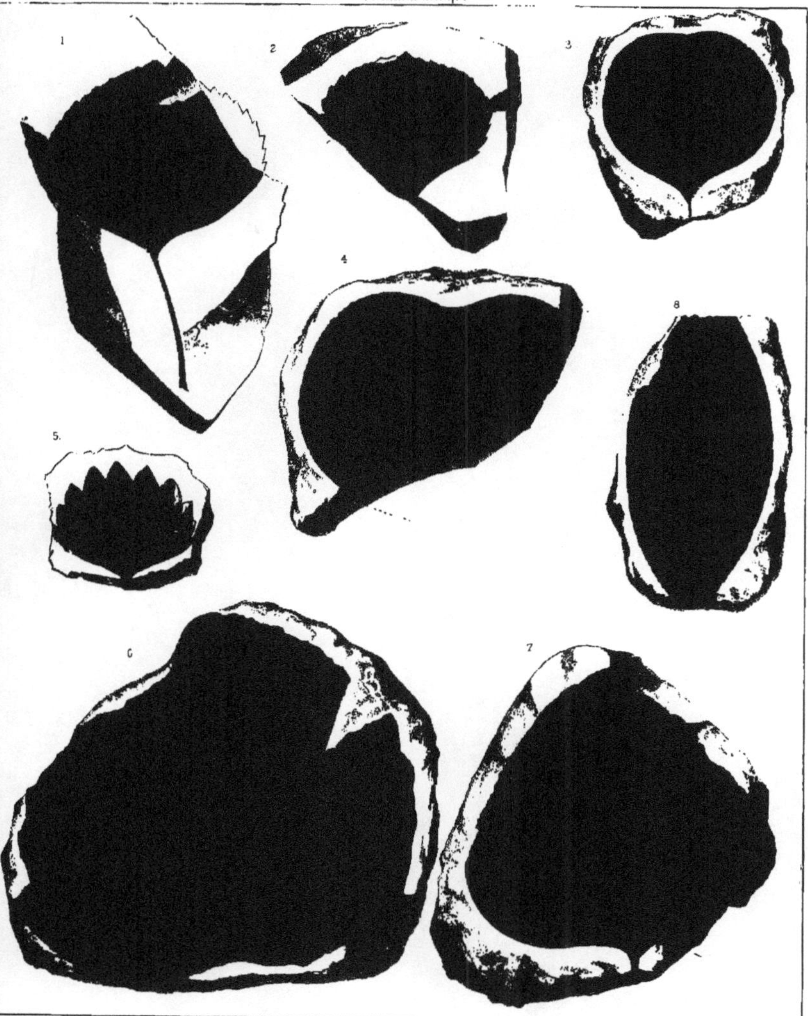

Thos. Sinclair & Son, Lith.

PLATE IV.

PLATE IV.

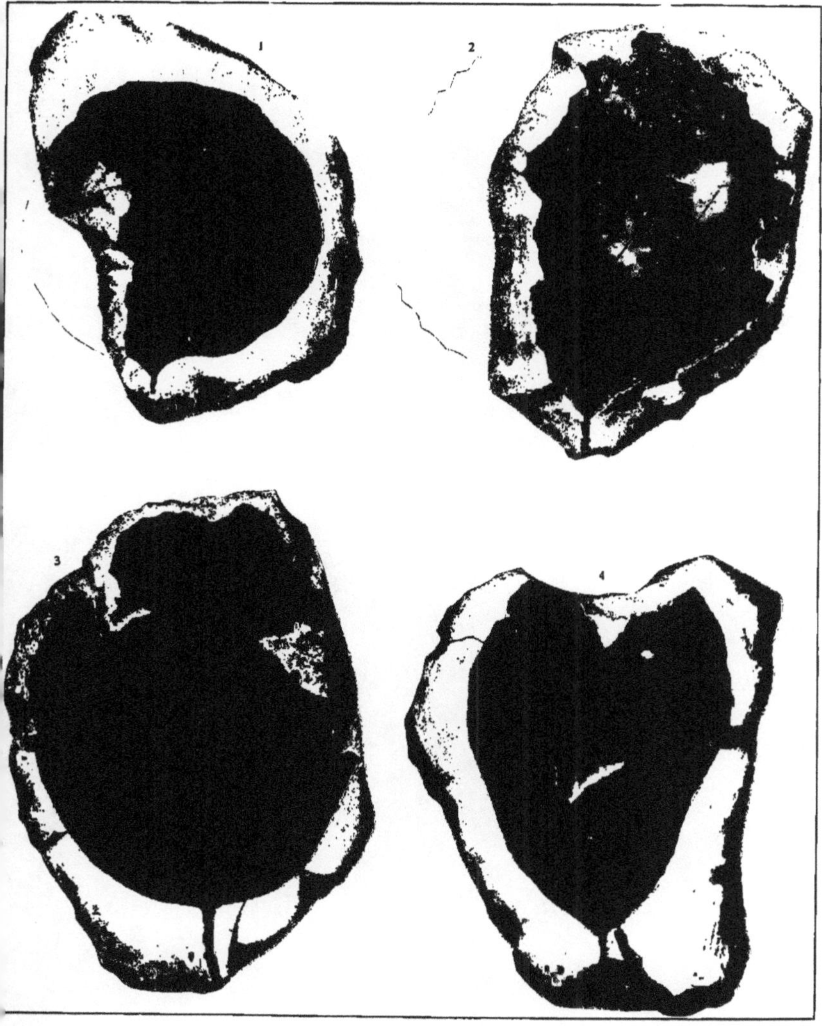
1
2
3
4

PLATE V.

PLATE V.

		Page.
Fig. 1	Sapotacites Haydenii Heer	126
2.	Phyllites obcordatus Heer	136
3.	Leguminosites Marcouanus Heer	113
4.	Liquidambar obtusilobatus (Heer) Hollick	101
5.	Populus (?) cordifolia Newb.	40
6.	Magnolia alternans Heer?	94
7.	Populus (?) Debeyana Heer	42

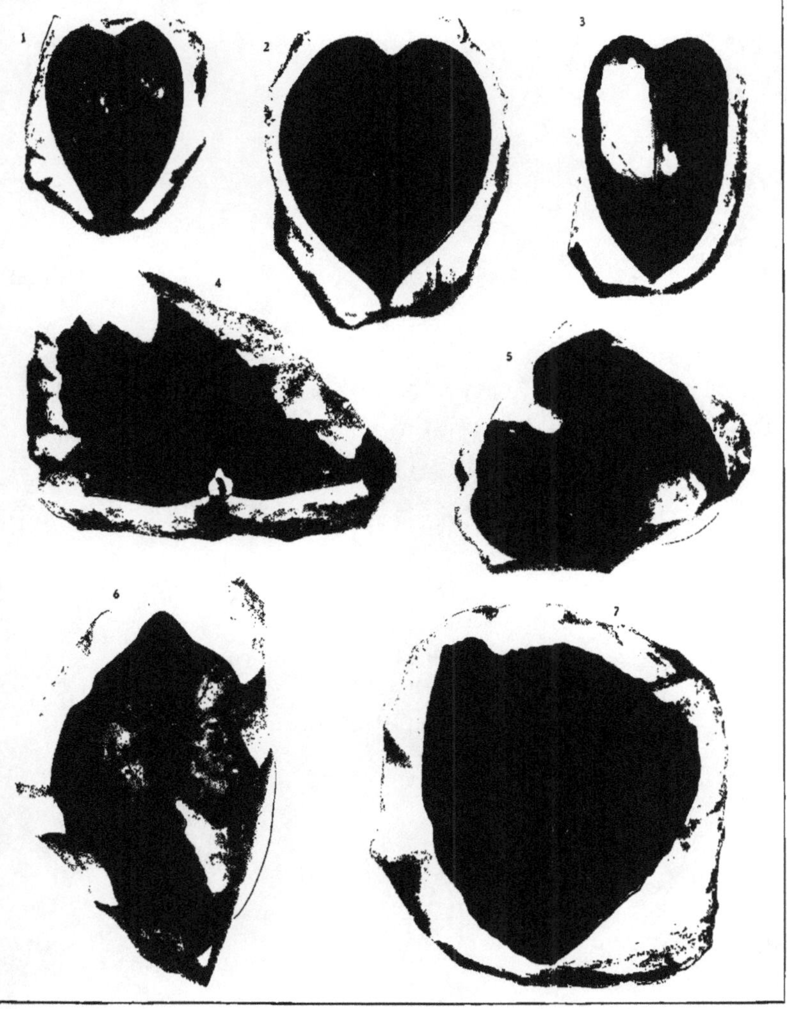
1
2
3
4
5
6
7

PLATE VI.

PLATE VI.

CRETACEOUS
(Dakota Group.)

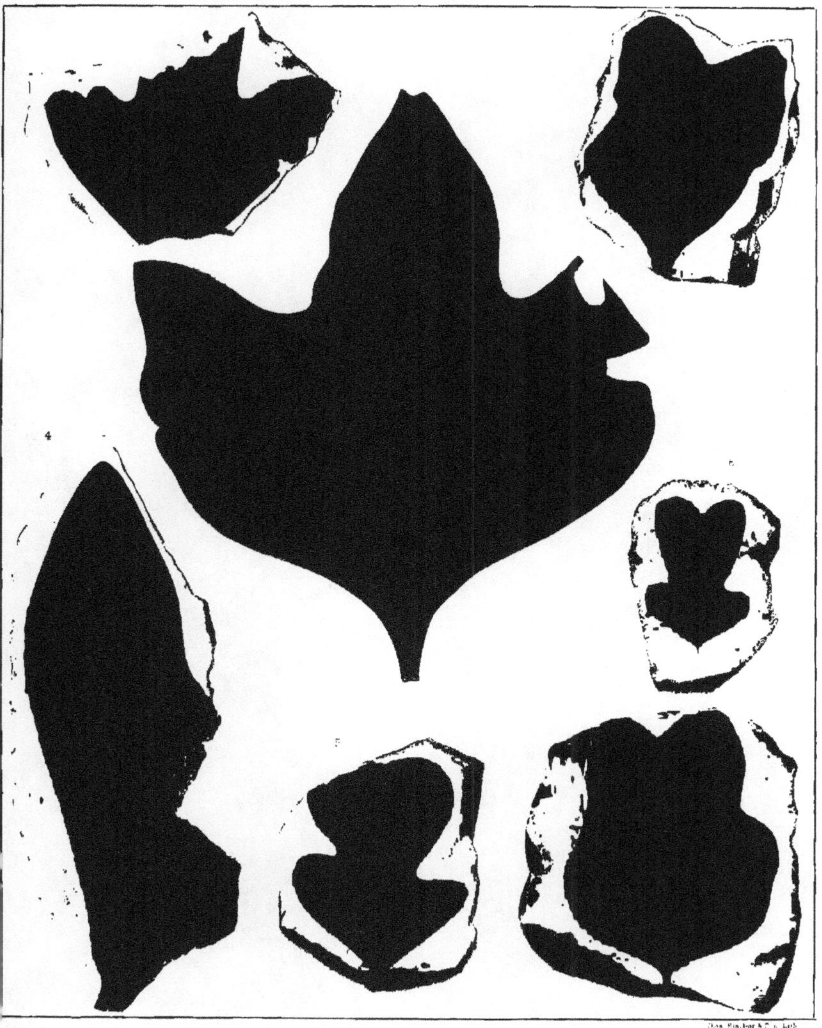

Thos. Sinclair & Co. Lith.

PLATE VII.

PLATE VII.

Page.

Figs. 1–3. Sassafras cretaceum Newb. 98

4. Protophyllum multinerve Lesq 132

168

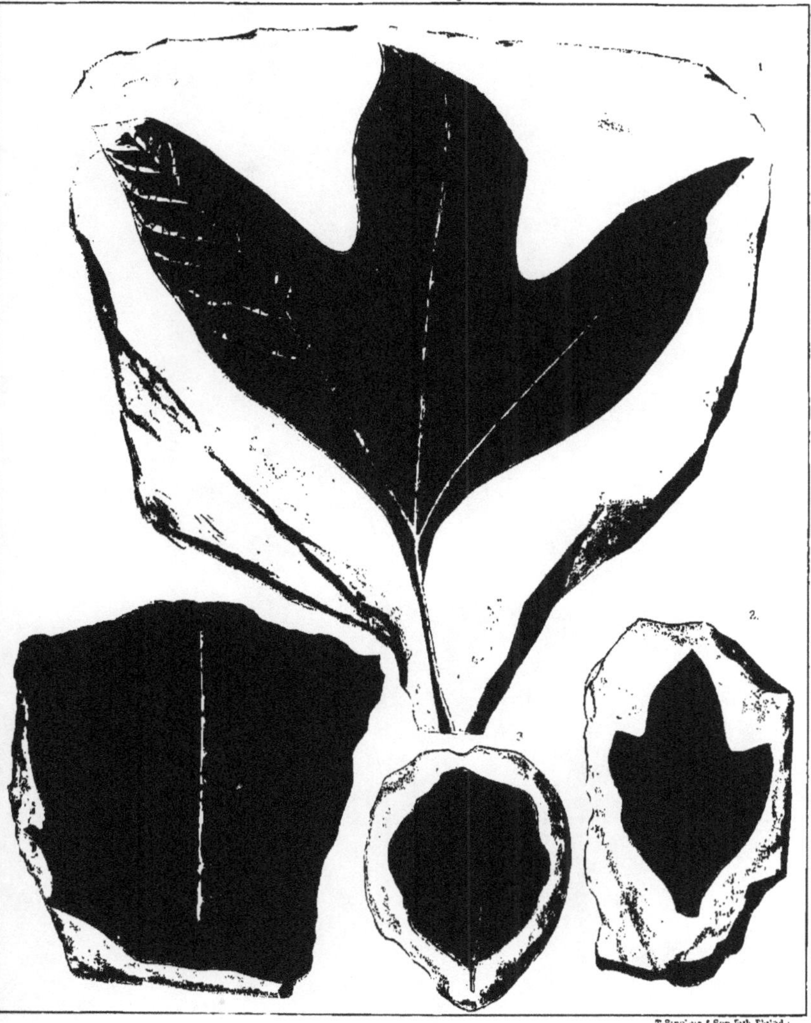

T. Sinclair & Son, Lith. Philada.

PLATE VIII.

PLATE VIII.

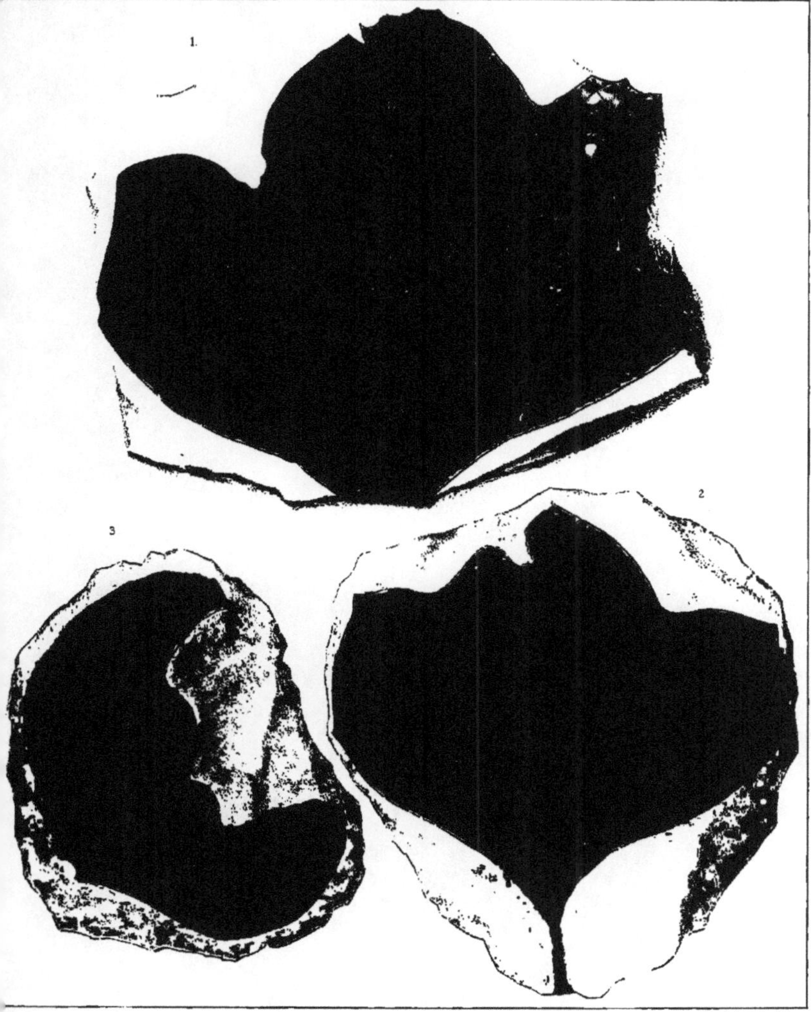

Thos. Sinclair & Son, Lith.

PLATE IX.

PLATE IX.

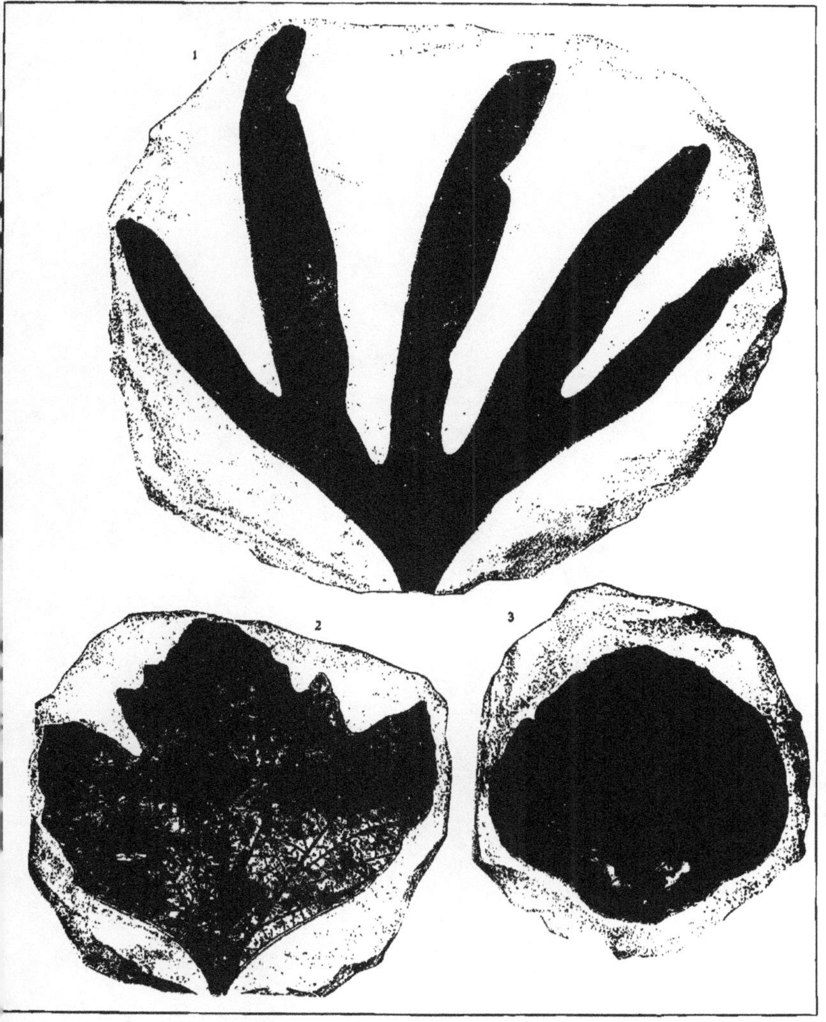
1
2
3

PLATE X.

PLATE X.

PLATE XI.

PLATE XI.

T Sinclair & Son Lith Philada

PLATE XII.

PLATE XII.

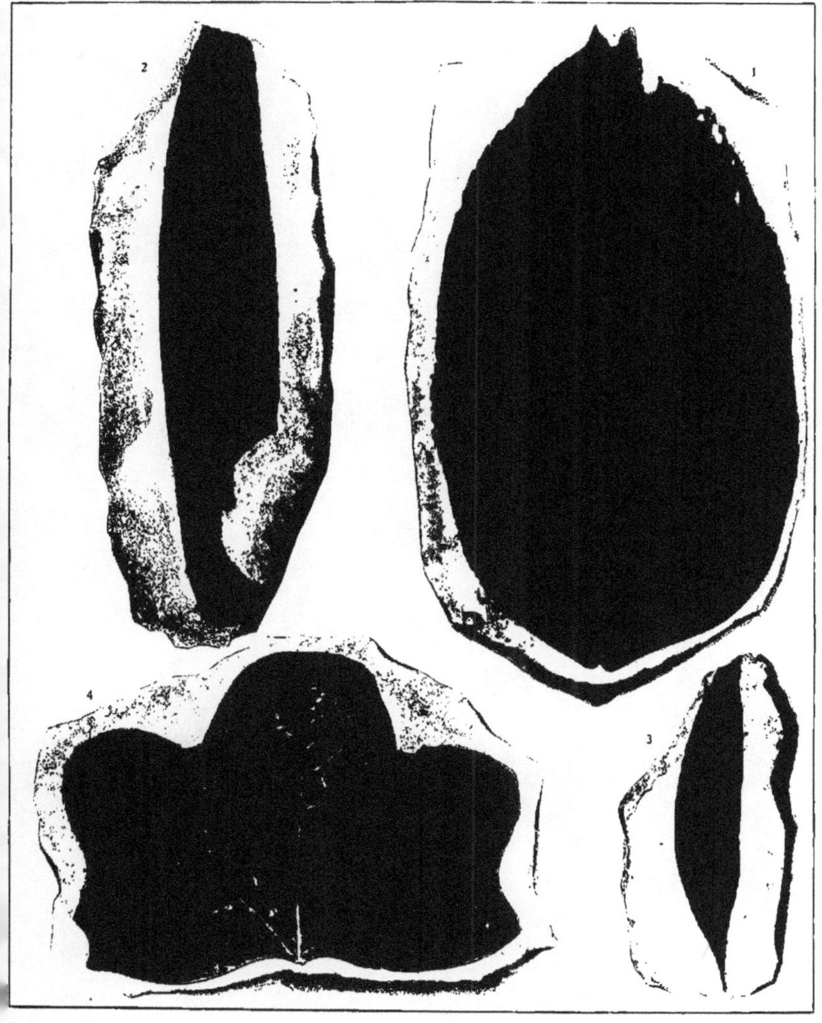
2
1
4
3

PLATE XIII.

PLATE XIII.

		Page.
FIG. 1.	Quercus sinuata Newb	78
2.	Quercus antiqua Newb	69
3, 4.	Salix flexuosa Newb	56
5, 6.	Salix foliosa Newb	57

1
5
3
6
4
2

PLATE XIV.

PLATE XIV.

	Page.
Fig. 1. Salix flexuosa Newb	56
2. Myrica (?) trifoliata Newb	37
3–4a. Sequoia cuneata Newb	18
5 Abietites cretacea Newb	18
6. Sequoia gracillima (Lesq.) Newb	19

CRETACEOUS
(Dakota Group.)

PLATE XV.

PLATE XV.

	Page.
Figs. 1, 1a. Anemia perplexa Hollick	3
2, 2a. Nilssonia Gibbsii (Newb.) Hollick	16

184

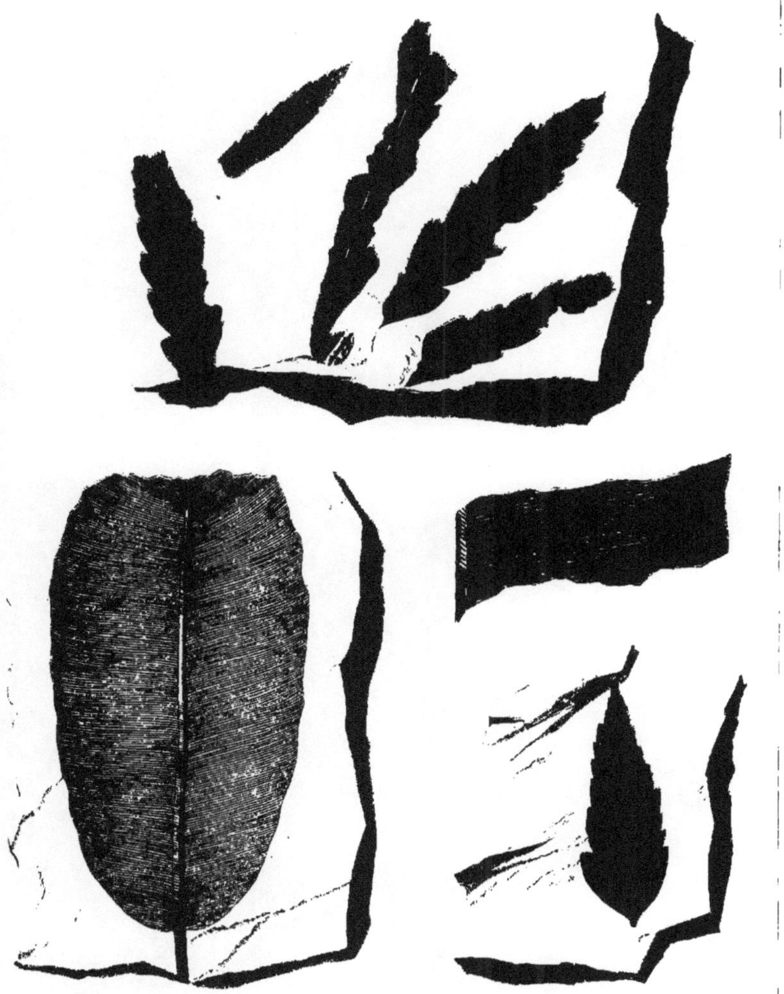

PLATE XVI.

PLATE XVI.

	Page.
FIGS. 1, 2. Equisetum robustum Newb	15
3. Anemia perplexa Hollick	8
4, 5. Aspidium Kennerlyi Newb	11
6, 6a. Sabal imperialis Dn	30

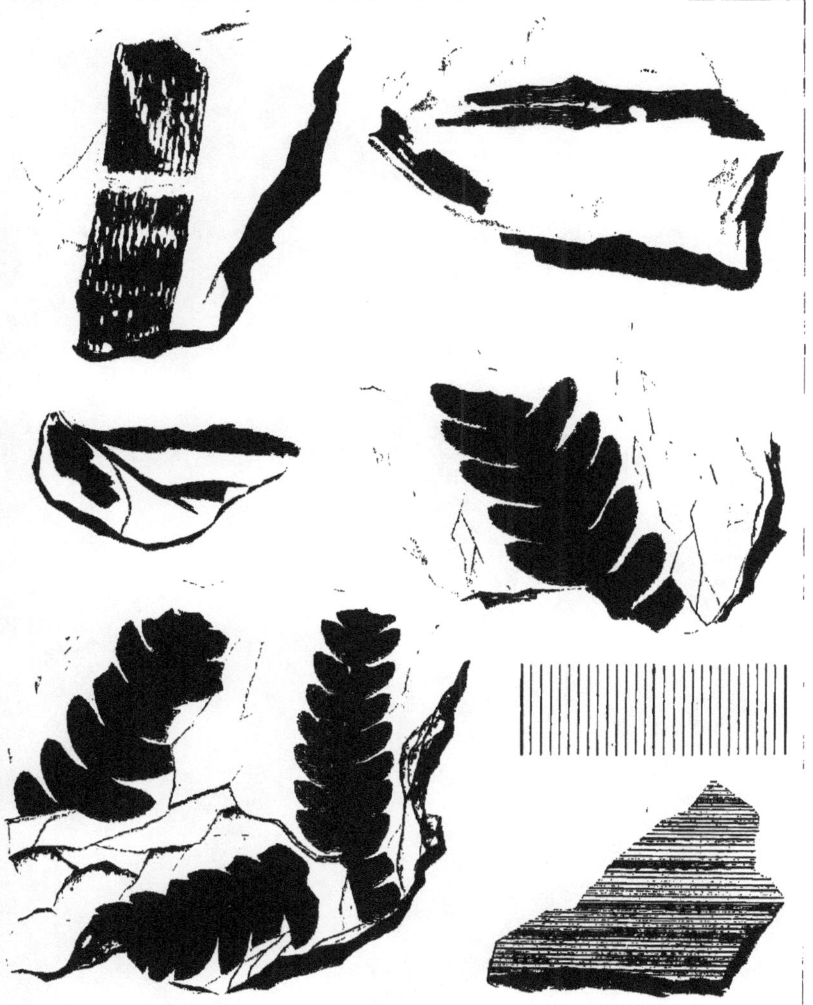

PLATE XVII.

PLATE XVII.

	Page.
FIGS. 1–3. Cinnamomum Heerii Lesq	100
4–6. Nyssa (?) cuneata Newb	125

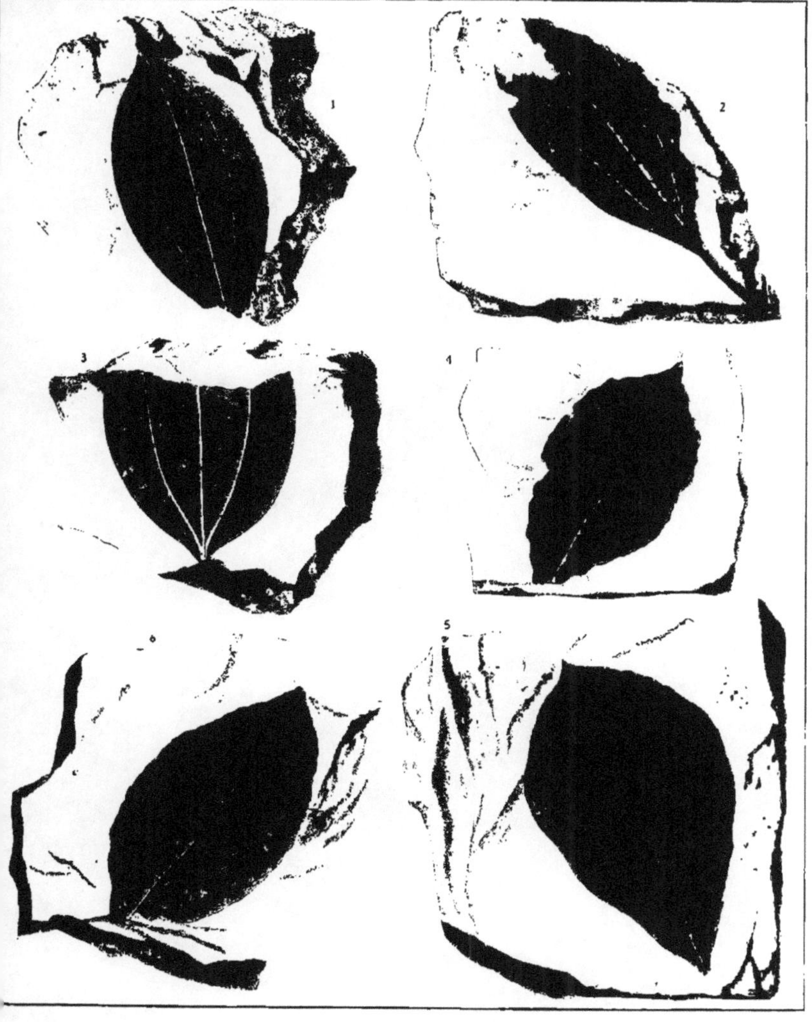
1
2
3
4
6
5

PLATE XVIII.

PLATE XVIII.

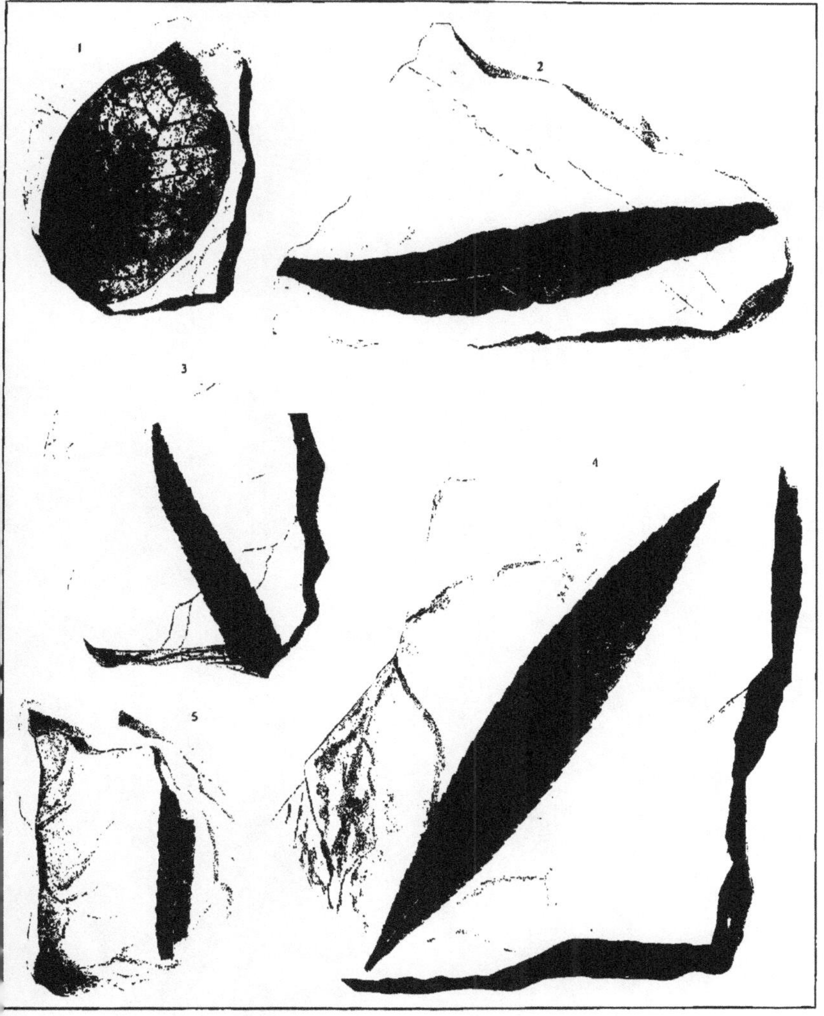
1
2
3
4
5

PLATE XIX.

PLATE XIX.

	Page.
FIGS. 1-3. Quercus coriacea Newb	73
4-6. Quercus flexuosa Newb	74

Thos. Sinclair & Son, Lith.

PLATE XX.

PLATE XX.

194

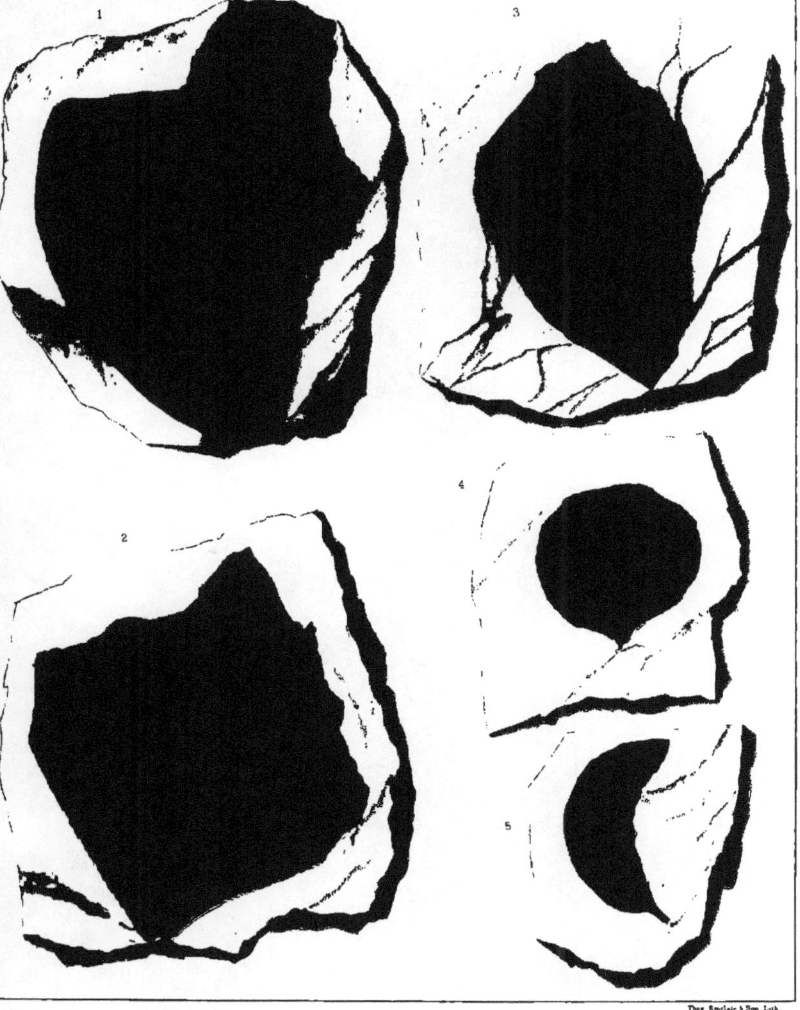

Thos Sinclair & Son, Lith

PLATE XXI.

PLATE XXI.

	Page.
FIGS. 1, 2. Sabal Campbelli Newb.....	27

PLATE XXII.

PLATE XXII.

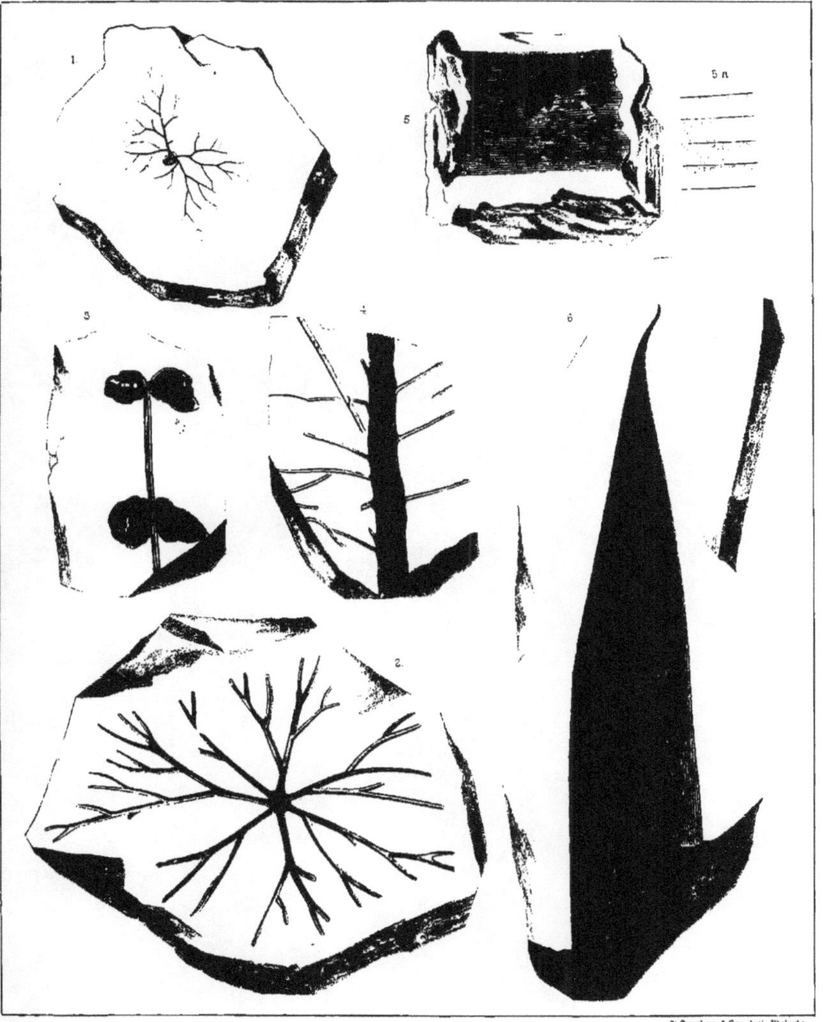

T Sinclair & Son, Lith Philada

PLATE XXIII.

PLATE XXIII.

	Page.
Fig. 1. Cabomba(?) gracilis Newb	91
2. Cabomba inermis (Newb.) Hollick	92
3. Onoclea sensibilis fossilis Newb	8
4. Onoclea sensibilis L. (introduced for comparison)	9
5, 6. Onoclea sensibilis obtusilobatus Torr. (introduced for comparison)	9

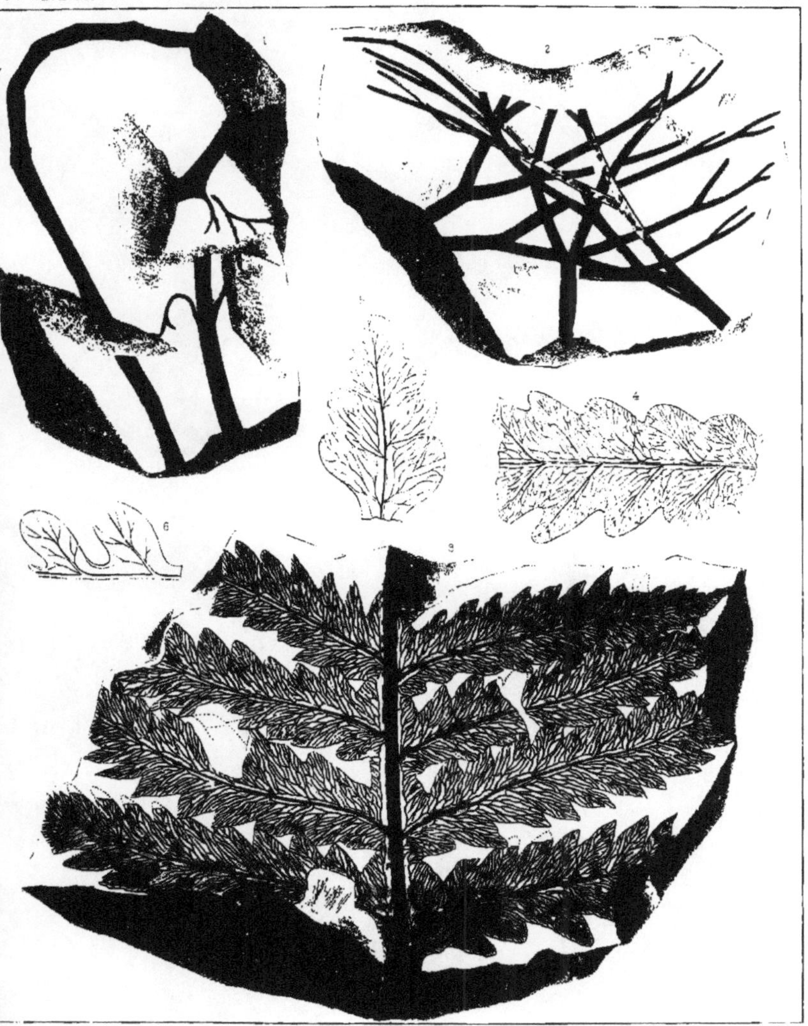

T. Sinclair & Son Lith. Philada.

PLATE XXIV.

PLATE XXIV.

	Page.
FIG. 1. Onoclea sensibilis fossilis Newb.	8
2. Onoclea sensibilis fossilis Newb. (summit of frond)	8
3. Onoclea sensibilis fossilis Newb. (pinna deeply cut, with elongated areolæ)	8
4, 5. Onoclea sensibilis fossilis Newb. (bases of upper and lower pinnæ)	8

3
4
5
1

PLATE XXV.

PLATE XXV.

	Page.
Sabal grandifolia Newb.	28

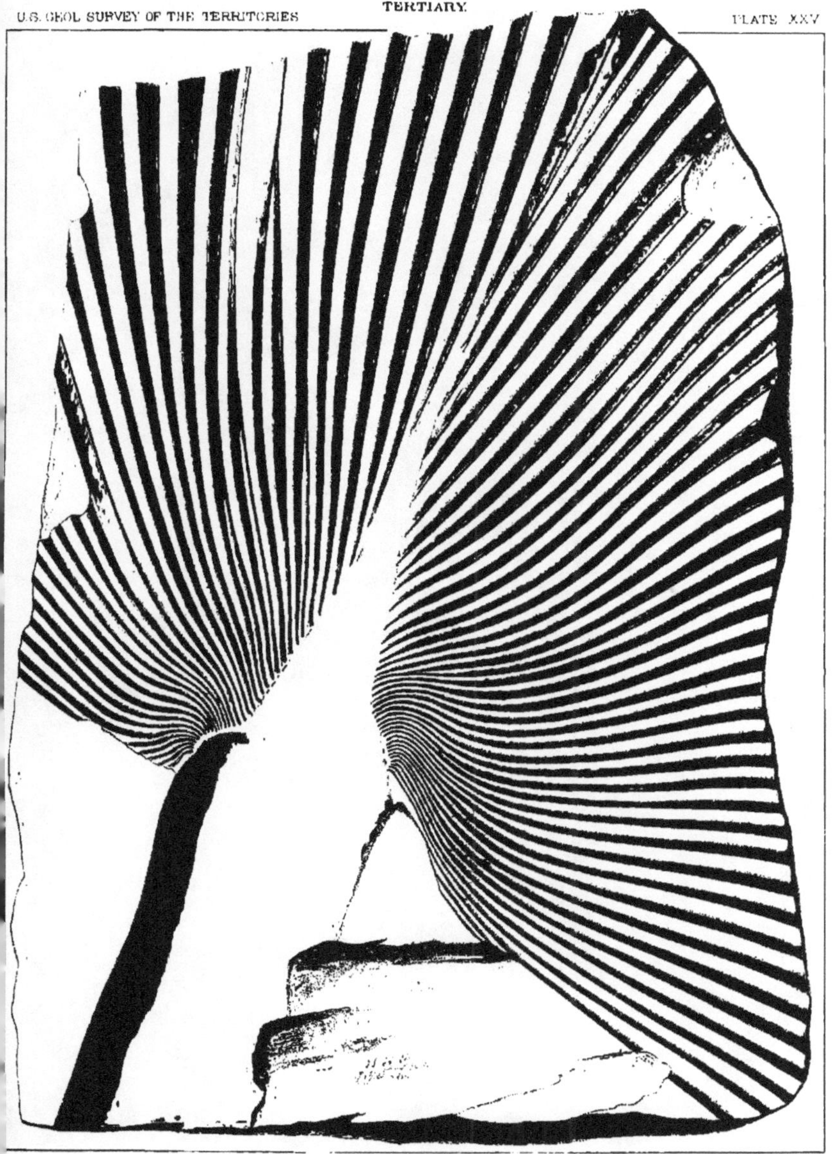

Thos Sinclair & Son, Lith

PLATE XXVI.

PLATE XXVI.

T. Sinclair & Son, Lith. Phila.

PLATE XXVII.

PLATE XXVII.

	Page.
FIG. 1. Populus genetrix Newb	44
2, 3. Populus nervosa Newb	48
4, 5. Populus Nebrascensis Newb	47

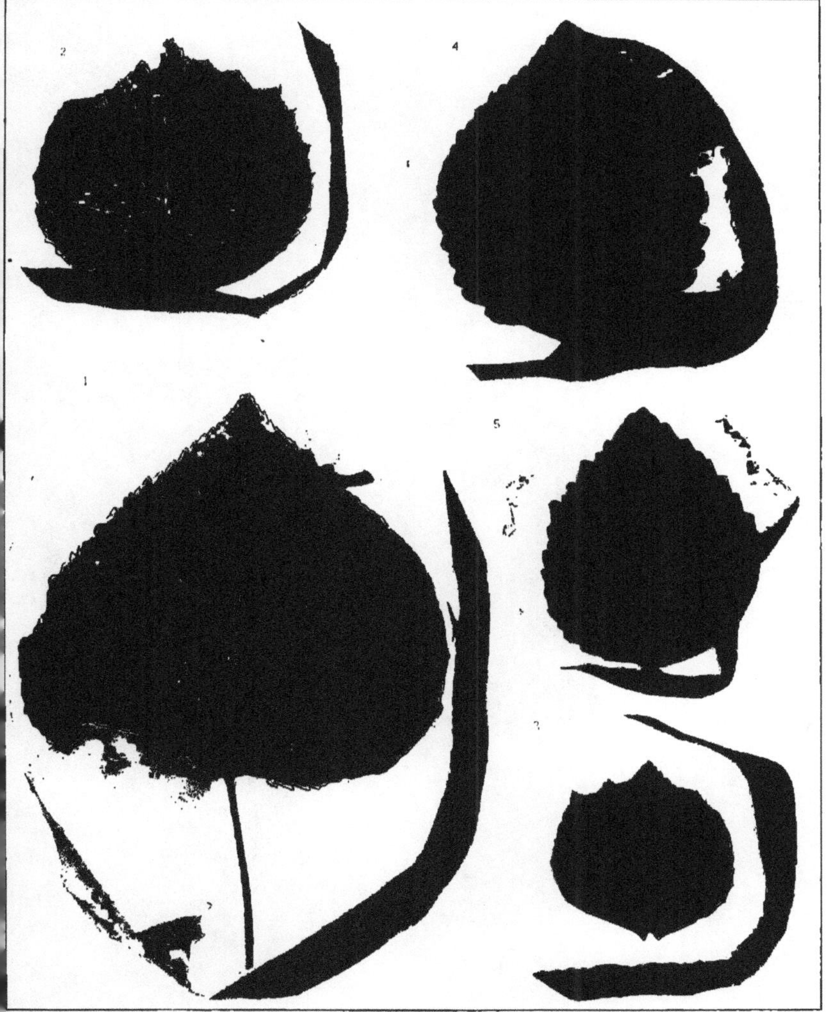

Thos Sinclair & Son, Lith

PLATE XXVIII.

PLATE XXVIII.

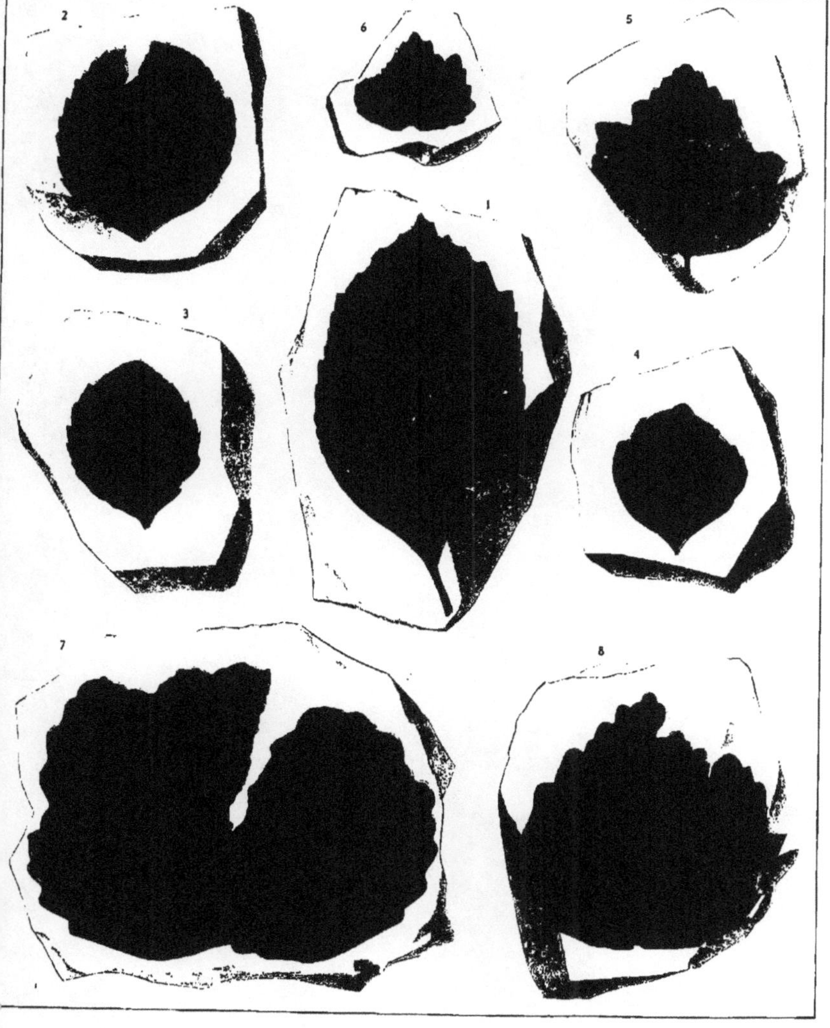
2
6
5
1
3
4
7
8

PLATE XXIX.

211

PLATE XXIX.

	Page.
Figs. 1–4. Populus rotundifolia Newb	51
5. Populus smilacifolia Newb	53
6. Populus cordata Newb	38
7. Populus cuneata Newb	41
8–10. Corylus Americana fossilis Newb	60

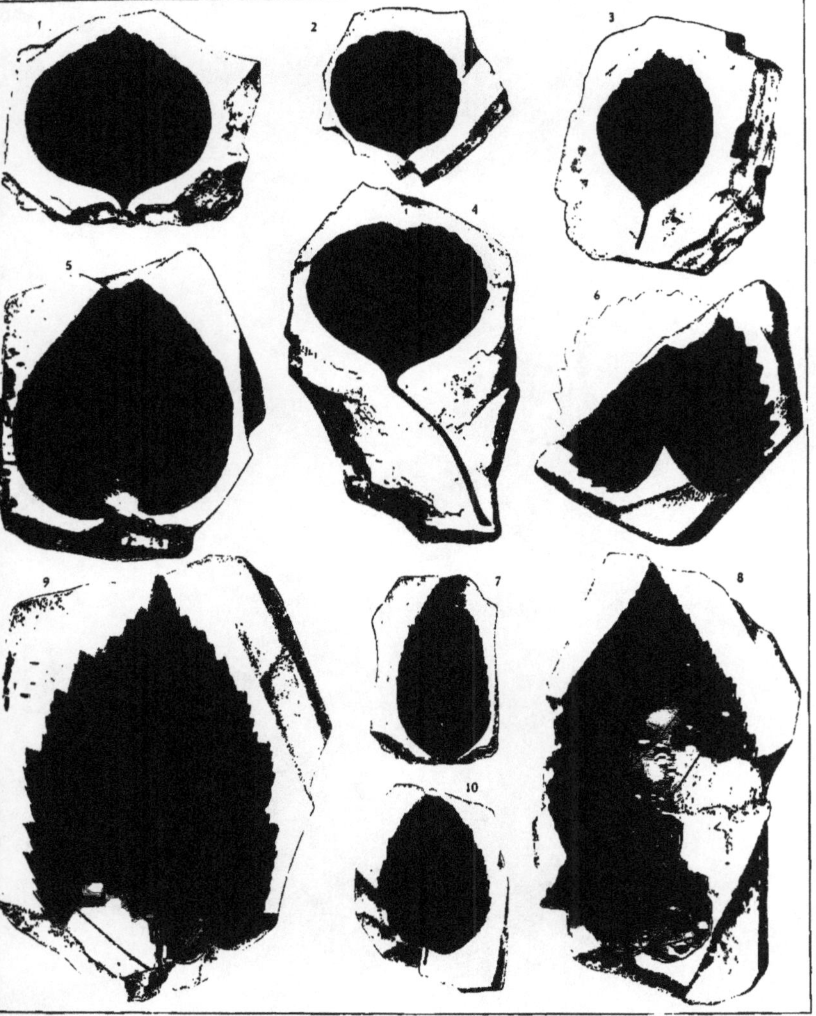
1
2
3
4
5
6
7
8
9
10

PLATE XXX.

PLATE XXX.

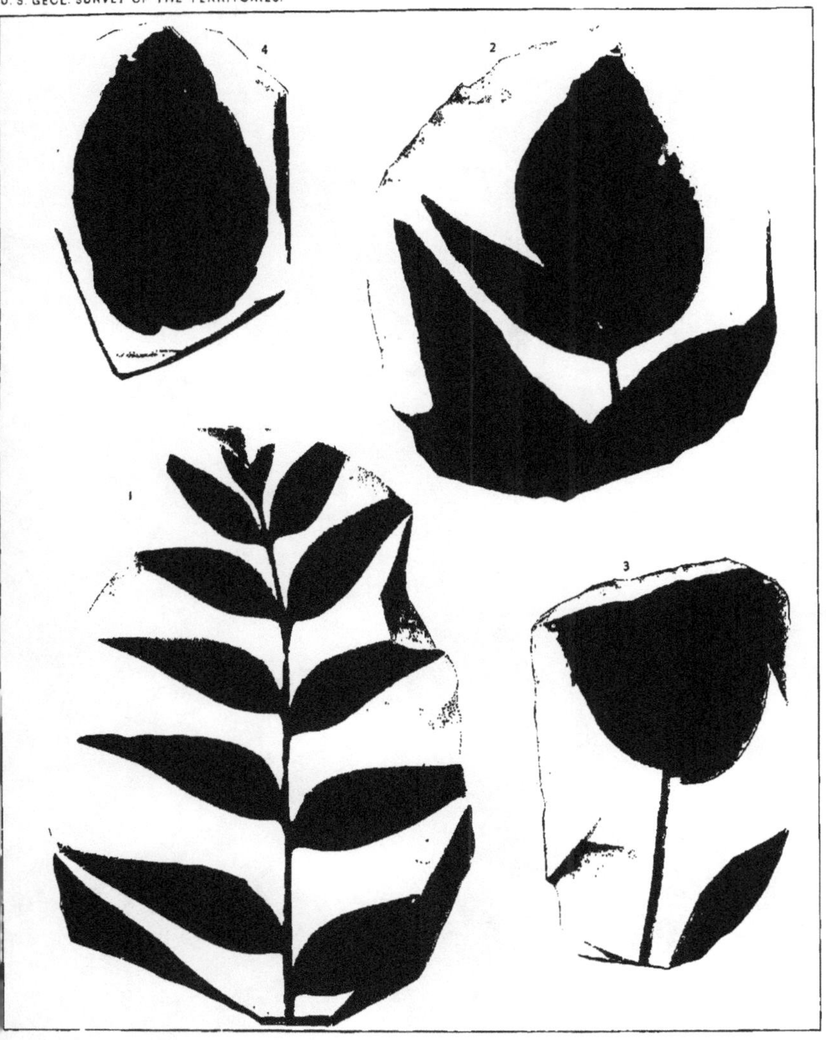
4
2
1
3

PLATE XXXI.

PLATE XXXI.

	Page.
Figs. 1–4. Carya antiquorum Newb	35
5. Negundo triloba Newb	115

216

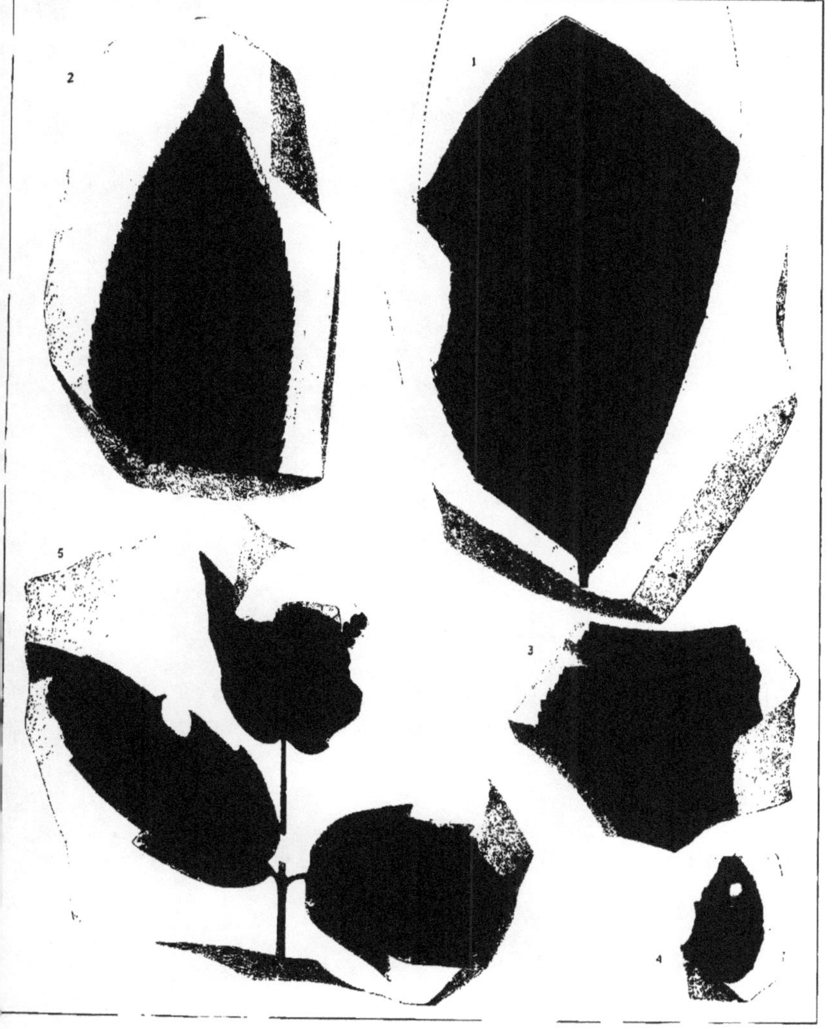
1
2
3
4
5

PLATE XXXII.

PLATE XXXII.

	Page.
Figs. 1–3. Corylus rostrata fossilis Newb	63
4. Corylus orbiculata Newb	62
5. Corylus MacQuarrii (Forbes) Heer	61

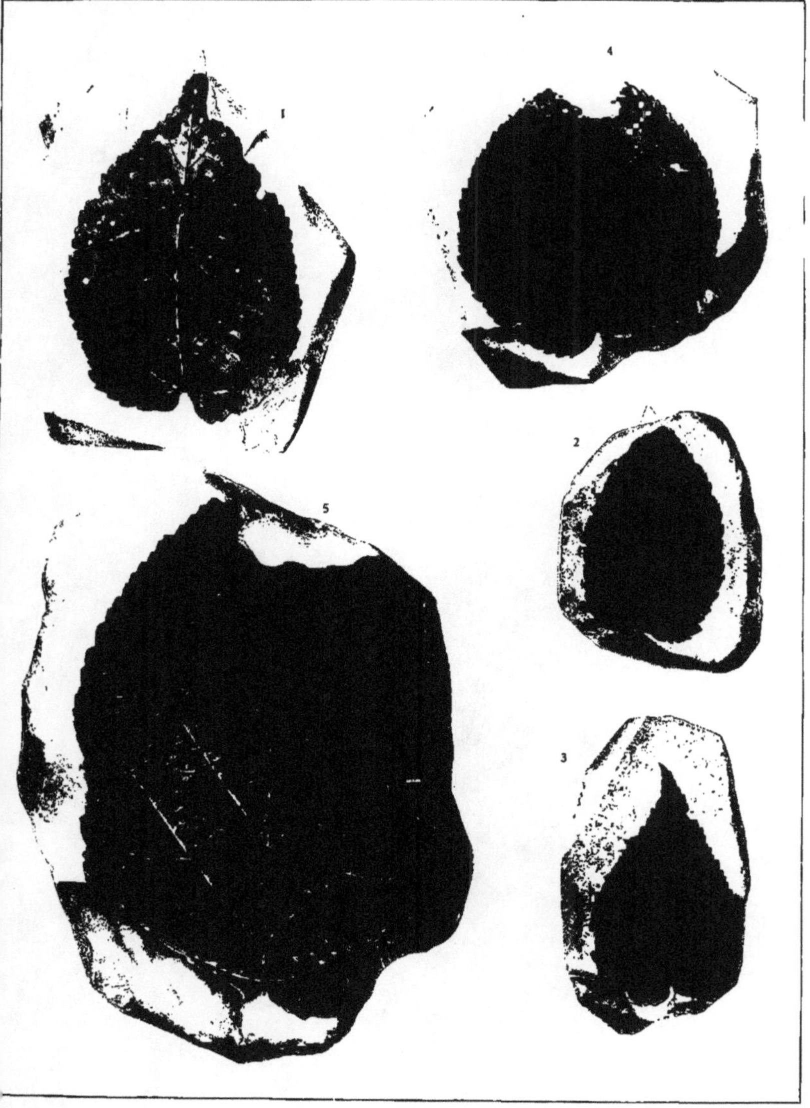
1
4
2
5
3

PLATE XXXIII.

PLATE XXXIII.

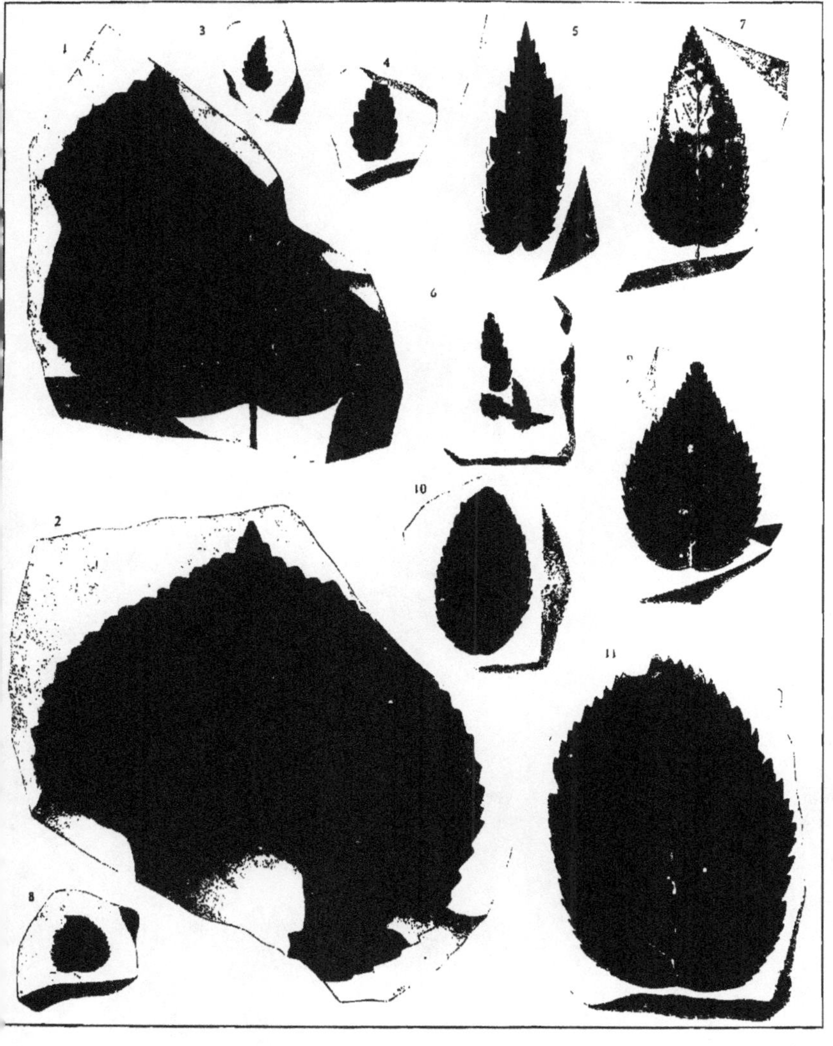
1
2
3
4
5
6
7
8
9
10
11

PLATE XXXIV.

PLATE XXXIV.

PLATE XXXV.

PLATE XXXV.

Page.

Platanus Raynoldsii Newb. 109

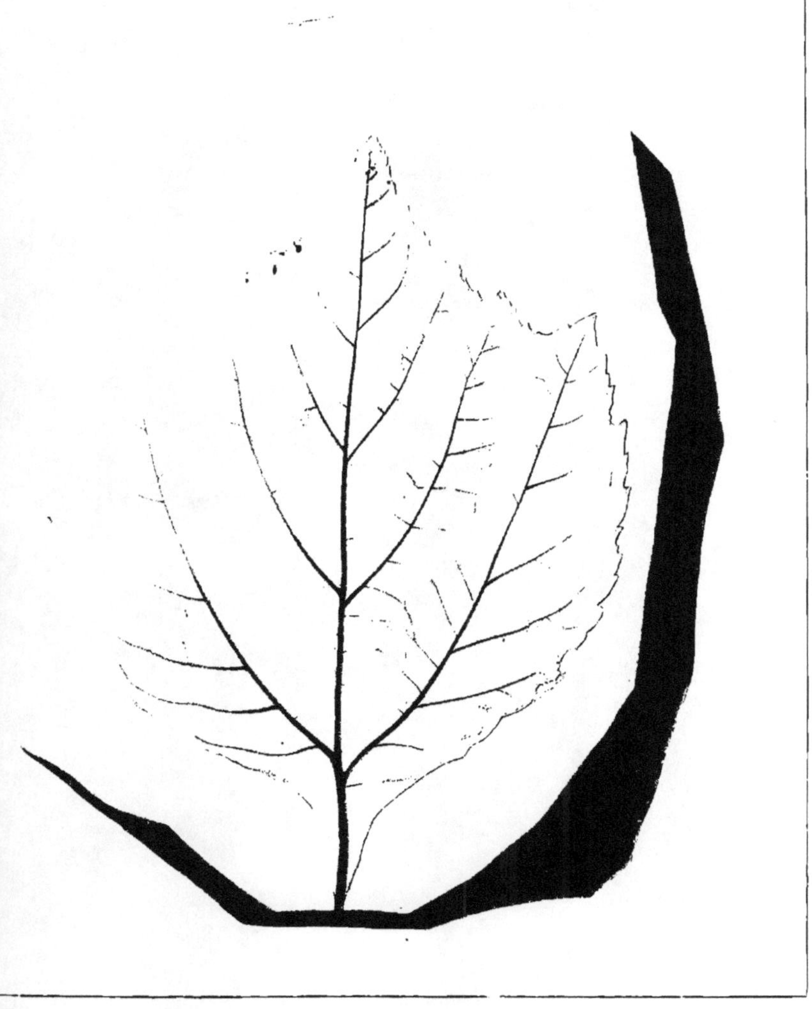

PLATE XXXVI.

PLATE XXXVI.

	Page.
Platanus Haydenii Newb.	103

PLATE XXXVII.

PLATE XXXVII.

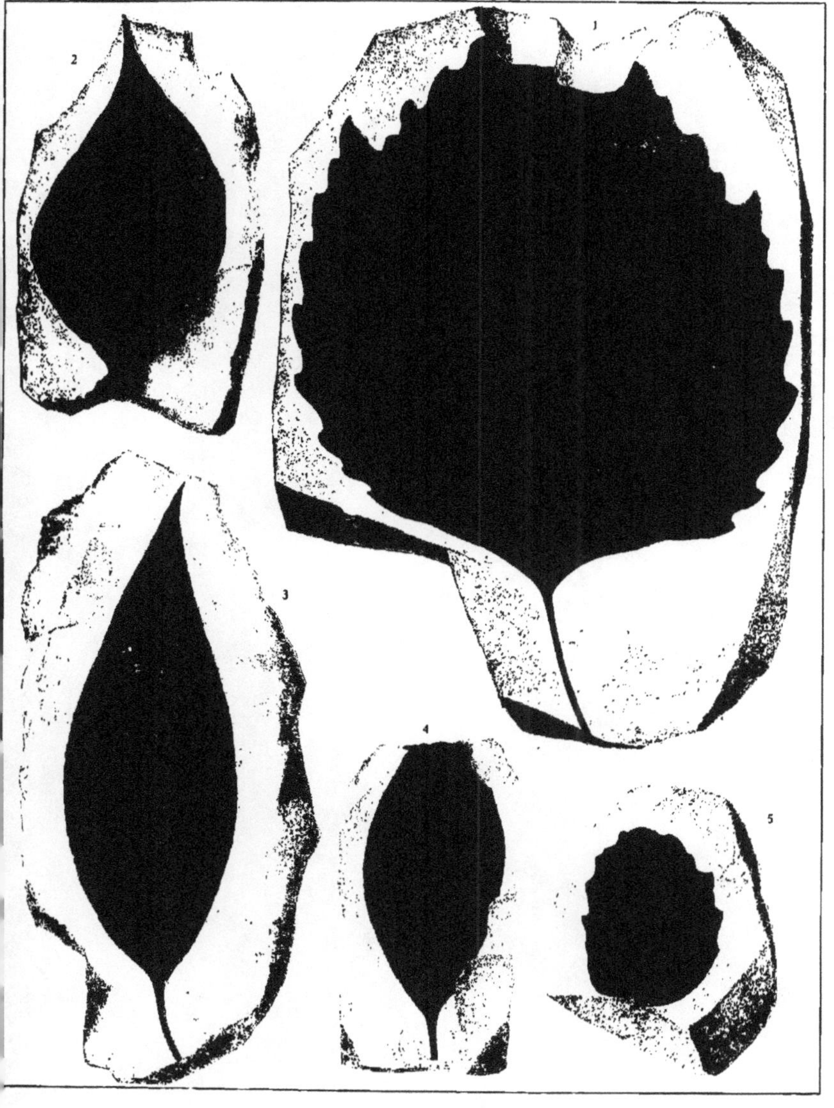
2
1
3
4
5

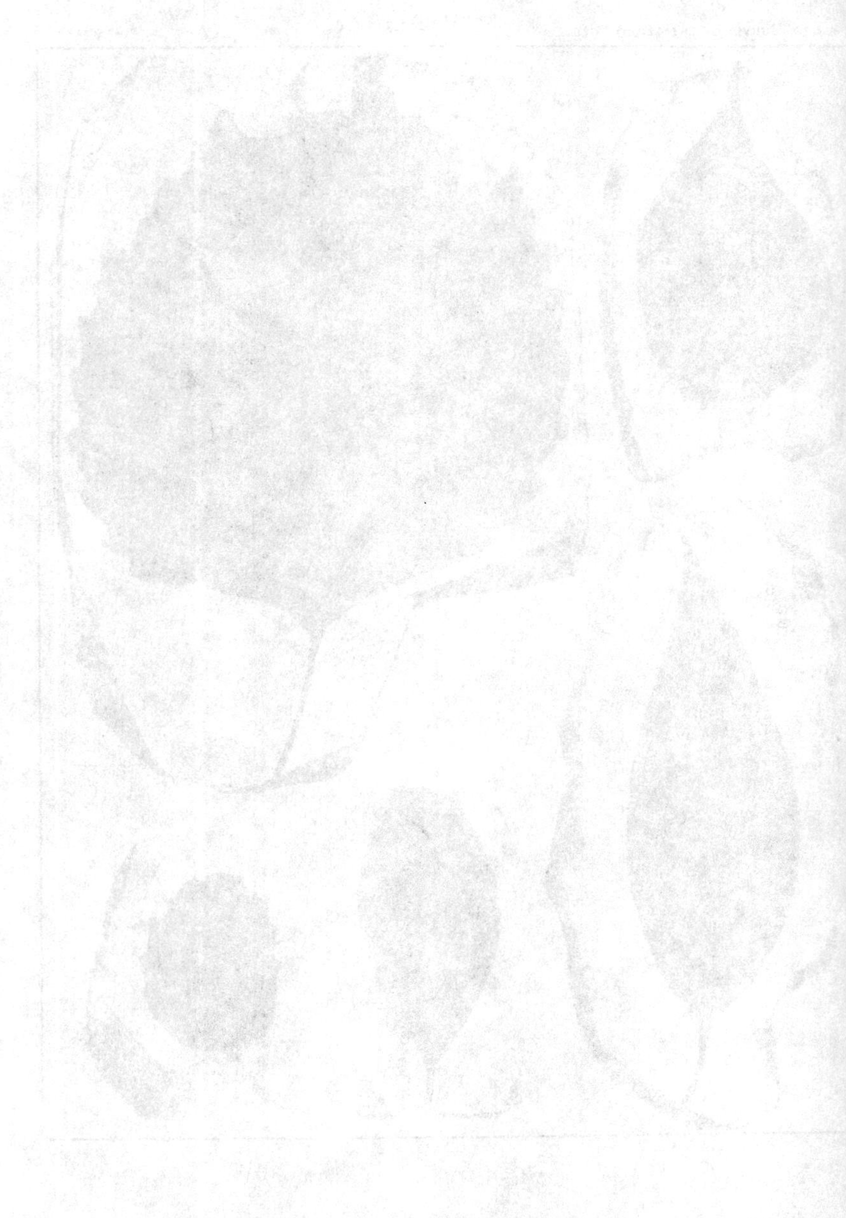

PLATE XXXVIII.

PLATE XXXVIII.

Page.

Platanus Haydenii Newb. (young leaf) 103

230

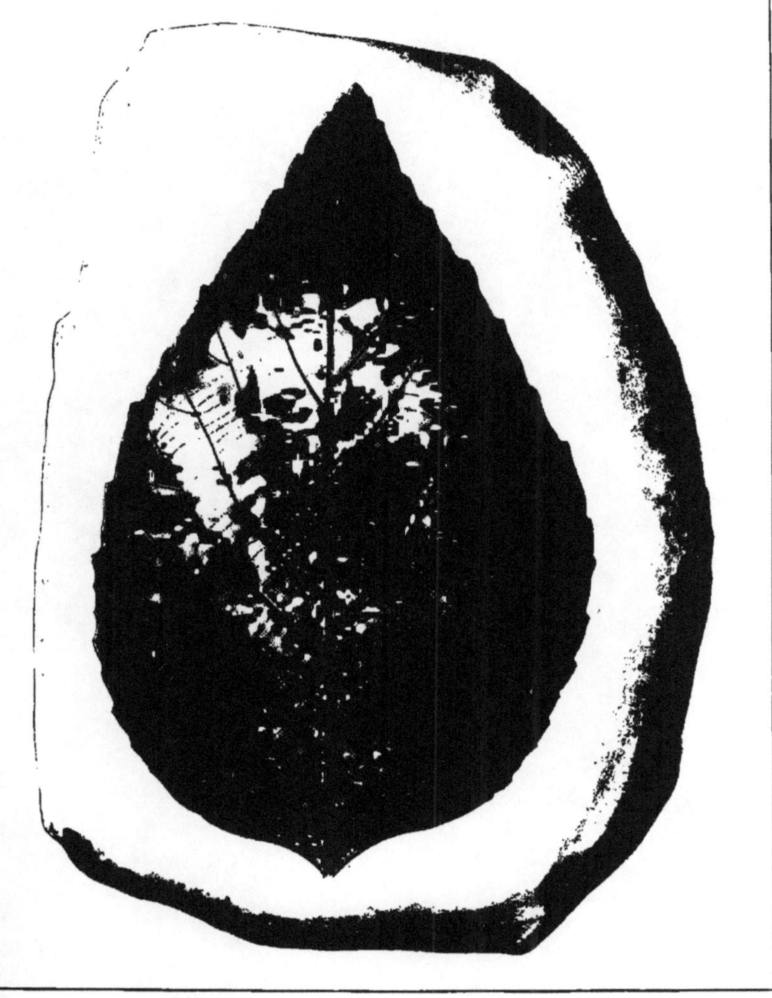

PLATE XXXIX.

PLATE XXXIX.

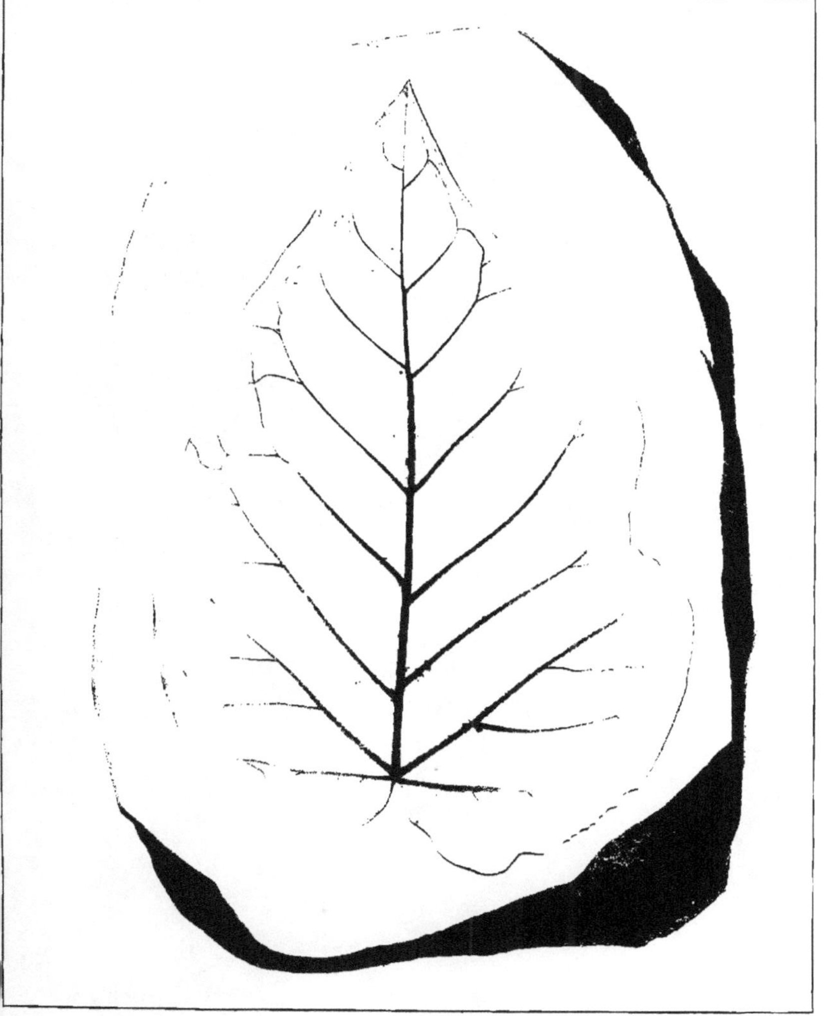

PLATE XL.

PLATE XL.

	Page.
FIG. 1. Carpolithes lineatus Newb	138
2. Sapindus affinis Newb	116
3. Calycites polysepala Newb	139
4, 5. Aralia triloba Newb	123
6. Amelanchier similis Newb	111
7. Aristolochia cordifolia Newb	90

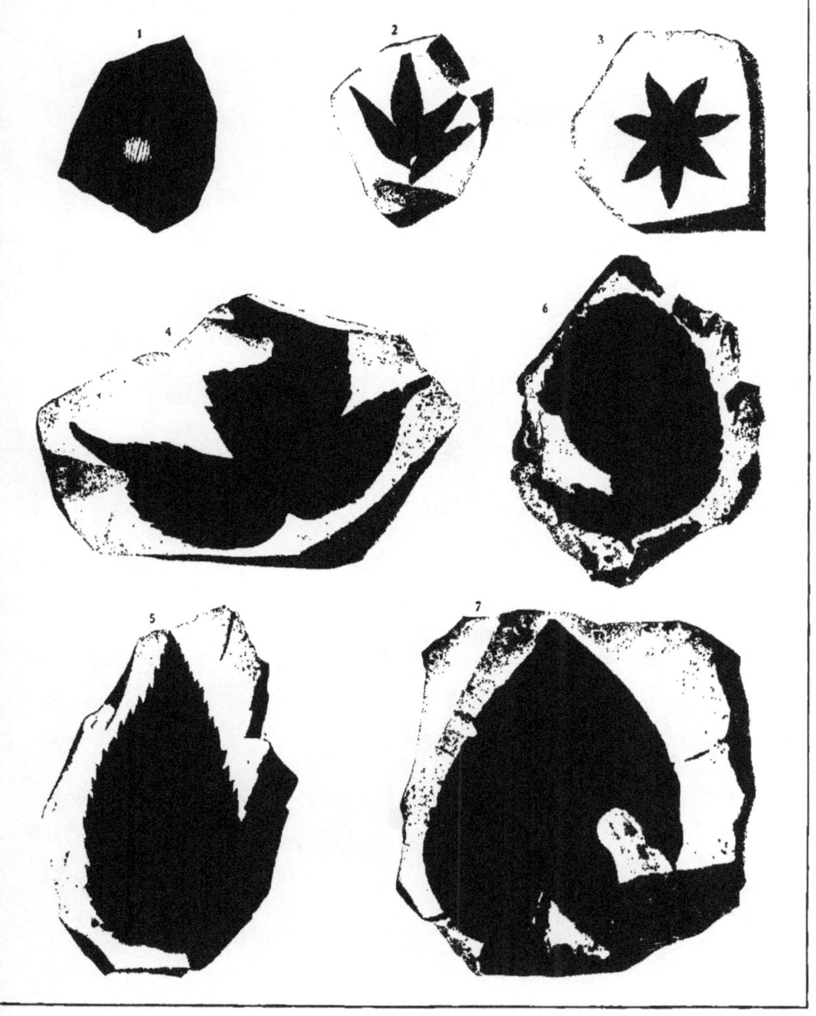
1
2
3
4
6
5
7

PLATE XLI.

PLATE XLI.

		Page.
FIGS. 1, 2.	Phyllites carneosus Newb	134
3, 4.	Phyllites cupanioides Newb	135

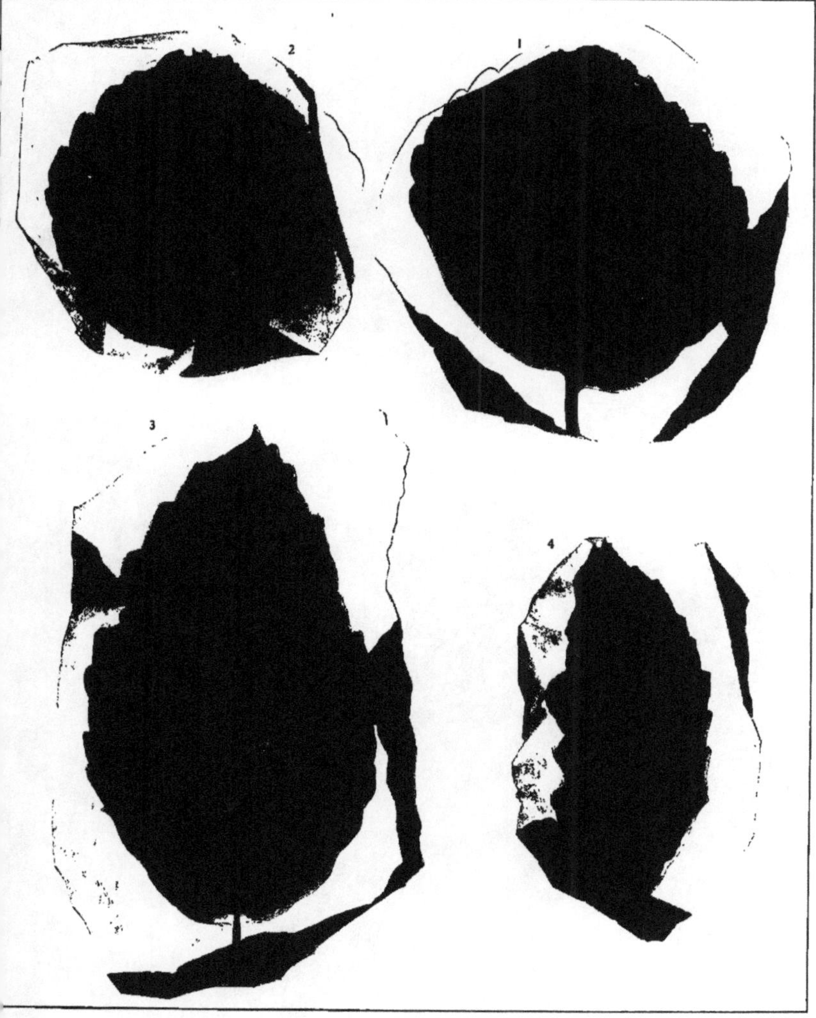
2
1
3
4

PLATE XLII.

PLATE XLII.

	Page.
Figs. 1-3. Platanus aspera Newb	102

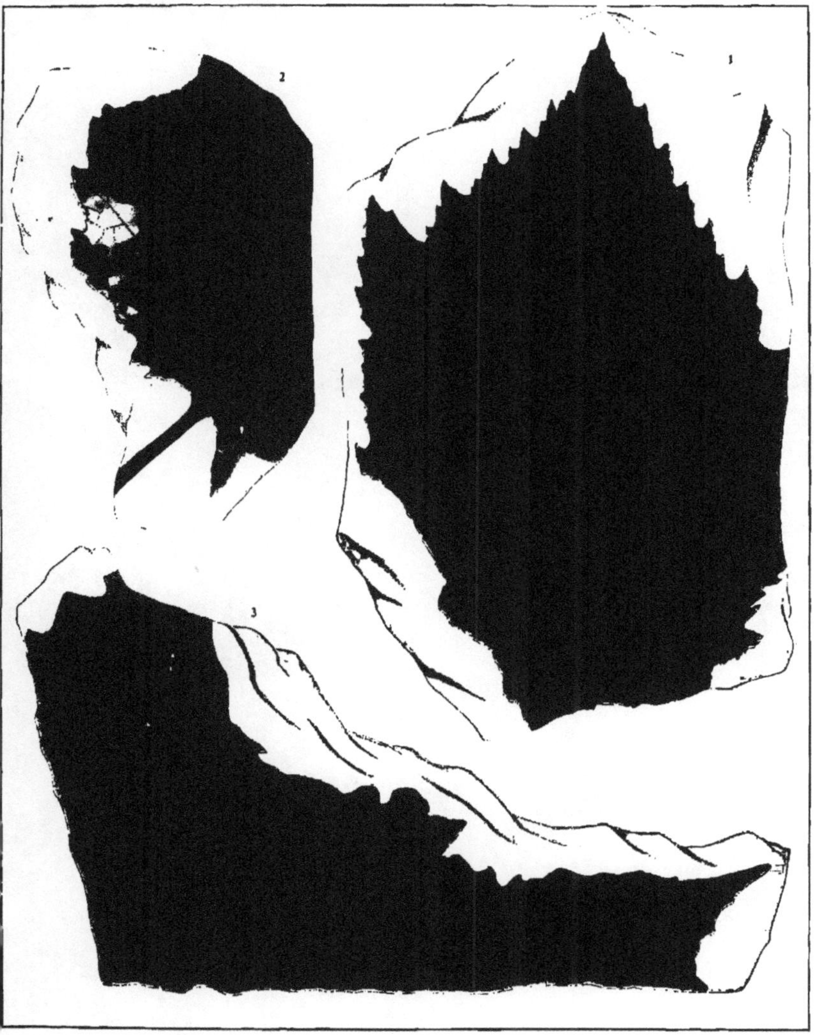

1

2

3

PLATE XLIII.

PLATE XLIII.

		Page.
Fig.	1. Quercus paucidentata Newb	76
	2-5. Quercus consimilis Newb	71
	6. Quercus simplex Newb	78
	7. Acorn of Quercus consimilis Newb	72
	8. Cupule of Quercus consimilis Newb	72
	9. Base of acorn of Quercus consimilis Newb	72
	10. Interior of cupule of Quercus consimilis Newb	72

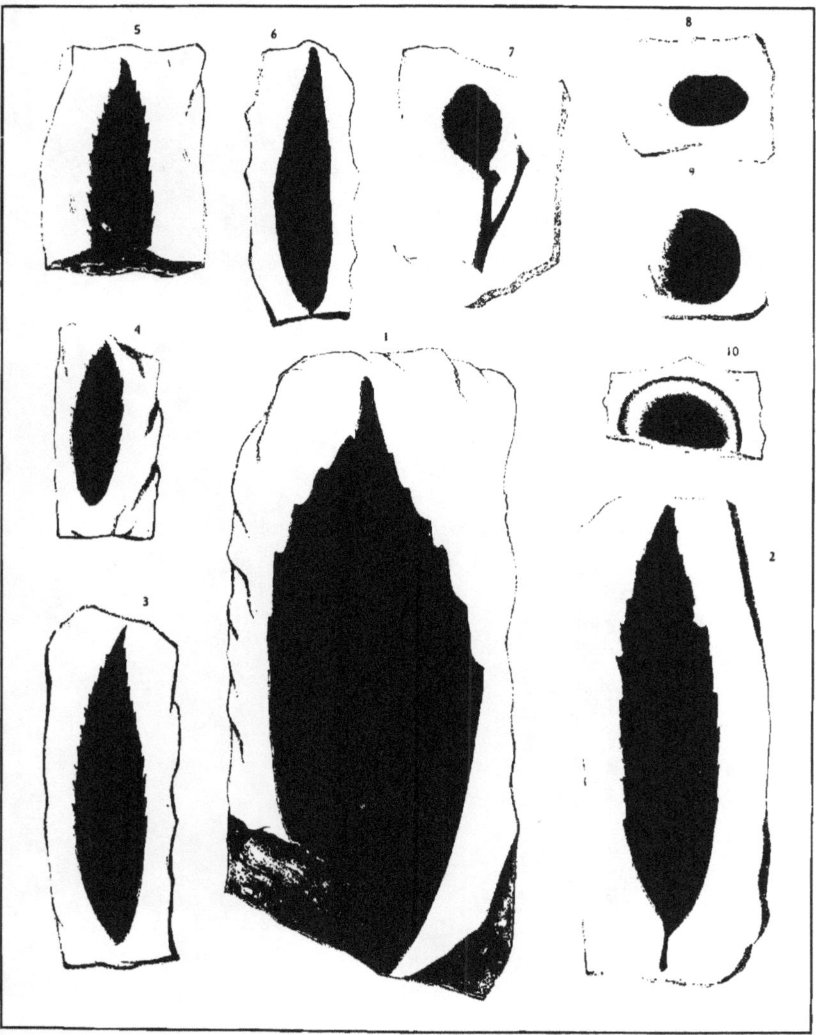
5
6
7
8
9
4
1
10
2
3

PLATE XLIV.

PLATE XLIV.

		Page.
Figs. 1–4.	Betula heterodonta Newb.	64
5.	Platanus aspera Newb.	102

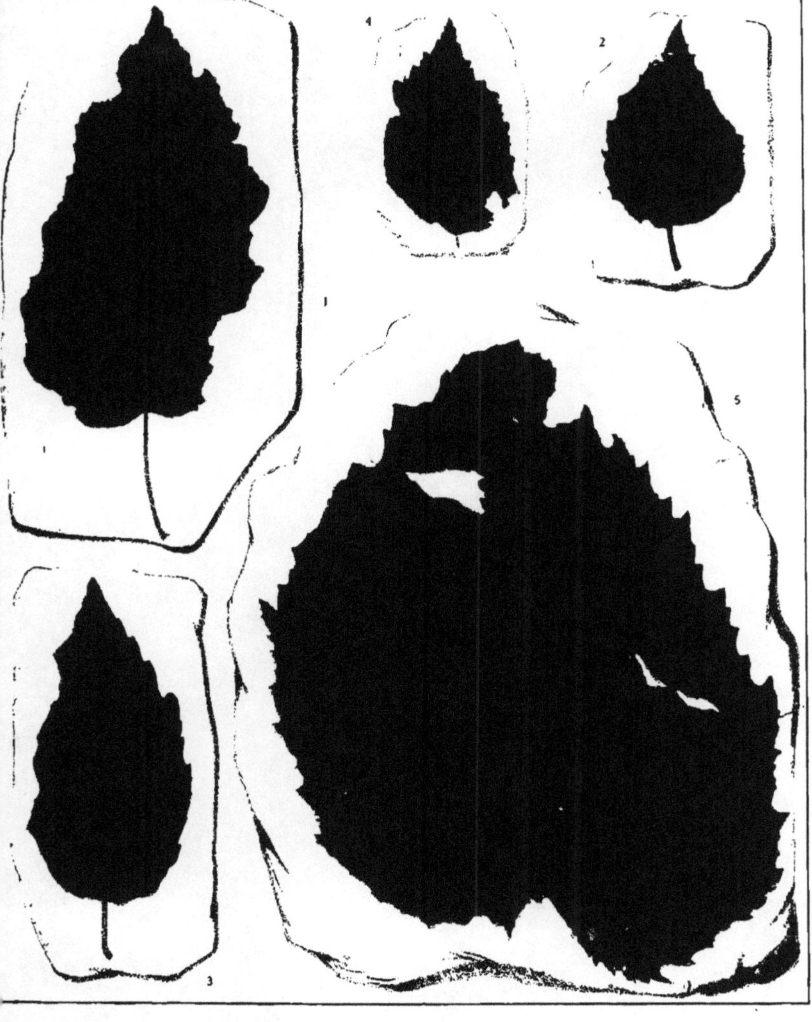
1
4
2
1
5
3

PLATE XLV.

PLATE XLV.

	Page.
Fig. 1. Betula heterodonta Newb	64
2–5, 8. Ulmus speciosa Newb	80
6. Fruit of Betula heterodonta Newb	65
7. Fruit of Ulmus speciosa Newb	80

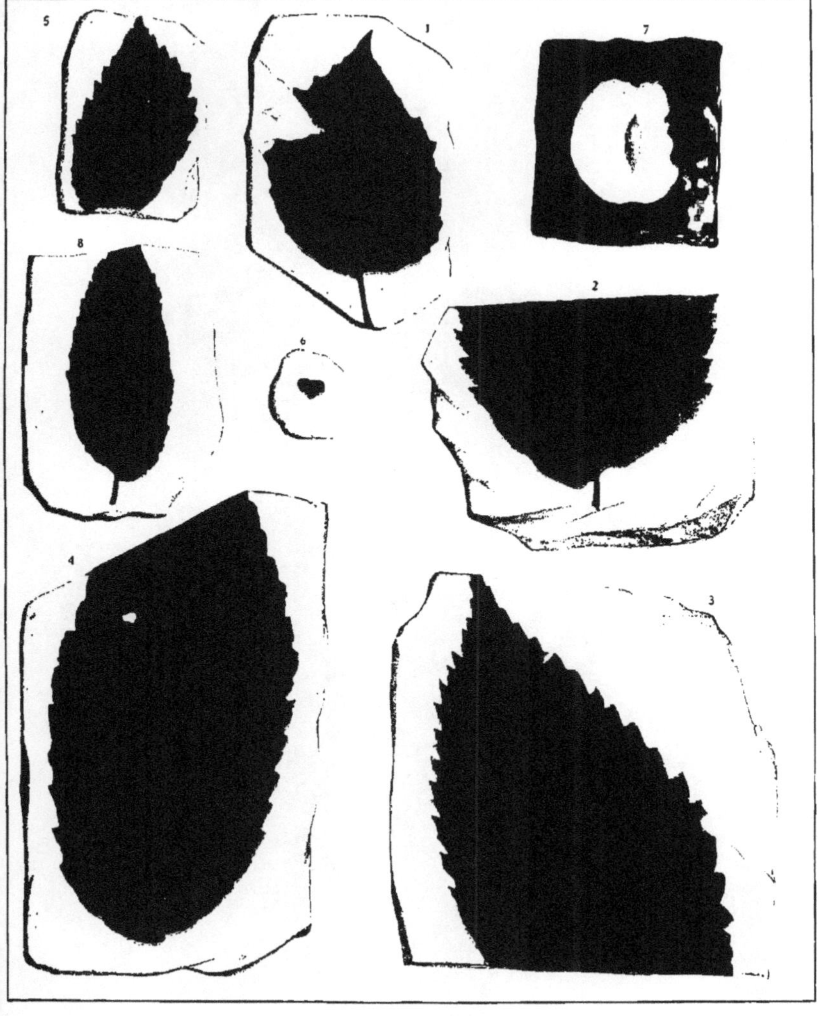
5
1
7
8
2
6
4
3

PLATE XLVI.

PLATE XLVI.

	Page.
Fig. 1. Ficus planicostata Lesq	88
2. Grewia crenata (Ung.) Heer	120
3, 4. Populus polymorpha Newb	50
5. Betula angustifolia Newb	63
6. Alnus serrulata fossilis Newb	66
7. Fruit of Alnus sp. ? Newb	67
8. Fruit of Acer sp. ? Newb	115
9. Monocotyledon gen. et sp. ? Hollick	33
10. Fruit of Cassia sp. ? Newb	113

246

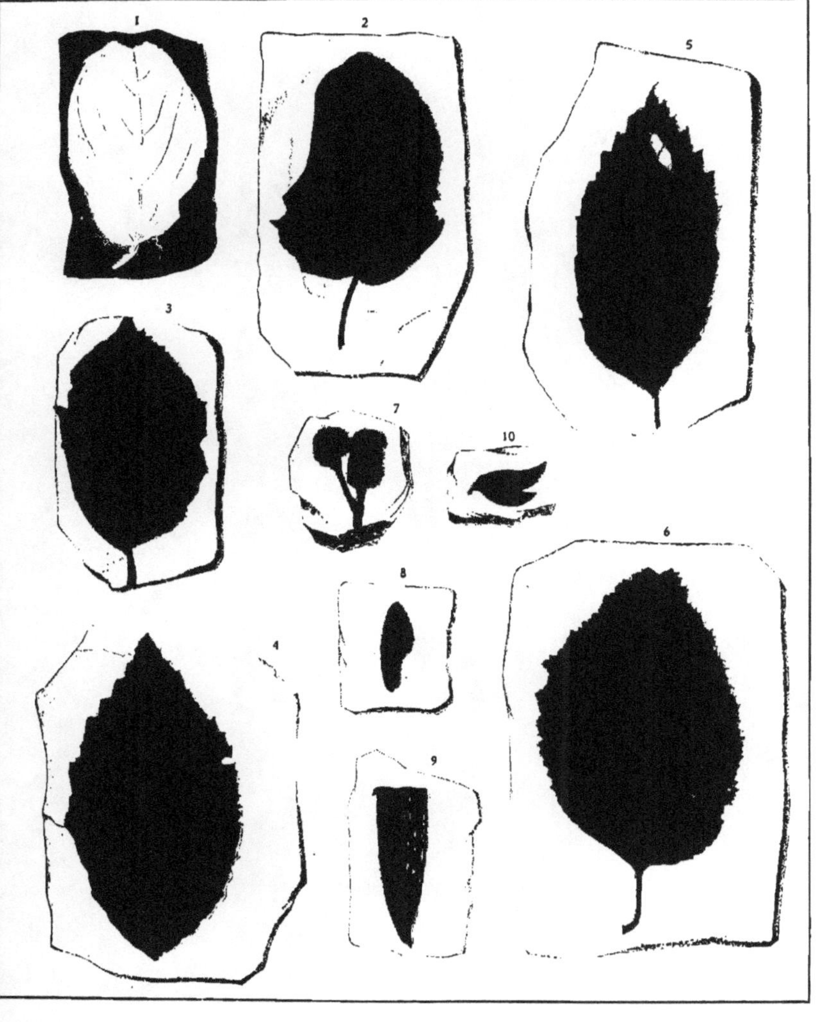
1
2
5
3
7
10
6
8
4
9

PLATE XLVII.

PLATE XLVII.

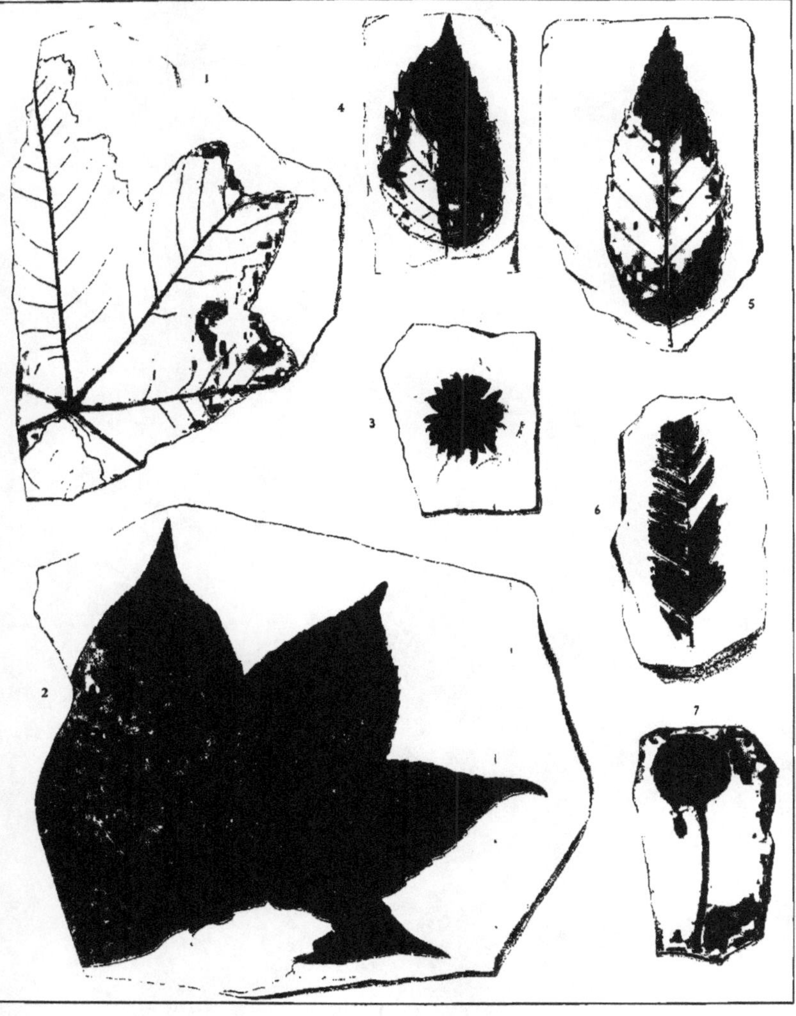
1
2
3
4
5
6
7

PLATE XLVIII.

PLATE XLVIII.

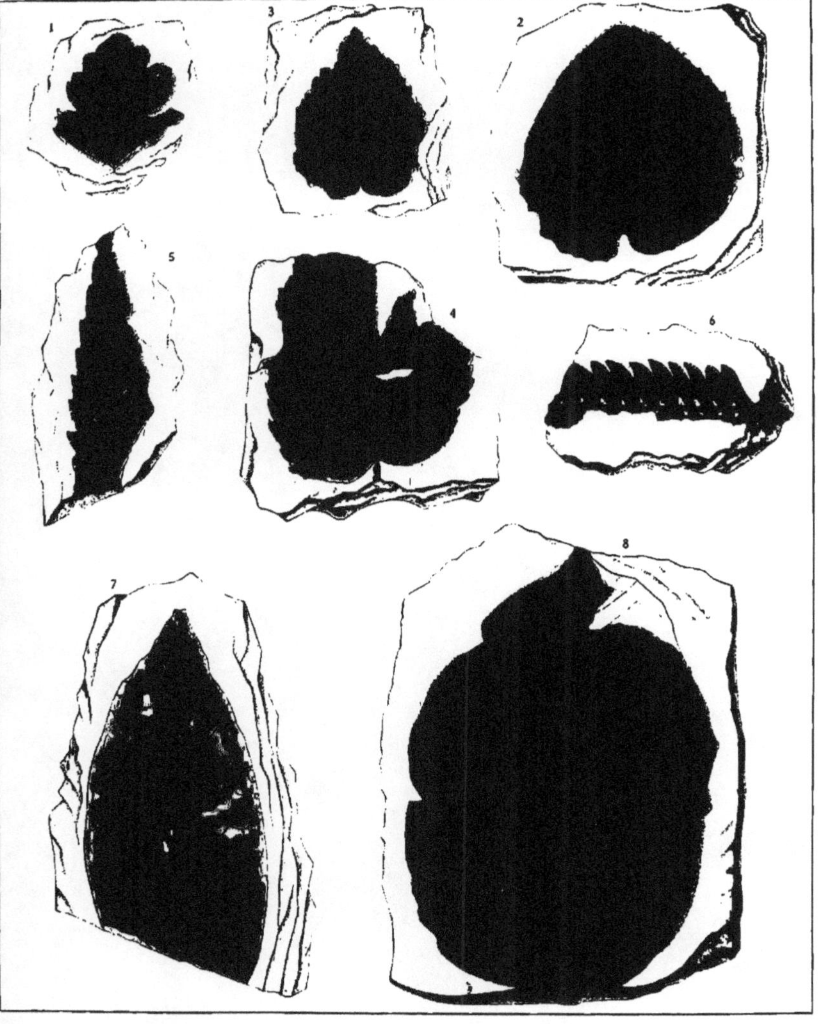
1
3
2
5
4
6
8
7

PLATE XLIX.

PLATE XLIX.

		Page.
Figs. 1-3.	Fraxinus integrifolia Newb	128
4, 7-9.	[fig. 9 misprinted fig. 1] Populus polymorpha Newb	50
5.	Fraxinus affinis Newb	127
6.	Fraxinus denticulata Heer?	128

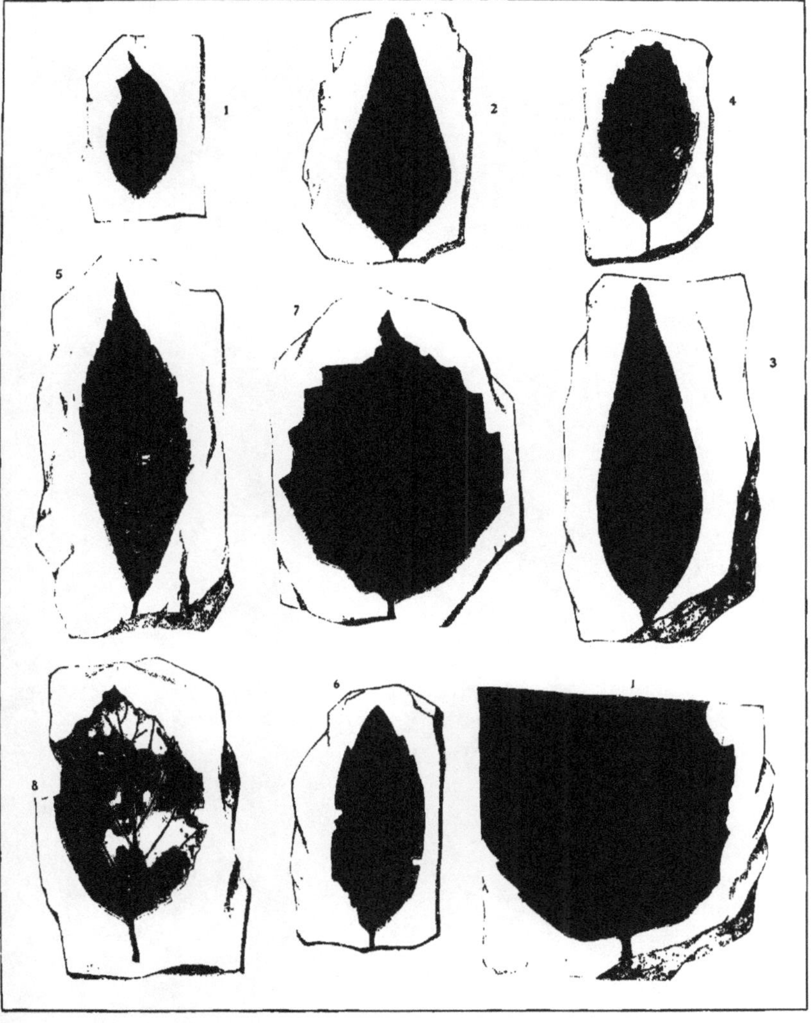
1
2
4
5
7
3
8
6
1

PLATE L.

PLATE L.

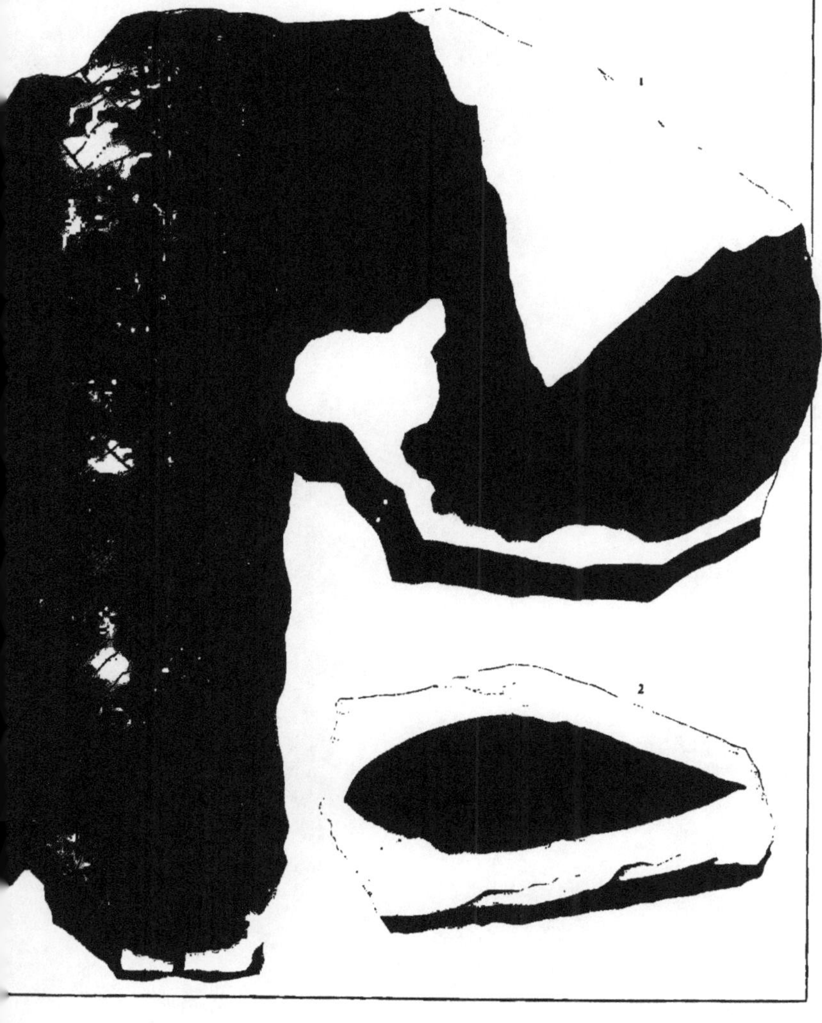
1
2

PLATE LI.

PLATE LI.

	Page.
Fig. 1. Ficus (?) Alaskana Newb	84
2, in part. Vitis rotundifolia Newb	120
2, in part. Juglans nigella Heer	33
3, in part. Quercus Grönlandica Heer	75
3, in part. Taxodium distichum miocenum Heer	22
4. Juglans nigella Heer	33

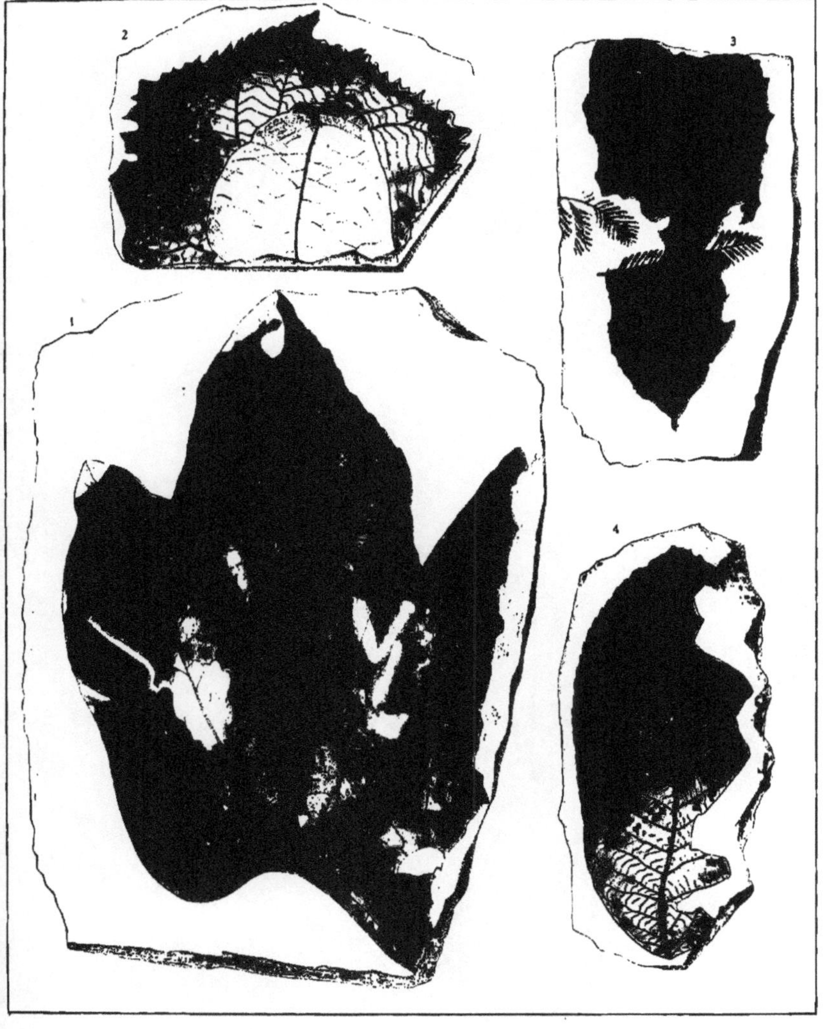
2
3
1
4

PLATE LII.

PLATE LII.

	Page.
Fig. 1. Ficus (?) Alaskana Newb	84
2, 3 and 4 in part. Taxodium distichum miocenum Heer	22
3 and 4 in part, 5. Prunus variabilis Newb	112

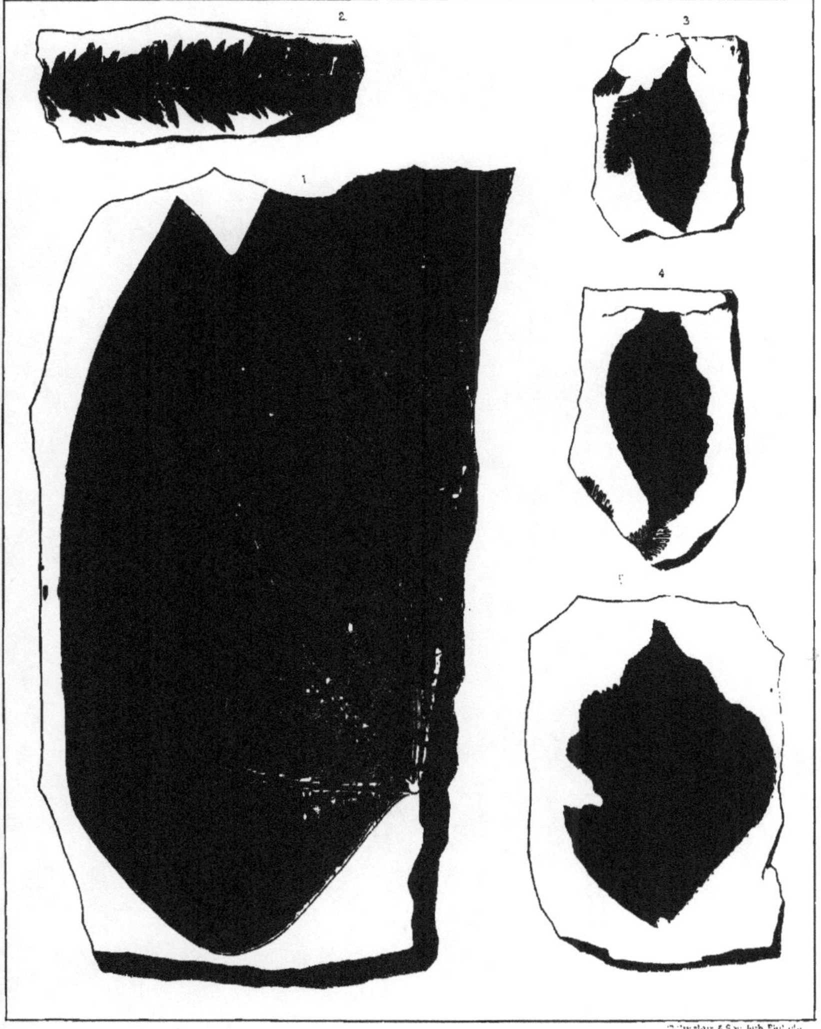

Thos. Sinclair & Son lith Phila.

PLATE LIII.

PLATE LIII.

	Page.
FIGS. 1, 2. Pterospermites dentatus Heer	133
3. Vitis rotundifolia Newb	120
4. Sequoia spinosa Newb	21
5. Fruit of Seqnoia spiuosa Newb	22
6. Celtis parvifolia Newb	84

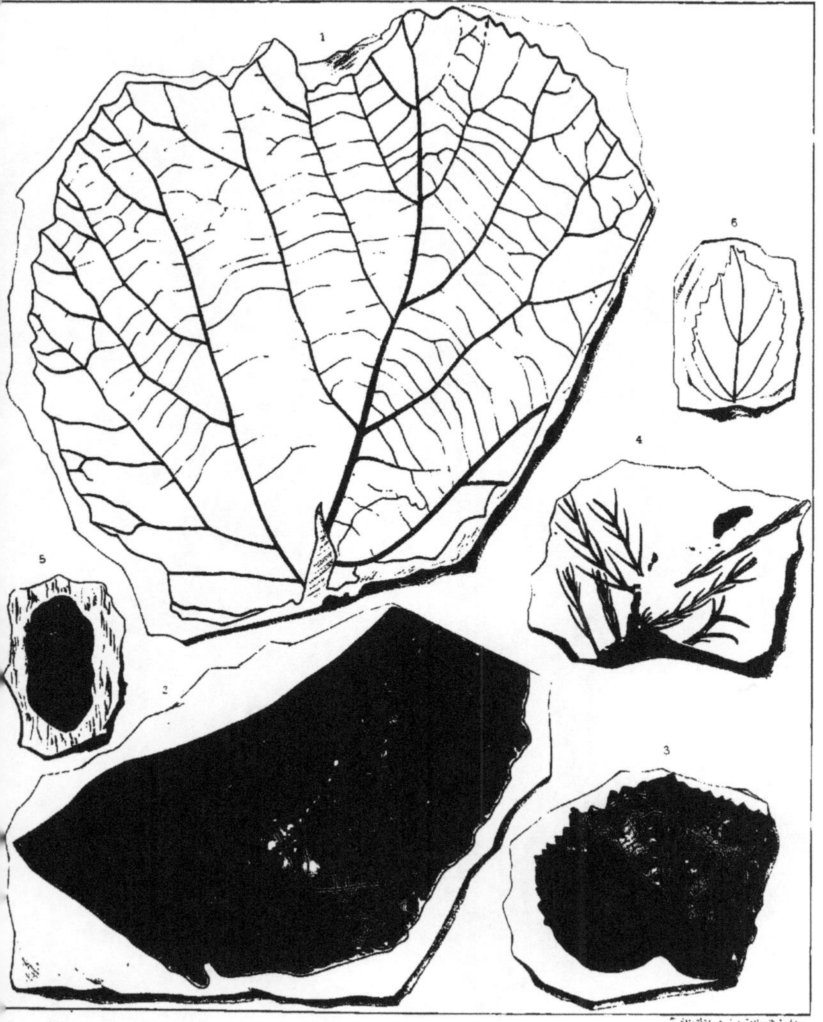
1
2
3
4
5
6

PLATE LIV.

PLATE LIV.

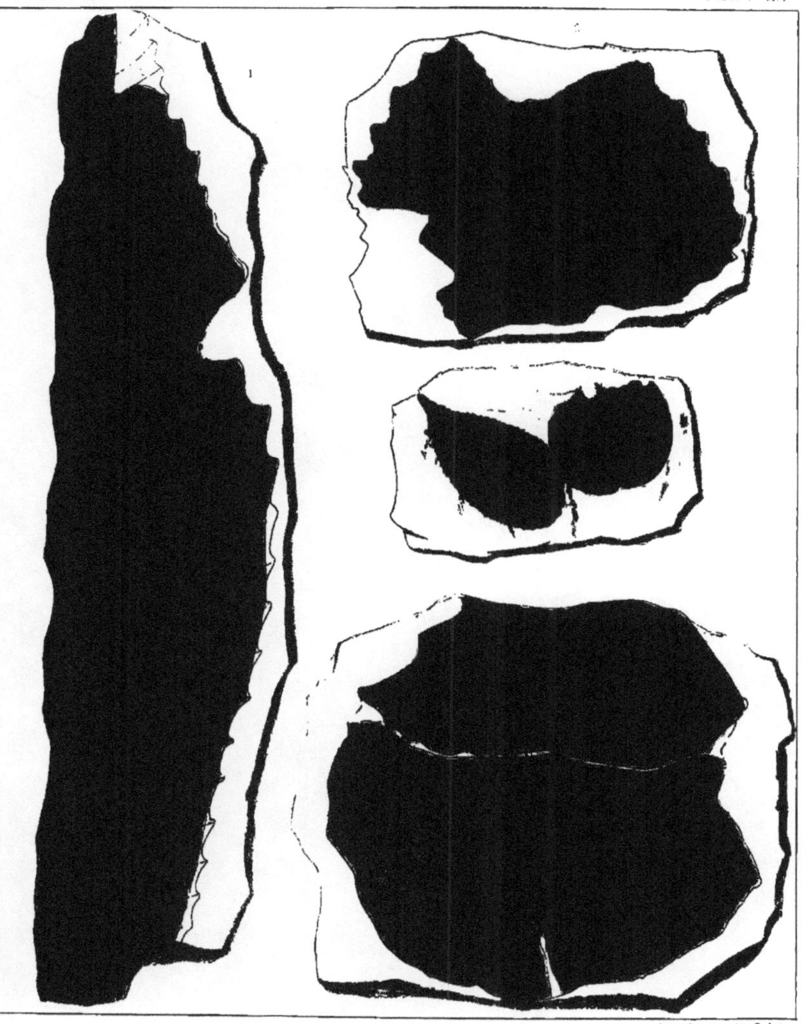

PLATE LV.

PLATE LV.

	Page.
FIGS. 1, 2. Ficus(?) Alaskana Newb	84
3, 4. Glyptostrobus Europæus (Brong.) Heer	24
5, in part. Taxodium occidentale Newb	23
5, in part. Taxodium distichum miocenum Heer	22
6. Carpinus grandis Ung	59

J T Sinclair & Son Lith Phila

PLATE LVI.

PLATE LVI.

	Page.
Fig. 1. Ficus (?) Condoni Newb	85
2. Berberis simplex Newb	97
3. Platanus Haydenii Newb	103
4. Quercus castanopsis Newb	71

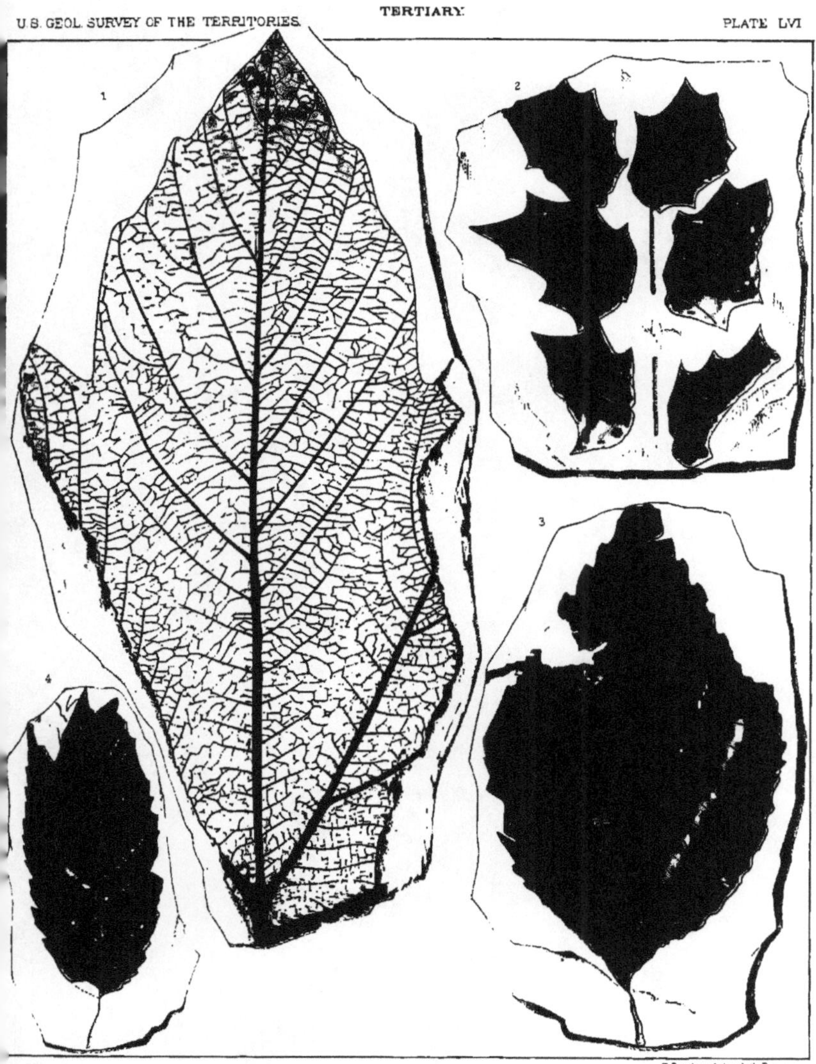

T Sinclair & Son Lith. Phila.

PLATE LVII.

PLATE LVII.

	Page.
Fig. 1. Ficus (?) Condoni Newb	85
2. Viburnum cuneatum Newb	130
3. Planera crenata Newb	81
4. Fruit of Betula sp. ? Newb	65

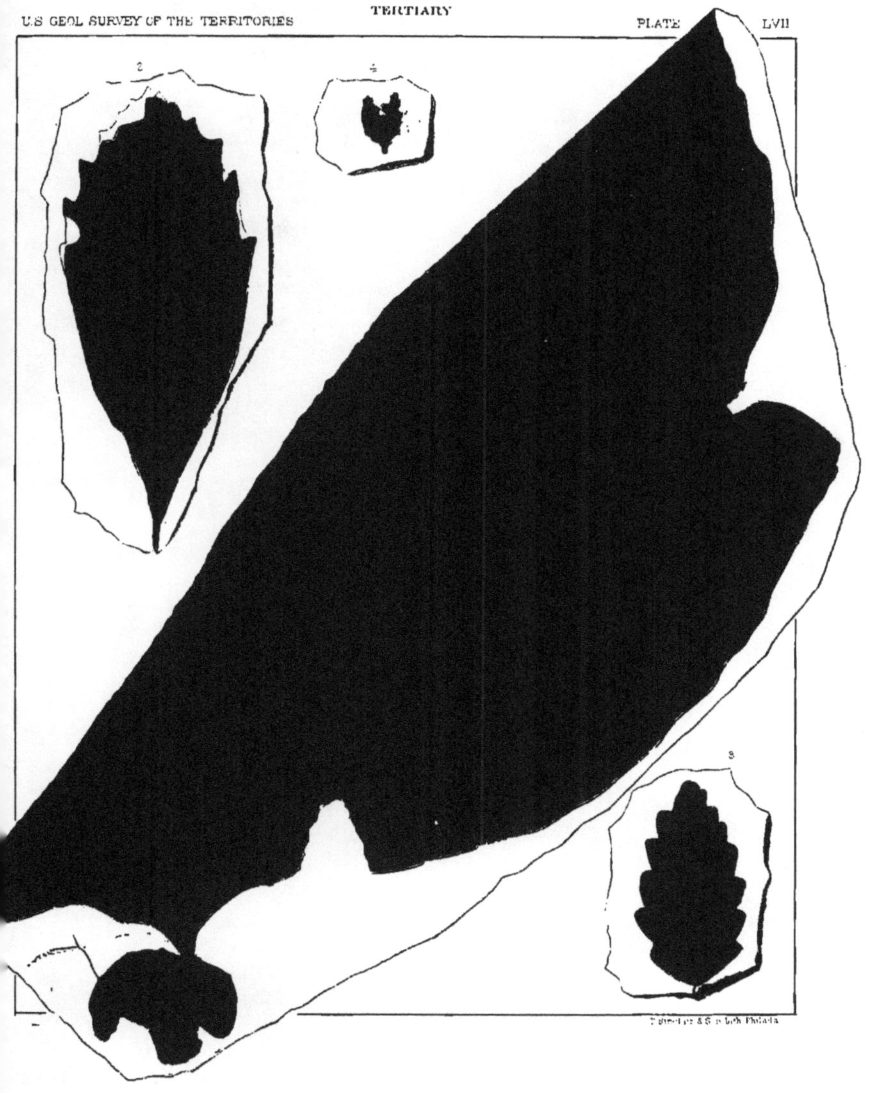

T. Sinclair & Son, lith. Philada.

PLATE LVIII.

PLATE LVIII.

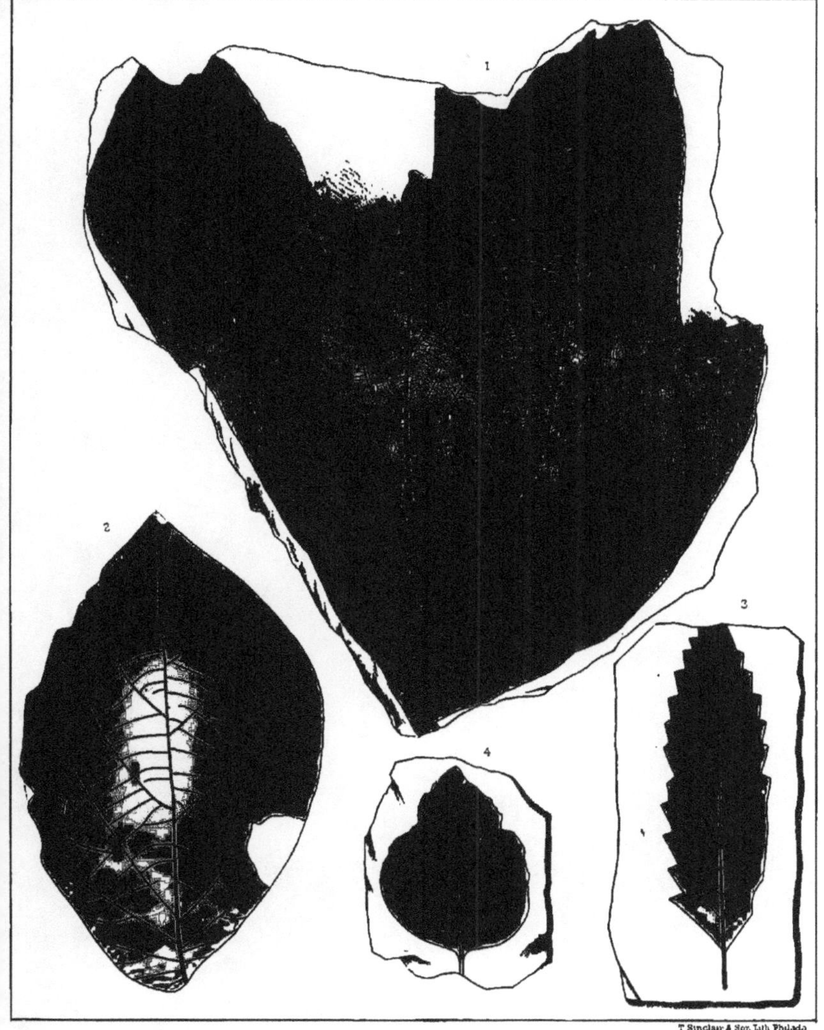

T Sinclair & Son, Lith Philada

PLATE LIX.

PLATE LIX.

	Page.
FIG. 1. Magnolia rotundifolia Newb	95
2. Ficus membranacea Newb	87
3. Platanus aspera Newb	102
4. Quercus laurifolia Newb	76

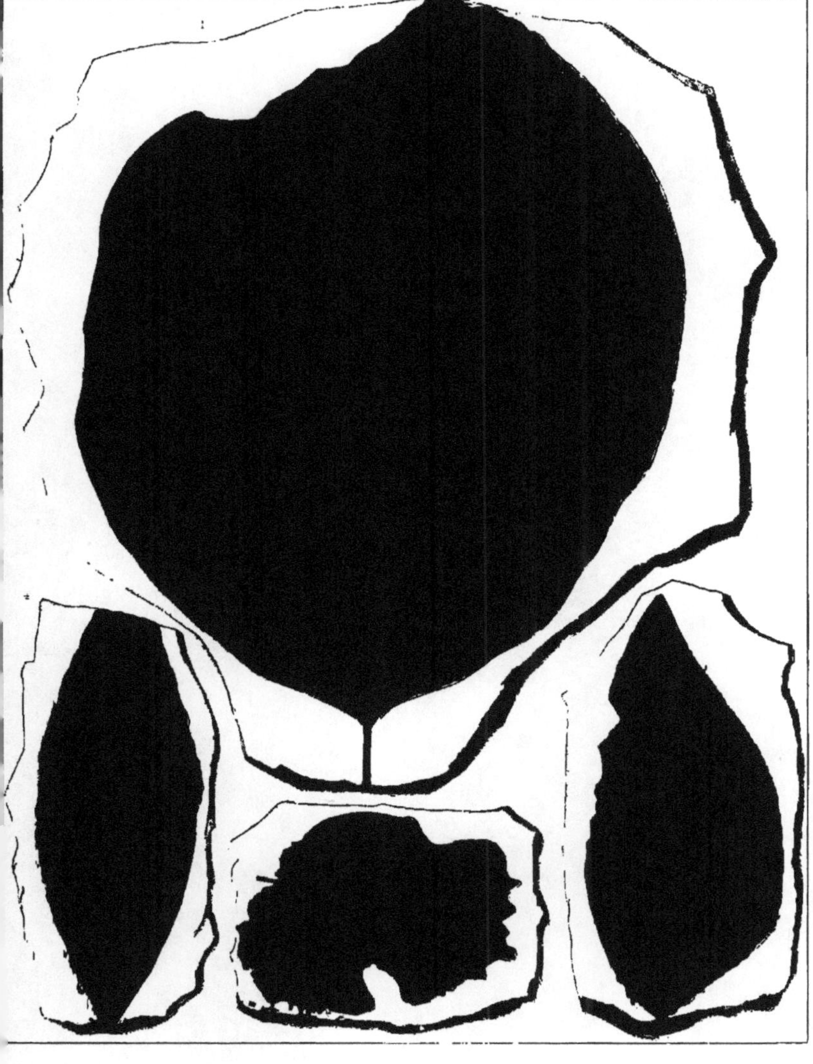

PLATE LX.

PLATE LX.

	Page.
Fig. 1. Protoficus inæqualis Newb	89
2. Quercus Sullyi Newb	79
3. Quercus laurifolia Newb	76
4. Aristolochia cordifolia Newb	90

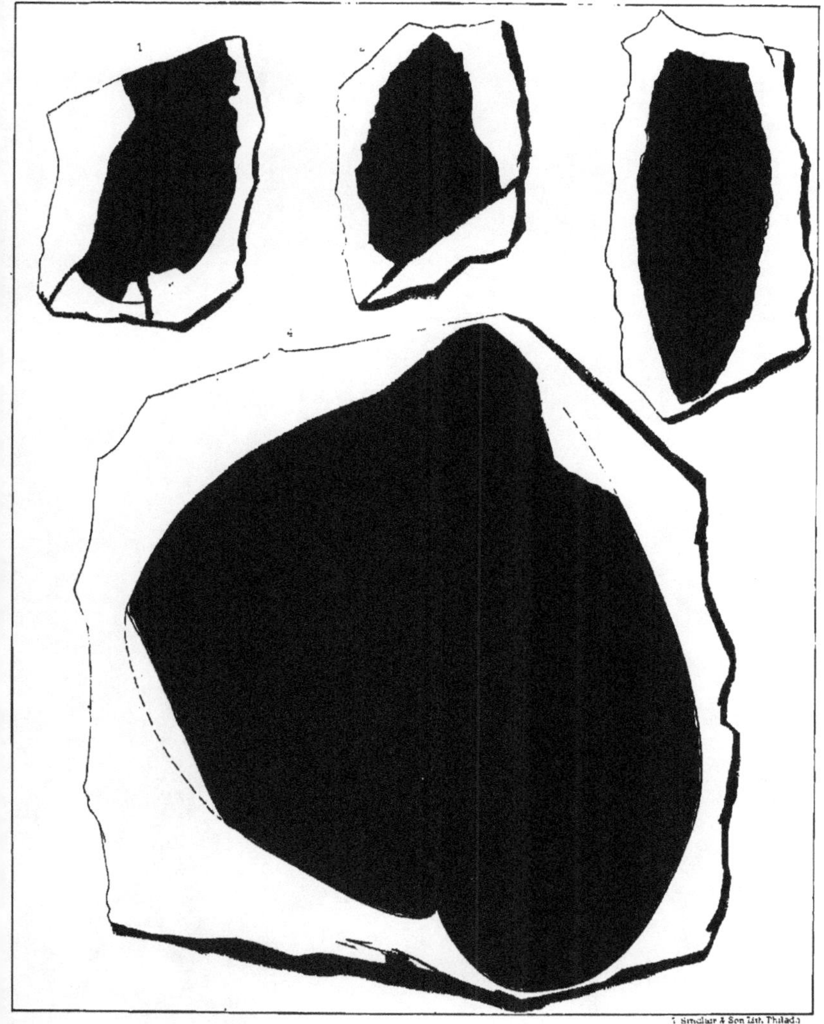

T. Sinclair & Son Lith. Philada

PLATE LXI.

PLATE LXI.

	Page.
FIGS. 1, 1a. Pteris Russellii Newb	7
2–5. Acrostichum hesperium Newb	6

T Sinclair & Son, Lith Phila

TERTIARY

PLATE LXII.

PLATE LXII.

		Page.
FIGS. 1–4.	Lygodium Kaulfussi Heer	1
5–6.	Pecopteris (Cheilanthes) sepulta Newb	12

T Sinclair & Son, Lith Phila

TERTIARY

PLATE LXIII.

PLATE LXIII.

		Page.
Figs. 1-4.	Anemia perplexa Hollick	3
5.	Sabal grandifolia Newb	28
6.	Sabal Powellii Newb	30

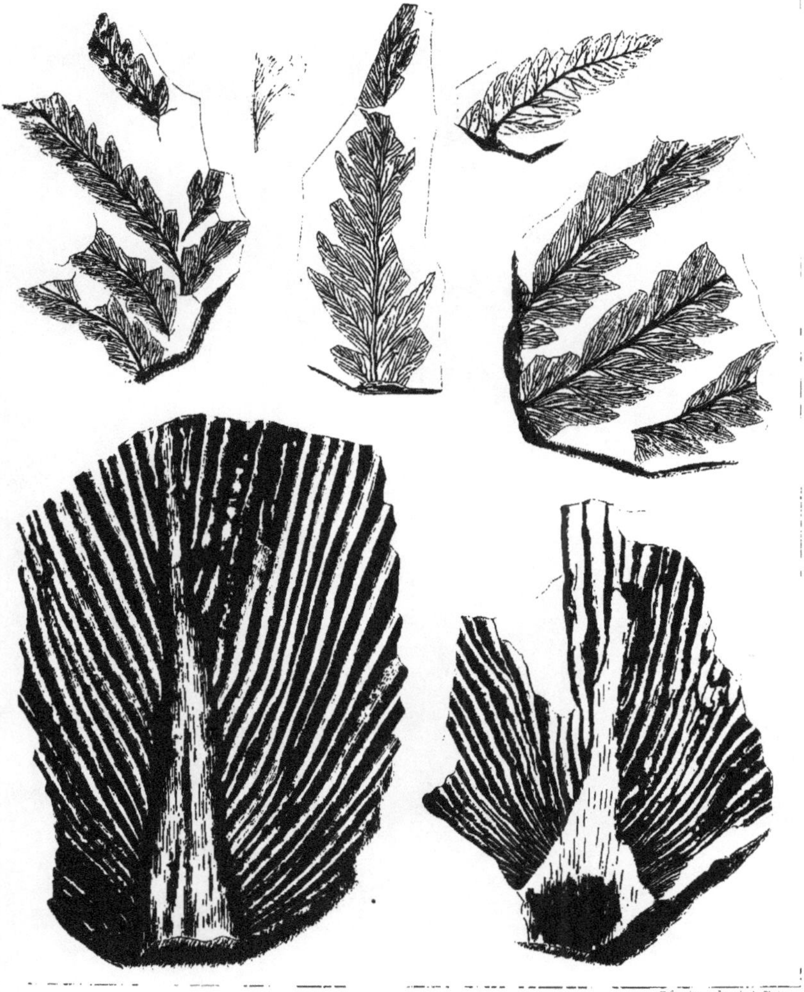

TERTIARY

PLATE LXIV.

PLATE LXIV.

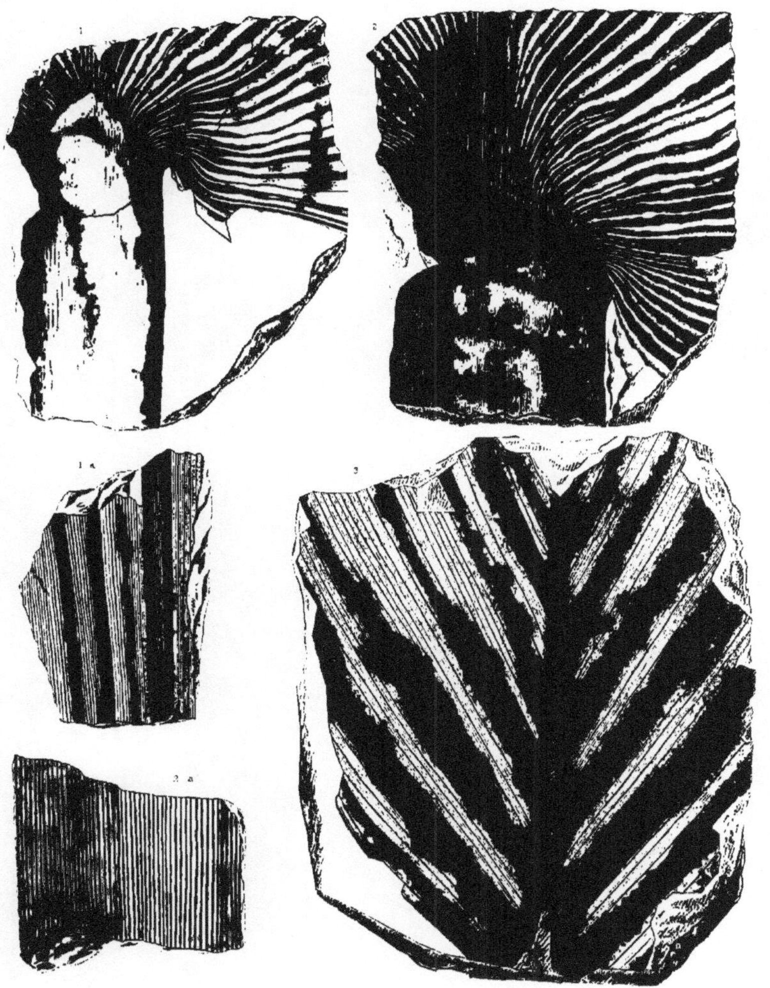

TERTIARY

PLATE LXV.

PLATE LXV.

		Page.
FIG. 1.	Juglans occidentalis Newb	34
2.	Salix angusta Al. Br.?	54
3–5.	Zizyphus longifolia Newb	119
6.	Quercus castanoides Newb	70
7.	Equisetum Oregonense Newb	14
8.	Equisetum Wyomingense Lesq	15

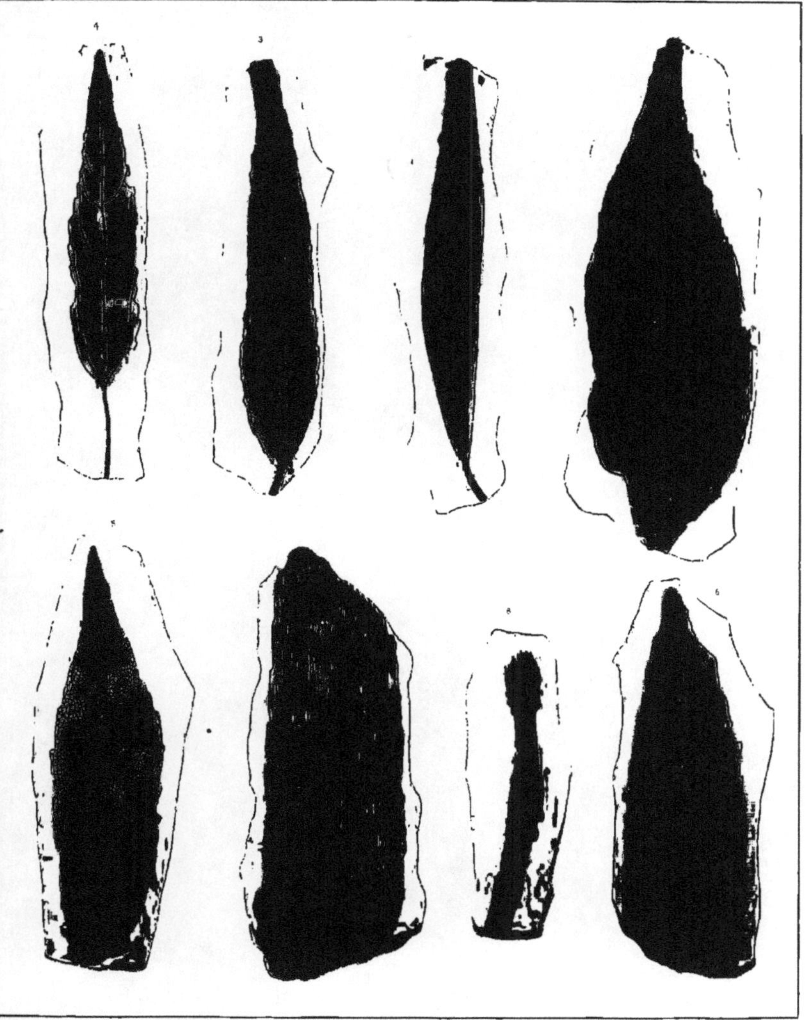

T. Sinclair & Son, Lith. Phila.

TERTIARY

PLATE LXVI.

PLATE LXVI.

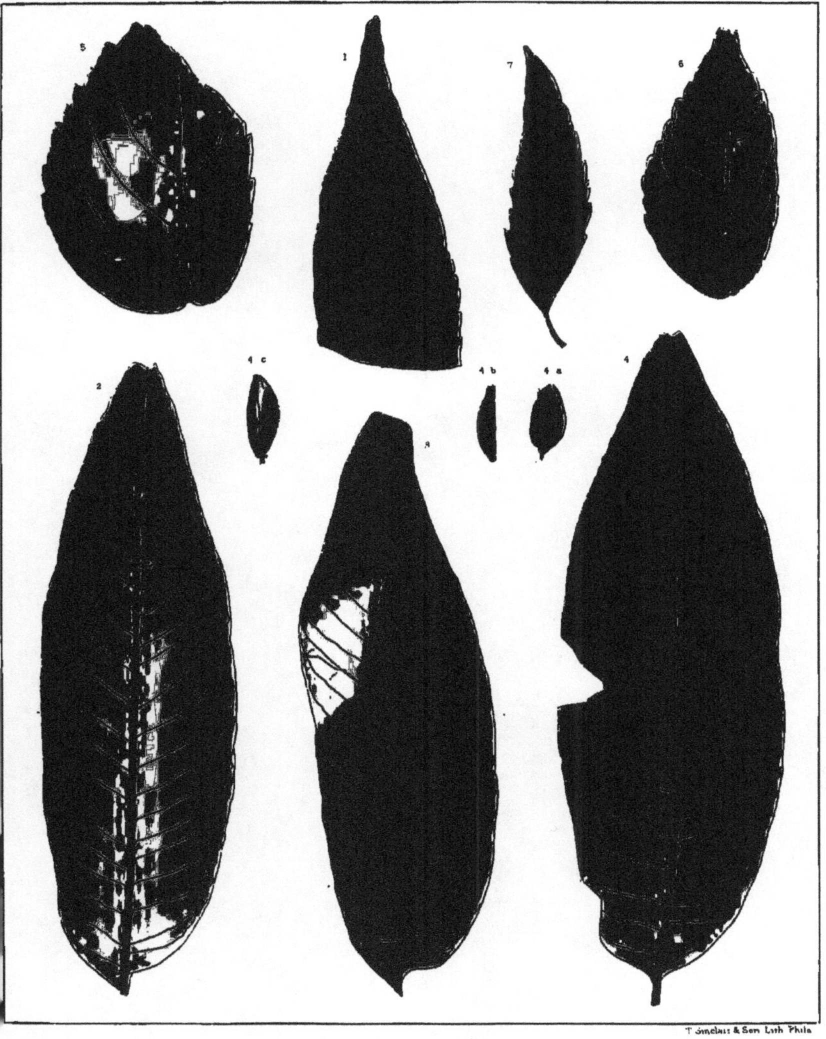

T. Sinclair & Son Lith Phila

TERTIARY

PLATE LXVII.

PLATE LXVII.

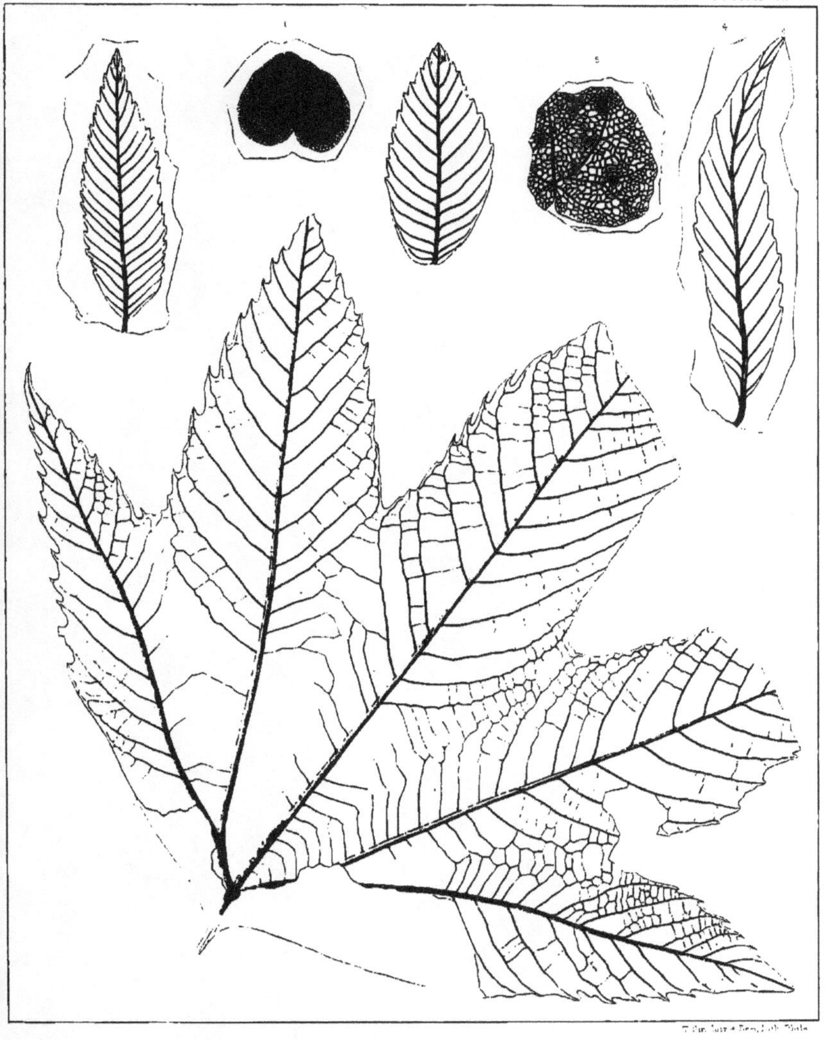

T Sinclair & Son, Lith. Phila.

TERTIARY

PLATE LXVIII.

PLATE LXVIII.

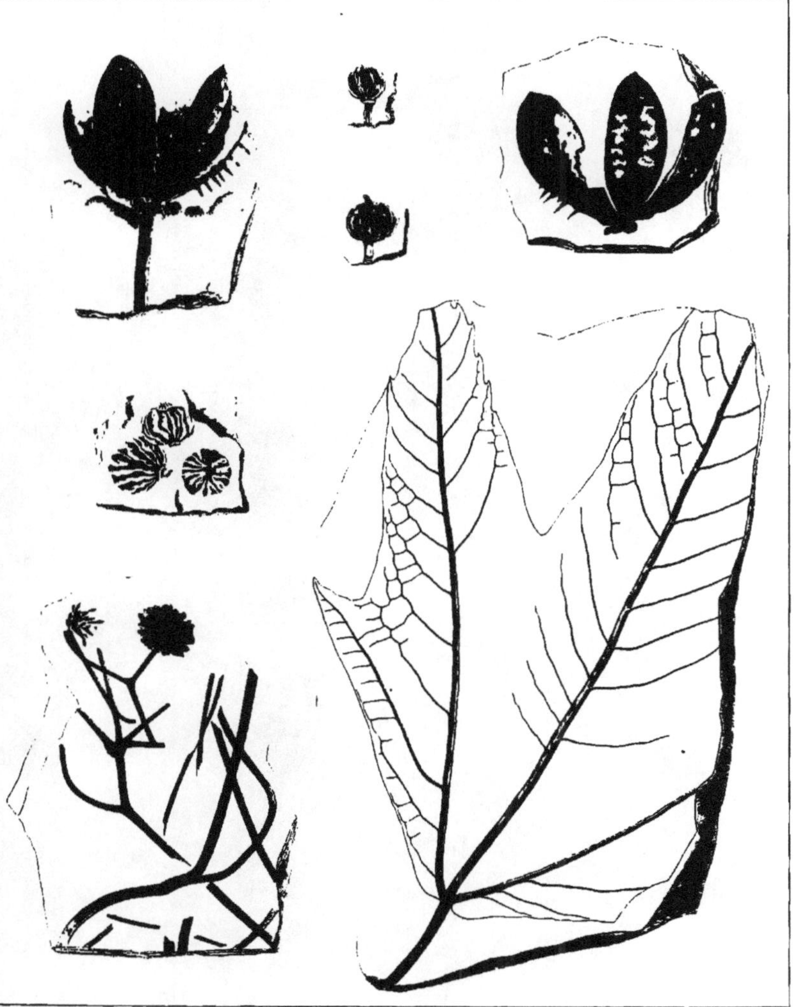

TERTIARY

INDEX.

[Genera and all divisions of higher rank are printed in SMALL CAPITALS; synonyms and names of species to which incidental reference is made, in *italics*. **Heavy-faced** figures refer to pages on which descriptions are given.]

ADVERTISEMENT.

[Monograph XXXV.]

The statute approved March 3, 1879, establishing the United States Geological Survey, contains the following provisions:

"The publications of the Geological Survey shall consist of the annual report of operations, geological and economic maps illustrating the resources and classification of the lands, and reports upon general and economic geology and paleontology. The annual report of operations of the Geological Survey shall accompany the annual report of the Secretary of the Interior. All special memoirs and reports of said Survey shall be issued in uniform quarto series if deemed necessary by the Director, but otherwise in ordinary octavos. Three thousand copies of each shall be published for scientific exchanges and for sale at the price of publication; and all literary and cartographic materials received in exchange shall be the property of the United States and form a part of the library of the organization: And the money resulting from the sale of such publications shall be covered into the Treasury of the United States."

Except in those cases in which an extra number of any special memoir or report has been supplied to the Survey by special resolution of Congress or has been ordered by the Secretary of the Interior, this office has no copies for gratuitous distribution.

ANNUAL REPORTS.

I. First Annual Report of the United States Geological Survey, by Clarence King. 1880. 8°. 79 pp. 1 map.—A preliminary report describing plan of organization and publications.

II. Second Annual Report of the United States Geological Survey, 1880–'81, by J. W. Powell. 1882. 8°. lv, 588 pp. 62 pl. 1 map.

III. Third Annual Report of the United States Geological Survey, 1881–'82, by J. W. Powell. 1883. 8°. xviii, 564 pp. 67 pl. and maps.

IV. Fourth Annual Report of the United States Geological Survey, 1882–'83, by J. W. Powell. 1884. 8°. xxxii, 473 pp. 85 pl. and maps.

V. Fifth Annual Report of the United States Geological Survey, 1883–'84, by J. W. Powell. 1885. 8°. xxxvi, 469 pp. 58 pl. and maps.

VI. Sixth Annual Report of the United States Geological Survey, 1884–'85, by J. W. Powell. 1885. 8°. xxix, 570 pp. 65 pl. and maps.

VII. Seventh Annual Report of the United States Geological Survey, 1885–'86, by J. W. Powell. 1888. 8°. xx, 656 pp. 71 pl. and maps.

VIII. Eighth Annual Report of the United States Geological Survey, 1886–'87, by J. W. Powell. 1889. 8°. 2 pt. xix, 474, xii pp., 53 pl. and maps; 1 prel. leaf, 475–1063 pp., 54–76 pl. and maps.

IX. Ninth Annual Report of the United States Geological Survey, 1887–'88, by J. W. Powell. 1889. 8°. xiii, 717 pp. 88 pl. and maps.

X. Tenth Annual Report of the United States Geological Survey, 1888–'89, by J. W. Powell. 1890. 8°. 2 pt. xv, 774 pp., 98 pl. and maps; viii, 123 pp.

XI. Eleventh Annual Report of the United States Geological Survey, 1889–'90, by J. W. Powell. 1891. 8°. 2 pt. xv, 757 pp., 66 pl. and maps; ix, 351 pp., 30 pl. and maps.

XII. Twelfth Annual Report of the United States Geological Survey, 1890–'91, by J. W. Powell. 1891. 8°. 2 pt., xiii, 675 pp., 53 pl. and maps; xviii, 576 pp., 146 pl. and maps.

XIII. Thirteenth Annual Report of the United States Geological Survey, 1891–'92, by J. W. Powell. 1893. 8°. 3 pt. vii, 240 pp., 2 maps; x, 372 pp., 105 pl. and maps; xi, 486 pp., 77 pl. and maps.

XIV. Fourteenth Annual Report of the United States Geological Survey, 1892–'93, by J. W. Powell. 1893. 8°. 2 pt. vi, 321 pp., 1 pl.; xx, 597 pp., 74 pl. and maps.

XV. Fifteenth Annual Report of the United States Geological Survey, 1893–'94, by J. W. Powell. 1895. 8°. xiv, 755 pp., 48 pl. and maps.

XVI. Sixteenth Annual Report of the United States Geological Survey, 1894–'95, Charles D. Walcott, Director. 1895. (Part I, 1896.) 8°. 4 pt. xxii, 910 pp., 117 pl. and maps; xix, 598 pp., 43 pl. and maps; xv, 646 pp., 23 pl.; xix, 735 pp., 6 pl.

XVII. Seventeenth Annual Report of the United States Geological Survey, 1895–'96, Charles D. Walcott, Director. 1896. 8°. 3 pt. in 4 vol. xxii, 1076 pp., 67 pl. and maps; xxv, 864 pp., 113 pl. and maps; xxiii, 542 pp., 8 pl. and maps; iii, 543–1058 pp., 9–13 pl.

XVIII. Eighteenth Annual Report of the United States Geological Survey, 1896–'97, Charles D. Walcott, Director 1897. (Parts II and III, 1898.) 8°. 5 pt. in 6 vol. 1–440 pp., 4 pl. and maps; i–v,

29. On the Fresh-Water Invertebrates of the North American Jurassic, by Charles A. White. 1886. 8°. 41 pp. 4 pl. Price 5 cents.
30. Second Contribution to the Studies on the Cambrian Faunas of North America, by Charles Doolittle Walcott. 1886. 8°. 369 pp. 33 pl. Price 25 cents.
31. Systematic Review of our Present Knowledge of Fossil Insects, including Myriapods and Arachnids, by Samuel Hubbard Scudder. 1886. 8°. 128 pp. Price 15 cents.
32. Lists and Analyses of the Mineral Springs of the United States; a Preliminary Study, by Albert C. Peale. 1886. 8°. 235 pp. Price 20 cents.
33. Notes on the Geology of Northern California, by J. S. Diller. 1886. 8°. 23 pp. Price 5 cents.
34. On the Relation of the Laramie Molluscan Fauna to that of the Succeeding Fresh-Water Eocene and Other Groups, by Charles A. White. 1886. 8°. 54 pp. 5 pl. Price 10 cents.
35. Physical Properties of the Iron-Carburets, by Carl Barus and Vincent Strouhal. 1886. 8°. 62 pp. Price 10 cents.
36. Subsidence of Fine Solid Particles in Liquids, by Carl Barus. 1886. 8°. 58 pp. Price 10 cents.
37. Types of the Laramie Flora, by Lester F. Ward. 1887. 8°. 354 pp. 57 pl. Price 25 cents.
38. Peridotite of Elliott County, Kentucky, by J. S. Diller. 1887. 8°. 31 pp. 1 pl. Price 5 cents.
39. The Upper Beaches and Deltas of the Glacial Lake Agassiz, by Warren Upham. 1887. 8°. 84 pp. 1 pl. Price 10 cents.
40. Changes in River Courses in Washington Territory due to Glaciation, by Bailey Willis. 1887. 8°. 10 pp. 4 pl. Price 5 cents.
41. On the Fossil Faunas of the Upper Devonian—the Genesee Section, New York, by Henry S. Williams. 1887. 8°. 121 pp. 4 pl. Price 15 cents.
42. Report of Work done in the Division of Chemistry and Physics, mainly during the Fiscal Year 1885-'86. F. W. Clarke, Chief Chemist. 1887. 8°. 152 pp. 1 pl. Price 15 cents.
43. Tertiary and Cretaceous Strata of the Tuscaloosa, Tombigbee, and Alabama Rivers, by Eugene A. Smith and Lawrence C. Johnson. 1887. 8°. 189 pp. 21 pl. Price 15 cents.
44. Bibliography of North American Geology for 1886, by Nelson H. Darton. 1887. 8°. 35 pp. Price 5 cents.
45. The Present Condition of Knowledge of the Geology of Texas, by Robert T. Hill. 1887. 8°. 94 pp. Price 10 cents.
46. Nature and Origin of Deposits of Phosphate of Lime, by R. A. F. Penrose, jr., with an Introduction by N. S. Shaler. 1888. 8°. 143 pp. Price 15 cents.
47. Analyses of Waters of the Yellowstone National Park, with an Account of the Methods of Analysis employed, by Frank Austin Gooch and James Edward Whitfield. 1888. 8°. 84 pp. Price 10 cents.
48. On the Form and Position of the Sea Level, by Robert Simpson Woodward. 1888. 8°. 88 pp. Price 10 cents.
49. Latitudes and Longitudes of Certain Points in Missouri, Kansas, and New Mexico, by Robert Simpson Woodward. 1889. 8°. 133 pp. Price 15 cents.
50. Formulas and Tables to Facilitate the Construction and Use of Maps, by Robert Simpson Woodward. 1889. 8°. 124 pp. Price 15 cents.
51. On Invertebrate Fossils from the Pacific Coast, by Charles Abiathar White. 1889. 8°. 102 pp. 14 pl. Price 15 cents.
52. Subaërial Decay of Rocks and Origin of the Red Color of Certain Formations, by Israel Cook Russell. 1889. 8°. 65 pp. 5 pl. Price 10 cents.
53. The Geology of Nantucket, by Nathaniel Southgate Shaler. 1889. 8°. 55 pp. 10 pl. Price 10 cents.
54. On the Thermo-Electric Measurement of High Temperatures, by Carl Barus. 1889. 8°. 313 pp., incl. 1 pl. 11 pl. Price 25 cents.
55. Report of Work done in the Division of Chemistry and Physics, mainly during the Fiscal Year 1886-'87. Frank Wigglesworth Clarke, Chief Chemist. 1889. 8°. 96 pp. Price 10 cents.
56. Fossil Wood and Lignite of the Potomac Formation, by Frank Hall Knowlton. 1889. 8°. 72 pp. 7 pl. Price 10 cents.
57. A Geological Reconnoissance in Southwestern Kansas, by Robert Hay. 1890. 8°. 49 pp. 2 pl. Price 5 cents.
58. The Glacial Boundary in Western Pennsylvania, Ohio, Kentucky, Indiana, and Illinois, by George Frederick Wright, with an Introduction by Thomas Chrowder Chamberlin. 1890. 8°. 112 pp., incl. 1 pl. 8 pl. Price 15 cents.
59. The Gabbros and Associated Rocks in Delaware, by Frederick D. Chester. 1890. 8°. 45 pp. 1 pl. Price 10 cents.
60. Report of Work done in the Division of Chemistry and Physics, mainly during the Fiscal Year 1887-'88. F. W. Clarke, Chief Chemist. 1890. 8°. 174 pp. Price 15 cents.
61. Contributions to the Mineralogy of the Pacific Coast, by William Harlow Melville and Waldemar Lindgren. 1890. 8°. 40 pp. 3 pl. Price 5 cents.
62. The Greenstone Schist Areas of the Menominee and Marquette Regions of Michigan, a Contribution to the Subject of Dynamic Metamorphism in Eruptive Rocks, by George Huntington Williams, with an Introduction by Roland Duer Irving. 1890. 8°. 241 pp. 16 pl. Price 30 cents.
63. A Bibliography of Paleozoic Crustacea from 1698 to 1889, including a List of North American Species and a Systematic Arrangement of Genera, by Anthony W. Vogdes. 1890. 8°. 177 pp. Price 15 cents.
64. A Report of Work done in the Division of Chemistry and Physics, mainly during the Fiscal Year 1888-'89. F. W. Clarke, Chief Chemist. 1890. 8°. 60 pp. Price 10 cents.

65. Stratigraphy of the Bituminous Coal Field of Pennsylvania, Ohio, and West Virginia, by Israel C. White. 1891. 8°. 212 pp. 11 pl. Price 20 cents.

66. On a Group of Volcanic Rocks from the Tewan Mountains, New Mexico, and on the Occurrence of Primary Quartz in Certain Basalts, by Joseph Paxson Iddings. 1890. 8°. 34 pp. Price 5 cents.

67. The Relations of the Traps of the Newark System in the New Jersey Region, by Nelson Horatio Darton. 1890. 8°. 82 pp. Price 10 cents.

68. Earthquakes in California in 1889, by James Edward Keeler. 1890. 8°. 25 pp. Price 5 cents.

69. A Classed and Annotated Biography of Fossil Insects, by Samuel Howard Scudder. 1890. 8°. 101 pp. Price 15 cents.

70. A Report on Astronomical Work of 1889 and 1890, by Robert Simpson Woodward. 1890. 8°. 79 pp. Price 10 cents.

71. Index to the Known Fossil Insects of the World, including Myriapods and Arachnids, by Samuel Hubbard Scudder. 1891. 8°. 744 pp. Price 50 cents.

72. Altitudes between Lake Superior and the Rocky Mountains, by Warren Upham. 1891. 8°. 229 pp. Price 20 cents.

73. The Viscosity of Solids, by Carl Barus. 1891. 8°. xii, 139 pp. 6 pl. Price 15 cents.

74. The Minerals of North Carolina, by Frederick Augustus Genth. 1891. 8°. 119 pp. Price 15 cents.

75. Record of North American Geology for 1887 to 1889, inclusive, by Nelson Horatio Darton. 1891. 8°. 173 pp. Price 15 cents.

76. A Dictionary of Altitudes in the United States (Second Edition), compiled by Henry Gannett, Chief Topographer. 1891. 8°. 393 pp. Price 25 cents.

77. The Texan Permian and its Mesozoic Types of Fossils, by Charles A. White. 1891. 8°. 51 pp. 4 pl. Price 10 cents.

78. A Report of Work done in the Division of Chemistry and Physics, mainly during the Fiscal Year 1889-'90. F. W. Clarke, Chief Chemist. 1891. 8°. 131 pp. Price 15 cents.

79. A Late Volcanic Eruption in Northern California and its Peculiar Lava, by J. S. Diller.

80. Correlation Papers—Devonian and Carboniferous, by Henry Shaler Williams. 1891. 8°. 279 pp. Price 20 cents.

81. Correlation Papers—Cambrian, by Charles Doolittle Walcott. 1891. 8°. 547 pp. 3 pl. Price 25 cents.

82. Correlation Papers—Cretaceous, by Charles A. White. 1891. 8°. 273 pp. 3 pl. Price 20 cents.

83. Correlation Papers—Eocene, by William Bullock Clark. 1891. 8°. 173 pp. 2 pl. Price 15 cents.

84. Correlation Papers—Neocene, by W. H. Dall and G. D. Harris. 1892. 8°. 349 pp. 3 pl. Price 25 cents.

85. Correlation Papers—The Newark System, by Israel Cook Russell. 1892. 8°. 344 pp. 13 pl. Price 25 cents.

86. Correlation Papers—Archean and Algonkian, by C. R. Van Hise. 1892. 8°. 549 pp. 12 pl. Price 25 cents.

87. A Synopsis of American Fossil. Brachiopoda, including Bibliography and Synonymy, by Charles Schuchert. 1897. 8°. 464 pp. Price 30 cents.

88. The Cretaceous Foraminifera of New Jersey, by Rufus Mather Bagg, Jr. 1898. 8°. 89 pp. 6 pl. Price 10 cents.

89. Some Lava Flows of the Western Slope of the Sierra Nevada, California, by F. Leslie Ransome. 1898. 8°. 74 pp. 11 pl. Price 15 cents.

90. A Report of Work done in the Division of Chemistry and Physics, mainly during the Fiscal Year 1890-'91. F. W. Clarke, Chief Chemist. 1892. 8°. 77 pp. Price 10 cents.

91. Record of North American Geology for 1890, by Nelson Horatio Darton. 1891. 8°. 88 pp. Price 10 cents.

92. The Compressibility of Liquids, by Carl Barus. 1892. 8°. 96 pp. 29 pl. Price 10 cents.

93. Some Insects of Special Interest from Florissant, Colorado, and Other Points in the Tertiaries of Colorado and Utah, by Samuel Hubbard Scudder. 1892. 8°. 35 pp. 3 pl. Price 5 cents.

94. The Mechanism of Solid Viscosity, by Carl Barus. 1892. 8°. 138 pp. Price 15 cents.

95. Earthquakes in California in 1890 and 1891, by Edward Singleton Holden. 1892. 8°. 31 pp. Price 5 cents.

96. The Volume Thermodynamics of Liquids, by Carl Barus. 1892. 8°. 100 pp. Price 10 cents.

97. The Mesozoic Echinodermata of the United States, by W. B. Clark. 1893. 8°. 207 pp. 50 pl. Price 20 cents.

98. Flora of the Outlying Carboniferous Basins of Southwestern Missouri, by David White. 1893. 8°. 139 pp. 5 pl. Price 15 cents.

99. Record of North American Geology for 1891, by Nelson Horatio Darton. 1892. 8°. 73 pp. Price 10 cents.

100. Bibliography and Index of the Publications of the U. S. Geological Survey, 1879-1892, by Philip Creveling Warman. 1893. 8°. 495 pp. Price 25 cents.

101. Insect Fauna of the Rhode Island Coal Field, by Samuel Hubbard Scudder. 1893. 8°. 27 pp. 2 pl. Price 5 cents.

102. A Catalogue and Bibliography of North American Mesozoic Invertebrata, by Cornelius Breckinridge Boyle. 1892. 8°. 315 pp. Price 25 cents.

103. High Temperature Work in Igneous Fusion and Ebullition, chiefly in Relation to Pressure, by Carl Barus. 1893. 8°. 57 pp. 9 pl. Price 10 cents.

104. Glaciation of the Yellowstone Valley north of the Park, by Walter Harvey Weed. 1893. 8°. 41 pp. 4 pl. Price 5 cents.

105. The Laramie and the Overlying Livingstone Formation in Montana, by Walter Harvey Weed, with Report on Flora, by Frank Hall Knowlton. 1893. 8°. 68 pp. 6 pl. Price 10 cents.

106. The Colorado Formation and its Invertebrate Fauna, by T. W. Stanton. 1893. 8°. 288 pp. 45 pl. Price 20 cents.

107. The Trap Dikes of the Lake Champlain Region, by James Furman Kemp and Vernon Freeman Marsters. 1893. 8°. 62 pp. 4 pl. Price 10 cents.

108. A Geological Reconnoissance in Central Washington, by Israel Cook Russell. 1893. 8°. 108 pp. 12 pl. Price 15 cents.

109. The Eruptive and Sedimentary Rocks on Pigeon Point, Minnesota, and their Contact Phenomena, by William Shirley Bayley. 1893. 8°. 121 pp. 16 pl. Price 15 cents.

110. The Paleozoic Section in the Vicinity of Three Forks, Montana, by Albert Charles Peale. 1893. 8°. 56 pp. 6 pl. Price 10 cents.

111. Geology of the Big Stone Gap Coal Fields of Virginia and Kentucky, by Marius R. Campbell. 1893. 8°. 106 pp. 6 pl. Price 15 cents.

112. Earthquakes in California in 1892, by Charles D. Perrine. 1893. 8°. 57 pp. Price 10 cents.

113. A Report of Work done in the Division of Chemistry during the Fiscal Years 1891-'92 and 1892-'93. F. W. Clarke, Chief Chemist. 1893. 8°. 115 pp. Price 15 cents.

114. Earthquakes in California in 1893, by Charles D. Perrine. 1894. 8°. 23 pp. Price 5 cents.

115. A Geographic Dictionary of Rhode Island, by Henry Gannett. 1894. 8°. 31 pp. Price 5 cents.

116. A Geographic Dictionary of Massachusetts, by Henry Gannett. 1894. 8°. 126 pp. Price 15 cents.

117. A Geographic Dictionary of Connecticut, by Henry Gannett. 1894. 8°. 67 pp. Price 10 cents.

118. A Geographic Dictionary of New Jersey, by Henry Gannett. 1894. 8°. 131 pp. Price 15 cents.

119. A Geological Reconnoissance in Northwest Wyoming, by George Homans Eldridge. 1894. 8°. 72 pp. Price 10 cents.

120. The Devonian System of Eastern Pennsylvania and New York, by Charles S. Prosser. 1894. 8°. 81 pp. 2 pl. Price 10 cents.

121. A Bibliography of North American Paleontology, by Charles Rollin Keyes. 1894. 8°. 251 pp. Price 20 cents.

122. Results of Primary Triangulation, by Henry Gannett. 1894. 8°. 412 pp. 17 pl. Price 25 cents.

123. A Dictionary of Geographic Positions, by Henry Gannett. 1895. 8°. 183 pp. 1 pl. Price 15 cents.

124. Revision of North American Fossil Cockroaches, by Samuel Hubbard Scudder. 1895. 8°. 176 pp. 12 pl. Price 15 cents.

125. The Constitution of the Silicates, by Frank Wigglesworth Clarke. 1895. 8°. 109 pp. Price 15 cents.

126. A Mineralogical Lexicon of Franklin, Hampshire, and Hampden counties, Massachusetts, by Benjamin Kendall Emerson. 1895. 8°. 180 pp. 1 pl. Price 15 cents.

127. Catalogue and Index of Contributions to North American Geology, 1732-1891, by Nelson Horatio Darton. 1896. 8°. 1045 pp. Price 60 cents.

128. The Bear River Formation and its Characteristic Fauna, by Charles A. White. 1895. 8°. 108 pp. 11 pl. Price 15 cents.

129. Earthquakes in California in 1894, by Charles D. Perrine. 1895. 8°. 25 pp. Price 5 cents.

130. Bibliography and Index of North American Geology, Paleontology, Petrology, and Mineralogy for 1892 and 1893, by Fred Boughton Weeks. 1896. 8°. 210 pp. Price 20 cents.

131. Report of Progress of the Division of Hydrography for the Calendar Years 1893 and 1894, by Frederick Haynes Newell, Topographer in Charge. 1895. 8°. 126 pp. Price 15 cents.

132. The Disseminated Lead Ores of Southeastern Missouri, by Arthur Winslow. 1896. 8°. 31 pp. Price 5 cents.

133. Contributions to the Cretaceous Paleontology of the Pacific Coast: The Fauna of the Knoxville Beds, by T. W. Stanton. 1895. 8°. 132 pp. 20 pl. Price 15 cents.

134. The Cambrian Rocks of Pennsylvania, by Charles Doolittle Walcott. 1896. 8°. 43 pp. 15 pl. Price 5 cents.

135. Bibliography and Index of North American Geology, Paleontology, Petrology, and Mineralogy for the Year 1894, by F. B. Weeks. 1896. 8°. 141 pp. Price 15 cents.

136. Volcanic Rocks of South Mountain, Pennsylvania, by Florence Bascom. 1896. 8°. 124 pp. 28 pl. Price 15 cents.

137. The Geology of the Fort Riley Military Reservation and Vicinity, Kansas, by Robert Hay. 1896. 8°. 35 pp. 8 pl. Price 5 cents.

138. Artesian-Well Prospects in the Atlantic Coastal Plain Region, by N. H. Darton. 1896. 8°. 228 pp. 19 pl. Price 20 cents.

139. Geology of the Castle Mountain Mining District, Montana, by W. H. Weed and L. V. Pirsson. 1896. 8°. 164 pp. 17 pl. Price 15 cents.

140. Report of Progress of the Division of Hydrography for the Calendar Year 1895, by Frederick Haynes Newell, Hydrographer in Charge. 1896. 8°. 356 pp. Price 25 cents.
141. The Eocene Deposits of the Middle Atlantic Slope in Delaware, Maryland, and Virginia, by William Bullock Clark. 1896. 8°. 167 pp. 40 pl. Price 15 cents.
142. A Brief Contribution to the Geology and Paleontology of Northwestern Louisiana, by T. Wayland Vaughan. 1896. 8°. 65 pp. 4 pl. Price 10 cents.
143. A Bibliography of Clays and the Ceramic Arts, by John C. Branner. 1896. 8°. 114 pp. Price 15 cents.
144. The Moraines of the Missouri Coteau and their Attendant Deposits, by James Edward Todd. 1896. 8°. 71 pp. 21 pl. Price 10 cents.
145. The Potomac Formation in Virginia, by W. M. Fontaine. 1896. 8°. 149 pp. 2 pl. Price 15 cents.
146. Bibliography and Index of North American Geology, Paleontology, Petrology, and Mineralogy for the Year 1895, by F. B. Weeks. 1896. 8°. 130 pp. Price 15 cents.
147. Earthquakes in California in 1895, by Charles D. Perrine, Assistant Astronomer in Charge of Earthquake Observations at the Lick Observatory. 1896. 8°. 23 pp. Price 5 cents.
148. Analyses of Rocks, with a Chapter on Analytical Methods, Laboratory of the United States Geological Survey, 1880 to 1896, by F. W. Clarke and W. F. Hillebrand. 1897. 8°. 306 pp. Price 20 cents.
149. Bibliography and Index of North American Geology, Paleontology, Petrology, and Mineralogy for the Year 1896, by Fred Boughton Weeks. 1897. 8°. 152 pp. Price 15 cents.
150. The Educational Series of Rock Specimens Collected and Distributed by the United States Geological Survey, by Joseph Silas Diller. 1898. 8°. 398 pp. 47 pl. Price 25 cents.
151. The Lower Cretaceous Gryphæas of the Texas Region, by R. T. Hill and T. Wayland Vaughan. 1898. 8°. 139 pp. 25 pl. Price 15 cents.
152. A Catalogue of the Cretaceous and Tertiary Plants of North America, by F. H. Knowlton. 1898. 8°. 247 pp. Price 20 cents.
153. A Bibliographic Index of North American Carboniferous Invertebrates, by Stuart Weller. 1898. 8°. 653 pp. Price 35 cents.
154. A Gazetteer of Kansas, by Henry Gannett. 1898. 8°. 246 pp. 6 pl. Price 20 cents.
155. Earthquakes in California in 1896 and 1897, by Charles D. Perrine, Assistant Astronomer in Charge of Earthquake Observations at the Lick Observatory. 1898. 8°. 47 pp. Price 5 cents.
156. Bibliography and Index of North American Geology, Paleontology, Petrology, and Mineralogy for the Year 1897, by Fred Boughton Weeks. 1898. 8°. 130 pp. Price 15 cents.

In preparation:

157. The Gneisses, Gabbro-Schists, and Associated Rocks of Southeastern Minnesota, by C. W. Hall.
— The Geology of Eastern Berkshire County, Massachusetts, by B. K. Emerson.
— The Moraines of Southeastern Dakota and their Attendant Deposits, by J. E. Todd.

WATER-SUPPLY AND IRRIGATION PAPERS.

By act of Congress approved June 11, 1896, the following provision was made:

"*Provided*, That hereafter the reports of the Geological Survey in relation to the gauging of streams and to the methods of utilizing the water resources may be printed in octavo form, not to exceed one hundred pages in length and five thousand copies in number; one thousand copies of which shall be for the official use of the Geological Survey, one thousand five hundred copies shall be delivered to the Senate, and two thousand five hundred copies shall be delivered to the House of Representatives, for distribution."

Under this law the following papers have been issued:

1. Pumping Water for Irrigation, by Herbert M. Wilson. 1896. 8°. 57 pp. 9 pl.
2. Irrigation near Phœnix, Arizona, by Arthur P. Davis. 1897. 8°. 97 pp. 31 pl.
3. Sewage Irrigation, by George W. Rafter. 1897. 8°. 100 pp. 4 pl.
4. A Reconnoissance in Southeastern Washington, by Israel Cook Russell. 1897. 8°. 96 pp. 7 pl.
5. Irrigation Practice on the Great Plains, by Elias Branson Cowgill. 1897. 8°. 39 pp. 12 pl.
6. Underground Waters of Southwestern Kansas, by Erasmus Haworth. 1897. 8°. 65 pp. 12 pl.
7. Seepage Waters of Northern Utah, by Samuel Fortier. 1897. 8°. 50 pp. 3 pl.
8. Windmills for Irrigation, by Edward Charles Murphy. 1897. 8°. 49 pp. 8 pl.
9. Irrigation near Greeley, Colorado, by David Boyd. 1897. 8°. 90 pp. 21 pl.
10. Irrigation in Mesilla Valley, New Mexico, by F. C. Barker. 1898. 8°. 51 pp. 11 pl.
11. River Heights for 1896, by Arthur P. Davis. 1897. 8°. 100 pp.
12. Water Resources of Southeastern Nebraska, by Nelson H. Darton. 1898. 8°. 55 pp. 21 pl.
13. Irrigation Systems in Texas, by William Ferguson Hutson. 1898. 8°. 67 pp. 10 pl.
14. New Tests of Certain Pumps and Water-Lifts used in Irrigation, by Ozni P. Hood. 1889. 8°. 91 pp. 1 pl.
15. Operations at River Stations, 1897, Part I. 1898. 8°. 100 pp.
16. Operations at River Stations, 1897, Part II. 1898. 8°. 101–200 pp.

In press:

17. Irrigation near Bakersfield, California, by C. E. Grunsky. 1898. 8°. 96 pp. 16 pl.
18. Irrigation near Fresno, California, by C. E. Grunsky. 1898. 8°. 94 pp. 14 pl.

In preparation:

19. Irrigation near Merced, California, by C. E. Grunsky.
20. Experiments with Windmills, by T. O. Perry.
21. Wells of Indiana, by Frank Leverett.
22. Sewage Irrigation, Part II, by George W. Rafter.
23. Water-Right Problems of Bighorn Mountains, by Elwood Mead.

TOPOGRAPHIC MAP OF THE UNITED STATES.

When, in 1882, the Geological Survey was directed by law to make a geologic map of the United States there was in existence no suitable topographic map to serve as a base for the geologic map. The preparation of such a topographic map was therefore immediately begun. About one-fifth of the area of the country, excluding Alaska, has now been thus mapped. The map is published in atlas sheets, each sheet representing a small quadrangular district, as explained under the following heading. The separate sheets are sold at 5 cents each when fewer than 100 copies are purchased, but when they are ordered in lots of 100 or more copies, whether of the same sheet or of different sheets, the price is 2 cents each. The mapped areas are widely scattered, nearly every State being represented. More than 800 sheets have been engraved and printed; they are tabulated by States in the Survey's "List of Publications," a pamphlet which may be had on application.

The map sheets represent a great variety of topographic features, and with the aid of descriptive text they can be used to illustrate topographic forms. This has led to the projection of an educational series of topographic folios, for use wherever geography is taught in high schools, academies, and colleges. Of this series the first folio has been issued, viz:

1. Physiographic types, by Henry Gannett, 1898, folio, consisting of the following sheets and 4 pages of descriptive text: Fargo (N. Dak.-Minn.), a region in youth; Charleston (W. Va.), a region in maturity; Caldwell (Kans.), a region in old age; Palmyra (Va.), a rejuvenated region; Mount Shasta, (Cal.), a young volcanic mountain; Eagle (Wis.), moraines; Sun Prairie (Wis.), drumlins; Donaldsonville (La.), river flood plains; Boothbay (Me.), a fiord coast; Atlantic City (N. J.), a barrier-beach coast.

GEOLOGIC ATLAS OF THE UNITED STATES.

The Geologic Atlas of the United States is the final form of publication of the topographic and geologic maps. The atlas is issued in parts, progressively as the surveys are extended, and is designed ultimately to cover the entire country.

Under the plan adopted the entire area of the country is divided into small rectangular districts (designated *quadrangles*), bounded by certain meridians and parallels. The unit of survey is also the unit of publication, and the maps and descriptions of each rectangular district are issued as a folio of the Geologic Atlas.

Each folio contains topographic, geologic, economic, and structural maps, together with textual descriptions and explanations, and is designated by the name of a principal town or of a prominent natural feature within the district.

Two forms of issue have been adopted, a "library edition" and a "field edition." In both the sheets are bound between heavy paper covers, but the library copies are permanently bound, while the sheets and covers of the field copies are only temporarily wired together.

Under the law a copy of each folio is sent to certain public libraries and educational institutions. The remainder are sold at 25 cents each, except such as contain an unusual amount of matter, which are priced accordingly. Prepayment is obligatory The folios ready for distribution are listed below.

No.	Name of sheet.	State.	Limiting meridians.	Limiting parallels.	Area, in square miles.	Price, in cents.
1	Livingston	Montana	110°-111°	45°-46°	3,354	25
2	Ringgold	Georgia Tennessee	85°-85° 30′	34° 30′-35°	980	25
3	Placerville	California	120° 30′-121°	38° 30′-39°	932	25
4	Kingston	Tennessee	84° 30′-85°	35° 30′-36°	969	25
5	Sacramento	California	121°-121° 30′	38° 30′-39°	932	25
6	Chattanooga	Tennessee	85°-85° 30′	35°-35° 30′	975	25
7	Pikes Peak (out of stock)	Colorado	105°-105° 30′	38° 30′-39°	932	25
8	Sewanee	Tennessee	85° 30′-86°	35°-35° 30′	975	25
9	Anthracite-Crested Butte	Colorado	106° 45′-107° 15′	38° 45′-39°	465	50
10	Harpers Ferry	Virginia West Virginia Maryland	77° 30′-78°	39°-39° 30′	925	25
11	Jackson	California	120° 30′-121°	38°-38° 30′	938	25
12	Estillville	Virginia Kentucky Tennessee	82° 30′-83°	36° 30′-37°	957	25
13	Fredericksburg	Maryland Virginia	77°-77° 30′	38°-38° 30′	938	25
14	Staunton	Virginia West Virginia	79°-79° 30′	38°-38° 30′	938	25
15	Lassen Peak	California	121°-122°	40°-41°	3,634	25
16	Knoxville	Tennessee North Carolina	83° 30′-84°	35° 30′-36°	925	25

No.	Name of sheet.	State.	Limiting meridians.	Limiting parallels.	Area, in square miles.	Price, in cents.
17	Marysville	California	121° 30′–122°	39°–39° 30′	925	25
18	Smartsville	California	121°–121° 30′	39°–39° 30′	925	25
19	Stevenson	Alabama Georgia Tennessee	85° 30′–86°	34° 30′–35°	980	25
20	Cleveland	Tennessee	84° 30′–85°	35°–35° 30′	975	25
21	Pikeville	Tennessee	85°–85° 30′	35° 30′–36°	969	25
22	McMinnville	Tennessee	85° 30′–86°	35° 30′–36°	969	25
23	Nomini	Maryland Virginia	76° 30′–77°	38°–38° 30′	938	25
24	Three Forks	Montana	111°–112°	45°–46°	3,354	50
25	Loudon	Tennessee	84°–84° 30′	35° 30′–36°	969	25
26	Pocahontas	Virginia West Virginia	81°–81° 30′	37°–37° 30′	951	25
27	Morristown	Tennessee	83°–83° 30′	36°–36° 30′	963	25
28	Piedmont	Virginia Maryland West Virginia	79°–79° 30′	39°–39° 30′	925	25
29	Nevada City: Nevada City Grass Valley Banner Hill	California	121° 00′ 25″–121° 03′ 45″ 121° 01′ 35″–121° 05′ 04″ 120° 57′ 05″–121° 00′ 25″	39° 13′ 50″–39° 17′ 16″ 39° 10′ 22″–39° 13′ 50″ 39° 13′ 50″–39° 17′ 16″	11.65 12.09 11.65	50
30	Yellowstone National Park: Gallatin Canyon Shoshone Lake	Wyoming	110°–111°	44°–45°	3,412	75
31	Pyramid Peak	California	120°–120° 30′	38° 30′–39°	932	25
32	Franklin	Virginia West Virginia	79°–79° 30′	38° 30′–39°	932	25
33	Briceville	Tennessee	84°–84° 30′	36°–36° 30′	963	25
34	Buckhannon	West Virginia	80°–80° 30′	38° 30′–39°	932	25
35	Gadsden	Alabama	86°–86° 30′	34°–34° 30′	986	25
36	Pueblo	Colorado	104° 30′–105°	38°–38° 30′	938	50
37	Downieville	California	120° 30′–121°	39° 30′–40°	919	25
39	Truckee	California	120°–120° 30′	39°–39° 30′	925	25
40	Wartburg	Tennessee	84° 30′–85°	36°–36° 30′	963	25
41	Sonora	California	120°–120° 30′	37° 30′–38°	944	25
42	Nueces	Texas	100°–100° 30′	29° 30′–30°	1,035	25
43	Bidwell Bar	California	121°–121° 30′	39° 30′–40°	918	25
44	Tazewell	Virginia West Virginia	81° 30′–82°	37°–37° 30′	950	25

STATISTICAL PAPERS.

Mineral Resources of the United States [1882], by Albert Williams, jr. 1883. 8°. xvii, 813 pp. Price 50 cents.

Mineral Resources of the United States, 1883 and 1884, by Albert Williams, jr. 1885. 8°. xiv, 1016 pp. Price 60 cents.

Mineral Resources of the United States, 1885. Division of Mining Statistics and Technology. 1886. 8°. vii, 576 pp. Price 40 cents.

Mineral Resources of the United States, 1886, by David T. Day. 1887. 8°. viii, 813 pp. Price 60 cents.

Mineral Resources of the United States, 1887, by David T. Day. 1888. 8°. vii, 832 pp. Price 50 cents.

Mineral Resources of the United States, 1888, by David T. Day. 1890. 8°. vii, 652 pp. Price 50 cents.

Mineral Resources of the United States, 1889 and 1890, by David T. Day. 1892. 8°. viii, 671 pp. Price 50 cents.

Mineral Resources of the United States, 1891, by David T. Day. 1893. 8°. vii, 630 pp. Price 50 cents.

Mineral Resources of the United States, 1892, by David T. Day. 1893. 8°. vii, 850 pp. Price 50 cents.

Mineral Resources of the United States, 1893, by David T. Day. 1894. 8°. viii, 810 pp. Price 50 cents.

On March 2, 1895, the following provision was included in an act of Congress:

"*Provided*, That hereafter the report of the mineral resources of the United States shall be issued as a part of the report of the Director of the Geological Survey."

In compliance with this legislation the following reports have been published:

Mineral Resources of the United States, 1894, David T. Day, Chief of Division. 1895. 8°. xv, 646 pp., 23 pl.; xix, 735 pp., 6 pl. Being Parts III and IV of the Sixteenth Annual Report.

Mineral Resources of the United States, 1895, David T. Day, Chief of Division. 1896. 8°. xxiii, 542 pp., 8 pl. and maps; iii, 543–1058 pp., 9–13 pl. Being Part III (in 2 vols.) of the Seventeenth Annual Report.

Mineral Resources of the United States, 1896, David T. Day, Chief of Division. 1897. 8°. xii, 642 pp., 1 pl.; 643–1400 pp. Being Part V (in 2 vols.) of the Eighteenth Annual Report.

Mineral Resources of the United States, 1897, David T. Day, Chief of Division. 1898. 8°. Being Part VI (in 2 vols.) of the Nineteenth Annual Report.

The money received from the sale of the Survey publications is deposited in the Treasury, and the Secretary of that Department declines to receive bank checks, drafts, or postage stamps; all remittances, therefore, must be by MONEY ORDER, made payable to the Director of the United States Geological Survey, or in CURRENCY—the exact amount. Correspondence relating to the publications of the Survey should be addressed to

THE DIRECTOR,
UNITED STATES GEOLOGICAL SURVEY,
WASHINGTON, D. C.

WASHINGTON, D. C., *October, 1898.*

[Take this leaf out and paste the separated titles upon three of your catalogue cards. The first and second titles need no addition; over the third write that subject under which you would place the book in your library.]

LIBRARY CATALOGUE SLIPS.

Series.

United States. *Department of the interior.* (*U. S. geological survey.*)
Department of the interior | — | Monographs | of the | United States geological survey | Volume XXXV | [Seal of the department] | Washington | government printing office | 1898
Second title: United States geological survey | Charles D. Walcott, director | — | The | later extinct floras of North America | by | John Strong Newberry | A | posthumous work | edited by | Arthur Hollick | [Vignette] |
Washington | government printing office | 1898
4°. xvii, 295 pp. 68 pl.

Author.

Newberry (John Strong).
United States geological survey | Charles D. Walcott, director | — | The | later extinct floras of North America | by | John Strong Newberry | A | posthumous work | edited by | Arthur Hollick | [Vignette] |
Washington | government printing office | 1898
4°. xvii, 295 pp. 68 pl.
[United States. *Department of the interior.* (*U. S. geological survey.*) Monograph XXXV.]

Subject.

United States geological survey | Charles D. Walcott, director | — | The | later extinct floras of North America | by | John Strong Newberry | A | posthumous work | edited by | Arthur Hollick | [Vignette] |
Washington | government printing office | 1898
4°. xvii, 295 pp. 68 pl.
[United States. *Department of the interior.* (*U. S. geological survey.*) Monograph XXXV.]

R. G. (Robert Griffith) Hatfield

The American house-carpenter; a treatise on the art of building, and the strength of materials

R. G. (Robert Griffith) Hatfield

The American house-carpenter; a treatise on the art of building, and the strength of materials

ISBN/EAN: 9783337057343

Printed in Europe, USA, Canada, Australia, Japan

Cover: Foto ©berggeist007 / pixelio.de

More available books at **www.hansebooks.com**

THE
AMERICAN HOUSE-CARPENTER.

A TREATISE

ON

THE ART OF BUILDING,

AND

THE STRENGTH OF MATERIALS.

BY

R. G. HATFIELD, ARCHITECT,

MEM. AM. INST. OF ARCHITECTS.

SEVENTH EDITION, REVISED AND ENLARGED

WITH ADDITIONAL ILLUSTRATIONS.

NEW YORK:

JOHN WILEY & SON,

15 ASTOR PLACE.

1873.

PREFACE.

This book is intended for carpenters—for masters, journeymen and apprentices. It has long been the complaint of this class that architectural books, intended for their instruction, are of a price so high as to be placed beyond their reach. This is owing, in a great measure, to the costliness of the plates with which they are illustrated: an unnecessary expense, as illustrations upon wood, printed on good paper, answer every useful purpose. Wood engravings, too, can be distributed among the letter-press; an advantage which plates but partially possess, and one of great importance to the reader.

Considerations of this kind induced the author to undertake the preparation of this volume. The subject matter has been gleaned from works of the first authority, and subjected to the most careful examination. The explanations have all been written out from the figures themselves, and not taken from any other work; and the figures have all been drawn expressly for this book. In doing this, the utmost care has been taken to make everything as plain as the nature of the case would admit.

The attention of the reader is particularly directed to the following new inventions, viz; an easy method of describing the curves of mouldings through three given points; a rule to determine the projection of eave cornices; a new method of proportioning a cornice to a larger given one; a way to determine the lengths and bevils of rafters for hip-roofs; a way to proportion the rise to the tread in stairs; to determine the true position of butt-joints in hand-rails; to find the bevils for splayed-work; a general rule for scrolls, &c. Many problems in geometry, also, have been simplified, and new ones introduced. Much labour has been bestowed upon the section on stairs, in which the subject of hand-railing is presented, in many respects, in a new, and it is hoped, more practical form than in previous treatises on that subject.

The author has endeavoured to present a fund of useful information to the *American house-carpenter* that would enable him to excel in his vocation; how far he has been successful in that object, the book itself must determine.

New York, Oct. 15, 1844.

FIFTH EDITION.

Since the first edition of this work was published, I have received numerous testimonials of its excellent practical value, from the very best sources, viz. from the workmen themselves who have used it, and who have profited by it. As a convenient manual for reference in respect to every question relating either to the simpler operations of Carpentry or the more intricate and abstruse problems of Geometry, those who have tried it assure me that they have been greatly assisted in using it. And, indeed, to the true workman, there is, in the study of the subjects of which this volume treats, a continual source of profitable and pleasurable interest. Gentlemen, in numerous instances, have placed it in the hands of their sons, who have manifested a taste for practical studies; and have also procured it for the use of the workmen upon their estates, as a guide in their mechanical operations. I was not, then, mistaken in my impressions, that a work of this kind was wanted; and this evidence of its usefulness rewards me in a measure for the pains taken in its preparation.

New York, Oct. 1, 1852.

SEVENTH EDITION.

It is now thirteen years since the first edition of the American House Carpenter was published. The attempt to furnish the recipients of this book with a fund of useful information in a compact and accessible form, has been so far successful that the sixth edition was exhausted nearly a year ago. At that time it was determined, before issuing another edition, to make a thorough revision of the work. The time occupied in this labour has been unexpectedly prolonged by at least six months, and this has resulted from various causes, but more especially from the absorbing nature of my professional duties. A large portion of the work has been rewritten, about 130 pages of new matter introduced and many new cuts inserted.

The most important additions to the work will be found in the section on Framing or Construction. Here will be found, now first published, the results of experiments on such building materials as are in common use in this country, and an extended series of rules for the application of this experimental knowledge to the practical purposes of building. Some of the rules are new, while others heretofore in use have been simplified. This section has been much improved, and it is hoped that it will be of service, not only to the house carpenter but also to the architect and civil engineer.

In preparing the original work, a desire to state the subjects treated of in terms suited to the comprehension of all classes of workmen, precluded the use of algebraical symbols and formulæ. In this edition, however, it has been deemed best to introduce them wherever they would contribute to the clearer elucidation of the subject; but care has been taken to state them in a simple form at first, and so to explain the symbols as they are introduced that those heretofore uninstructed in regard to them, may comprehend what little is here exhibited, and at the same time be induced to pursue the study more fully in works more strictly mathematical. But for those who may not succeed in comprehending the algebraical formulæ, it may be stated that all the practical deductions derived from them are written out in words at length, so as to be fully understood without their assistance.

New York, Sept. 1, 1857. R. G. H.

TABLE OF CONTENTS.

SECTION I.—PRACTICAL GEOMETRY

SECTION II.—ARCHITECTURE.

SECTION III.—MOULDINGS, CORNICES, &c.

SECTION IV.—FRAMING, OR CONSTRUCTION.

SECTION V.—DOORS, WINDOWS, &c.

SECTION VI.—STAIRS.

SECTION VII.—SHADOWS.

APPENDIX.

INTRODUCTION.

Art. 1.—A knowledge of the properties and principles of *lines* can best be acquired by practice. Although the various problems throughout this work may be understood by inspection, yet they will be impressed upon the mind with much greater force, if they are actually performed with pencil and paper by the student. Science is acquired by study—art by practice: he, therefore, who would have any thing more than a theoretical, (which must of necessity be a superficial,) knowledge of Carpentry, will attend to the following directions, provide himself with the articles here specified, and perform all the operations described in the following pages. Many of the problems may appear, at the first reading, somewhat confused and intricate; but by making one line at a time, according to the explanations, the student will not only succeed in copying the figures correctly, but by ordinary attention will learn the principles upon which they are based, and thus be able to make them available in any unexpected case to which they may apply.

2.—The following articles are necessary for drawing, viz: a drawing-board, paper, drawing-pins or mouth-glue, a sponge, a T-square, a set-square, two straight-edges, or flat rulers, a lead pencil, a piece of india-rubber, a cake of india-ink, a set of drawing-instruments, and a scale of equal parts.

3.—The size of the *drawing-board* must be regulated according to the size of the drawings which are to be made upon it. Yet for ordinary practice, in learning to draw, a board about 15 by 20 inches, and one inch thick, will be found large enough, and more convenient than a larger one. This board should be well-seasoned, perfectly square at the corners, and without clamps on the ends. A board is better without clamps, because the little service they are supposed to render by preventing the board from warping, is overbalanced by the consideration that the shrinking of the panel leaves the ends of the clamps projecting beyond the edge of the board, and thus interfering with the proper working of the stock of the T-square. When the stuff is well-seasoned, the warping of the board will be but trifling; and by exposing the rounding side to the fire, or to the sun, it may be brought back to its proper shape.

4.—For mere line drawings, it is unnecessary to use the *best* drawing-paper; and since, where much is used the expense will be considerable, it is desirable for economy to procure paper of as low a price as will be suitable for the purpose. The best paper is made in England and marked "Whatman." This is a hand-made paper. There is also a machine-made paper at about half-price, and the Manilla paper, of various tints of russet color, is still less in price. These papers are of the various sizes needed, and are quite sufficient for ordinary drawings.

5.—A *drawing-pin* is a small brass button, having a steel pin projecting from the under side. By having one of these at each corner, the paper can be fixed to the board; but this can be done in a much better manner with *mouth-glue.* The pins will prevent the paper from changing its position on the board; but, more than this, the glue keeps the paper perfectly tight and smooth, thus making it so much the more pleasant to work on.

To attach the paper with mouth-glue, lay it with the bottom side up, on the board; and with a straight-edge and penknife, cut off the rough and uneven edge. With a sponge moderately wet, rub all the surface of the paper, except a strip around the edge about half an inch wide. As soon as the glistening of the water disappears, turn the sheet over, and place it upon the board just where you wish it glued. Commence upon one of the longest sides, and proceed thus: lay a flat ruler upon the paper, parallel to the edge, and within a quarter of an inch of it With a knife, or any thing similar, turn up the edge of the paper against the edge of the ruler, and put one end of the cake of mouth-glue between your lips to dampen it. Then holding it upright, rub it against and along the entire edge of the paper that is turned up against the ruler, bearing moderately against the edge of the ruler, which must be held firmly with the left hand. Moisten the glue as often as it becomes dry, until a sufficiency of it is rubbed on the edge of the paper. Take away the ruler, restore the turned-up edge to the level of the board, and lay upon it a strip of pretty stiff paper. By rubbing upon this, not very hard but pretty rapidly, with the thumb nail of the right hand, so as to cause a gentle friction, and heat to be imparted to the glue that is on the edge of the paper, you will make it adhere to the board. The other edges in succession must be treated in the same manner.

Some short distances along one or more of the edges, may afterwards be found loose: if so, the glue must again be applied, and the paper rubbed until it adheres. The board must then be laid away in a warm or dry place; and in a short time, the surface of the paper will be drawn out, perfectly tight and smooth, and ready for use. The paper dries best when the board is laid level. When the drawing is finished, lay a straight-edge upon the paper, and cut it from the board, leaving the glued strip still attached. This may afterwards be taken off by wetting it freely with the sponge; which will soak the glue, and loosen the paper. Do this as soon as the drawing is taken off, in order that the board may be dry when it is wanted for use again. Care must be taken that, in applying the glue, the edge of the paper does not become damper than the rest: if it should, the paper must be laid aside to dry, (to use at another time,) and another sheet be used in its place.

Sometimes, especially when the drawing board is new, the paper will not stick very readily; but by persevering, this difficulty may be overcome. In the place of the mouth-glue, a strong solution of gum-arabic may be used, and on some accounts is to be preferred; for the edges of the paper need not be kept dry, and it adheres more readily. Dissolve the gum in a sufficiency of warm water to make it of the consistency of linseed oil. It must be applied to the paper with a brush, when the edge is turned up against the ruler, as was described for the mouth-glue. If two drawing-boards are used, one may be in use while the other is laid away to dry; and as they may be cheaply made, it is advisable to have two. The drawing-board having a frame around it, commonly called a panel-board, may afford rather more facility in attaching the paper when this is of the size to suit; yet it has objections which overbalance that consideration.

6.—A *T-square* of mahogany, at once simple in its construction, and affording all necessary service, may be thus made. Let the stock or handle be seven inches long, two and a quarter inches wide, and three-eighths of an inch thick: the blade, twenty inches long, (exclusive of the stock,) two inches wide, and one-eighth of an inch thick. In joining the blade to the stock, a very firm and simple joint may be made by dovetailing it—as shown at *Fig.* 1.

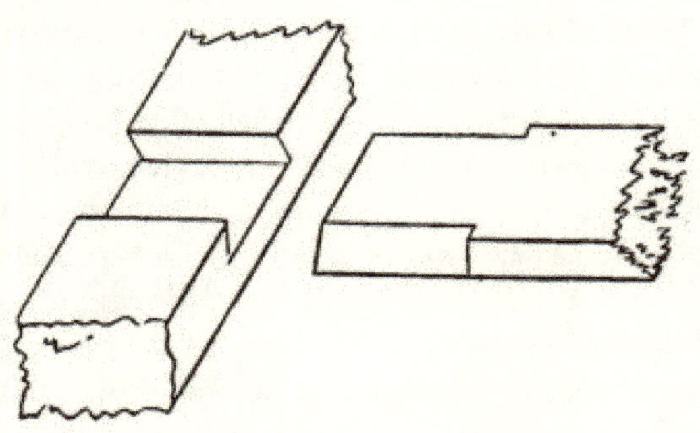

7.—The *set-square* is in the form of a right-angled triangle; and is commonly made of mahogany, one-eighth of an inch in thickness. The size that is most convenient for general use, is six inches and three inches respectively for the sides which contain the right angle; although a particular length for the sides is by no means necessary. Care should be taken to have the square corner exactly true. This, as also the T-square and rulers, should have a hole bored through them, by which to hang them upon a nail when not in use.

8.—One of the *rulers* may be about twenty inches long, and the other six inches. The *pencil* ought to be hard enough to retain a fine point, and yet not so hard as to leave ineffaceable marks. It should be used lightly, so that the extra marks that are not needed when the drawing is inked, may be easily rubbed off with the rubber. The best kind of *india-ink* is that which will easily rub off upon the plate; and, when the cake is rubbed against the teeth, will be free from grit.

9.—The *drawing-instruments* may be purchased of mathematical instrument makers at various prices: from one to one hundred dollars a set. In choosing a set, remember that the lowest price articles are not always the cheapest. A set, comprising a sufficient number of instruments for ordinary use, well made and fitted in a mahogany box, may be purchased of the mathematical instrument-makers in New York for four or five dollars. But for permanent use those which come at ten or twelve dollars will be found to be the best.

10.—The best *scale of equal parts* for carpenters' use, is one that has one-eighth, three-sixteenths, one-fourth, three-eighths, one-half, five-eighths, three-fourths, and seven-eighths of an inch, and one inch, severally divided into *twelfths*, instead of being divided, as they usually are, into tenths. By this, if it be required to proportion a drawing so that every foot of the object represented will upon the paper measure one-fourth of an inch, use that part of the scale which is divided into one-fourths of an inch taking for every foot one of those divisions, and for every inch one of the subdivisions into twelfths; and proceed in like manner in proportioning a drawing to any of the other divisions of the scale. An instrument in the form of a semi-circle, called a *protractor*, and used for laying down and measuring angles, is of much service to surveyors, but not much to carpenters.

11.—In drawing parallel lines, when they are to be parallel to either side of the board, use the T-square; but when it is required to draw lines parallel to a line which is drawn in a direction oblique to either side of the board, the set-square must be used. Let *a b*, (*Fig.* 2,) be a line, parallel to which it is

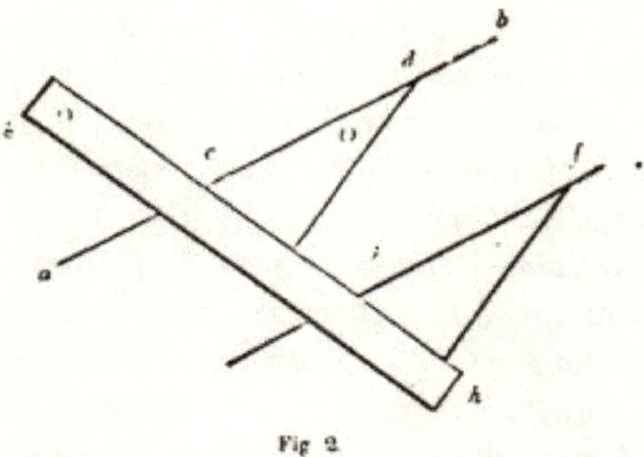

Fig 2

desired to draw one or more lines. Place any edge, as *c d*, of the set-square even with said line; then place the ruler, *g h*, against one of the other sides, as *c e*, and hold it firmly; slide the set-square along the edge of the ruler as far as it is desired, as at *f*; and a line drawn by the edge, *i f*, will be parallel to *a b*.

12.—To draw a line, as *k l*, (*Fig.* 3,) perpendicular to another, as *a b*, set the shortest edge of the set-square at the line, *a b*; place the ruler against the longest side, (the hypothenuse of the right-angled triangle;) hold the ruler firmly, and slide the set-square along until the side, *e d* touches the point, *k*; then the line, *l k*, drawn by it, will be perpendicular to *a b*. In like manner the drawing of other problems may be facilitated, as will be discovered in using the instruments.

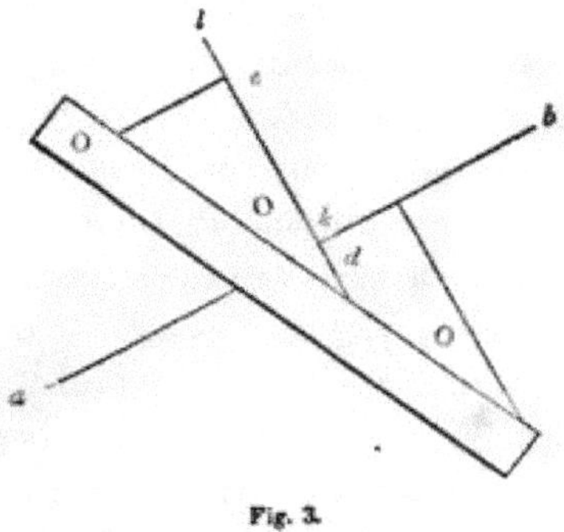

Fig. 3.

13.—In drawing a problem, proceed, with the pencil sharpened to a point, to lay down the several lines until the whole figure is completed; observing to let the lines cross each other at the several angles, instead of merely meeting. By this, the length of every line will be clearly defined. With a drop or two of water, rub one end of the cake of ink upon a plate or saucer, until a sufficiency adheres to it. Be careful to dry the cake of ink; because if it is left wet, it will crack and crumble in pieces. With an inferior camel's-hair pencil, add a little water to the ink that was rubbed on the plate, and mix it well. It should be diluted sufficiently to flow freely from the pen, and yet be thick enough to make a *black* line. With the hair pencil, place a little of the ink between the nibs of the drawing-pen, and screw the nibs together until the pen makes a fine line. Beginning with the curved lines, proceed to ink *all* the lines of the figure; being careful now to make every line of its requisite length. If they are a trifle too short or too long, the drawing will have a ragged appearance; and this is opposed to that neatness and accuracy which is indispensable to a good drawing. When the ink is dry, efface the pencil-marks with the india-rubber. If the pencil is used lightly, they will all rub off, leaving those lines only that were inked.

14.—In problems, all auxiliary lines are drawn light; while the lines given and those sought, in order to be distinguished at a glance, are made much heavier. The heavy lines are made so, by passing over them a second time, having the nibs of the pen separated far enough to make the lines as heavy as desired. If the heavy lines are made before the drawing is cleaned with the rubber, they will not appear so black and neat; because the india-rubber takes away part of the ink. If the drawing is a ground-plan or elevation of a house, the shade-lines, as they are termed, should not be put in until the drawing is shaded; as there is danger of the heavy lines spreading, when the brush, in shading or coloring, passes over them. If the lines are inked with common writing-ink, they will, however fine they may be made, be subject to the same evil; for which reason, india-ink is the only kind to be used.

THE

AMERICAN HOUSE-CARPENTER.

SECTION I.—PRACTICAL GEOMETRY.

DEFINITIONS.

15.—*Geometry* treats of the properties of magnitudes.

16.—A *point* has neither length, breadth, nor thickness.

17.—A *line* has length only.

18.—*Superficies* has length and breadth only.

19.—A *plane* is a surface, perfectly straight and even in every direction; as the face of a panel when not warped nor winding.

20.—A *solid* has length, breadth and thickness.

21.—A *right*, or *straight*, line is the shortest that can be drawn between two points.

22.—*Parallel lines* are equi-distant throughout their length.

23.—An *angle* is the inclination of two lines towards one another. (*Fig.* 4.)

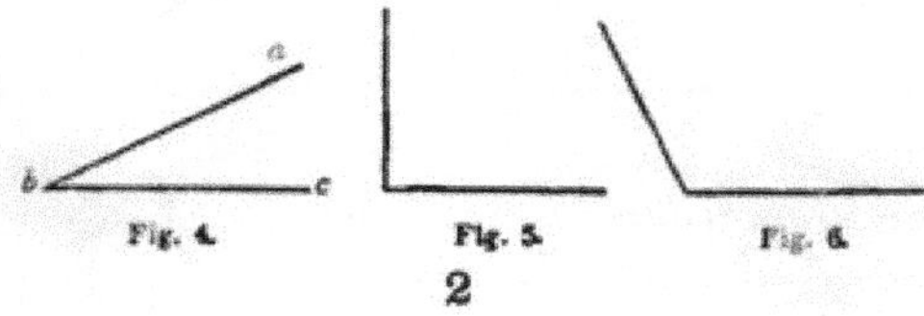

Fig. 4. Fig. 5. Fig. 6.

24.—A *right angle* has one line perpendicular to the other. (*Fig.* 5.)

25.—An *oblique angle* is either greater or less than a right angle. (*Fig.* 4 and 6.)

26.—An *acute angle* is less than a right angle. (*Fig.* 4.)

27.—An *obtuse angle* is greater than a right angle. (*Fig.* 6.)

When an angle is denoted by three letters, the middle one, in the order they stand, denotes the angular point, and the other two the sides containing the angle; thus, let *a b c*, (*Fig.* 4,) be the angle, then *b* will be the angular point, and *a b* and *b c* will be the two sides containing that angle.

28.—A *triangle* is a superficies having three sides and angles. (*Fig.* 7, 8, 9 and 10.)

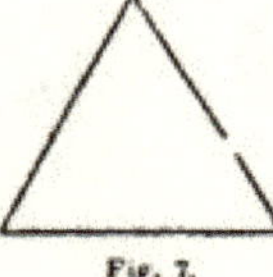
Fig. 7.

Fig. 8.

29.—An *equi-lateral triangle* has its three sides equal. (*Fig.* 7.)

30.—An *isosceles triangle* has only two sides equal. (*Fig.* 8.)

31.—A *scalene triangle* has all its sides unequal. (*Fig.* 9)

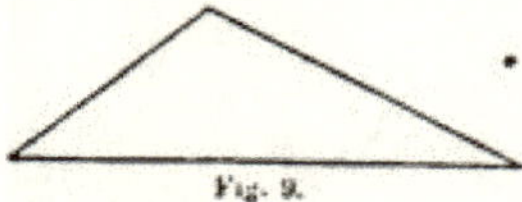
Fig. 9.

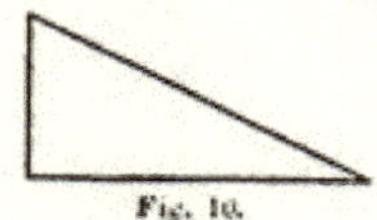
Fig. 10.

32.—A *right-angled triangle* has one right angle. (*Fig.* 10.)

33.—An *acute-angled triangle* has all its angles acute. (*Fig.* 7 and 8.)

34.—An *obtuse-angled triangle* has one obtuse angle. (*Fig.* 9.)

35.—A *quadrangle* has four sides and four angles. (*Fig.* 11 to 16.)

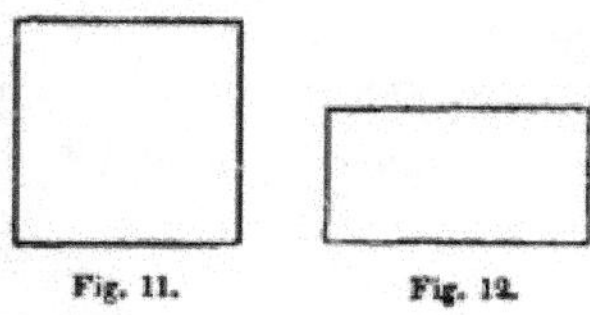

Fig. 11. Fig. 12.

36.—A *parallelogram* is a quadrangle having its opposite sides parallel. (*Fig.* 11 to 14.)

37.—A *rectangle* is a parallelogram, its angles being right angles. (*Fig.* 11 and 12.)

38.—A *square* is a rectangle having equal sides. (*Fig.* 11.)

39.—A *rhombus* is an equi-lateral parallelogram having oblique angles. (*Fig.* 13.)

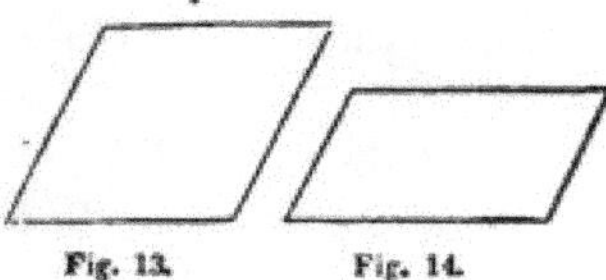

Fig. 13. Fig. 14.

40.—A *rhomboid* is a parallelogram having oblique angles. (*Fig.* 14.)

41.—A *trapezoid* is a quadrangle having only two of its sides parallel. (*Fig.* 15.)

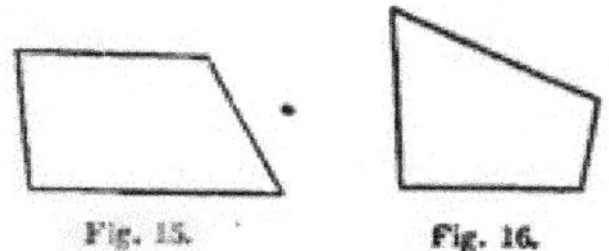

Fig. 15. Fig. 16.

42.—A *trapezium* is a quadrangle which has no two of its sides parallel. (*Fig.* 16.)

43.—A *polygon* is a figure bounded by right lines.

44.—A *regular polygon* has its sides and angles equal.

45.—An *irregular polygon* has its sides and angles unequal.

46.—A *trigon* is a polygon of three sides, (*Fig.* 7 to 10,) a *tetragon* has four sides, (*Fig.* 11 to 16;) a *pentagon* has five, (*Fig.* 17;) a *hexagon* six, (*Fig.* 18;) a *heptagon* seven, (*Fig.* 19;) an *octagon* eight, (*Fig.* 20;) a *nonagon* nine; a *decagon* ten; an *undecagon* eleven; and a *dodecagon* twelve sides.

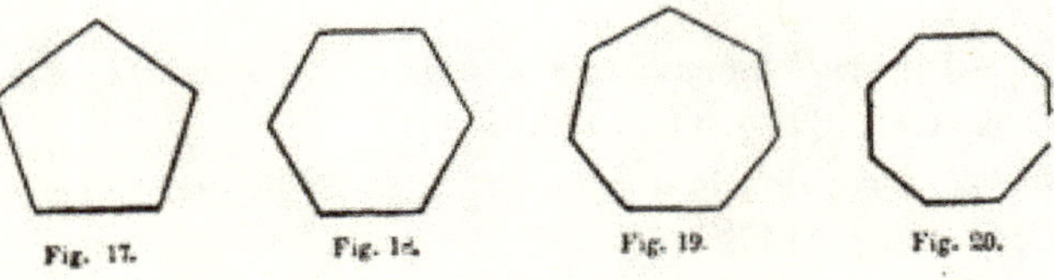
Fig. 17. Fig. 18. Fig. 19. Fig. 20.

47.—A *circle* is a figure bounded by a curved line, called the *circumference;* which is every where equi-distant from a certain point within, called its *centre.*

The circumference is also called the *periphery*, and sometimes the *circle.*

48.—The *radius* of a circle is a right line drawn from the centre to any point in the circumference. (*a b*, *Fig.* 21.)

All the *radii* of a circle are equal.

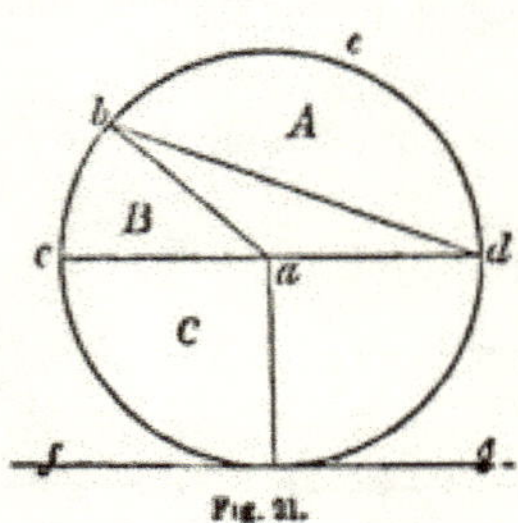

Fig. 21.

49.—The *diameter* is a right line passing through the centre, and terminating at two opposite points in the circumference. Hence it is twice the length of the radius. (*c d*, *Fig.* 21.)

50.—An *arc* of a circle is a part of the circumference. (*c b* or *b e d*, *Fig.* 21.)

51.—A *chord* is a right line joining the extremities of an arc. (*b d*, *Fig.* 21.)

52.—A *segment* is any part of a circle bounded by an arc and its chord. (*A*, *Fig.* 21.)

53.—A *sector* is any part of a circle bounded by an arc and two radii, drawn to its extremities. (*B*, *Fig.* 21.)

54.—A *quadrant*, or quarter of a circle, is a sector having a quarter of the circumference for its arc. (*C*, *Fig.* 21.)

55.—A *tangent* is a right line, which in passing a curve, touches, without cutting it. (*f g*, *Fig.* 21.)

56.—A *cone* is a solid figure standing upon a circular base diminishing in straight lines to a point at the top, called its vertex. (*Fig.* 22.)

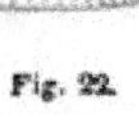

Fig. 22.

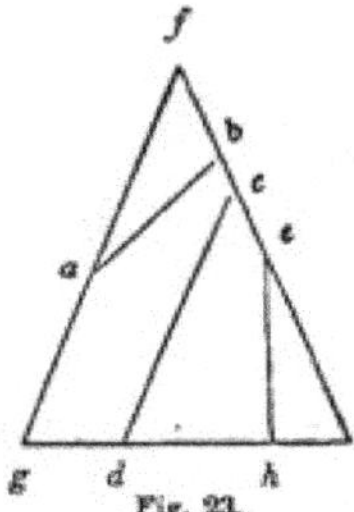

Fig. 23.

57.—The *axis* of a cone is a right line passing through it, from the vertex to the centre of the circle at the base.

58.—An *ellipsis* is described if a cone be cut by a plane, not parallel to its base, passing quite through the curved surface. (*a b*, *Fig.* 23.)

59.—A *parabola* is described if a cone be cut by a plane, parallel to a plane touching the curved surface. (*c d*, *Fig.* 23—*c d* being parallel to *f g*.)

60.—An *hyperbola* is described if a cone be cut by a plane, parallel to any plane within the cone that passes through its vertex. (*e h*, *Fig.* 23.)

61.—*Foci* are the points at which the pins are placed in describing an ellipse. (See *Art.* 115, and *f*, *f*, *Fig.* 24.)

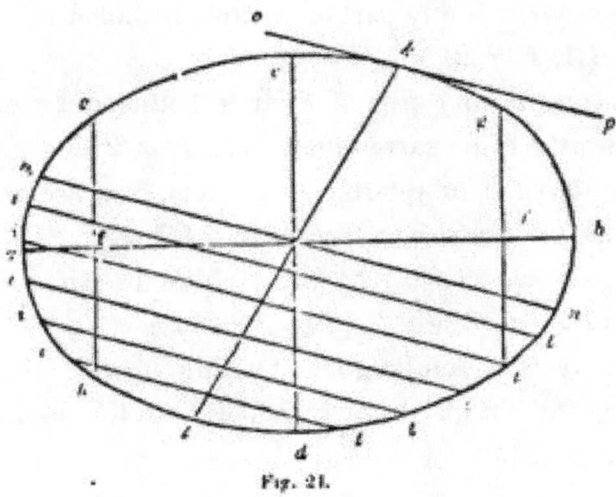

Fig. 24.

62.—The *transverse axis* is the longest diameter of the ellipsis. (*a b*, *Fig*. 24.)

63.—The *conjugate axis* is the shortest diameter of the ellipsis; and is, therefore, at right angles to the transverse axis. (*c d*, *Fig*. 24.)

64.—The *parameter* is a right line passing through the focus of an ellipsis, at right angles to the transverse axis, and terminated by the curve. (*g h* and *g t*, *Fig*. 24.)

65.—A *diameter of an ellipsis* is any right line passing through the centre, and terminated by the curve. (*k l*, or *m n*, *Fig*. 24.)

66.—A diameter is *conjugate* to another when it is parallel to a tangent drawn at the extremity of that other—thus, the diameter, *m n*, (*Fig*. 24,) being parallel to the tangent, *o p*, is therefore conjugate to the diameter, *k l*.

67.—A *double ordinate* is any right line, crossing a diameter of an ellipsis, and drawn parallel to a tangent at the extremity of that diameter. (*i t*, *Fig*. 24.)

68.—A *cylinder* is a solid generated by the revolution of a right-angled parallelogram, or rectangle, about one of its sides; and consequently the ends of the cylinder are equal circles. (*Fig*. 25.)

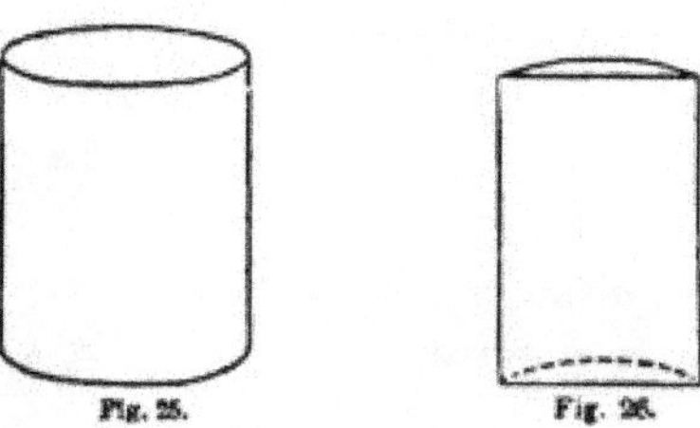

Fig. 25. Fig. 26.

69.—The *axis* of a cylinder is a right line passing through it, from the centres of the two circles which form the ends.

70.—A *segment* of a cylinder is comprehended under three planes, and the curved surface of the cylinder. Two of these are segments of circles: the other plane is a parallelogram, called by way of distinction, the *plane of the segment.* The circular segments are called, the ends of the cylinder. (*Fig.* 26.)

N. B.—For Algebraical Signs, Trigonometrical Terms, &c., see Appendix.

PROBLEMS.

RIGHT LINES AND ANGLES.

71.—*To bisect a line.* Upon the ends of the line, *a b*, (*Fig.* 27,) as centres, with any distance for radius greater than half

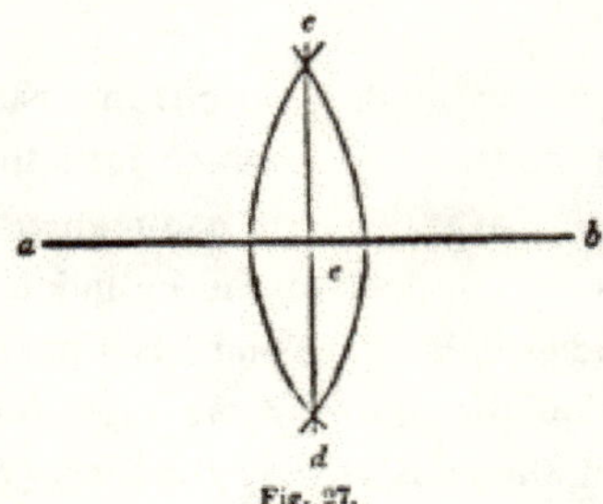

Fig. 27.

a b, describe arcs cutting each other in *c* and *d*; draw the line, *c d*, and the point, *e*, where it cuts *a b*, will be the middle of the line, *a b*.

In practice, a line is generally divided with the compasses, or dividers; but this problem is useful where it is desired to draw, at the middle of another line, one at right angles to it. (See *Art.* 85.)

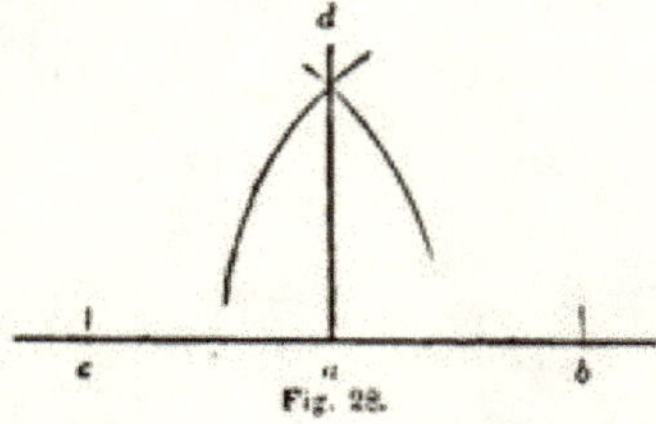

Fig. 28.

72.—*To erect a perpendicular.* From the point, *a*, (*Fig.* 28,) set oft any distance, as *a b*, and the same distance from *a* to *c*, upon *c*, as a centre, with any distance for radius greater than *c a*, describe an arc at *d*; upon *b*, with the same radius, describe another at *d*; join *d* and *a*, and the line, *d a*, will be the perpendicular required.

This, and the three following problems, are more easily performed by the use of the set-square—(see *Art.* 12.) Yet they are useful when the operation is so large that a set-square cannot be used.

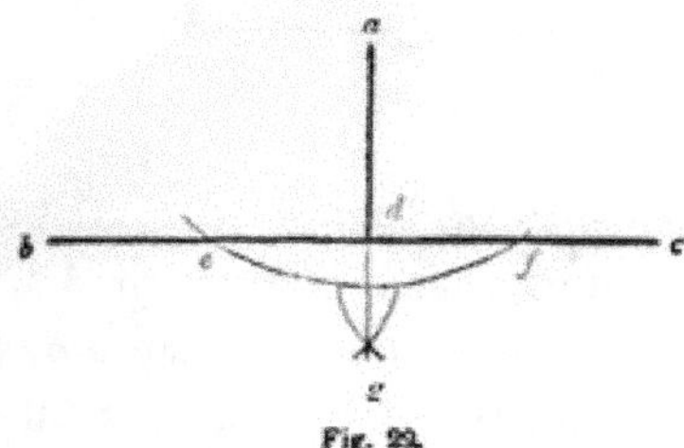

Fig. 29.

73.—*To let fall a perpendicular.* Let *a*, (*Fig.* 29,) be the point, above the line, *b c*, from which the perpendicular is required to fall. Upon *a*, with any radius greater than *a d*, describe an arc, cutting *b c* at *e* and *f*; upon the points, *e* and *f* with any radius greater than *e d*, describe arcs, cutting each other at *g*; join *a* and *g*, and the line, *a d*, will be the perpendicular required.

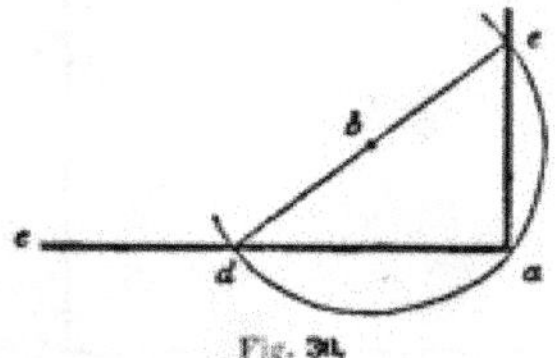

Fig. 30.

74.—*To erect a perpendicular at the end of a line.* Let *a*, (*Fig.* 30,) at the end of the line, *c a*, be the point at which the perpendicular is to be erected. Take any point, as *b*, above the line, *c a*, and with the radius, *b a*, describe the arc, *d a e*. through *d* and *b*, draw the line, *d e*; join *e* and *a*, then *e a* will be the perpendicular required.

The principle here made use of, is a very important one; and is applied in many other cases—(see *Art.* 81, *b.* and *Art.* 84. For proof of its correctness, see *Art.* 156.)

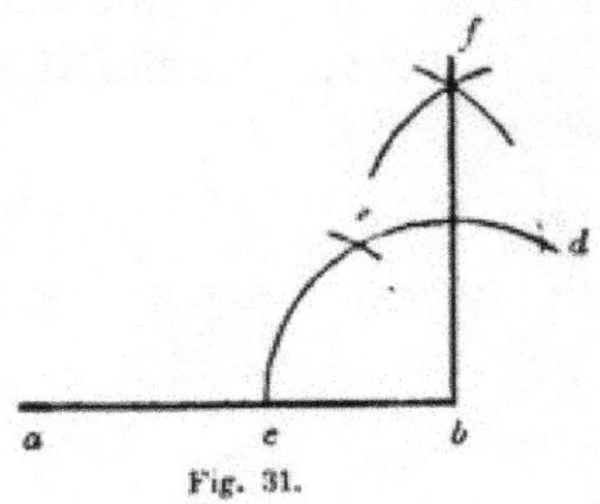

Fig. 31.

74, *a.*—*A second method.* Let *b*, (*Fig.* 31,) at the end of the line, *a b*, be the point at which it is required to erect a perpendicular. Upon *b*, with any radius less than *b a*, describe the arc, *c e d*; upon c, with the same radius, describe the small arc at *e*, and upon *e*, another at *d*; upon *e* and *d*, with the same or any other radius greater than half *e d*, describe arcs intersecting at *f*; join *f* and *b*, and the line, *f b*, will be the perpendicular required. This method of erecting a perpendicular and that of the following article, depend for accuracy upon the fact that the side of a hexagon is equal to the radius of the circumscribing circle.

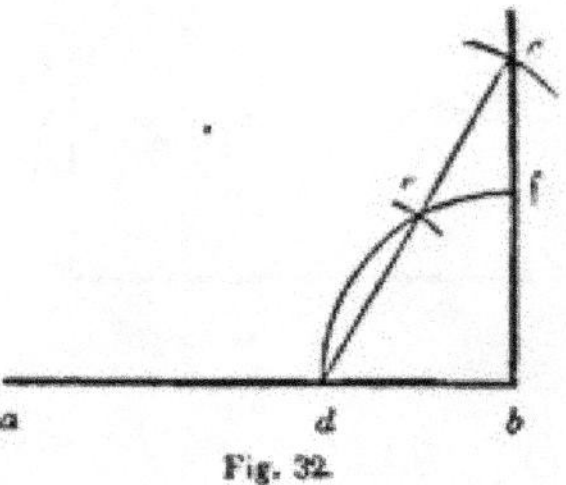

Fig. 32.

74, *b.*—*A third method.* Let *b*, (*Fig.* 32,) be the given point at which it is required to erect a perpendicular. Upon *b*, with any radius less than *b a*, describe the quadrant, *d e f*; upon *d*, with the same radius, describe an arc at *e*, and upon *e*, another at c, through *d* and *e*, draw *d c*, cutting the arc in *c*; join *c* and *b*, then *c b* will be the perpendicular required.

This problem can be solved by the *six, eight and ten* rule as it is called; which is founded upon the same principle as the problems at *Art.* 103, 104; and is applied as follows. Let *a d*, (*Fig.* 30,) equal eight, and *a e*, six; then, if *d e* equals ten, the angle, *e a d*, is a right angle. Because the square of six and that of eight, added together, equal the square of ten, thus: 6 × 6 = 36, and 8 × 8 = 64; 36 + 64 = 100, and 10 × 10 = 100. Any sizes, taken in the same proportion, as six, eight and ten, will produce the same effect: as 3, 4 and 5, or 12, 16 and 20. (See *Art.* 103.)

By the process shown at *Fig.* 30, the end of a board may be squared without a carpenters'-square. All that is necessary is a pair of compasses and a ruler. Let *c a* be the edge of the board, and *a* the point at which it is required to be squared. Take the point, *b*, as near as possible at an angle of forty-five degrees, or on a *mitre*-line, from *a*, and at about the middle of the board. This is not necessary to the working of the problem, nor does it affect its accuracy, but the result is more easily obtained. Stretch the compasses from *b* to *a*, and then bring the leg at *a* around to *d*; draw a line from *d*, through *b*, out indefinitely; take the distance, *d b*, and place it from *b* to *e*; join *e* and *a*; then *e a* will be at right angles to *c a*. In squaring the foundation of a building, or laying-out a garden, a rod and chalk-line may be used instead of compasses and ruler.

75.—*To let fall a perpendicular near the end of a line.* Let *e*, (*Fig.* 30,) be the point above the line, *c a*, from which the perpendicular is required to fall. From *e*, draw any line, as *e d*, obliquely to the line, *c a*; bisect *e d* at *b*; upon *b*, with the radius, *b e*, describe the arc, *e a d*; join *e* and *a*; then *e a* will be the perpendicular required.

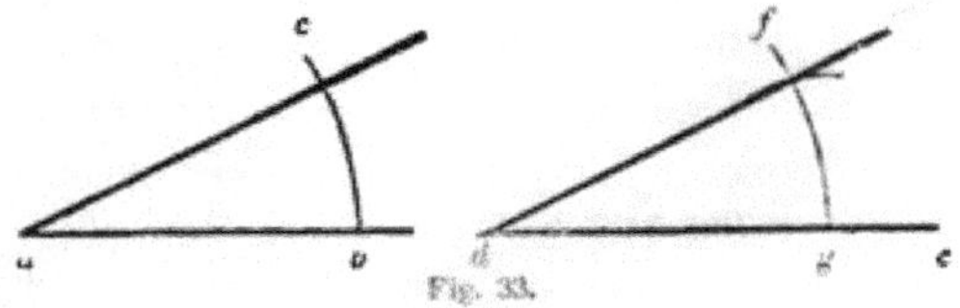

Fig. 33.

76.—*To make an angle,* (as *e d f*, *Fig.* 33,) *equal to a given angle,* (as *b a c*.) From the angular point, *a*, with any radius describe the arc, *b c*; and with the same radius, on the line, *d e*, and from the point, *d*, describe the arc, *f g*; take the distance, *b c*, and upon *g*, describe the small arc at *f*; join *f* and *d*; and the angle, *e d f*, will be equal to the angle, *b a c*.

If the given line upon which the angle is to be made, is situated parallel to the similar line of the given angle, this may be performed more readily with the set-square. (See *Art.* 11.)

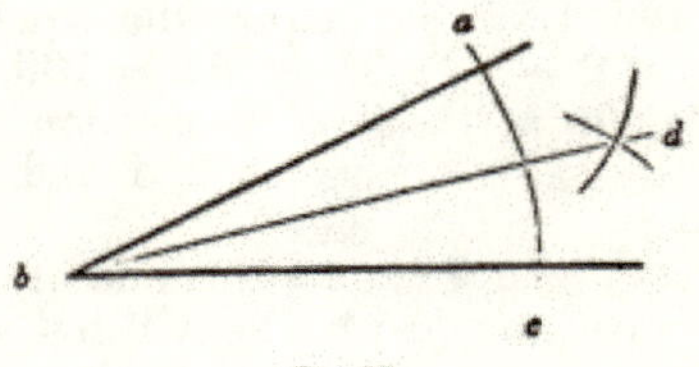

Fig. 34.

77.—*To bisect an angle.* Let *a b c*, (*Fig.* 34,) be the angle to be bisected. Upon *b*, with any radius, describe the arc, *a c*; upon *a* and *c*, with a radius greater than half *a c*, describe arcs cutting each other at *d*; join *b* and *d*; and *b d* will bisect the angle, *a b c*, as was required.

This problem is frequently made use of in solving other problems; it should therefore be well impressed upon the memory.

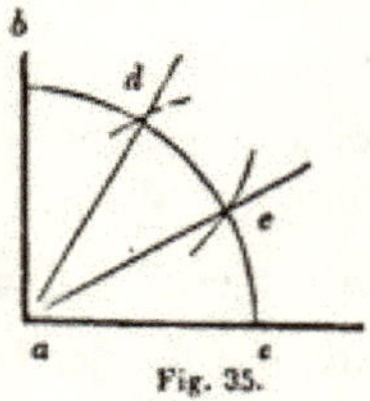

Fig. 35.

78.—*To trisect a right angle.* Upon *a*, (*Fig.* 35,) with any radius, describe the arc, *b c*; upon *b* and *c*, with the same radius, describe arcs cutting the arc, *b c*, at *d* and *e*; from *d* and *e*, draw lines to *a*, and they will trisect the angle as was required.

The truth of this is made evident by the following operation. Divide a circle into quadrants: also, take the radius in the dividers, and space off the circumference. This will divide the circumference into just six parts. A semi-circumference, there-tore, is equal to three, and a quadrant to one and a half of those parts. The radius, therefore, is equal to ⅔ of a quadrant; and this is equal to a right angle.

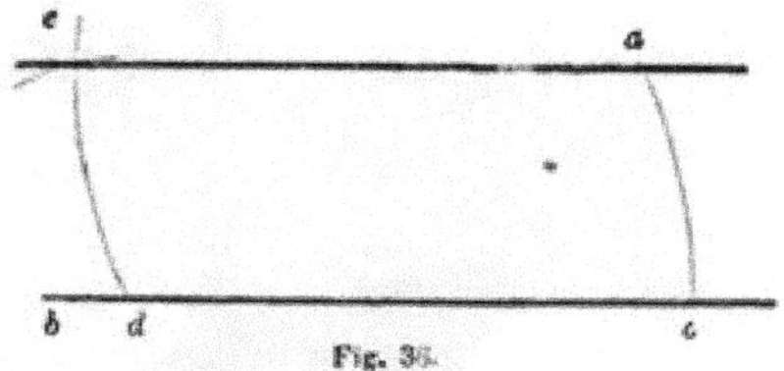

Fig. 36.

79.—*Through a given point, to draw a line parallel to a given line.* Let *a*, (*Fig.* 36,) be the given point, and *b c* the given line. Upon any point, as *d*, in the line, *b c*, with the radius, *d a*, describe the arc, *a c;* upon *a*, with the same radius, describe the arc, *d e ;* make *d e* equal to *a c ;* through *e* and *a* draw the line, *e a ;* which will be the line required.

This is upon the same principle as *Art.* 76.

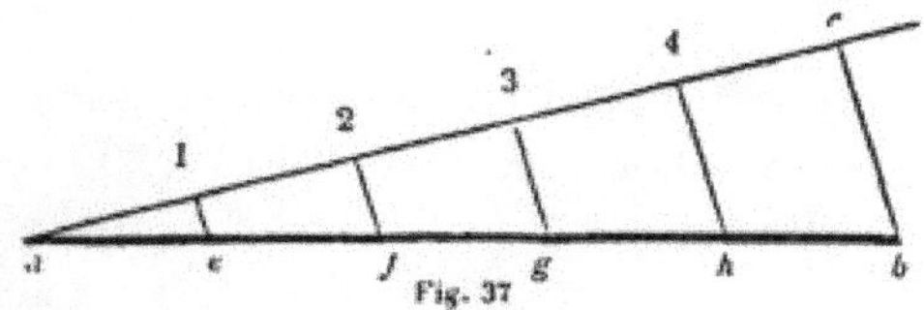

Fig. 37

80.—*To divide a given line into any number of equal parts.* Let *a b*, (*Fig.* 37,) be the given line, and 5 the number of parts. Draw *a c*, at any angle to *a b ;* on *a c*, from *a*, set off 5 equal parts of any length, as at 1, 2, 3, 4 and *c ;* join *c* and *b ;* through the points, 1, 2, 3 and 4, draw 1 *e*, 2 *f*, 3 *g* and 4 *h*, parallel to *c b ;* which will divide the line, *a b*, as was required.

The lines, *a b* and *a c*, are divided in the same proportion. (See *Art.* 109.)

THE CIRCLE.

81.—*To find the centre of a circle.* Draw any chord, as *a b*

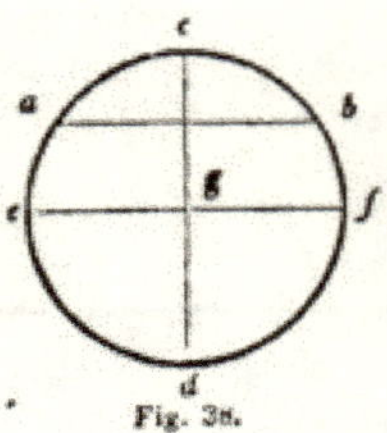

Fig. 38.

(*Fig.* 38,) and bisect it with the perpendicular, *c d*; bisect *c a* with the line, *e f*, as at *g*; then *g* is the centre as was required.

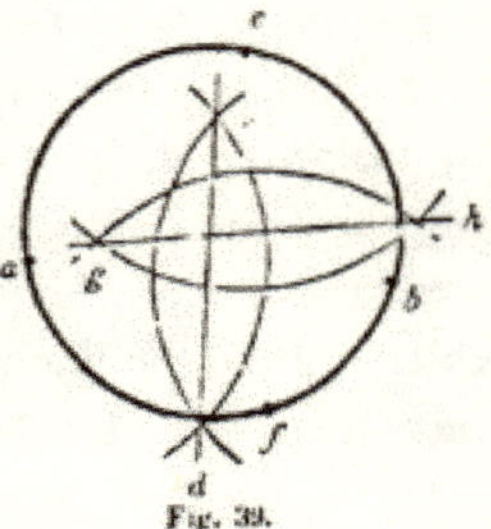

Fig. 39.

81, *a.*—*A second method.* Upon any two points in the circumference nearly opposite, as *a* and *b*. (*Fig.* 39,) describe arcs cutting each other at c and *d*: take any other two points, as *e* and *f*, and describe arcs intersecting as at *g* and *h*; join *g* and *h*, and *c* and *d*; the intersection, *o*, is the centre.

This is upon the same principle as *Art.* 85.

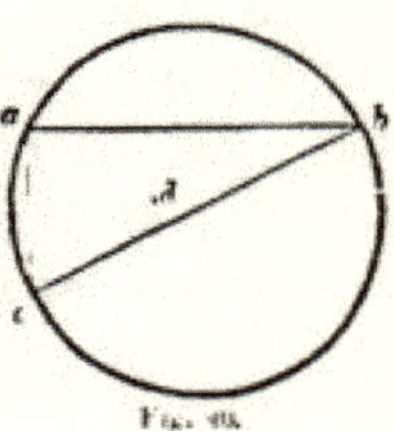

Fig. 40.

81, *b.*—*A third method.* Draw any chord, as *a b*. (*Fig.* 40,) and from the point, *a*, draw *a c*, at right angles to *a b*; join *c* and *b*; bisect *c b* at *d*—which will be the centre of the circle.

If a circle be not too large for the purpose, its centre may very readily be ascertained by the help of a carpenters'-square, thus: app'y the corner of the square to any point in the circumference, as at *a*; by the edges of the square, (which the lines, *a b* and *a c*, represent,) draw lines cutting the circle, as at *b* and *c*; join *b* and *c*; then if *b c* is bisected, as at *d*, the point, *d*, will be the centre. (See *Art.* 156.)

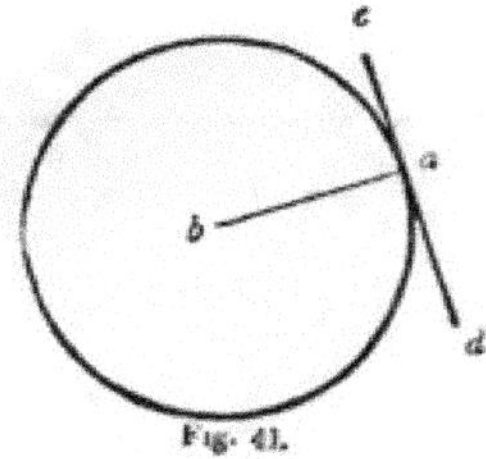

Fig. 41.

82.—*At a given point in a circle, to draw a tangent thereto.* Let *a*, (*Fig.* 41,) be the given point, and *b* the centre of the circle. Join *a* and *b*; through the point, *a*, and at right angles to *a b*, draw *c d*; then *c d* is the tangent required.

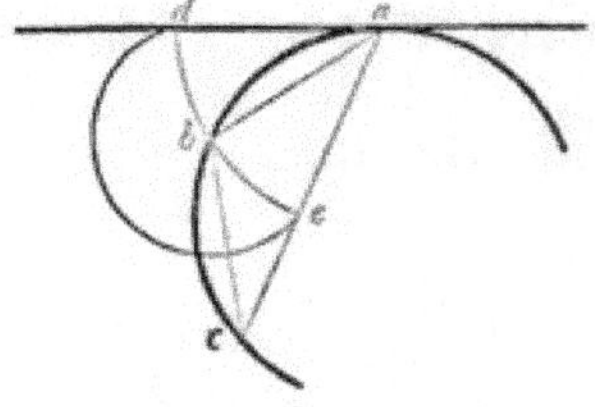

Fig. 42

83.—*The same, without making use of the centre of the circle.* Let *a*, (*Fig.* 42,) be the given point. From *a*, set off any distance to *b*, and the same from *b* to *c*; join *a* and *c*, upon *a*, with *a b* for radius, describe the arc, *d b e*; make *d b* equal to *b e*; through *a* and *d*, draw a line; this will be the tangent required.

The correctness of this method depends upon the fact that the angle formed by a chord and tangent is equal to any inscribed angle in the opposite segment of the circle, (*Art.* 163;) *a b* being the chord, and *b c a* the angle in the opposite segment of the circle. Now, the angles *d a b* and *b c a* are equal, because the angles *d a b* and *b a c* are, by construction, equal; and the angles *b a c* and *b c a* are equal, because the triangle *a b c* is an isosceles triangle, having its two sides, *a b* and *b c*, by construction equal; therefore the angles *d a b* and *b c a* are equal.

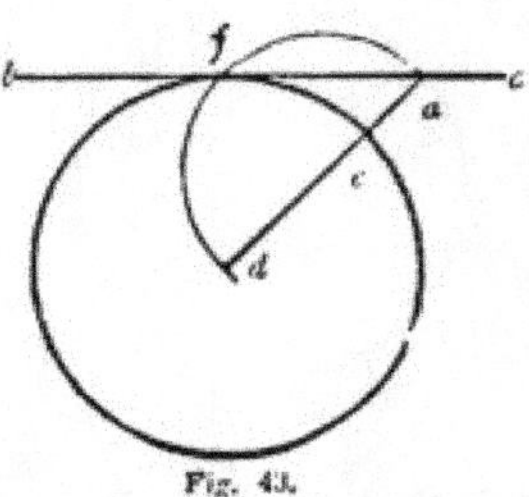

Fig. 43.

84.—*A circle and a tangent given, to find the point of contact.* From any point, as *a*, (*Fig.* 43,) in the tangent, *b c*, draw a line to the centre *d*; bisect *a d* at *e*; upon *e*, with the radius, *e a*, describe the arc, *a f d*; *f* is the point of contact required.

If *f* and *d* were joined, the line would form right angles with the tangent, *b c*. (See *Art.* 156.)

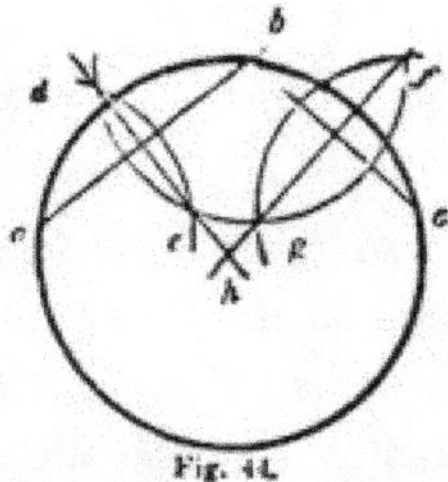

Fig. 44.

85.—*Through any three points not in a straight line, to draw a circle.* Let *a*, *b* and *c*, (*Fig.* 44,) be the three given points. Upon *a* and *b*, with any radius greater than half *a b*, describe arcs intersecting at *d* and *e*; upon *b* and *c*, with any radius greater than half *b c*, describe arcs intersecting at *f* and *g*; through *d* and *e*, draw a right line, also another through *f* and *g*; upon the intersection, *h*, with the radius, *h a*, describe the circle, *a b c*, and it will be the one required.

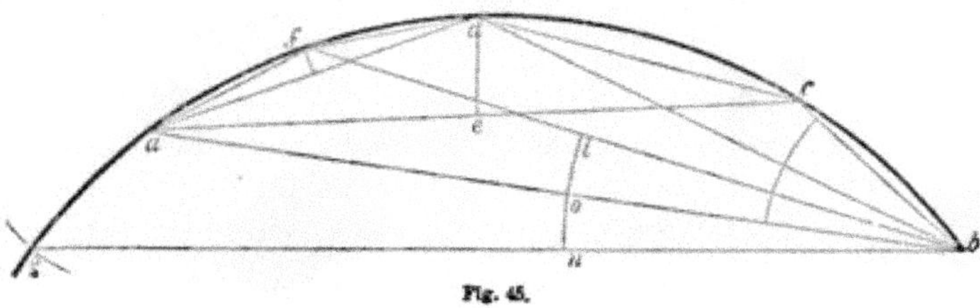

Fig. 45.

86.—*Three points not in a straight line being given, to find a fourth that shall, with the three, lie in the circumference of a circle.* Let *a b c*, (*Fig.* 45,) be the given points. Connect them with right lines, forming the triangle, *a c b*; bisect the angle, *c b a*, (*Art.* 77,) with the line *b d*; also bisect *c a* in *e*, and erect *e d*, perpendicular to *a c*, cutting *b d* in *d*; then *d* is the *fourth* point required.

A fifth point may be found, as at *f*, by assuming *a*, *d* and *b*, as the three given points, and proceeding as before. So, also, any number of points may be found; simply by using any three already found. This problem will be serviceable in obtaining short pieces of very flat sweeps. (See *Art.* 397.)

The proof of the correctness of this method is found in the fact that equal chords subtend equal angles, (*Art.* 162.) Join *d* and *c*; then since *a e* and *e c* are, by construction, equal, therefore the chords *a d* and *d c* are equal; hence the angles they subtend, *d b a* and *d b c*, are equal. So likewise chords drawn from *a* to *f*, and from *f* to *d*, are equal, and subtend the equal angles, *d b f* and *f b a*. Additional points, *beyond a* or *b*, may be obtained on the same principle. To obtain a point beyond *a*, on *b*, as a centre, describe with any radius the arc *i o n*, make *o n* equal to *o i*; through *b* and *n* draw *b g*; on *a* as

4

centre and with $a f$ for radius, describe the arc, cutting $g b$ at g, then g is the point sought.

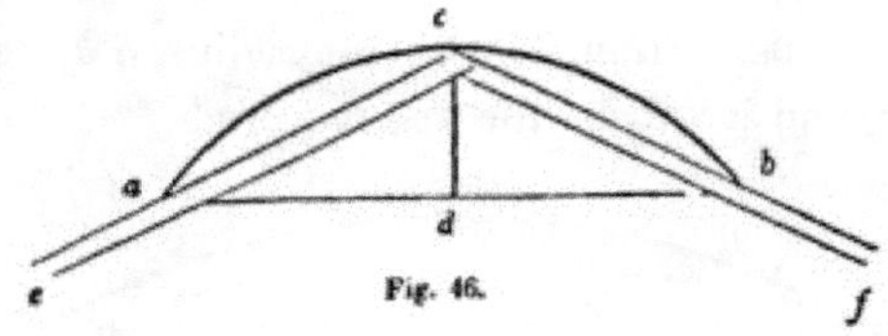

Fig. 46.

87.—*To describe a segment of a circle by a set-triangle.* Let $a b$, (*Fig.* 46,) be the chord, and $c d$ the height of the segment. Secure two straight-edges, or rulers, in the position, $c e$ and $c f$, by nailing them together at c, and affixing a brace from e to f; put in pins at a and b; move the angular point, c, in the direction, $a c b$; keeping the edges of the triangle hard against the pins, a and b; a pencil held at c will describe the arc, $a c b$.

A curve described by this process is accurately *circular*, and is not a mere approximation to a circular arc, as some may suppose. This method produces a circular curve, because all inscribed angles on one side of a chord line are equal. (*Art.* 161.) To obtain the radius from a chord and its versed sine, see *Art.* 165.

If the angle formed by the rulers at c be a right angle, the segment described will be a semi-circle. This problem is useful in describing centres for brick arches, when they are required to be rather flat. Also, for the head hanging-stile of a window-frame, where a brick arch, instead of a stone lintel, is to be placed over it.

87 *a.*—*To find the radius of an arc of a circle when the chord and versed sine are given.* The radius is equal to the sum of the squares of half the chord and of the versed sine, divided by twice the versed sine. This is expressed, algebraically, thus—$r = \frac{(\frac{c}{2})^2 + v^2}{2 v}$, where r is the radius, c the chord, and v the versed sine. (*Art.* 165.)

Example.—In a given arc of a circle, a chord of 12 feet has the rise at the middle, or the versed sine, equal to 2 feet, what is the radius?

Half the chord equals 6, the square of 6 is, $6 \times 6 = 36$
The square of the versed sine is, $2 \times 2 = \quad 4$

Their sum equals, 40

Twice the versed sine equals 4, and 40 divided by 4 equals 10 Therefore the radius, in this case, is 10 feet. This result is shown in less space and more neatly by using the above algebraical formula. For the letters, substituting their value, the formula $r = \frac{(\frac{c}{2})^2 + v^2}{2v}$ becomes $r = \frac{(\frac{12}{2})^2 + 2^2}{2 \times 2}$, and performing the arithmetical operations here indicated equals

$$\frac{6^2 + 2^2}{4} = \frac{36 + 4}{4} = \frac{40}{4} = 10$$

87 *b.*—*To find the versed sine of an arc of a circle when the radius and chord are given.* The versed sine is equal to the radius, less the square root of the difference of the squares of the radius and half chord: expressed algebraically thus—$v = r - \sqrt{r^2 - (\frac{c}{2})^2}$, where r is the radius, v the versed sine, and c the chord. (*Art.* 158.)

Example.—In an arc of a circle whose radius is 75 feet, what is the versed sine to a chord of 120 feet? By the table in the Appendix it will be seen that—

The square of the radius, 75, equals, . .	5625
The square of half the chord, 60, equals, .	3600
The difference is, ...	2025
The square root of this is, ...	45
This deducted from the radius, ...	75
The remainder is the versed sine, =	30

This is expressed by the formula thus—

$$v = 75 - \sqrt{75^2 - (\tfrac{120}{2})^2} = 75 - \sqrt{5625 - 3600} = 75 - 45 = 30$$

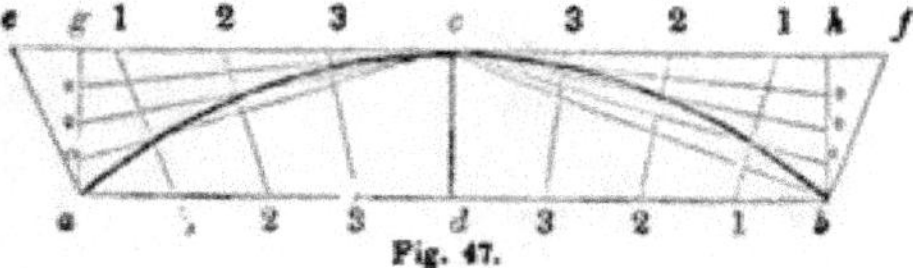

Fig. 47.

88.—*To describe the segment of a circle by intersection of lines.* Let *a b*, (*Fig.* 47,) be the chord, and *c d* the height of the segment. Through *c*, draw *c f*, parallel to *a b*; draw *b f* at right angles to *c b*; make *c e* equal to *c f*; draw *a g* and *b h*, at right angles to *a b*; divide *c e*, *c f*, *d a*, *d b*, *a g*, and *b h*, each into a like number of equal parts, as four; draw the lines, 1 1, 2 2, &c., and from the points, *o*, *o* and *o*, draw lines to *c*; at the intersection of these lines, trace the curve, *a c b*, which will be the segment required.

In very large work, or in laying out ornamented gardens, &c., this will be found useful; and where the centre of the proposed arc of a circle is inaccessible it will be invaluable. (To trace the curve, see note at *Art.* 117.)

The lines *e a*, *c d* and *f b*, would, were they extended, meet in a point, and that point would be in the opposite side of the circumference of the circle of which *a c b* is a segment. The lines 1 1, 2 2, 3 3, would likewise, if extended, meet in the same point. The line, *c d*, if extended to the opposite side of the circle, would become a diameter. The line, *f b*, forms, by construction, a right angle with *b c*, and hence the extension of *f b* would also form a right angle with *b c*, on the opposite side of *b c*; and this right angle would be the inscribed angle in the semicircle; and since this is required to be a *right* angle, (*Art.* 156,) therefore the construction thus far is correct, and it will be found likewise that at each point in the curve formed by the intersection of the radiating lines, these intersecting lines are at right angles.

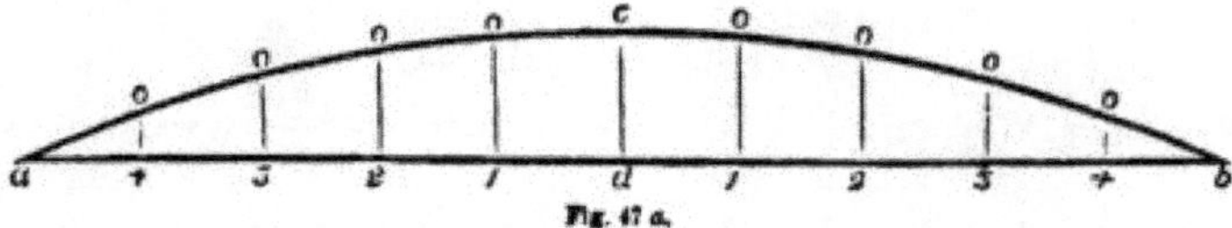

Fig. 47 a.

88 *a*.—Points in the circumference of a circle may be obtained arithmetically, and positively accurate, by the calculation of *ordinates*, or the parallel lines, 0 1, 0 2, 0 3, 0 4. (*Fig.*

47 *a*.) These ordinates are drawn at right angles to the chord line, *a b*, and they may be drawn at any distance apart, either equally distant or unequally, and there may be as many of them as is desirable; the more there are the more points in the curve will be obtained. If they are located in pairs, equally distant from the versed sine, *c d*, calculation need be made only for those on one side of *c d*, as those on the opposite side will be of equal lengths, respectively; for example, 0 1, on the left-hand side of *c d*, is equal to 0 1 on the right-hand side, 0 2 on the right equals 0 2 on the left, and in like manner for the others.

The length of any ordinate is equal to the square root of the difference of the squares of the radius and abscissa, less the difference between the radius and versed sine. (*Art.* 166.) The abscissa being the distance from the foot of the versed sine to the foot of the ordinate. Algebraically, $y = \sqrt{r^2 - x^2} - (r - v)$, where y is put to represent the ordinate; x, the abscissa; v, the versed sine; and r, the radius.

Example.—An arc of a circle has its chord, *a b*, (*Fig.* 47 *a*,) 100 feet long, and its versed sine, *c d*, 5 feet. It is required to ascertain the length of ordinates for a sufficient number of points through which to describe the curve. To this end it is requisite, first, to ascertain the radius. This is readily done in accordance with *Art.* 87 *a*. For, $\frac{(\frac{c}{2})^2 + v^2}{2v}$, becomes $\frac{50^2 + 5^2}{2 \times 5} =$ 252·5 = radius. Having the radius, the curve might at once be described without the ordinate points, but for the impracticability that usually occurs, in large, flat segments of the circle, of getting a location for the centre; the centre usually being inaccessible. The ordinates are, therefore, to be calculated. In *Fig.* 47 *a* the ordinates are located equidistant, and are 10 feet apart. It will only be requisite, therefore, to calculate those on one side of the versed sine, *c d*. For the first ordinate, 0 1, the formula, $y = \sqrt{r^2 - x^2} - (r - v)$ becomes

$$
\begin{aligned}
y &= \sqrt{252{\cdot}5^2 - 10^2} - (252{\cdot}5 - 5).\\
&= \sqrt{63756{\cdot}25 - 100} - 247{\cdot}5.\\
&= 252{\cdot}3019 - 247{\cdot}5.\\
&= 4{\cdot}8019 = \text{the first ordinate, 0 1.}
\end{aligned}
$$

For the second—

$$y = \sqrt{252{\cdot}5^2 - 20^2} - (252{\cdot}5 - 5).$$
$$= \quad 251{\cdot}7066 - 247{\cdot}5.$$
$$= \quad 4{\cdot}2066 = \text{the second ordinate, } 0\,2.$$

For the third—

$$y = \sqrt{252{\cdot}5^2 - 30^2} - 247{\cdot}5.$$
$$= \quad 250{\cdot}7115 - 247{\cdot}5.$$
$$= \quad 3{\cdot}2115 = \text{the third ordinate, } 0\,3.$$

For the fourth—

$$y = \sqrt{252{\cdot}5^2 - 40^2} - 247{\cdot}5.$$
$$= \quad 249{\cdot}3115 - 247{\cdot}5.$$
$$= \quad 1{\cdot}8115 = \text{the fourth ordinate, } 0\,4.$$

The results here obtained are in feet and decimals of a foot. To reduce these to feet, inches, and eighths of an inch, proceed as at Reduction of Decimals in the Appendix. If the two-feet rule, used by carpenters and others, were decimally divided, there would be no necessity of this reduction, and it is to be hoped that the rule will yet be thus divided, as such a reform would much lessen the labor of computations, and insure more accurate measurements.

Versed sine, *c d*, =	ft. 5·0		= ft. 5·0 inches.	
Ordinates,	0 1, =	4·8019	=	4·9⅝ inches nearly.
"	0 2, =	4·2066	=	4·2½ inches nearly.
"	0 3, =	3·2115	=	3·2½ inches nearly.
"	0 4, =	1·8115	=	1·9¾ inches nearly.

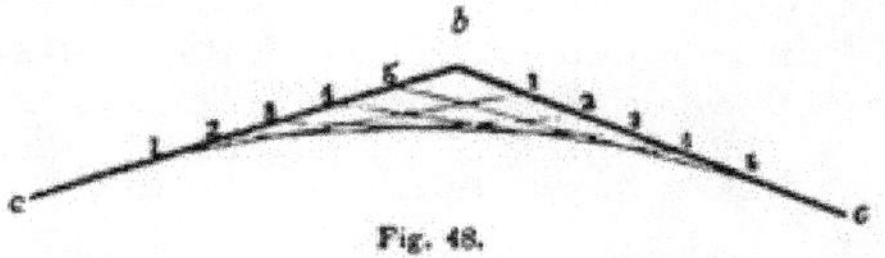

Fig. 48.

89.—*In a given angle, to describe a tanged curve.* Let *a b c*, (*Fig.* 48,) be the given angle, and 1 in the line, *a b*, and 5 in the line, *b c*, the termination of the curve. Divide 1 *b* and *b* 5 into a like number of equal parts, as at 1, 2, 3, 4 and 5; join 1 and 1, 2 and 2, 3 and 3, &c.; and a regular curve will be formed that will be tangical to the line, *a b*, at the point, 1, and to *b c* at 5.

This is of much use in stair-building, in easing the angles formed between the wall-string and the base of the hall, also between the front string and level facia, and in many other instances. The curve is not circular, but of the form of the parabola, (*Fig.* 93;) yet in large angles the difference is not perceptible. This problem can be applied to describing the

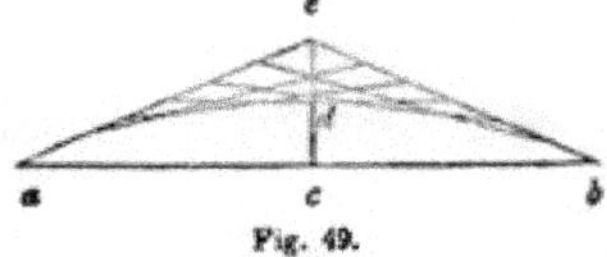

Fig. 49.

curve for door heads, window-heads, &c., to rather better advantage than *Art.* 87. For instance, let *a b*, (*Fig.* 49,) be the width of the opening, and *c d* the height of the arc. Extend *c d*, and make *d e* equal to *c d*; join *a* and *e*, also *e* and *b*; and proceed as directed above.

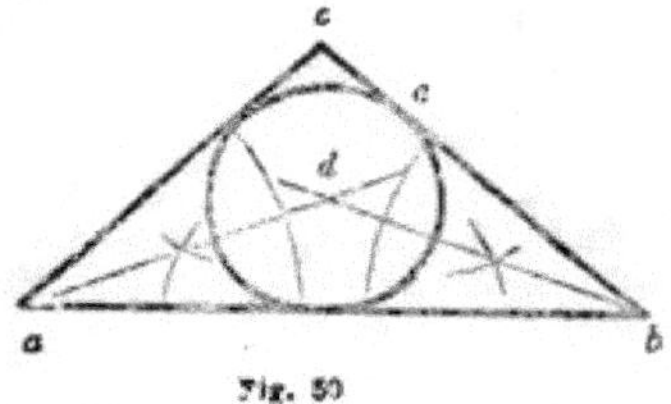

Fig. 50

90.—*To describe a circle within any given triangle, so that the sides of the triangle shall be tangical.* Let *a b c*, (*Fig.* 50,) be the given triangle. Bisect the angles *a* and *b*, according to *Art.* 77; upon *d*, the point of intersection of the bisecting lines, with the radius, *d e*, describe the required circle.

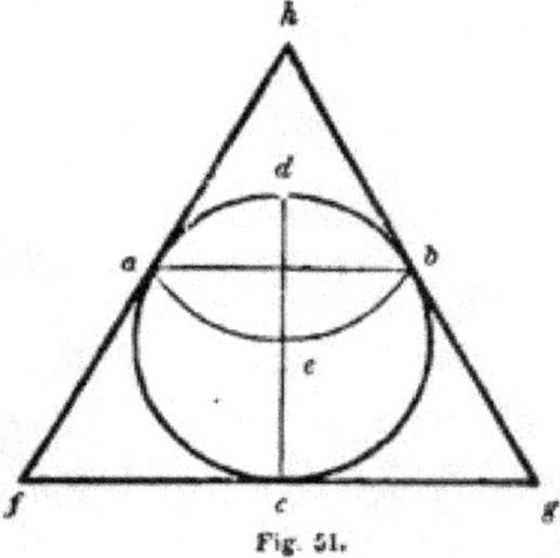

Fig. 51.

91.—*About a given circle, to describe an equi-lateral triangle.* Let *a d b c*, (*Fig.* 51,) be the given circle. Draw the diameter, *c d*; upon *d*, with the radius of the given circle, describe the arc, *a e b*; join *a* and *b*; draw *f g*, at right angles to *d c*; make *f c* and *c g*, each equal to *a b*; from *f*, through *a*, draw *f h*, also from *g*, through *b*, draw *g h*; then *f g h* will be the triangle required.

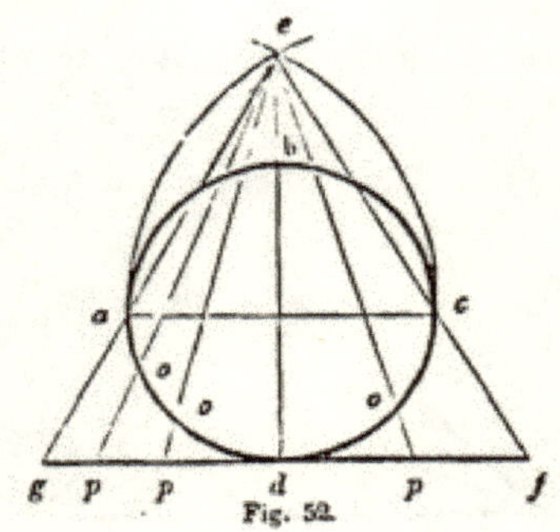

Fig. 52.

92.—*To find a right line nearly equal to the circumference of a circle.* Let *a b c d*. (*Fig.* 52,) be the given circle. Draw the diameter, *a c*; on this erect an equi-lateral triangle, *a e c*, according to *Art.* 93; draw *g f*, parallel to *a c*; extend *e c* to *f*, also *e a* to *g*; then *g f* will be nearly the length of the semi-circle, *a d c*; and twice *g f* will nearly equal the circumference of the circle, *a b c d*, as was required.

Lines drawn from *e*, through any points in the circle, as *o*, *o* and *o*, to *p*, *p* and *p*, will divide *g f* in the same way as the semi-circle, *a d c*, is divided. So, any portion of a circle may be transferred to a straight line. This is a very useful problem, and should be well studied; as it is frequently used to solve problems on stairs, domes, &c.

92, *a.*—*Another method.* Let *a b f c*, (*Fig.* 53,) be the given circle. Draw the diameter, *a c*; from *d*, the centre, and at right angles to *a c*, draw *d b*; join *b* and *c*; bisect *b c* at *e*; from *d*, through *e*, draw *d f*; then *e f* added to three times the diameter, will equal the circumference of the circle sufficiently near for

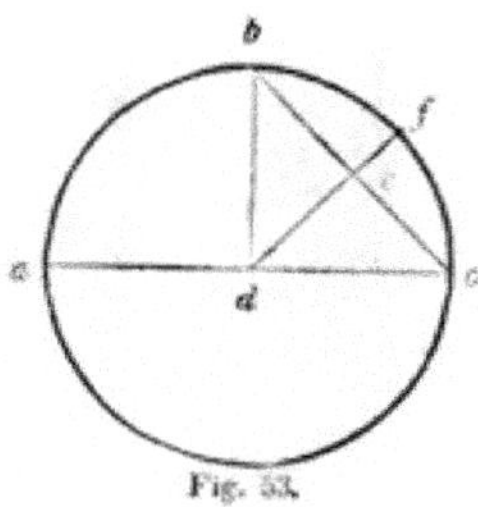

Fig. 53.

many uses. The result is a trifle too large, If the circumference found by this rule, be divided by 648·22, the quotient will be the excess. Deduct this excess, and the remainder will be the true circumference. This problem is rather more curious than useful, as it is less labor to perform the operation arithmetically: simply multiplying the given diameter by 3·1416, or where a great degree of accuracy is needed by 3·1415926.

POLYGONS, &C.

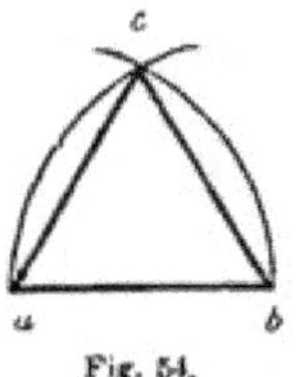

Fig. 54.

93.—*Upon a given line to construct an equi-lateral triangle.* Let *a b*, (*Fig.* 54,) be the given line. Upon *a* and *b*, with *a b* for radius, describe arcs, intersecting at *c;* join *a* and *c*, also *c* and *b;* then *a c b* will be the triangle required.

94.—*To describe an equi-lateral rectangle, or square.* Let *a b*, (*Fig.* 55,) be the length of a side of the proposed square. Upon *a* and *b*, with *a b* for radius, describe the arcs *a d* and *b c;* bisect the arc, *a e*, in *f;* upon *e*, with *e f* for radius, de-

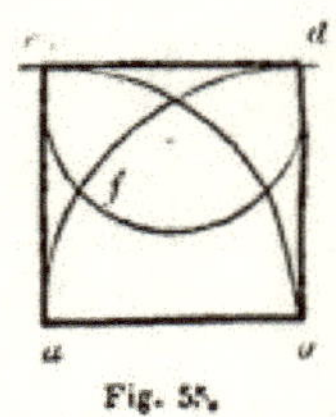

Fig. 55.

scribe the arc, *c f d;* join *a* and *c*, *c* and *d*, *d* and *b;* then *a c d b* will be the square required.

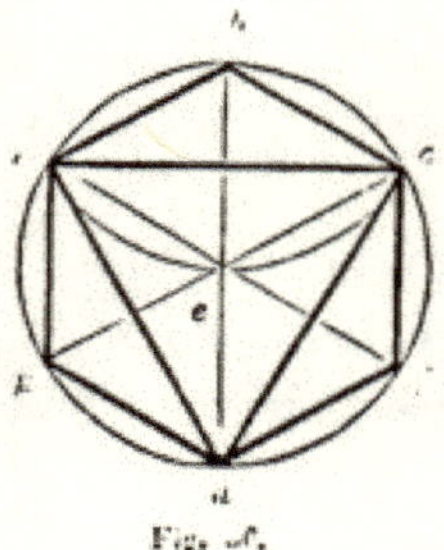

Fig. 56.

95.—*Within a given circle, to inscribe an equi-lateral triangle, hexagon or dodecagon.* Let *a b c d*, (*Fig.* 56,) be the given circle. Draw the diameter, *b d;* upon *b*, with the radius of the given circle, describe the arc, *a e c;* join *a* and *c*, also *a* and *d*, and *c* and *d*—and the triangle is completed. For the hexagon: from *a*, also from *c*, through *e*, draw the lines, *a f* and *c g;* join *a* and *b*, *b* and *c*, *c* and *f*, &c., and the hexagon is completed. The dodecagon may be formed by bisecting the sides of the hexagon.

Each side of a regular hexagon is exactly equal to the radius of the circle that circumscribes the figure. For the radius is equal to a chord of an arc of 60 degrees; and, as every circle is supposed to be divided into 360 degrees, there is just 6 times 60, or 6 arcs of 60 degrees, in the whole circumference. A line drawn from each angle of the hexagon to the centre, (as in the figure,) divides it into six equal, equi-lateral triangles.

96.—*Within a square to inscribe an octagon.* Let *a b c d*,

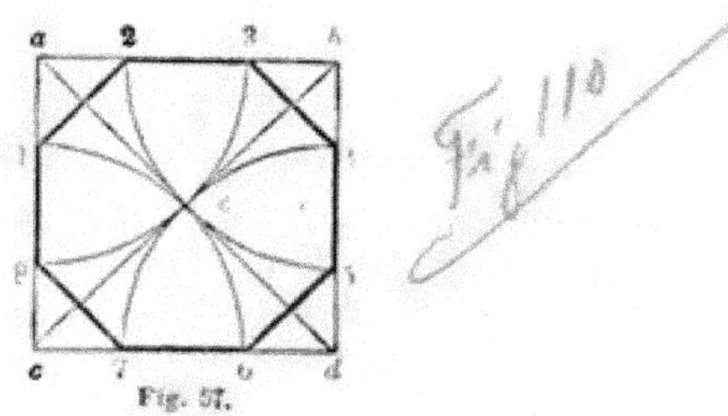

Fig. 57.

(*Fig.* 57,) be the given square. Draw the diagonals, *a d* and *b c;* upon *a*, *b*, *c* and *d*, with *a e* for radius, describe arcs cutting the sides of the square at 1, 2, 3, 4, 5, 6, 7 and 8; join 1 and 2, 3 and 4, 5 and 6, &c., and the figure is completed.

In order to eight-square a hand-rail, or any piece that is to be afterwards rounded, draw the diagonals, *a d* and *b c*, upon the end of it, after it has been squared-up. Set a gauge to the distance, *a e*, and run it upon the whole length of the stuff, from each corner both ways. This will show how much is to be chamfered off, in order to make the piece octagonal. (*Art.* 159.)

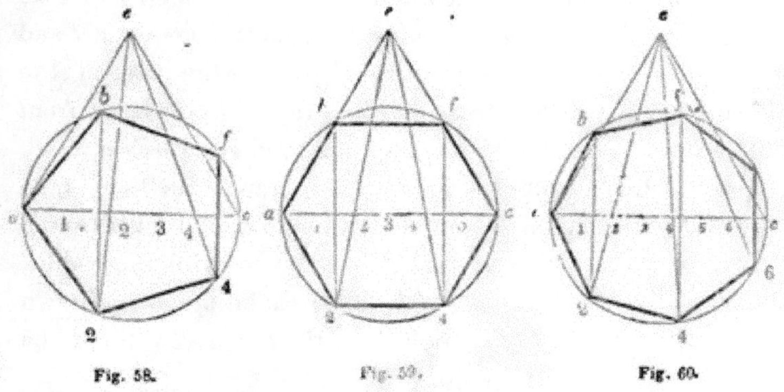

Fig. 58. Fig. 59. Fig. 60.

97.—*Within a given circle to inscribe any regular polygon.* Let *a b c* 2, (*Fig.* 58, 59 and 60,) be given circles. Draw the diameter, *a c;* upon this, erect an equi-lateral triangle, *a e c*, according to *Art.* 93; divide *a c* into as many equal parts as the polygon is to have sides, as at 1, 2, 3, 4, &c.; from *e*, through each even number, as 2, 4, 6, &c., draw lines cutting the circle in the points, 2, 4, &c.; from these points and at right angles to *a c*, draw lines to the opposite part of the circle; this will give the remaining points for the polygon, as *b*, *f*, &c.

In forming a hexagon, the sides of the triangle erected upon *a c*, (as at *Fig.* 59,) mark the points *b* and *f*. This method of locating the angles of a polygon is an approximation sufficiently near for many purposes; it is based upon the like principle with the method of obtaining a right line *nearly* equal to a circle. (*Art.* 92.) The method shown at *Art.* 98 is accurate.

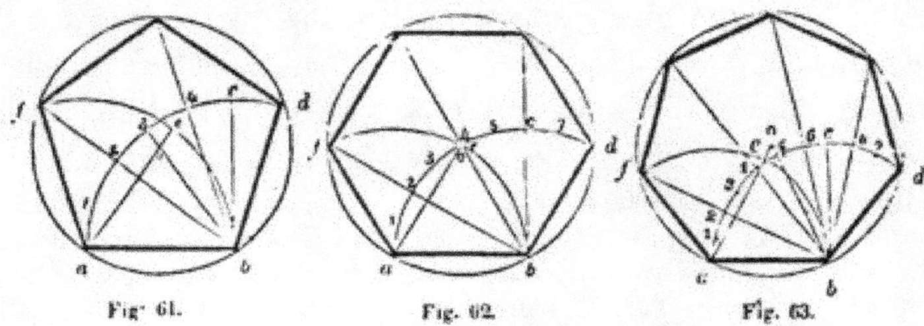

Fig. 61. Fig. 62. Fig. 63.

98.—*Upon a given line to describe any regular polygon.* Let *a b*, (*Fig.* 61, 62 and 63,) be given lines, equal to a side of the required figure. From *b*, draw *b c*, at right angles to *a b*; upon *a* and *b*, with *a b* for radius, describe the arcs, *a c d* and *f c b*; divide *a c* into as many equal parts as the polygon is to have sides, and extend those divisions from *c* towards *d*; from the second point of division counting from *c* towards *a*, as 3, (*Fig.* 61,) 4, (*Fig.* 62,) and 5, (*Fig.* 63,) draw a line to *b*; take the distance from said point of division to *a*, and set it from *b* to *e*; join *e* and *a*; upon the intersection, *o*, with the radius, *o a*, describe the circle *a f d b*; then radiating lines, drawn from *b* through the even numbers on the arc, *a d*, will cut the circle at the several angles of the required figure.

In the hexagon, (*Fig.* 62,) the divisions on the arc, *a d*, are not necessary; for the point, *o*, is at the intersection of the arcs, *a d* and *f b*, the points, *f* and *d*, are determined by the intersection of those arcs with the circle, and the points above, *g* and *h*, can be found by drawing lines from *a* and *b*, through the centre, *o*. In polygons of a greater number of sides than the hexagon, the intersection, *o*, comes above the arcs; in such case, therefore, the lines, *a e* and *b* 5, (*Fig.* 63,) have to be extended before they will intersect. This method of describing polygons is founded on correct principles, and is therefore accurate. In the circle equal arcs subtend equal angles, (*Arts.* 86 and 162.) Although this method is accurate, yet polygons may be described as accurately and more simply in the following manner. It will be observed that much of the process in this method is for the purpose of ascertaining the centre of a circle that will circumscribe the proposed polygon. By reference to the Table of Polygons in the Appendix it will be seen how this centre may be obtained arithmetically. This is the *Rule.*—Multiply the given side by the tabular radius for polygons of a like number of sides with the proposed figure, and the product will be the radius of the required circumscribing circle. Divide this circle into as many equal parts as the polygon is to have sides, connect the points of division by straight lines, and the figure is complete. For example: It is desired to describe a polygon of 7 sides, and 20 inches a side. The tabular radius is 1·1523824. This multiplied by 20, the product, 23·047648 is the required radius in inches. The Rules for the Reduction of Decimals, also in the Appendix, show how to change decimals to the fractions of a foot or an inch. From this, 23·047648 is equal to $23\frac{1}{16}$ inches nearly. It is not needed to take all the decimals in the table, three or four of them will give a result sufficiently near for all ordinary practice.

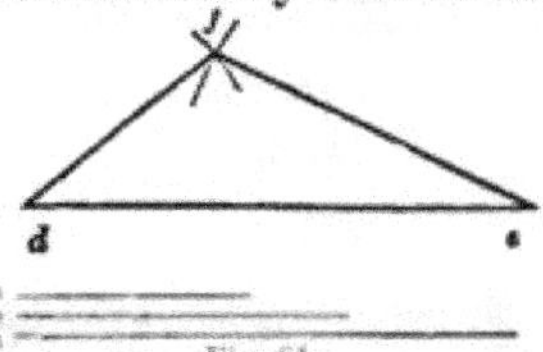

Fig. 64.

99.—*To construct a triangle whose sides shall be severally equal to three given lines.* Let *a*, *b* and *c*, (*Fig.* 64,) be the given lines. Draw the line, *d e*, and make it equal to *c*; upon *e*, with *b* for radius, describe an arc at *f*; upon *d*, with *a* for radius, describe an arc intersecting the other at *f*; join *d* and *f*, also *f* and *e*; then *d f e* will be the triangle required.

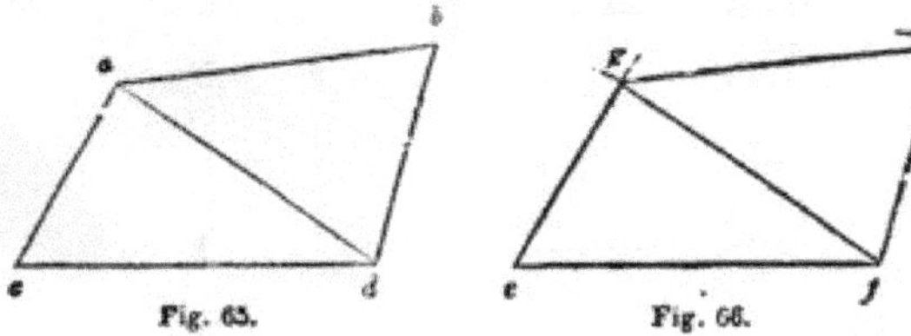

Fig. 65. Fig. 66.

100.—*To construct a figure equal to a given, right-lined figure.* Let *a b c d*, (*Fig.* 65,) be the given figure. Make *e f*, (*Fig.* 66,) equal to *c d*; upon *f*, with *d a* for radius, describe an arc at *g*; upon *e*, with *c a* for radius, describe an arc intersecting the other at *g*; join *g* and *e*; upon *f* and g, with *d b* and *a b* for radius, describe arcs intersecting at *h*; join *g* and *h*, also *h* and *f*; then *Fig.* 66 will every way equal *Fig.* 65.

So, right-lined figures of any number of sides may be copied, by first dividing them into triangles, and then proceeding as above. The shape of the floor of any room, or of any piece of land, &c., may be accurately laid out by this problem, at a scale upon paper; and the contents in square feet be ascertained by the next.

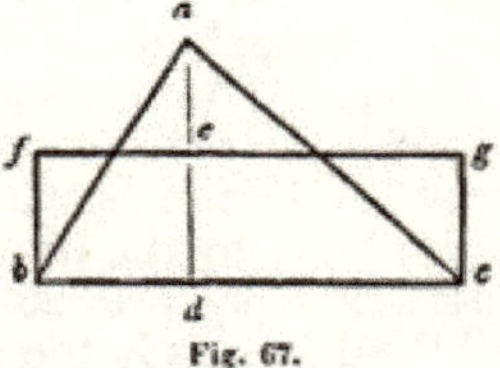

Fig. 67.

101.—*To make a parallelogram equal to a given triangle.* Let *a b c*, (*Fig.* 67,) be the given triangle. From *a*, draw *a d*, at right angles to *b c*; bisect *a d* in *e*; through *e*, draw *f g*, parallel to *b c*; from *b* and *c*, draw *b f* and *c g*, parallel to *d e*; then *b f g c* will be a parallelogram containing a surface exactly equal to that of the triangle, *a b c*.

Unless the parallelogram is required to be a rectangle, the lines, *b f* and *c g*, need not be drawn parallel to *d e*. If a rhomboid is desired, they may be drawn at an oblique angle, provided they be parallel to one another. To ascertain the area of a triangle, multiply the base, *b c*, by half the perpendicular height, *d a*. In doing this, it matters not which side is taken for base.

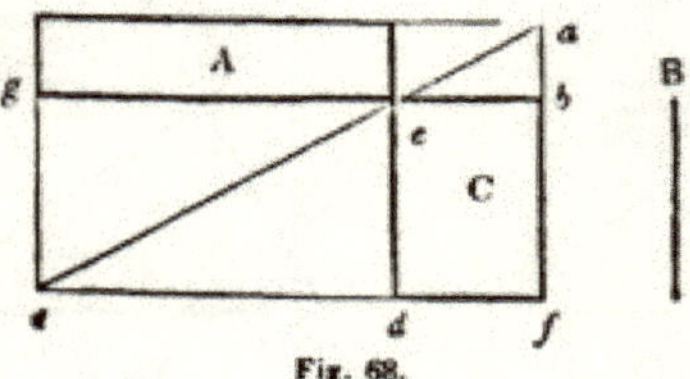

Fig. 68.

102.—*A parallelogram being given, to construct another equal to it, and having a side equal to a given line.* Let *A* (*Fig.* 68,) be the given parallelogram, and *B* the given line Produce the sides of the parallelogram, as at *a*, *b*, *c* and *d*; make *e d* equal to *B*; through *d*, draw *c f*, parallel to *g b*; through *e*, draw the diagonal, *c a*; from *a*, draw *a f*, parallel to *e d* then *C* will be equal to *A*. (See *Art.* 144.)

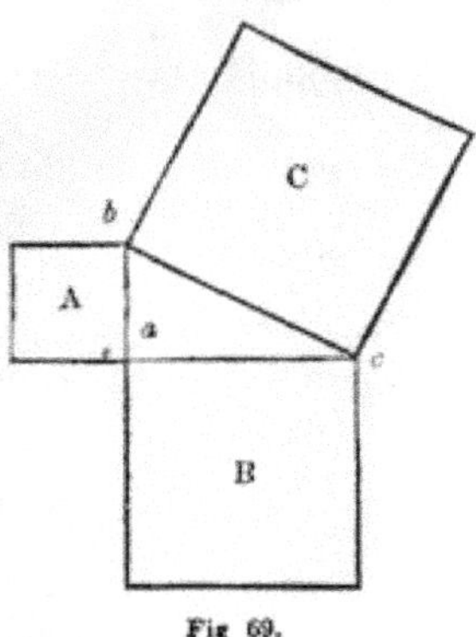

Fig 69.

103.—*To make a square equal to two or more given squares.* Let *A* and *B*, (*Fig.* 69,) be two given squares. Place them so as to form a right angle, as at *a*; join *b* and *c*; then the square, *C*, formed upon the line, *b c*, will be equal in extent to the squares, *A* and *B*, added together. Again: if *a b*, (*Fig.* 70,) be equal to

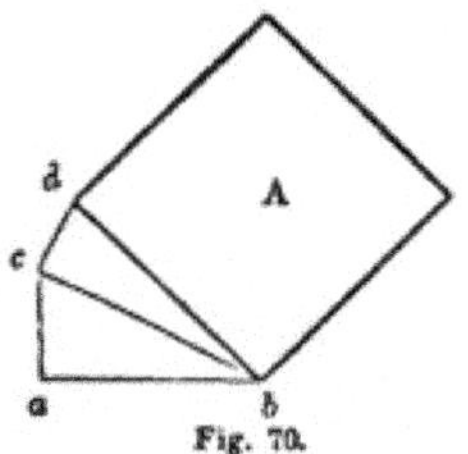

Fig. 70.

the side of a given square, *c a*, placed at right angles to *a b*, be the side of another given square, and *c d*, placed at right angles to *c b*, be the side of a third given square; then the square, *A*, formed upon the line, *d b*, will be equal to the three given squares. (See *Art.* 157.)

The usefulness and importance of this problem are proverbial. To ascertain the length of braces and of rafters in framing, the length of stair-strings, &c., are some of the purposes to which it may be applied in carpentry. (See note to *Art.* 74, *b.*) If the length of any two sides of a right-angled triangle is known, that of the third can be ascertained. Because the square of the hypothenuse is equal to the united squares of the two sides that contain the right angle.

(1.)—The two sides containing the right angle being known, to find the hypothenuse. *Rule.*—Square each given side, add the squares together, and from the product extract the square-root: this will be the answer. For instance, suppose it were required to find the length of a rafter for a house, 34 feet wide,—the ridge of the roof to be 9 feet high, above the level of the wall-plates. Then 17 feet, half of the span, is one, and 9 feet, the height, is the other of the sides that contain the right angle Proceed as directed by the rule:

```
 17                          9
 17                          9
----                       ----
119                         81 = square of 9.
 17                        289 = square of 17.
----                       ----
289 = square of 17.        370 Product.
```

```
1 ) 370 ( 19·235 + = square-root of 370; equal 19 feet, 2[illegible] in.
1   1                    nearly: which would be the required
-  ----                  length of the rafter.
29 ) 270
 9   261
---  ----
382) ··900
  2    764
----  -----
3843 ) 13600
   3   11529
-----  ------
38465)· 207100
        192325
        ------
```

(By reference to the table of square-roots in the Appendix, the root of almost any number may be found ready calculated; also, to change the decimals of a foot to inches and parts, see Rules for the Reduction of Decimals in the Appendix.)

Again: suppose it be required, in a frame building, to find the length of a brace, having a run of three feet each way from the point of the right angle. The length of the sides containing the right angle will be each 3 feet: then, as before—

```
               3
               3
              —
               9 = square of one side.
3 times 3 =    9 = square of the other side.
              —
              18 Product:
```

18 Product: the square-root of which is 4·2426 + ft., or 4 feet, 2 inches and $\frac{7}{8}$ths. full.

(2.)—The hypothenuse and one side being known, to find the other side. *Rule.*—Subtract the square of the given side from the square of the hypothenuse, and the square-root of the product will be the answer. Suppose it were required to ascertain the greatest perpendicular height a roof of a given span may have, when pieces of timber of a given length are to be used as rafters. Let the span be 20 feet, and the rafters of 3×4 hemlock joist. These come about 13 feet long. The known hypothenuse, then, is 13 feet, and the known side, 10 feet—that being half the span of the building.

```
                  13
                  13
                 ———
                  39
                 13
                 ———
                 169 = square of hypothenuse.
10 times 10 =    100 = square of the given side.
                 ———
                  69 Product:
```

69 Product: the square-root of which is 8·3066 + feet, or 8 feet, 3 inches and $\frac{5}{8}$ths. full. This will be the greatest perpendicular height, as required. Again: suppose that in a story of 8 feet, from floor to floor, a step-ladder is required, the strings of which are to be of plank, 12 feet long; and it is desirable to know the greatest run such a length of string will afford. In this case, the two given sides are—hypothenuse 12, perpendicular 8 feet.

```
12 times 12 =  144 = square of hypothenuse.
 8 times  8 =   64 = square of perpendicular.
               ———
                80 Product:
```

80 Product: the square-root of which is 8·9442 + feet. or 8 feet, 11 inches and $\frac{5}{16}$ths.—the answer, as required.

Many other cases might be adduced to show the utility of this problem. A practical and ready method of ascertaining the length of braces, rafters, &c., when not of a great length, is to apply a rule across the carpenters'-square. Suppose, for the length of a rafter, the base be 12 feet and the height 7. Apply the rule diagonally on the square, so that it touches 12 inches from the corner on one side, and 7 inches from the corner on the other. The number of inches on the rule, which are intercepted by the sides of the square, 13⅞ nearly, will be the length of the rafter in feet; viz, 13 feet and ⅞ths of a foot. If the dimensions are large, as 30 feet and 20, take the half of each on the sides of the square, viz, 15 and 10 inches; then the length in inches across, will be one-half the number of feet the rafter is long. This method is just as accurate as the preceding; but when the length of a very long rafter is sought, it requires great care and precision to ascertain the fractions. For the least variation on the square, or in the length taken on the rule, would make perhaps several inches difference in the length of the rafter. For shorter dimensions, however, the result will be true enough.

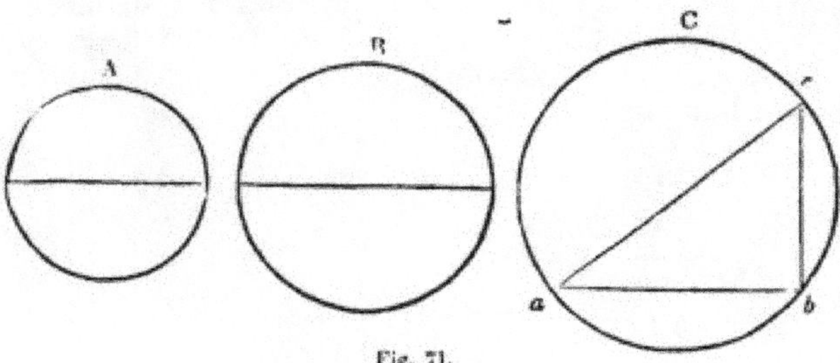

Fig. 71.

104.—*To make a circle equal to two given circles.* Let *A* and *B*, (*Fig.* 71,) be the given circles. In the right-angled triangle, *a b c*, make *a b* equal to the diameter of the circle, *B*, and *c b* equal to the diameter of the circle, *A*; then the hypothenuse,

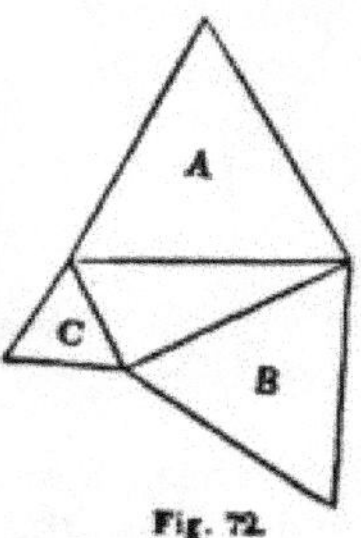

Fig. 72.

a c, will be the diameter of a circle, *C*, which will be equal in area to the two circles, *A* and *B*, added together.

Any polygonal figure, as *A*, (*Fig.* 72,) formed on the hypothenuse of a right-angled triangle, will be equal to two similar figures,* as *B* and *C*, formed on the two legs of the triangle.

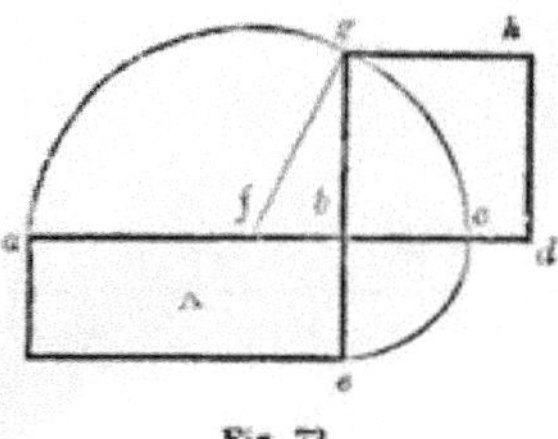

Fig. 73

105.—*To construct a square equal to a given rectangle.* Let *A*, (*Fig.* 73,) be the given rectangle. Extend the side, *a b*, and make *b c* equal to *b e*; bisect *a c* in *f*, and upon *f*, with the radius, *f a*, describe the semi-circle, *a g c*; extend *e b*, till it cuts the curve in *g*; then a square, *b g h d*, formed on the line, *b g*, will be equal in area to the rectangle, *A*.

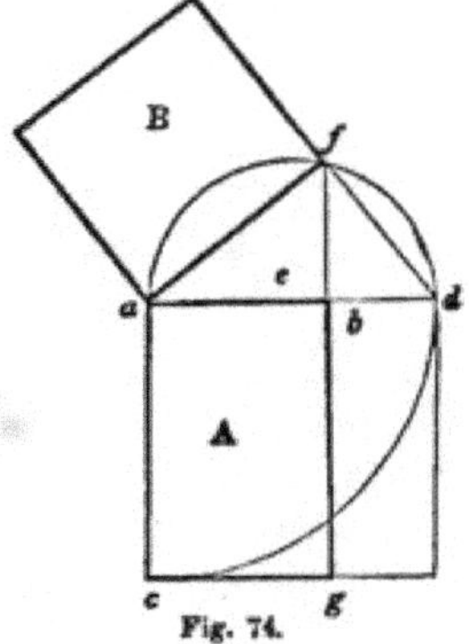

Fig. 74.

105, *a.*—*Another method.* Let *A*, (*Fig.* 74,) be the given rectangle. Extend the side, *a b*, and make *a d* equal to *a c*,

* Similar figures are such as have their several angles respectively equal, and their sides respectively proportionate.

bisect *a d* in *e*; upon *e*, with the radius, *e a*, describe the semicircle, *a f d*; extend *g b* till it cuts the curve in *f*; join *a* and *f*; then the square, *B*, formed on the line, *a f*, will be equal in area to the rectangle, *A*. (See *Art.* 156 and 157.)

106.—*To form a square equal to a given triangle.* Let *a b*, (*Fig.* 73,) equal the base of the given triangle, and *b e* equal half its perpendicular height, (see *Fig.* 67;) then proceed as directed at *Art.* 105.

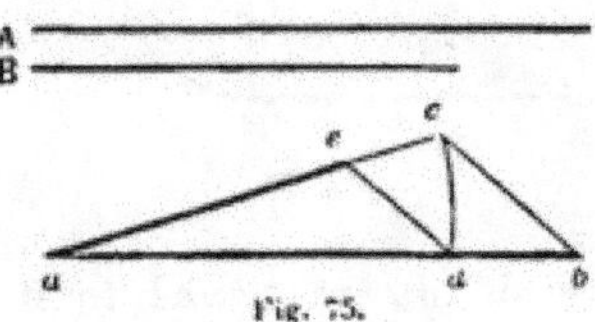

Fig. 75.

107.—*Two right lines being given, to find a third proportional thereto.* Let *A* and *B*, (*Fig.* 75,) be the given lines. Make *a b* equal to *A*; from *a*, draw *a c*, at any angle with *a b*; make *a c* and *a d* each equal to *B*; join *c* and *b*; from *d*, draw *d e*, parallel to *c b*; then *a e* will be the third proportional required. That is, *a e* bears the same proportion to *B*, as *B* does to *A*.

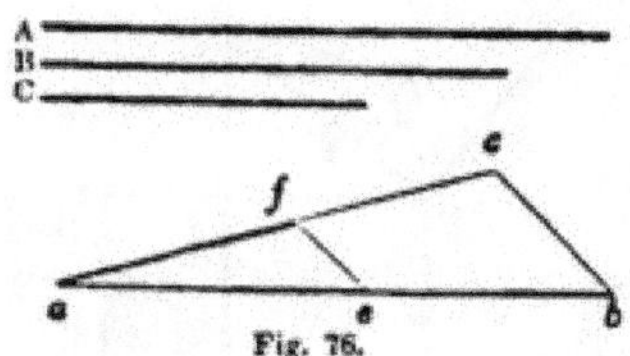

Fig. 76.

108.—*Three right lines being given, to find a fourth proportional thereto.* Let *A*, *B* and *C*, (*Fig.* 76,) be the given lines. Make *a b* equal to *A*; from *a*, draw *a c*, at any angle with *a b*; make *a c* equal to *B*, and *a e* equal to *C*; join *c* and *b*; from *e*, draw *e f*, parallel to *c b*; then *a f* will be the fourth proportional required. That is, *a f* bears the same proportion to *C*, as *B* does to *A*.

To apply this problem, suppose the two axes of a given ellipsis and the longer axis of a proposed ellipsis are given. Then, by this problem, the length of the shorter axis to the proposed ellipsis, can be found; so that it will bear the same proportion to the longer axis, as the shorter of the given ellipsis does to its longer. (See also, *Art.* 126.)

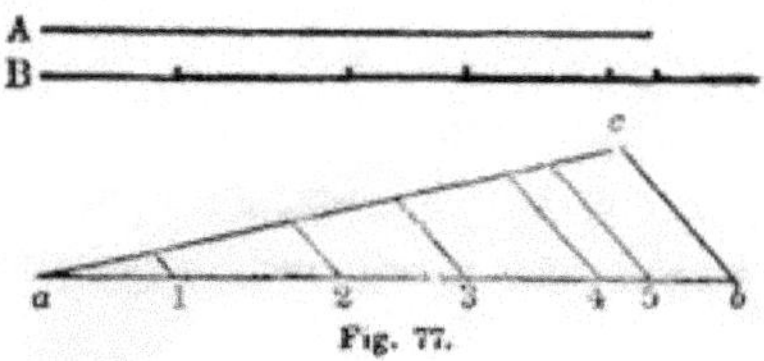

Fig. 77.

109.—*A line with certain divisions being given, to divide another, longer or shorter, given line in the same proportion.* Let *A*, (*Fig.* 77,) be the line to be divided, and *B* the line with its divisions. Make *a b* equal to *B*, with all its divisions, as at 1, 2, 3, &c.; from *a*, draw *a c*, at any angle with *a b*; make *a c* equal to *A*; join *c* and *b*; from the points, 1, 2, 3, &c., draw lines, parallel to *c b*; then these will divide the line, *a c*, in the same proportion as *B* is divided—as was required.

This problem will be found useful in proportioning the members of a proposed cornice, in the same proportion as those of a given cornice of another size. (See *Art.* 253 and 254.) So of a pilaster, architrave, &c.

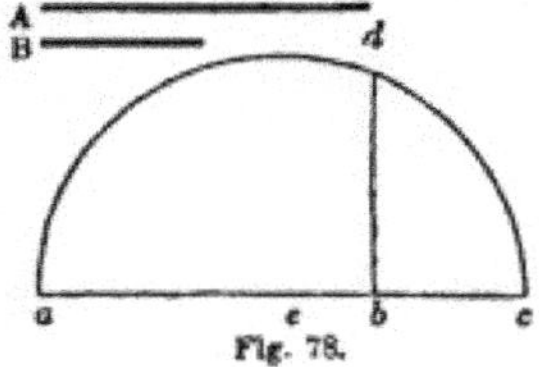

Fig. 78.

110.—*Between two given right lines, to find a mean proportional.* Let *A* and *B*, (*Fig.* 78,) be the given lines. On the line, *a c*, make *a b* equal to *A*, and *b c* equal to *B*; bisect *a c* in *e*; upon *e*, with *e a* for radius, describe the semi-circle, *a d* c; at b, erect b d, at right angles to a c; then b d will be the mean proportional between A and B. That is, a b is to b d as b d is to b c. This is usually stated thus—$a\,b : b\,b :: b\,d : b\,c$, and since the product of the means equals the product of the extremes, therefore, $a\,b \times b\,c = \overline{b\,d}^2$. This is shown geometrically at *Art.* 105.

CONIC SECTIONS.

111.—If a cone, standing upon a base that is at right angles with its axis, be cut by a plane, perpendicular to its base and passing through its axis, the section will be an isosceles triangle;

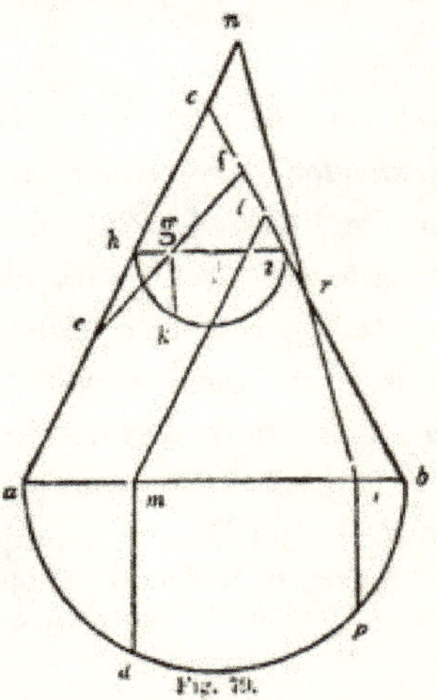

Fig. 79.

(as *a b c*, *Fig.* 79;) and the base will be a semi-circle. If a cone be cut by a plane in the direction, *e f*, the section will be an *ellipsis;* if in the direction, *m l*, the section will be a *parabola;* and if in the direction, *r o*, an *hyperbola.* (See *Art.* 56 to 60.) If the cutting planes be at right angles with the plane, *a b c*, then—

112.—*To find the axes of the ellipsis*, bisect *e f*, (*Fig.* 79,) in *g;* through *g*, draw *h i*, parallel to *a b;* bisect *h i* in *j;* upon *j*, with *j h* for radius, describe the semi-circle, *h k i;* from *g*, draw *g k*, at right angles to *h i;* then twice *g k* will be the conjugate axis, and *e f* the transverse.

113.—*To find the axis and base of the parabola.* Let *m l*, (*Fig.* 79,) parallel to *a c*, be the direction of the cutting plane From *m*, draw *m d*, at right angles to *a b*; then *l m* will be the axis and height, and *m d* an ordinate and half the base; as at *Fig.* 92, 93.

114.—*To find the height, base and transverse axis of an hyperbola.* Let *o r*, (*Fig.* 79,) be the direction of the cutting plane. Extend *o r* and *a c* till they meet at *n*; from *o*, draw *o p*, at right angles to *a b*; then *r o* will be the height, *n r* the transverse axis, and *o p* half the base; as at *Fig.* 94.

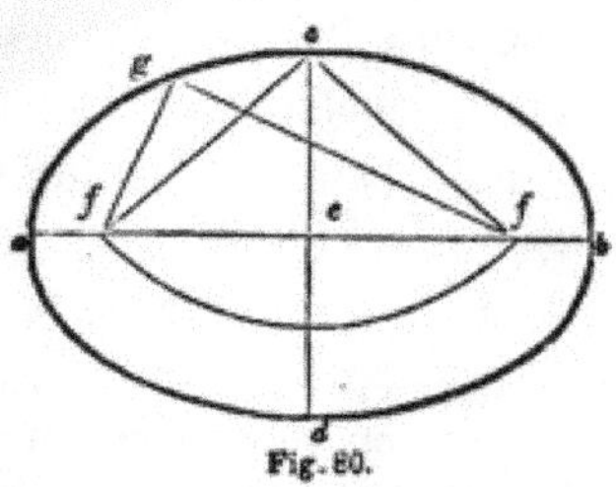

Fig. 80.

115.—*The axes being given, to find the foci, and to describe an ellipsis with a string.* Let *a b*, (*Fig.* 80,) and *c d*, be the given axes. Upon *c*, with *a e* or *b e* for radius, describe the arc, *f f*; then *f* and *f*, the points at which the arc cuts the transverse axis, will be the *foci.* At *f* and *f* place two pins, and another at *c*; tie a string about the three pins, so as to form the triangle, *f f c*; remove the pin from *c*, and place a pencil in its stead; keeping the string taut, move the pencil in the direction, *c g a*; it will then describe the required ellipsis. The lines, *f g* and *g f*, show the position of the string when the pencil arrives at *g*.

This method, when performed correctly, is perfectly accurate; but the string is liable to stretch, and is, therefore, not so good to use as the trammel. In making an ellipse by a string or twine, that kind should be used which has the least tendency to elasticity. For this reason, a cotton cord, such as chalk-lines are commonly made of, is not proper for the purpose: a linen, or flaxen cord is much better.

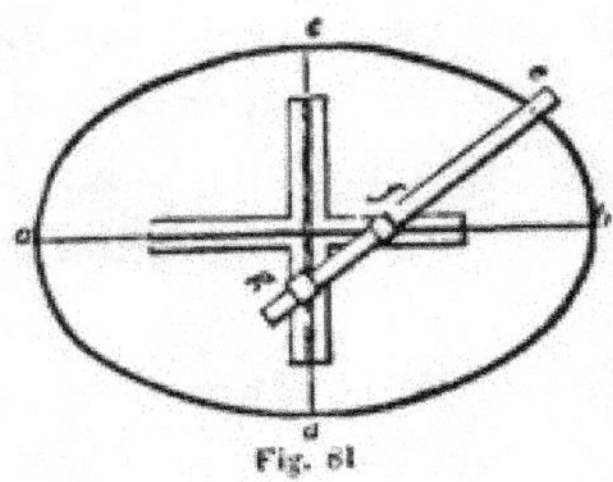

Fig. 81

116.—*The axes being given, to describe an ellipsis with a trammel.* Let *a b* and *c d*, (*Fig.* 81,) be the given axes. Place the trammel so that a line passing through the centre of the grooves, would coincide with the axes; make the distance from the pencil, *e*, to the nut, *f*, equal to half *c d*; also, from the pencil, *e*, to the nut, *g*, equal to half *a b*; letting the pins under the nuts slide in the grooves, move the trammel, *e g*, in the direction, *c b d*; then the pencil at *e* will describe the required ellipse.

A trammel may be constructed thus: take two straight strips of board, and make a groove on their face, in the centre of their width; join them together, in the middle of their length, at right angles to one another; as is seen at *Fig.* 81. A rod is then to be prepared, having two moveable nuts made of wood, with a mortice through them of the size of the rod, and pins under them large enough to fill the grooves. Make a hole at one end of the rod, in which to place a pencil. In the absence of a regular trammel, a temporary one may be made, which, for any short job, will answer every purpose. Fasten two straight-edges at right angles to one another. Lay them so as to coincide with the axes of the proposed ellipse, having the angular point at the centre. Then, in a rod having a hole for the pencil at one end, place two brad-awls at the distances described at *Art.* 116. While the pencil is moved in the direction of the curve, keep the brad-awls hard against the straight-edges, as directed for using the trammel-rod, and one-quarter of the ellipse will be drawn. Then, by shifting the straight-edges, the other three quarters in succession may be drawn. If the required ellipse be not too large, a carpenters'-square may be made use of, in place of the straight-edges.

An improved method of constructing the trammel, is as follows: make the sides of the grooves bevilling from the face of the stuff, or dove-tailing instead of square. Prepare two slips of wood, each about two inches long, which shall be of a shape to just fill the groove when slipped in at the end. These, instead of pins, are to be attached one to each of the moveable nuts with a screw, loose enough for the nut to move freely about the screw as an axis. The advantage of this contrivance is, in preventing the nuts from slipping out of their places, during the operation of describing the curve.

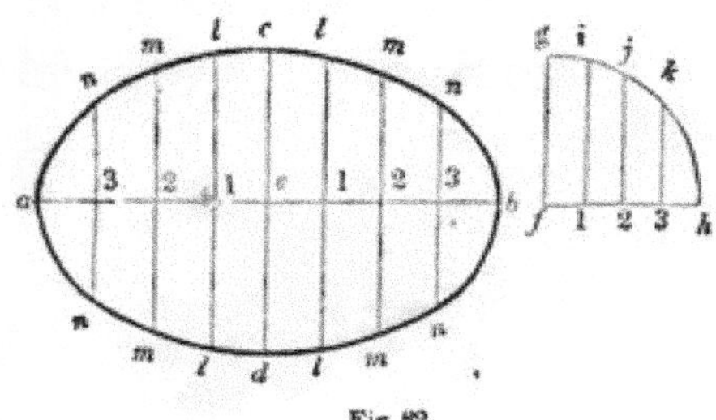

Fig. 82.

117.—*To describe an ellipsis by ordinates.* Let *a b* and *c a*, (*Fig.* 82,) be given axes. With *c e* or *e d* for radius, describe the quadrant, *f g h*; divide *f h*, *a e* and *e b*, each into a like number of equal parts, as at 1, 2 and 3; through these points, draw ordinates, parallel to *c d* and *f g*; take the distance, 1 *i*, and place it at 1 *l*, transfer 2 *j* to 2 *m*, and 3 *k* to 3 *n*; through the points, *a*, *n*, *m*, *l* and *c*, trace a curve, and the ellipsis will be completed.

The greater the number of divisions on *a e*, &c., in this and the following problem, the more points in the curve can be found, and the more accurate the curve can be traced. If pins are placed in the points, *n*, *m*, *l*, &c., and a thin slip of wood bent around by them, the curve can be made quite correct. This method is mostly used in tracing face-moulds for stair hand-railing.

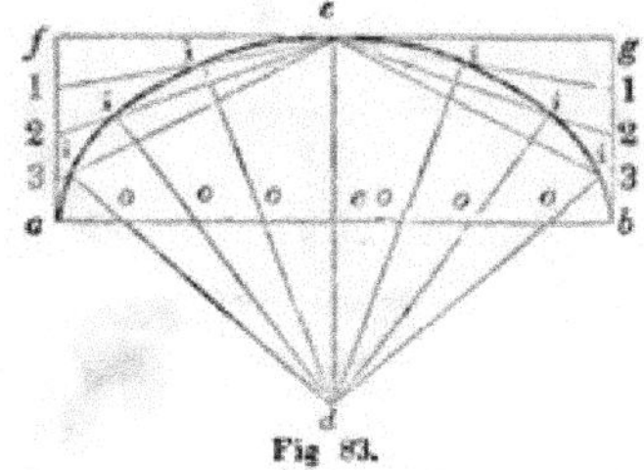

Fig. 83.

118.—*To describe an ellipsis by intersection of lines.* Let *a b* and *c d*, (*Fig.* 83,) be given axes. Through *e* draw *f g*, parallel to *a b*; from *a* and *b*, draw *a f* and *b g*, at right angles to *a b*; divide *f a*, *g b*, *a e* and *e b*, each into a like number of equal parts, as at 1, 2, 3 and *o*, *o*, *o*; from 1, 2 and 3, draw lines to *c*; through *o*, *o* and *o*, draw lines from *d*, intersecting those drawn to *c*; then a curve, traced through the points, *i*, *i*, *i*, will be that of an ellipsis.

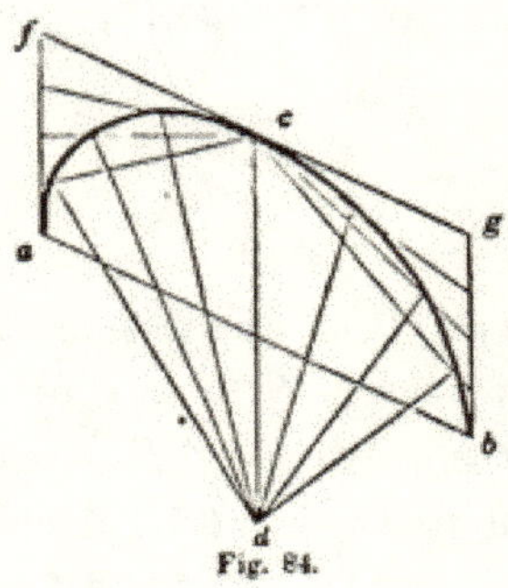

Fig. 84.

Where neither trammel nor string is at hand, this, perhaps, is the most ready method of drawing an ellipsis. The divisions should be small, where accuracy is desirable. By this method, an ellipsis may be traced without the axes, provided that a diameter and its conjugate be given. Thus, *a b* and *c d*, (*Fig.* 84,) are conjugate diameters: *f g* is drawn parallel to *a b*, instead of being at right angles to *c d*; also, *f a* and *g b* are drawn parallel to *c d*, instead of being at right angles to *a b*.

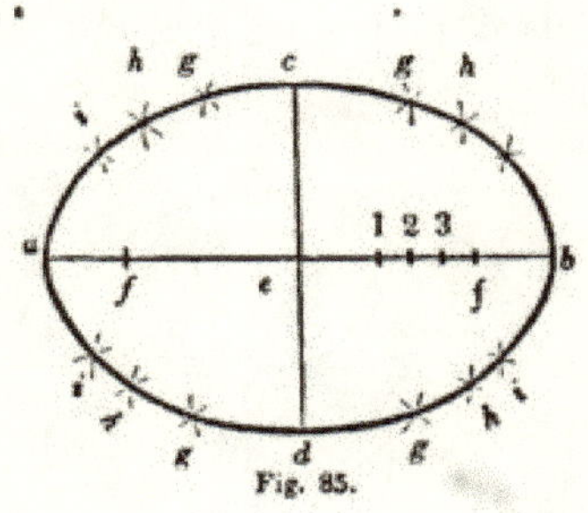

Fig. 85.

119.—*To describe an ellipsis by intersecting arcs.* Let *a b* and *c d*, (*Fig.* 85,) be given axes. Between one of the foci, *f* and *f*, and the centre, *e*, mark any number of points, at random, as 1, 2 and 3; upon *f* and *f*, with *b* 1 for radius, describe arcs at *g*, *g*, *g* and *g*; upon *f* and *f*, with *a* 1 for radius, describe arcs intersecting the others at *g*, *g*, *g* and *g*; then these points of intersection will be in the curve of the ellipsis. The other points, *h* and *i*, are found in like manner, viz: *h* is found by taking *b* 2 for one radius, and *a* 2 for the other; *i* is found by taking *b* 3 for one radius, and *a* 3 for the other, always using the foci for centres. Then by tracing a curve through the points, *c*, *g*, *h*, *i*, *b*, &c., the ellipse will be completed.

This problem is founded upon the same principle as that of the string. This is obvious, when we reflect that the length of the string is equal to the transverse axis, added to the distance between the foci. See *Fig.* 80; in which *c f* equals *a e*, the half of the transverse axis.

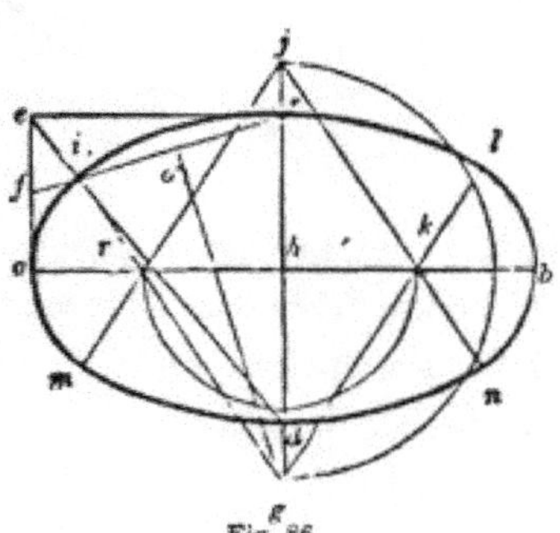

Fig. 86.

120.—*To describe a figure nearly in the shape of an ellipsis, by a pair of compasses.* Let *a b* and *c d*, (*Fig.* 86,) be given axes. From *c*, draw *c e*, parallel to *a b*; from *a*, draw *a e*, parallel to *c d*; join *e* and *d*; bisect *e a* in *f*; join *f* and *c*, intersecting *e d* in *i*; bisect *i c* in *o*; from *o*, draw *o g*, at right angles to *i c*, meeting *c d* extended to *g*; join *i* and *g*, cutting the transverse axis in *r*; make *h j* equal to *h g*, and *h k* equal to *h r*, from *j*, through *r* and *k*, draw *j m* and *j n*; also, from *g*, through *k*, draw *g l*; upon *g* and *j*, with *g c* for radius, describe the arcs, *i l* and *m n*: upon *r* and *k*, with *r a* for radius, describe the arcs, *m i* and *l n*, this will complete the figure.

When the axes are proportioned to one another as 2 to 3, the extremities, *c* and *d*, of the shortest axis, will be the centres for describing the arcs, *i l* and *m n*; and the intersection of *e d* with the transverse axis, will be the centre for describing the arc, *m i*, &c. As the elliptic curve is continually changing its course from that of a circle, a true ellipsis cannot be described with a pair of compasses. The above, therefore, is only an approximation.

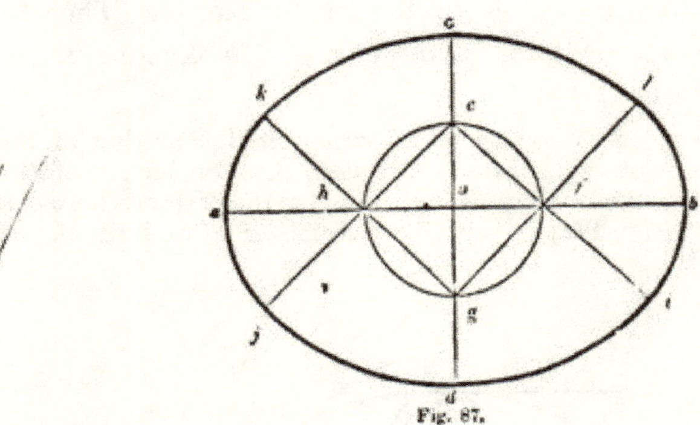

Fig. 87.

121.—*To draw an oval in the proportion, seven by nine.* Let *c d*, (*Fig.* 87,) be the given conjugate axis. Bisect *c d* in *o*, and through *o*, draw *a b*, at right angles to *c d*; bisect *c o* in *e*, upon *o*, with *o e* for radius, describe the circle, *e f g h*; from *e*, through *h* and *f*, draw *e j* and *e i*; also, from *g*, through *h* and *f*, draw *g k* and *g l*; upon *g*, with *g c* for radius, describe the arc, *k l*; upon *e*, with *e d* for radius, describe the arc, *j i*; upon *h* and *f*, with *h k* for radius, describe the arcs, *j k* and *l i*; this will complete the figure.

This is an approximation to an ellipsis; and perhaps no method can be found, by which a well-shaped oval can be drawn with greater facility. By a little variation in the process, ovals of different proportions may be obtained. If quarter of the transverse axis is taken for the radius of the circle, *e f g h*, one will be drawn in the proportion, five by seven.

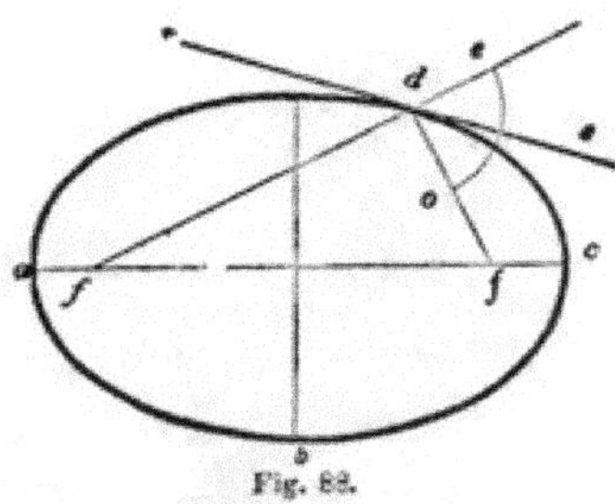

Fig. 88.

122.—*To draw a tangent to an ellipsis.* Let *a b c d*, (*Fig.* 88,) be the given ellipsis, and *d* the point of contact. Find the foci, (*Art.* 115,) *f* and *f*, and from them, through *d*, draw *f e* and *f d*; bisect the angle, (*Art.* 77,) *e d o*, with the line, *s r*; then *s r* will be the tangent required.

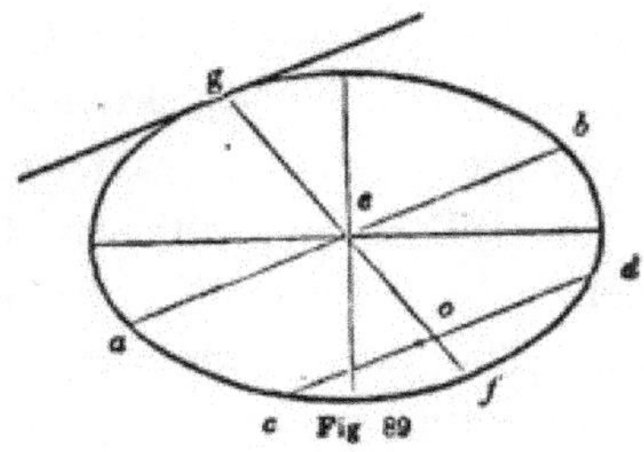

Fig 89

123.—*An ellipsis with a tangent given, to detect the point of contact.* Let *a g b f*, (*Fig.* 89,) be the given ellipsis and tangent. Through the centre, *e*, draw *a b*, parallel to the tangent; any where between *e* and *f*, draw *c d*, parallel to *a b*; bisect *c d* in *o*; through *o* and *e*, draw *f g*; then *g* will be the point of contact required.

124.—*A diameter of an ellipsis given, to find its conjugate.* Let *a b*, (*Fig.* 89,) be the given diameter. Find the line, *f g*, by the last problem; then *f g* will be the diameter required.

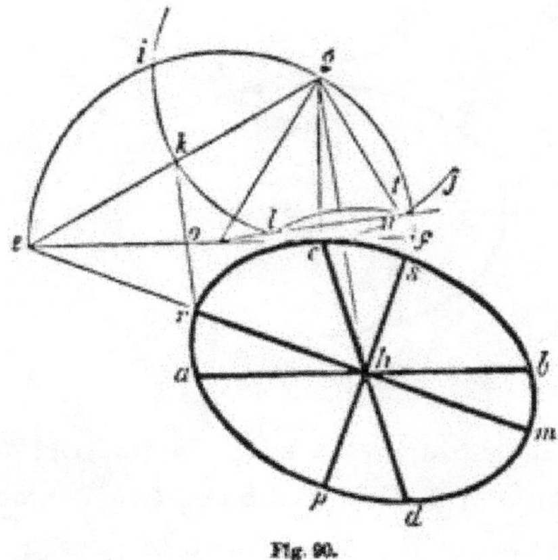

Fig. 90.

125.—*Any diameter and its conjugate being given, to ascertain the two axes, and thence to describe the ellipsis.* Let *a b* and *c d*, (*Fig.* 90,) be the given diameters, conjugate to one another. Through *c*, draw *e f*, parallel to *a b*; from *c*, draw *c g*, at right angles to *e f*; make *c g* equal to *a h* or *h b*; join *g* and *h*; upon *g*, with *g c* for radius, describe the arc, *i k c j*; upon *h*, with the same radius, describe the arc, *l n*; through the intersections, *l* and *n*, draw *n o*, cutting the tangent, *e f*, in *o*; upon *o*, with *o g* for radius, describe the semi-circle, *e i g f*; join *e* and *g*, also *g* and *f*, cutting the arc, *i c j*, in *k* and *t*; from *e*, through *h*, draw *e m*, also from *f*, through *h*, draw *f p*; from *k* and *t*, draw *k r* and *t s*, parallel to *g h*, cutting *e m* in *r*, and *f p* in *s*; make *h m* equal to *h r*, and *h p* equal to *h s*; then *r m* and *s p* will be the axes required, by which the ellipsis may be drawn in the usual way.

126.—*To describe an ellipsis, whose axes shall be proportionate to the axes of a larger or smaller given one.* Let *a c b d*, (*Fig.* 91,) be the given ellipsis and axes, and *i j* the transverse axis of a proposed smaller one. Join *a* and *c*; from *i* draw *i e*, parallel to *a c*; make *o f* equal to *o e*; then *e f* will be

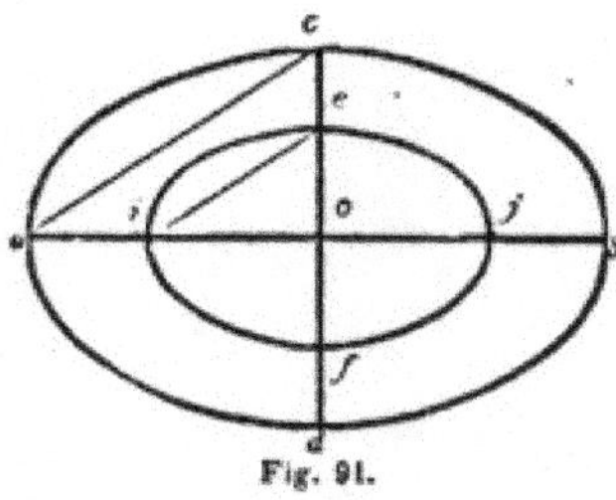
Fig. 91.

the conjugate axis required, and will bear the same proportion to *i j*, as *c d* does to *a b*. (See *Art.* 108.)

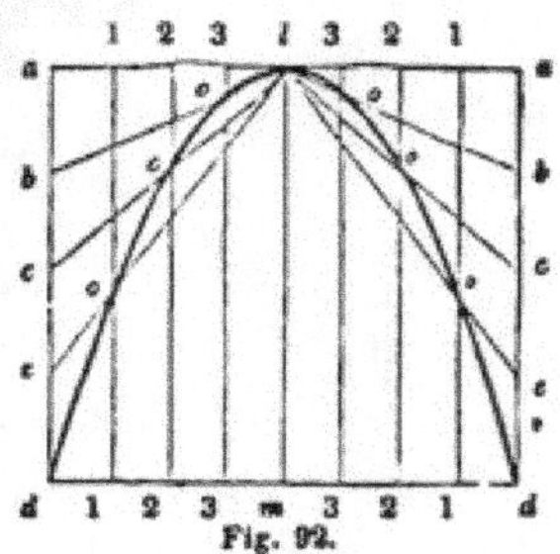
Fig. 92.

127.—*To describe a parabola by intersection of lines.* Let *m l*, (*Fig.* 92,) be the axis and height, (see *Fig.* 79,) and *d d*, a double ordinate and base of the proposed parabola. Through *l*, draw *a a*, parallel to *d d*; through *d* and *d*, draw *d a* and *d a*, parallel to *m l*; divide *a d* and *d m*, each into a like number of equal parts; from each point of division in *d m*, draw the lines, 1 1, 2 2, &c., parallel to *m l*; from each point of division in *d a*, draw lines to *l*; then a curve traced through the points of intersection, *o*, *o* and *o*, will be that of a parabola.

127, *a*.—*Another method.* Let *m l*, (*Fig.* 93,) be the axis and height, and *d d* the base. Extend *m l*, and make *l a* equal to *m l*; join *a* and *d*, and *a* and *d*; divide *a d* and *a d*, each into a like number of equal parts, as at 1, 2, 3, &c.; join 1 and 1, 2 and 2, &c., and the parabola will be completed

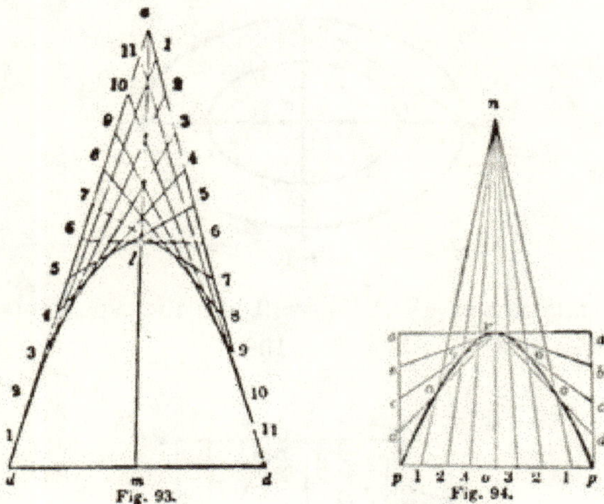

Fig. 93. Fig. 94.

128.—*To describe an hyperbola by intersection of lines.* Let *r o*, (*Fig.* 94,) be the height, *p p* the base, and *n r* the transverse axis. (See *Fig.* 79.) Through *r*, draw *a a*, parallel to *p p*; trom *p*, draw *a p*, parallel to *r o*; divide *a p* and *p o*, each into a like number of equal parts; from each of the points of divisions in the base, draw lines to *n*; from each of the points of division in *a p*, draw lines to *r*; then a curve traced through the points of intersection, *o*, *o*, &c., will be that of an hyperbola.

The parabola and hyperbola afford handsome curves for various mo ildings.

DEMONSTRATIONS.

129.—To impress more deeply upon the mind of the learner some of the more important of the preceding problems, and to indulge a very common and praiseworthy curiosity to discover the cause of things, are some of the reasons why the following exercises are introduced. In all reasoning, definitions are necessary; in order to insure, in the minds of the proponent and respondent, identity of ideas. A *corollary* is an inference deduced from a previous course of reasoning. An *axiom* is a proposition evident at first sight. In the following demonstrations, there are many axioms taken for granted; (such as, things equal to the same thing are equal to one another, &c.;) these it was thought not necessary to introduce in form.

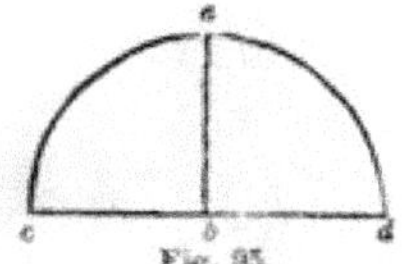

Fig. 95.

130.—*Definition.* If a straight line, as *a b*, (*Fig.* 95,) stand upon another straight line, as *c d*, so that the two angles made at the point, *b*, are equal—*a b c* to *a b d*, (see note to *Art.* 27,) then each of the two angles is called *a right angle.*

131.—*Definition.* The circumference of every circle is supposed to be divided into 360 equal parts, called *degrees ;* hence a semi-circle contains 180 degrees, a quadrant 90, &c.

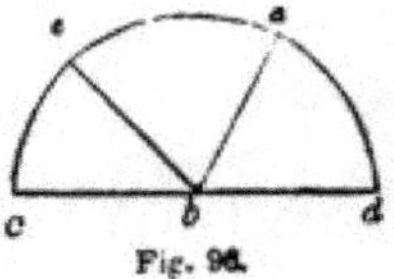

Fig. 96.

132.—*Definition.* The *measure of an angle* is the number of degrees contained between its two sides, using the angular point as a centre upon which to describe the arc. Thus the arc, *c e*, (*Fig.* 96,) is the measure of the angle, *c b e ; e a*, of the angle, *e b a ;* and *a d*, of the angle, *a b d.*

133.—*Corollary.* As the two angles at *b*, (*Fig.* 95,) are right angles, and as the semi-circle, *c a d*, contains 180 degrees, (*Art.* 131,) the measure of two right angles, therefore, is 180 degrees ; of one right angle, 90 degrees ; of half a right angle, 45 ; of one-third of a right angle, 30, &c.

134.—*Definition.* In measuring an angle, (*Art.* 132,) no regard is to be had to the length of its sides, but only to the degree of their inclination. Hence *equal angles* are such as have the same degree of inclination, without regard to the length of their sides.

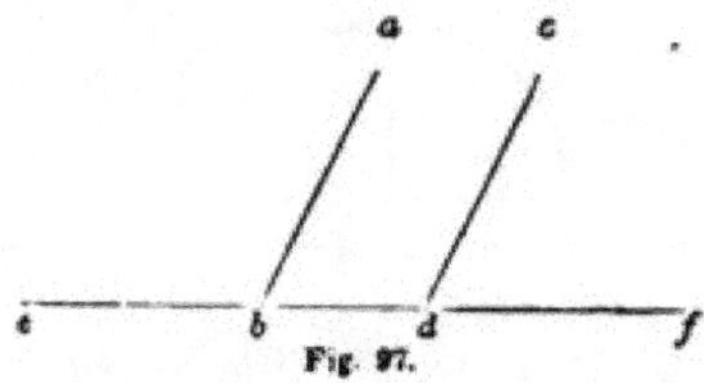

Fig. 97.

135.—*Axiom.* If two straight lines, parallel to one another, as *a b* and *c d*, (*Fig.* 97,) stand upon another straight line, as *e f*, the angles, *a b f* and *c d f*, are equal; and the angle, *a b e*, is equal to the angle, *c d e*.

136.—*Definition.* If a straight line, as *a b*, (*Fig.* 96,) stand obliquely upon another straight line, as *c d*, then one of the angles, as *a b c*, is called ***an obtuse angle***, and the other, as *a b d*, ***an acute angle.***

137.—*Axiom.* The two angles, *a b d* and *a b c*, (*Fig.* 96,) are together equal to two right angles, (*Art.* 130, 133;) also, the three angles, *a b d*, *e b a* and *c b e*, are together equal to two right angles.

138.—*Corollary.* Hence all the angles that can be made upon one side of a line, meeting in a point in that line, are together equal to two right angles.

139.—*Corollary.* Hence all the angles that can be made on both sides of a line, at a point in that line, or all the angles that can be made about a point, are together equal to four right angles.

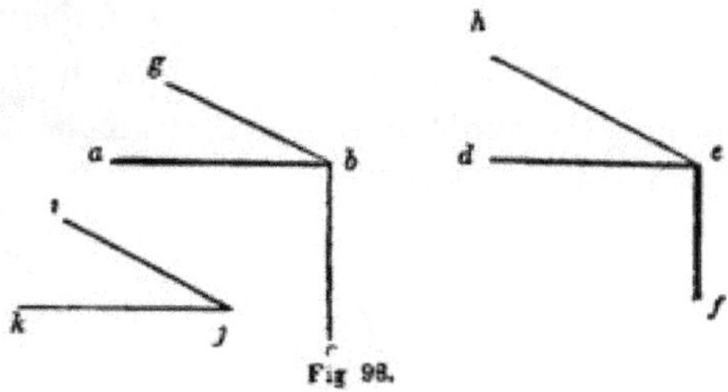

Fig 98.

140.—*Proposition.* If to each of two equal angles a third angle be added, their sums will be equal. Let *a b c* and *d e f*, (*Fig.* 98,) be equal angles, and the angle, *i j k*, the one to be added. Make the angles, *g b a* and *h e d*, each equal to the given angle, *i j k*; then the angle, *g b c*, will be equal to the angle, *h e f*; for, if *a b c* and *d e f* be angles of 90 degrees, and *i j k*, 30, then the angles, *g b c* and *h e f*, will be each equal to 90 and 30 added, viz: 120 degrees.

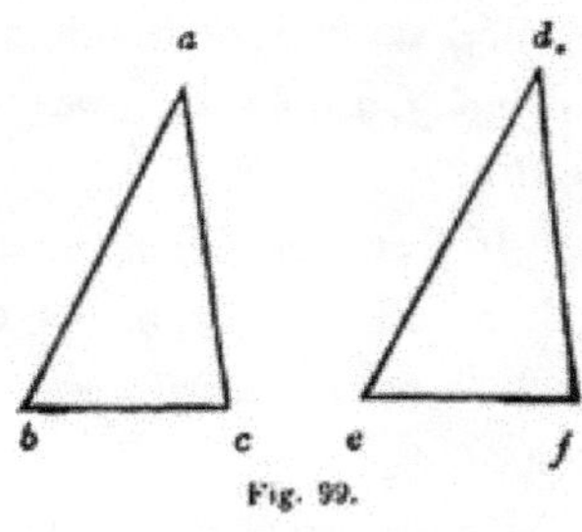

Fig. 99.

141.—*Proposition.* Triangles that have two of their sides and the angle contained between them respectively equal, have also their third sides and the two remaining angles equal; and consequently one triangle will every way equal the other. Let *a b c*, (*Fig.* 99,) and *d e f* be two given triangles, having the angle at *a* equal to the angle at *d*, the side, *a b*, equal to the side, *d e*, and the side, *a c*, equal to the side, *d f;* then the third side of one, *b c*, is equal to the third side of the other, *e f;* the angle at *b* is equal to the angle at *e*, and the angle at *c* is equal to the angle at *f*. For, if one triangle be applied to the other, the three points, *b*, *a*, *c*, coinciding with the three points, *e*, *d*, *f*, the line, *b c*, must coincide with the line, *e f;* the angle at *b* with the angle at *e;* the angle at *c* with the angle at *f*; and the triangle, *b a c*, be every way equal to the triangle, *e d f*.

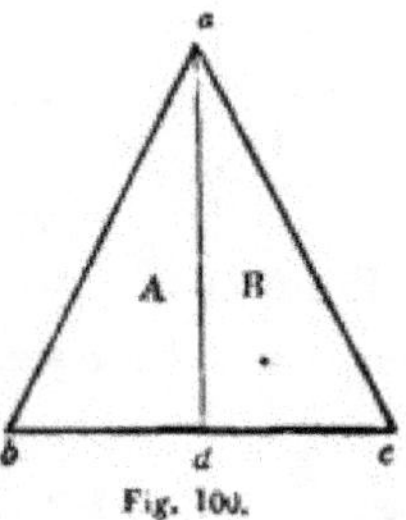

Fig. 100.

142.—*Proposition.* The two angles at the base of an isoceles triangle are equal. Let *a b c*, (*Fig.* 100,) be an isoceles triangle, of which the sides, *a b* and *a c*, are equal. Bisect the angle, (*Art.*

77,) *b a c*, by the line, *a d*. Then the line, *b a*, being equal to the line, *a c*; the line, *a d*, of the triangle, *A*, being equal to the line, *a d*, of the triangle, *B*, being common to each; the angle, *b a d*, being equal to the angle, *d a c*; the line, *b d*, must, according to *Art.* 141, be equal to the line, *d c*; and the angle at *b* must be equal to the angle at *c*.

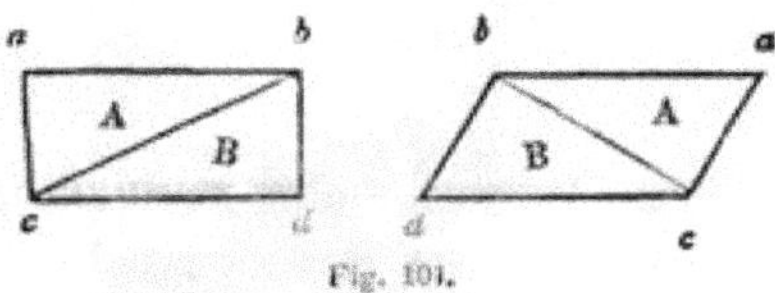

Fig. 101.

143.—*Proposition.* A diagonal crossing a parallelogram divides it into two equal triangles. Let *a b c d*, (*Fig.* 101,) be a given parallelogram, and *b c*, a line crossing it diagonally. Then, as *a c* is equal to *b d*, and *a b* to *c d*, the angle at *a* to the angle at *d*, the triangle, *A*, must, according to *Art.* 141, be equal to the triangle, *B*.

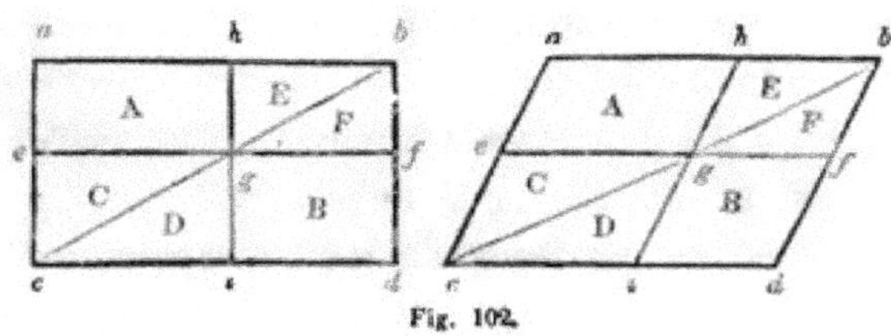

Fig. 102.

144.—*Proposition.* Let *a b c d*, (*Fig.* 102,) be a given parallelogram, and *b c* a diagonal. At any distance between *a b* and *c d*, draw *e f*, parallel to *a b*; through the point, *g*, the intersection of the lines, *b c* and *e f*, draw *h i*, parallel to *b d*. In every parallelogram thus divided, the parallelogram, *A*, is equal to the parallelogram, *B*. According to *Art.* 143, the triangle, *a b c*, is equal to the triangle, *b c d*; the triangle, *C*, to the triangle, *D*; and *E* to *F*; this being the case, take *D* and *F* from the triangle, *b c d*, and *C* and *E* from the triangle, *a b c*, and what remains in one must be equal to what remains in the other; therefore, the parallelogram, *A*, is equal to the parallelogram, *B*.

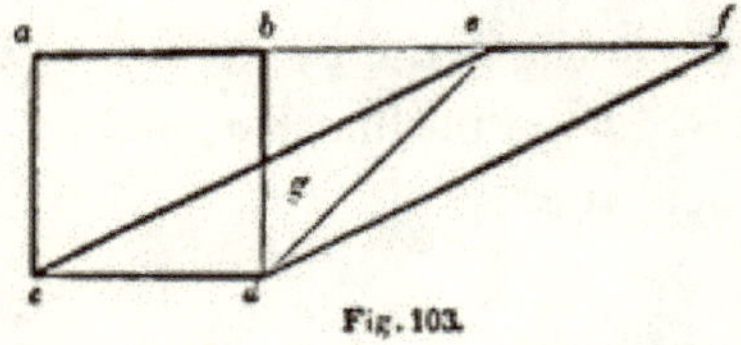

Fig. 103.

145.—*Proposition.* Parallelograms standing upon the same base and between the same parallels, are equal. Let *a b c d* and *e f c d*, (*Fig.* 103,) be given parallelograms, standing upon the same base, *c d*, and between the same parallels, *a f* and *c d*. Then, *a b* and *e f* being equal to *c d*, are equal to one another; *b e* being added to both *a b* and *e f*, *a e* equals *b f;* the line, *a c*, being equal to *b d*, and *a e* to *b f*, and the angle, *c a e*, being equal, (*Art.* 135,) to the angle, *d b f*, the triangle, *a e c*, must be equal, (*Art.* 141,) to the triangle, *b f d;* these two triangles being equal, take the same amount, the triangle, *b e g*, from each, and what remains in one, *a b g c*, must be equal to what remains in the other, *e f d g;* these two quadrangles being equal, add the same amount, the triangle, *c g d*, to each, and they must still be equal; therefore, the parallelogram, *a b c d*, is equal to the parallelogram, *e f c d*.

146.—*Corollary.* Hence, if a parallelogram and triangle stand upon the same base and between the same parallels, the parallelogram will be equal to double the triangle. Thus, the parallelogram, *a d*, (*Fig.* 103,) is double, (*Art.* 143,) the triangle, *c e d*.

147.—*Proposition.* Let *a b c d*, (*Fig.* 104,) be a given quadrangle with the diagonal, *a d*. From *b*, draw *b e*, parallel to *a d*, extend *c d* to *e;* join *a* and *e;* then the triangle, *a e c*, will be equal in area to the quadrangle, *a b c d*. Since the triangles, *a d b* and *a d e*, stand upon the same base, *a d*, and between the same paral-

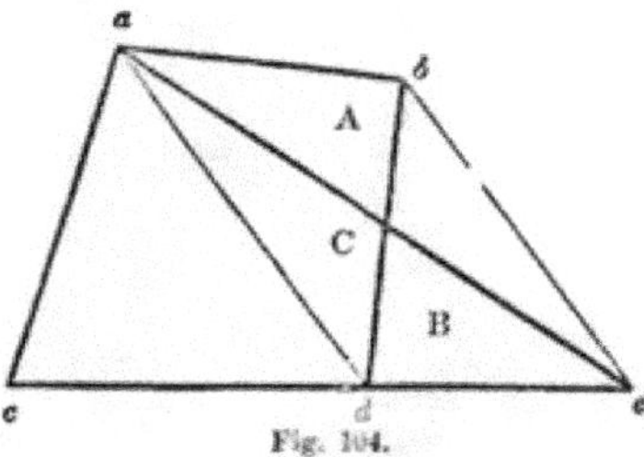

Fig. 104.

lels, *a d* and *b e*, they are therefore equal, (*Art.* 145, 146;) and since the triangle, *C*, is common to both, the remaining triangles, *A* and *B*, are therefore equal; then *B* being equal to *A*, the triangle, *a e c*, is equal to the quadrangle, *a b c d*.

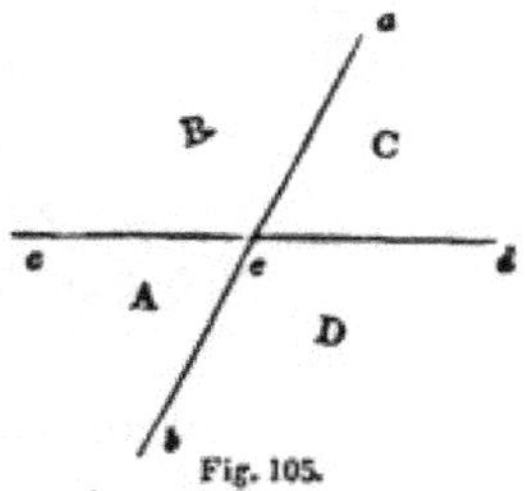

Fig. 105.

148.—*Proposition.* If two straight lines cut each other, as *a b* and *c d*, (*Fig.* 105,) the vertical, or opposite angles, *A* and *C*, are equal. Thus, *a e*, standing upon *c d*, forms the angles, *B* and *C*, which together amount, (*Art.* 137,) to two right angles; in the same manner, the angles, *A* and *B*, form two right angles; since the angles, *A* and *B*, are equal to *B* and *C*, take the same amount, the angle, *B*, from each pair, and what remains of one pair is equal to what remains of the other; therefore, the angle, *A*, is equal to the angle, *C*. The same can be proved of the opposite angles, *B* and *D*.

149.—*Proposition.* The three angles of any triangle are equal to two right angles. Let *a b c*, (*Fig.* 106,) be a given triangle, with its sides extended to *f*, *e*, and *d*, and the line, *c g*,

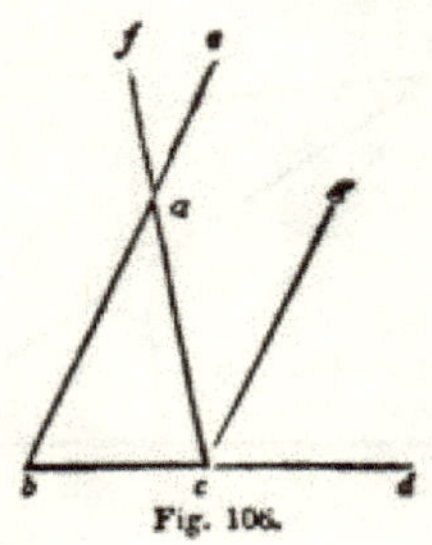

Fig. 106.

drawn parallel to *b e*. As *g c* is parallel to *e b*, the angle, *g c d*, is, equal, (*Art.* 135,) to the angle, *e b d*; as the lines, *f c* and *b e*, cut one another at *a*, the opposite angles, *f a e* and *b a c*, are equal, (*Art.* 148;) as the angle, *f a e*, is equal, (*Art.* 135,) to the angle, *a c g*, the angle, *a c g*, is equal to the angle, *b a c*; therefore, the three angles meeting at *c*, are equal to the three angles of the triangle, *a b c*; and since the three angles at *c* are equal, (*Art.* 137,) to two right angles, the three angles of the triangle, *a b c*, must likewise be equal to two right angles. Any triangle can be subjected to the same proof.

150.—*Corollary.* Hence, if one angle of a triangle be a right angle, the other two angles amount to just one right angle.

151.—*Corollary.* If one angle of a triangle be a right angle, and the two remaining angles are equal to one another, these are each equal to half a right angle.

152.—*Corollary.* If any two angles of a triangle amount to a right angle, the remaining angle is a right angle.

153.—*Corollary.* If any two angles of a triangle are together equal to the remaining angle. that remaining angle is a right angle.

154.—*Corollary.* If any two angles of a triangle are each equal to two-thirds of a right angle, the remaining angle is also equal to two-thirds of a right angle.

155.—*Corollary.* Hence, the angles of an equi-lateral triangle. are each equal to two-thirds of a right angle.

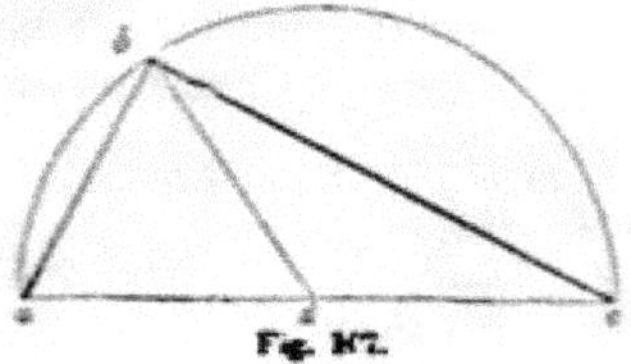

Fig. 107.

156.—*Proposition.* If from the extremities of the diameter of a semi-circle, two straight lines be drawn to any point in the circumference, the angle formed by them at that point will be a right angle. Let *a b c*, (*Fig.* 107,) be a given semi-circle, and *a b* and *b c*, lines drawn from the extremities of the diameter, *a c*, to the given point, *b*; the angle formed at that point by these lines, is a right angle. Join the point, *b*, and the centre, *d*; the lines, *d a*, *d b* and *d c*, being radii of the same circle, are equal; the angle at *a* is therefore equal, (*Art.* 142,) to the angle, *a b d*, also, the angle at *c* is, for the same reason, equal to the angle, *d b c*; the angle, *a b c*, being equal to the angles at *a* and *c* taken together, must therefore, (*Art.* 153,) be a right angle.

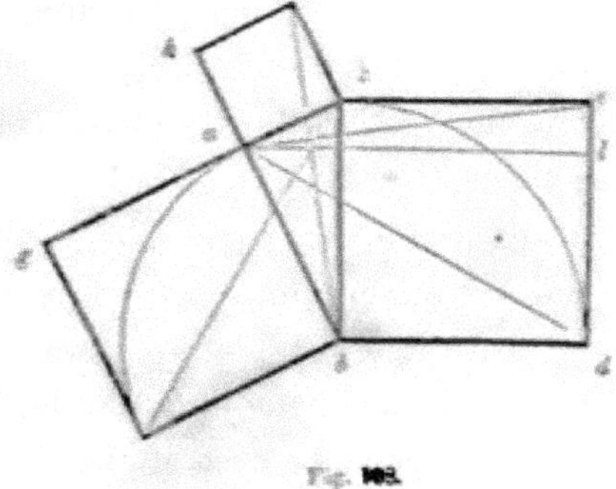

Fig. 108.

157.—*Proposition.* The square of the hypothenuse of a right-angled triangle, is equal to the squares of the two remaining sides. Let *a b c*, (*Fig.* 108,) be a given right-angled triangle, having a square formed on each of its sides: then, the square, *b e*, is equal to the squares, *h c* and *g b*, taken together. This can be proved by showing that the parallelogram, *b l*, is equal to the square, *g b*; and that the parallelogram, *c l*, is equal to the square, *h c*. The angle, *c b d*, is a right angle, and the angle, *a b f*, is a right angle; add to each of these the angle, *a b c*; then the angle, *f b c*, will evidently be equal, (*Art.* 140,) to the angle, *a b d*; the triangle, *f b c*, and the square, *g b*, being both upon the same base, *f b*, and between the same parallels, *f b* and *g c*, the square, *g b*, is equal, (*Art.* 146,) to twice the triangle, *f b c*; the triangle, *a b d*, and the parallelogram, *b l*, being both upon the same base, *b d*, and between the same parallels, *b d* and *a l*, the parallelogram, *b l*, is equal to twice the triangle, *a b d*; the triangles, *f b c* and *a b d*, being equal to one another, (*Art.* 141,) the square, *g b*, is equal to the parallelogram, *b l*, either being equal to twice the triangle, *f b c* or *a b d*. The method of proving *h c* equal to *c l* is exactly similar—thus proving the square, *b e*, equal to the squares, *h c* and *g b*, taken together.

This problem, which is the 47th of the First Book of Euclid is said to have been demonstrated first by Pythagoras. It is stated, (but the story is of doubtful authority,) that as a thank-offering for its discovery he sacrificed a hundred oxen to the gods. From this circumstance, it is sometimes called the *hecatomb* problem. It is of great value in the exact sciences, more especially in Mensuration and Astronomy, in which many otherwise intricate calculations are by it made easy of solution.

158.—*Proposition.* In a segment of a circle, the versed sine equals the radius, less the square root of the difference of the squares of the radius and half-chord. That is, the versed sine, *a c*, (*Fig.* 109,) equals *a b*, less *c b*. Now *a b* is radius, hence the radius, minus *c b*, equals *a c*, the versed sine. To find the value of *c b*, it will be observed that *c b* is the side of the square, *c f*, while the radius *b d* is the side of the square, *b h*, and the half-chord, *c d*, is the side of the square, *c e*; also, that these three squares are made upon the three sides of the right angled

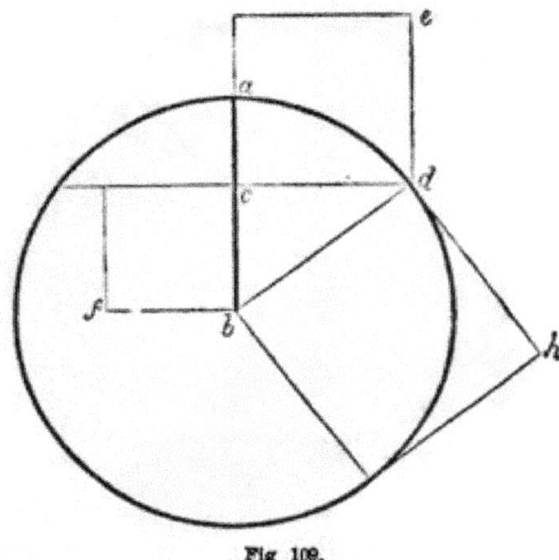

Fig 109.

triangle, *b c d*, and the square, *b h*, is therefore equal to the two squares, *c e* and *c f*, (*Art.* 157;) therefore, the square, *c f*, is equal to the square, *b h*, minus the square, *c e*;—or, is equal to the difference of the squares on *b d* and *c d*. Consequently the square *root* of *c f* is equal to the square *root* of the difference of the squares on *b d* and *c d*; and since *c b* is the square root of *c f*, therefore *c b* equals the square root of the difference of the squares on *b d* and *c d*—or, equals the square root of the difference of the squares of the radius and the half-chord. Having found an expression for the value of *c b*, it remains merely to deduct this value from the radius, and the residue equals the versed sine; for, as before stated, the versed sine, *a c*, equals the radius, *a b*, minus *c b*; therefore, the versed sine equals the radius, minus the square root of the difference of the squares on the radius and half-chord. The rule expressed algebraically is $v = r - \sqrt{r^2 - a^2}$, where v is the versed sine, r the radius, and a the half-chord. It is read, v equals r, minus the square root of the difference of the squares of r and a.

159.—*Proposition.* In an equilateral octagon the semi-diagonal of a circumscribed square, having its sides coincident with four of the sides of the octagon, equals the distance along

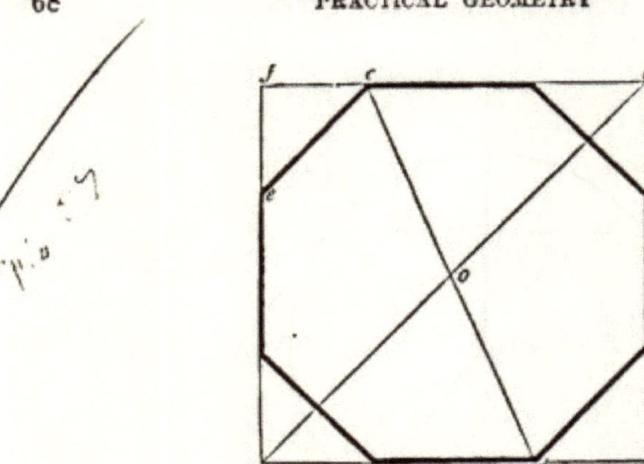

Fig. 110.

a side of the square from its corner to the more remote angle of the octagon occurring on that side of the square. To prove this, it need only to be shown that the triangle, *a o d*, (*Fig.* 110,) is an isosceles triangle having its sides *a o* and *a d*, equal. The octagon being equi-lateral, it is also equi-angular, therefore the angles, *b c o*, *e c o*, *a d o*, &c., are all equal. Of the right-angled triangle, *f e c*, *f c* and *f e* being equal, the two angles, *f e c* and *f c e* are equal, (*Art.* 142,) and are therefore, (*Art.* 151,) each equal to half a right angle. In like manner it may be shown that *f a b* and *f b a* are also each equal to half a right angle. And since *f e c* and *f a b* are equal angles, therefore the lines *e c* and *a b* are parallel, (*Art.* 135,) and hence the angles, *e c o* and *a o d*, are equal. These being equal, and the angles *e c o* and *a d o* being, by construction, equal, as before shown, therefore the angles *a o d* and *a d o* are equal, and consequently the lines *a o* and *a d* are equal. (*Art.* 142.)

160.—*Proposition.* An angle at the circumference of a circle is measured by half the arc that subtends it: that is, the angle *a b c*, (*Fig.* 111,) is equal to half the angle *a d c*. Through the centre, *d*, draw the diameter, *b e*. The triangle *a b d* is an isosceles triangle, *a d* and *b d* being radii, and therefore equal; hence the two angles, *d a b* and *d b a*, are equal.

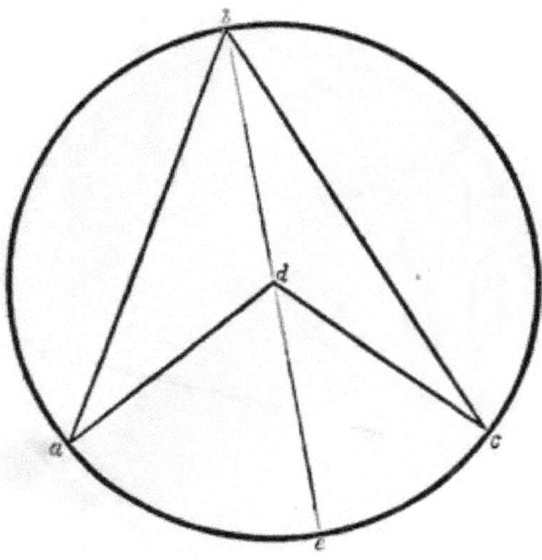

Fig. 111.

(*Art.* 142,) and the sum of these two angles is equal to the angle *a d e*, (*Art.* 149,) and therefore one of them, *a b d*, is equal to the half of *a d e*. The angles *a d e* and *a b d* (or *a b e*) are both subtended by the arc *a e*. Now, since the angle, *a d e*, is measured by the arc *a e*, which subtends it, therefore the half of the angle, *a d e*, would be measured by the half of the arc *a e;* and since *a b d* is equal to the half of *a d e*, therefore *a b d*, or *a b e*, is measured by the half of the arc *a e*. It may be shown in like manner that the angle *e b c* is measured by half the arc *e c*, and hence it follows that the angle, *a b c*, is measured by half the arc, *a c*, that subtends it.

161.—*Proposition.* In a circle, all the inscribed angles, *a b c*, (*Fig.* 112,) which stand upon the same side of the chord *d e*, are equal. For each angle is measured by half the arc *d f e*, (*Art.* 160,) hence the angles are all equal.

162.—*Corollary.* Equal chords, in the same circle, subtend equal angles.

163.—*Proposition.* The angle formed by a chord and tangent is equal to any inscribed angle in the opposite segment

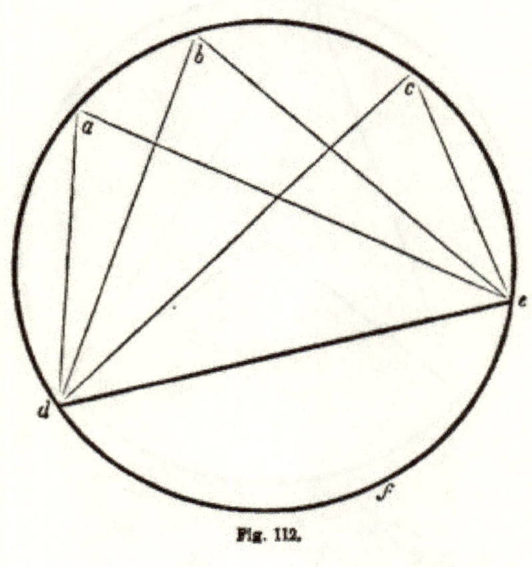

Fig. 112.

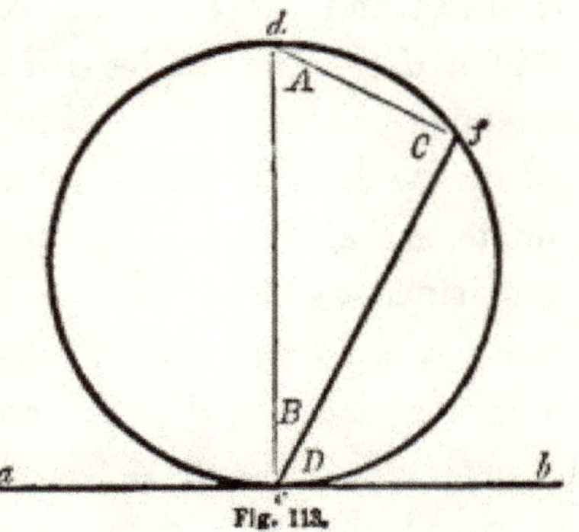

Fig. 113.

of the circle; that is, the angle *D*, (*Fig.* 113,) equals the angle *A*. Let *c f* be the chord, and *a b* the tangent; draw the diameter, *d c;* then *d c b* is a right angle, also *d f c* is a right angle. (*Art.* 156.) The angles *A* and *B* together equal a right angle, (*Art.* 150;) also the angles *B* and *D* together equal a right angle, (equal the angle *d c b;*) therefore the sum of *A* and *B* equals the sum of *B* and *D*. From each of these two equals, taking the like quantity *B*, the remainders, *A* and

D, are equal. Thus, it is proved for the angle at *d*; it is also true for any other angle; for, since all other inscribed angles on that side of the chord line, *c f*, equal the angle *A*, (*Art.* 161,) therefore the angle formed by a chord and tangent equals any angle in the opposite segment of the circle. This being proved for the acute angle, *D*, it is also true for the obtuse

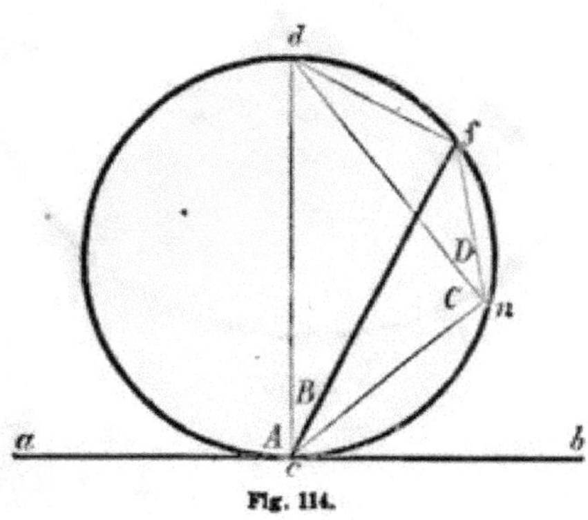

Fig. 114.

angle, *a c f*; for, from any point, *n*, (*Fig.* 114,) in the arc *c n f*, draw lines to *d*, *f* and *c*; now, if it can be proved that the angle *a c f* equals the angle *f n c*, the entire proposition is proved, for the angle *f n c* equals any of all the inscribed angles that can be drawn on that side of the chord. (*Art.* 161.) To prove, then, that *a c f* equals *c n f*: the angle *a c f* equals the sum of the angles *A* and *B*; also the angle *c n f* equals the sum of the angles *C* and *D*. The angles *B* and *D*, being inscribed angles on the same chord, *d f*, are equal. The angles *C* and *A* being right angles, (*Art.* 156,) are likewise equal. Now, since *A* equals *C*, and *B* equals *D*, therefore the sum of *A* and *B* equals the sum of *C* and *D*—or the angle *a c f* equals the angle *c n f*.

164.—*Proposition.* Two chords, *a b* and *c d*, (*Fig.* 115,) intersecting, the parallelogram or rectangle formed by the two parts of one is equal to the rectangle formed by the two parts of the other. That is, *c e* multiplied by *e d*, the product is

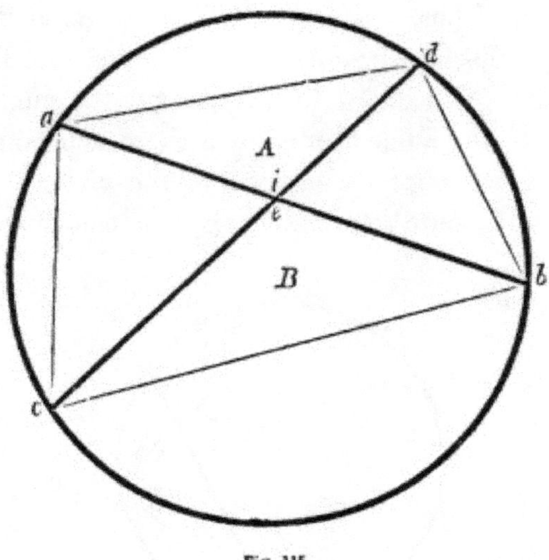

Fig. 115.

equal to the product of *a e* multiplied by *e b*. The triangle *A* is similar to the triangle *B*, because it has corresponding angles. The angle *i* equals the angle *e*, (*Art.* 148;) the angle at *c* equals the angle at *a* because they stand upon the same chord, *d b*, (*Art.* 161;) for the same reason the angle *b* equals the angle *d*, for each stands upon the same chord, *a c*. Therefore, the triangle *A* having the same angles as the triangle *B*, the length of the sides of one are in like proportion as the length of the sides in the other. So, *e d* : *a e* :: *e b* : *c e*. Hence, *a e* multiplied by *e b* is equal to *e d* mutiplied by *c e*— or the product of the means equals the product of the extremes.

165.—*Proposition.* In any circle, when a segment is given, the radius is equal to the sum of the squares of half the chord and of the versed sine, divided by twice the versed sine. Let *a b*, (*Fig.* 116,) be the chord line, and *v* the versed sine of the segment. By the preceding article the triangle *A* is shown to be like the triangle *B*, having equal angles and proportionate

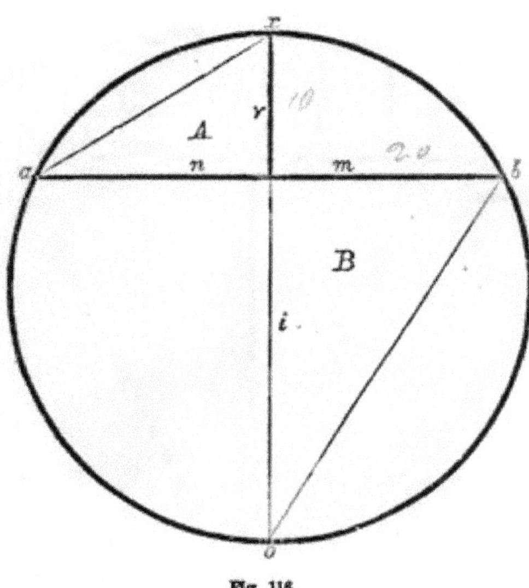

Fig. 116.

length of sides. Therefore, $v : n :: m : i$, or $\frac{n^2}{v} = i$; that is, i is equal to the square of n (or $n \times n$) divided by v. This result being added to v equals the diameter $o\,x$, which may be indicated by the letter d; thus, $\frac{n^2}{v} + v = i + v = d$; and the half of this, or $\frac{\frac{n^2}{v} + v}{2} = \frac{d}{2} = r =$ the radius. Reducing this expression by multiplying the numerator and denominator each by the like quantity, viz. v, there results, $\frac{n^2 + v^2}{2\,v} = r$; and where c represents the chord, the expression is, $\frac{(\frac{c}{2})^2 + v^2}{2\,v} = r$: that is, as stated above, the radius is equal to the sum of the squares of half the chord and of the versed sine, divided by twice the versed sine.

166.—*Proposition.* Any ordinate, $m\,n$, (*Fig.* 117,) in the segment of a circle, is equal to the square root of the difference

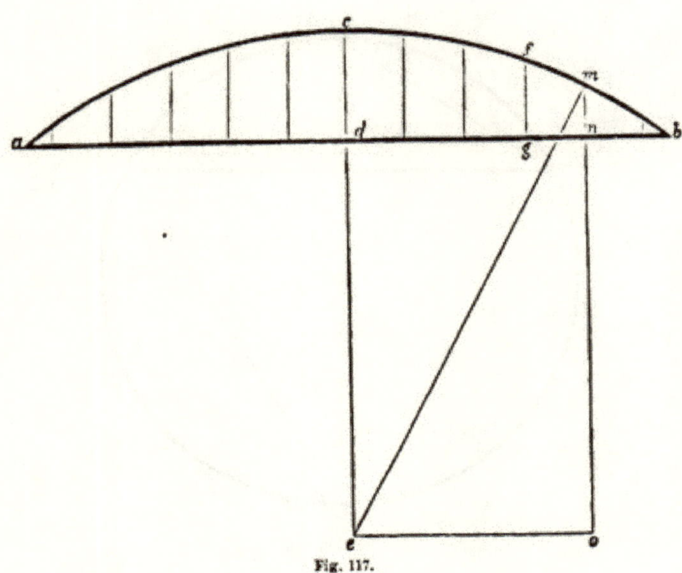

Fig. 117.

of the squares of the radius and abscissa, ($d\ n$,) less the difference of the radius and versed sine. So, if the chord $a\ b$, and the versed sine $c\ d$, be given, the length of any number of ordinates may be found by which to describe the arc. Find the radius, $c\ e$, by the preceding Article. It will be observed that $e\ m$ is also radius. Then, to find the length of the ordinate, $m\ n$, make $e\ o$ equal to $d\ n$: now, according to Article 157, the square of $e\ o$ taken from the square of $e\ m$, the residue equals the square of $o\ m$, and the square root of this residue will be the length of the line $o\ m$. Then from $o\ m$ take $o\ n$ equal to $e\ d$, and the result will be the length of $m\ n$. That is, the ordinate is equal to the square root of the difference of the squares of the radius and abscissa, less the difference of the radius and versed sine. This may be expressed algebraically thus: $y = \sqrt{r^2 - x^2} - (r - v)$, where y is the ordinate, r the radius, x the abscissa, and v the versed sine;—$d\ n$ being the abscissa of the ordinate $n\ m$, $d\ g$ the abscissa of the ordinate *g f*, &c.: the abscissa being in each case the distance from the foot of the versed sine, *c d*, to the foot of the ordinate whose length is sought.

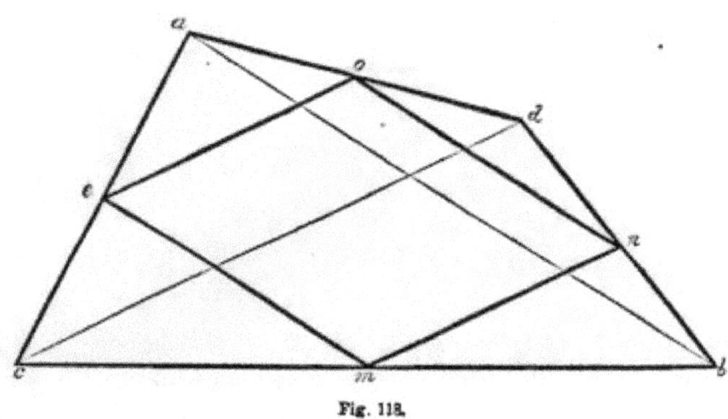

Fig. 118.

167.—*Proposition.* The sides of any quadrangle being bisected, and lines drawn joining the points of bisection in the adjacent sides, these lines will form a parallelogram. Draw the diagonals, *a b* and *c d*, (*Fig.* 118.) It will here be perceived that the two triangles, *a e o* and *a c d*, are homologous, having like angles and proportionate sides. Two of the sides of one triangle lie coincident with the two corresponding sides of the other triangle, therefore the contained angles between these sides in each triangle are identical. By construction, these corresponding sides are proportionate; *a c* being equal to twice *a e*, and *a d* being equal to twice *a o*; therefore the remaining sides are proportionate, *c d* being equal to twice *e o*, hence the remaining corresponding angles are equal. Since, then, the angles *a e o* and *a c d* are equal, therefore the line *e o* is parallel with the diagonal *c d*—so, likewise, the line *m n* is parallel to the same diagonal, *c d*. If, therefore, these two lines, *e o* and *m n*, are parallel to the same line, *c d*, they must be parallel to each other. In the same manner the lines *o n* and *e m* are proved parallel to the diagonal, *a b*, and to each other; therefore the inscribed figure, *m e o n*, is a parallelogram. It may be remarked also, that the parallelogram so formed will contain just one-half the area of the circumscribing quadrangle.

These demonstrations, which relate mostly to the problems previously given, are introduced to satisfy the learner in regard to their mathematical accuracy. By studying and thoroughly understanding them, he will soonest arrive at a knowledge of their importance, and be likely the longer to retain them in memory. Should he have a relish for such exercises, and wish to continue them farther, he may consult Euclid's Elements, in which the whole subject of theoretical geometry is treated of in a manner sufficiently intelligible to be understood by the young mechanic. The house-carpenter, especially, needs information of this kind, and were he thoroughly acquainted with the principles of geometry, he would be much less liable to commit mistakes, and be better qualified to excel in the execution of his often difficult undertakings.

HISTORY OF ARCHITECTURE.

168.—Architecture has been defined to be—"the art of build ing;" but, in its common acceptation, it is—"the art of designing and constructing buildings, in accordance with such principles as constitute stability, utility and beauty." The literal signification of the Greek word *archi-tecton*, from which the word *architect* is derived, is chief-carpenter; but the architect has always been known as the chief *designer* rather than the chief *builder*. Of the three classes into which architecture has been divided—viz., Civil, Military, and Naval, the first is that which refers to the construction of edifices known as dwellings, churches and other public buildings, bridges, &c., for the accommodation of civilized man—and is the subject of the remarks which follow.

169.—This is one of the most ancient of the arts: the scriptures inform us of its existence at a very early period. Cain, the son of Adam,—"builded a city, and called the name of the city after the name of his son, Enoch"—but of the peculiar style or manner of building we are not informed. It is presumed that it was not remarkable for beauty, but that utility and perhaps stability were its characteristics. Soon after the deluge—that me morable event, which removed from existence all traces of the works of man—the Tower of Babel was commenced. This was a work of such magnitude that the gathering of the materials, according to some writers, occupied three years; the period from its commencement until the work was abandoned, was twenty-two years; and the bricks were like blocks of stone, being twenty feet long, fifteen broad and seven thick. Learned men have given it as their opinion, that the tower in the temple of Belus at Babylon was the same as that which in the scriptures is called the Tower of Babel. The tower of the temple of Belus was square at its base, each side measuring one furlong, and consequently half a mile in circumference. Its form was that of a pyramid, and its height was 660 feet. It had a winding passage on the outside from the base to the summit, which was wide enough for two carriages.

170.—Historical accounts of ancient cities, of which there are now but few remains—such as Babylon, Palmyra and Ninevah of the Assyrians; Sidon, Tyre, Aradus and Serepta of the Phœnicians; and Jerusalem, with its splendid temple, of the Israelites —show that architecture among them had made great advances. Ancient monuments of the art are found also among other nations; the subterraneous temples of the Hindoos upon the islands, Elephanta and Salsetta; the ruins of Persepolis in Persia; pyramids, obelisks, temples, palaces and sepulchres in Egypt—all prove that the architects of those early times were possessed of skill and judgment highly cultivated. The principal characteristics of their works, are gigantic dimensions, immoveable solidity, and, in some instances, harmonious splendour. The extraordinary size of some is illustrated in the pyramids of Egypt. The largest of these stands not far from the city of Cairo: its base, which is square, covers about 11¼ acres, and its height is nearly 500 feet. The stones of which it is built are immense—the smallest being full thirty feet long.

171.—Among the Greeks, architecture was cultivated as a fine art, and rapidly advanced towards perfection. Dignity and grace were added to stability and magnificence. In the Doric order, their first style of building, this is fully exemplified. Phidias, Ictinus and Callicrates, are spoken of as masters in the art at this period: the encouragement and support of Pericles stimulated them to a noble emulation. The beautiful temple of Minerva, erected upon the acropolis of Athens, the Propyleum, the Odeum and others, were lasting monuments of their success. The Ionic and Corinthian orders were added to the Doric, and many magnificent edifices arose. These exemplified, in their chaste proportions, the elegant refinement of Grecian taste. Improvement in Grecian architecture continued to advance, until perfection seems to have been attained. The specimens which have been partially preserved, exhibit a combination of elegant proportion, dignified simplicity and majestic grandeur. Architecture among the Greeks was at the height of its glory at the period immediately preceding the Peloponnesian war; after which the art declined. An excess of enrichment succeeded its former simple grandeur; yet a strict regularity was maintained amid the profusion of ornament. After the death of Alexander, 323 B. C., a love of gaudy splendour increased: the consequent decline of the art was visible, and the Greeks afterwards paid but little attention to the science.

172.—While the Greeks were masters in architecture, which they applied mostly to their temples and other public buildings, the Romans gave their attention to the science in the construction of the many aqueducts and sewers with which Rome abounded; building no such splendid edifices as adorned Athens, Corinth and Ephesus, until about 200 years B. C., when their intercourse with the Greeks became more extended. Grecian architecture was introduced into Rome by Sylla; by whom, as also by Marius and Cæsar, many large edifices were erected in various cities of Italy. But under Cæsar Augustus, at about the beginning of the christian era, the art arose to the greatest perfection it ever attained in Italy. Under his patronage, Grecian artists were encouraged, and many emigrated to Rome. It was at about this time that Solomon's temple at Jerusalem was rebuilt by Herod—a Roman. This was 46 years in the erection, and was most probably of the Grecian style of building—perhaps of the Corinthian order. Some of the stones of which it was built were 46 feet long, 21 feet high and 14 thick; and others were of the astonishing length of 82 feet. The porch rose to a great height; the whole being built of white marble exquisitely polished. This is the building concerning which it was remarked—"Master, see what manner of stones, and what buildings are here." For the construction of private habitations also, finished artists were employed by the Romans: their dwellings being often built with the finest marble, and their villas splendidly adorned. After Augustus, his successors continued to beautify the city, until the reign of Constantine; who, having removed the imperial residence to Constantinople, neglected to add to the splendour of Rome; and the art, in consequence, soon fell from its high excellence.

Thus we find that Rome was indebted to Greece for what she possessed of architecture—not only for the knowledge of its principles, but also for many of the best buildings themselves; these having been originally erected in Greece, and stolen by the unprincipled conquerors—taken down and removed to Rome. Greece was thus robbed of her best monuments of architecture. Touched by the Romans, Grecian architecture lost much of its elegance and dignity. The Romans, though justly celebrated for their scientific knowledge as displayed in the construction of their various edifices, were not capable of appreciating the simple grandeur, the refined elegance of the Grecian style; but sought to improve upon it by the addition of luxurious enrichment, and thus deprived it of true elegance. In the days of Nero, whose palace of gold is so celebrated, buildings were lavishly adorned. Adrian did much to encourage the art; but not satisfied with the simplicity of the Grecian style, the artists of his time aimed at inventing new ones, and added to the already redundant embellishments of the previous age. Hence the origin of the pedestal, the great variety of intricate ornaments, the convex frieze, the round and the open pediments, &c. The rage for luxury continued until Alexander Severus, who made some improvement; but very soon after his reign, the art began rapidly to decline, as particularly evidenced in the mean and trifling character of the ornaments.

173.—The Goths and Vandals, when they overran the countries of Italy, Greece, Asia and Africa, destroyed most of the works of ancient architecture. Cultivating no art but that of war, these savage hordes could not be expected to take any interest in the beautiful forms and proportions of their habitations. From this time, architecture assumed an entirely different aspect. The celebrated styles of Greece were unappreciated and forgotten; and modern architecture took its first step on the platform of existence. The Goths, in their conquering invasions, gradually extended it over Italy, France, Spain, Portugal and Germany, into England. From the reign of Gallienus may be reckoned the total extinction of the arts among the Romans. From his time until the 6th or 7th century, architecture was almost entirely neglected. The buildings which were erected during this suspension of the arts, were very rude. Being constructed of the fragments of the edifices which had been demolished by the Visigoths in their unrestrained fury, and the builders being destitute of a proper knowledge of architecture, many sad blunders and extensive patchwork might have been seen in their construction—entablatures inverted, columns standing on their wrong ends, and other ridiculous arrangements characterized their clumsy work. The vast number of columns which the ruins around them afforded, they used as piers in the construction of arcades—which by some is thought, after having passed through various changes, to have been the origin of the plan of the Gothic cathedral. Buildings generally, which are not of the classical styles, and which were erected after the fall of the Roman empire, have by some been indiscriminately included under the term *Gothic.* But the changes which architecture underwent during the dark ages, show that there were several distinct modes of building.

174.—Theodoric, king of the Ostrogoths, a friend of the arts, who reigned in Italy from A. D. 493 to 525, endeavoured to restore and preserve some of the ancient buildings; and erected others, the ruins of which are still seen at Verona and Ravenna. Simplicity and strength are the characteristics of the structures erected by him; they are, however, devoid of grandeur and elegance, or fine proportions. These are properly of the GOTHIC style; by some called the *old* Gothic to distinguish it from the pointed style, which is generally called *modern* Gothic.

175.—The Lombards, who ruled in Italy from A. D. 568, had no taste for architecture nor respect for antiquities. Accordingly, they pulled down the splendid monuments of classic architecture which they found standing, and erected in their stead huge buildings of stone which were greatly destitute of proportion, elegance or utility—their characteristics being scarcely any thing more than stability and immensity combined with ornaments of a puerile character. Their churches were disfigured with rows of small columns along the cornice of the pediment, small doors and windows with circular heads, roofs supported by arches having arched buttresses to resist their thrust, and a lavish display of incongruous ornaments. This kind of architecture is called, the LOMBARD style, and was employed in the 7th century in Pavia, the chief city of the Lombards; at which city, as also at many other places, a great many edifices were erected in accordance with its inelegant forms.

176.—The Byzantine architects, from Byzantium, Constantinople, erected many spacious edifices; among which are included the cathedrals of Bamberg, Worms and Mentz, and the most ancient part of the minster at Strasburg; in all of these they combined the Roman-Ionic order with the Gothic of the Lombards.

This style is called the LOMBARD-BYZANTINE. To the last style there were afterwards added cupolas similar to those used in the east, together with numerous slender pillars with tasteless capitals, and the many minarets which are the characteristics of the proper *Byzantine*, or *Oriental* style.

177.—In the eighth century, when the Arabs and Moors destroyed the kingdom of the Goths, the arts and sciences were mostly in possession of the Musselmen-conquerors; at which time there were three kinds of architecture practised; viz: the Arabian, the Moorish and the modern-Gothic. The ARABIAN style was formed from Greek models, having circular arches added, and towers which terminated with globes and minarets. The MOORISH is very similar to the Arabian, being distinguished from it by arches in the form of a horse-shoe. It originated in Spain in the erection of buildings with the ruins of Roman architecture, and is seen in all its splendour in the ancient palace of the Mohammedan monarchs at Grenada, called the *Alhambra*, or *red-house*. The MODERN-GOTHIC was originated by the Visigoths in Spain by a combination of the Arabian and Moorish styles; and introduced by Charlemagne into Germany. On account of the changes and improvements it there underwent, it was, at about the 13th or 14th century, termed the *German*, or *romantic* style. It is exhibited in great perfection in the towers of the minster of Strasburgh, the cathedral of Cologne and other edifices. The most remarkable features of this lofty and aspiring style, are the lancet or pointed arch, clustered pillars, lofty towers and flying buttresses. It was principally employed in ecclesiastical architecture, and in this capacity introduced into France, Italy, Spain, and England.

178.—The Gothic architecture of England is divided into the *Norman*, the *Early-English*, the *Decorated*, and the *Perpendicular* styles. The Norman is principally distinguished by the character of its ornaments—the chevron, or *zigzag*, being the most common. Buildings in this style were erected in the 12th century. The Early-English is celebrated for the beauty of its edifices, the chaste simplicity and purity of design which they display, and the peculiarly graceful character of its foliage. This style is of the 13th century. The Decorated style, as its name implies, is characterized by a great profusion of enrichment, which consists principally of the crocket, or feathered-ornament, and ball-flower. It was mostly in use in the 14th century. The Perpendicular style, which dates from the 15th century, is distinguished by its high towers, and parapets surmounted with spires similar in number and grouping to oriental minarets.

179.—Thus these several styles, which have been erroneously termed *Gothic*, were distinguished by peculiar characteristics as well as by different names. The first symptoms of a desire to return to a pure style in architecture, after the ruin caused by the Goths, was manifested in the character of the art as displayed in the church of St. Sophia at Constantinople, which was erected by Justinian in the 6th century. The church of St. Mark at Venice, which arose in the 10th or 11th century, was the work of Grecian architects, and resembles in magnificence the forms of ancient architecture. The cathedral at Pisa, a wonderful structure for the age, was erected by a Grecian architect in 1016. The marble with which the walls of this building were faced, and of which the four rows of columns that support the roof are composed, is said to be of an excellent character. The Campanile, or leaning-tower as it is usually called, was erected near the cathedral in the 12th century. Its inclination is generally supposed to have arisen from a poor foundation; although by some it is said to have been thus constructed originally, in order to inspire in the minds of the beholder sensations of sublimity and awe. In the 13th century, the science in Italy was slowly progressing; many fine churches were erected, the style of which displayed a decided advance in the progress towards pure classical architecture. In other parts of Europe, the Gothic, or pointed style, was prevalent. The cathedral at Strasburg, designed by Irwin Steinbeck, was erected in the 13th and 14th centuries. In France and England during the 14th century, many very superior edifices were erected in this style.

180.—In the 14th and 15th centuries, and particularly in the latter, architecture in Italy was greatly revived. The masters began to study the remains of ancient Roman edifices; and many splendid buildings were erected, which displayed a purer taste in the science. Among others, St. Peter's of Rome, which was built about this time, is a lasting monument of the architectural skill of the age. Giocondo, Michael Angelo, Palladio, Vignola, and other celebrated architects, each in their turn, did much to restore the art to its former excellence. In the edifices which were erected under their direction, however, it is plainly to be seen that they studied not from the pure models of Greece, but from the remains of the deteriorated architecture of Rome. The high pedestal, the coupled columns, the rounded pediment, the many curved-and-twisted enrichments, and the convex frieze, were unknown to pure Grecian architecture. Yet their efforts were serviceable in correcting, to a good degree, the very impure taste that had prevailed since the overthrow of the Roman empire.

181.—At about this time, the Italian masters and numerous artists who had visited Italy for the purpose, spread the Roman style over various countries of Europe; which was gradually received into favor in place of the modern-Gothic. This fell into disuse; although it has of late years been again cultivated. It requires a building of great magnitude and complexity for a perfect display of its beauties. In America, the pure Grecian style was at first more or less studied; and perhaps the simplicity of its principles would be better adapted to a republican country, than the intricacy and extent of those of the Gothic; but at the present time the latter style is being introduced, especially for ecclesiastical structures.

STYLES OF ARCHITECTURE.

182.—It is generally acknowledged that the various styles in architecture, were originated in accordance with the different pursuits of the early inhabitants of the earth; and were brought by their descendants to their present state of perfection, through the propensity for imitation and desire of emulation which are found more or less among all nations. Those that followed agricultural pursuits, from being employed constantly upon the same piece of land, needed a permanent residence, and the wooden *hut* was the offspring of their wants; while the shepherd, who followed his flocks and was compelled to traverse large tracts of country for pasture, found the *tent* to be the most portable habitation; again, the man devoted to hunting and fishing—an idle and vagabond way of living—is naturally supposed to have been content with the *cavern* as a place of shelter. The latter is said to have been the origin of the Egyptian style; while the curved roof of Chinese structures gives a strong indication of their having had the tent for their model; and the simplicity of the original style of the Greeks, (the Doric,) shows quite conclusively, as is generally conceded, that its original was of wood. The modern-Gothic, or pointed style, which was most generally confined to ecclesiastical structures, is said by some to have originated in an attempt to imitate the bower, or grove of trees, in which the ancients performed their idol-worship.

183.—There are numerous styles, or orders, in architecture; and a knowledge of the peculiarities of each is important to the student in the art. An Order, in architecture, is composed of three principal parts, viz: the Stylobate, the Column and the Entablature.

184.—The Stylobate is the substructure, or basement, upon which the columns of an order are arranged. In Roman architecture—especially in the interior of an edifice—it frequently occurs, that each column has a separate substructure; this is called a *pedestal.* If possible, the pedestal should be avoided in all cases; because it gives to the column, the appearance of having been originally designed for a small building, and afterwards pieced-out to make it long enough for a larger one.

185.—The COLUMN is composed of the base, shaft and capital.

186.—The ENTABLATURE, above and supported by the columns, is horizontal; and is composed of the architrave, frieze and cornice. These principal parts are again divided into various members and mouldings. (See *Sect.* III.)

187.—The BASE of a column is so called from *basis*, a foundation, or footing.

188.—The SHAFT, the upright part of a column standing upon the base and crowned with the capital, is from *shafto*, to dig—in the manner of a well, whose inside is not unlike the form of a column.

189.—The CAPITAL, from *kephale* or *caput*, the head, is the uppermost and crowning part of the column.

190.—The ARCHITRAVE, from *archi*, chief or principal, and *trabs*, a beam, is that part of the entablature which lies in immediate connection with the column.

191.—The FRIEZE, from *fibron*, a fringe or border, is that part of the entablature which is immediately above the architrave and beneath the cornice. It was called by some of the ancients, *zophorus*, because it was usually enriched with sculptured animals.

192.—The CORNICE, from *corona*, a crown, is the upper and projecting part of the entablature—being also the uppermost and crowning part of the whole order.

193.—The PEDIMENT, above the entablature, is the triangular portion which is formed by the inclined edges of the roof at the end of the building. In Gothic architecture, the pediment is called, a *gable*.

194.—The TYMPANUM is the perpendicular triangular surface which is enclosed by the cornice of the pediment.

195.—The ATTIC is a small order, consisting of pilasters and entablature, raised above a larger order, instead of a pediment. An attic story is the upper story, its windows being usually square.

196.—An order, in architecture, has its several parts and members proportioned to one another by a scale of 60 equal parts, which are called minutes. If the height of buildings were always the same, the scale of equal parts would be a fixed quantity—an exact number of feet and inches. But as buildings are erected of different heights, the column and its accompaniments are required to be of different dimensions. To ascertain the scale of equal parts, it is necessary to know the height to which the whole order is to be erected. This must be divided by the number of diameters which is directed for the order under consideration. Then the quotient obtained by such division, is the length of the scale of equal parts—and is, also, the diameter of the column next above the base. For instance, in the Grecian Doric order the whole height, including column and entablature, is 8 diameters. Suppose now it were desirable to construct an example of this order, forty feet high. Then 40 feet divided by 8, gives 5 feet for the length of the scale; and this being divided by 60, the scale is completed. The upright columns of figures, marked *H* and *P*, by the side of the drawings illustrating the orders, designate the height and the projection of the members. The projection of each member is reckoned from a line passing through the axis of the column, and extending above it to the top of the entablature. The figures represent minutes, or 60ths, of the major diameter of the shaft of the column.

197.—GRECIAN STYLES. The original method of building among the Greeks, was in what is called the *Doric* order: to this were afterwards added the *Ionic* and the *Corinthian*. These three were the only styles known among them. Each is distinguished from the other two, by not only a peculiarity of some one or more of its principal parts, but also by a particular destination. The character of the Doric is robust, manly and Herculean-like; that of the Ionic is more delicate, feminine, matronly; while that of the Corinthian is extremely delicate, youthful and virgin-like. However they may differ in their general character, they are alike famous for grace and dignity, elegance and grandeur, to a high degree of perfection.

198.—The Doric Order, (*Fig.* 120,) is so ancient that its origin is unknown—although some have pretended to have discovered it. But the most general opinion is, that it is an improvement upon the original wooden buildings of the Grecians. These no doubt were very rude, and perhaps not unlike the following figure.

Fig. 119.

The trunks of trees, set perpendicularly to support the roof, may be taken for columns; the tree laid upon the tops of the perpendicular ones, the architrave; the ends of the cross-beams which rest upon the architrave, the triglyphs; the tree laid on the cross-beams as a support for the ends of the rafters, the bed-moulding of the cornice; the ends of the rafters which project beyond the bed-moulding, the mutules; and perhaps the projection of the roof in front, to screen the entrance from the weather, gave origin to the portico.

The peculiarities of the Doric order are the triglyphs—those parts of the frieze which have perpendicular channels cut in

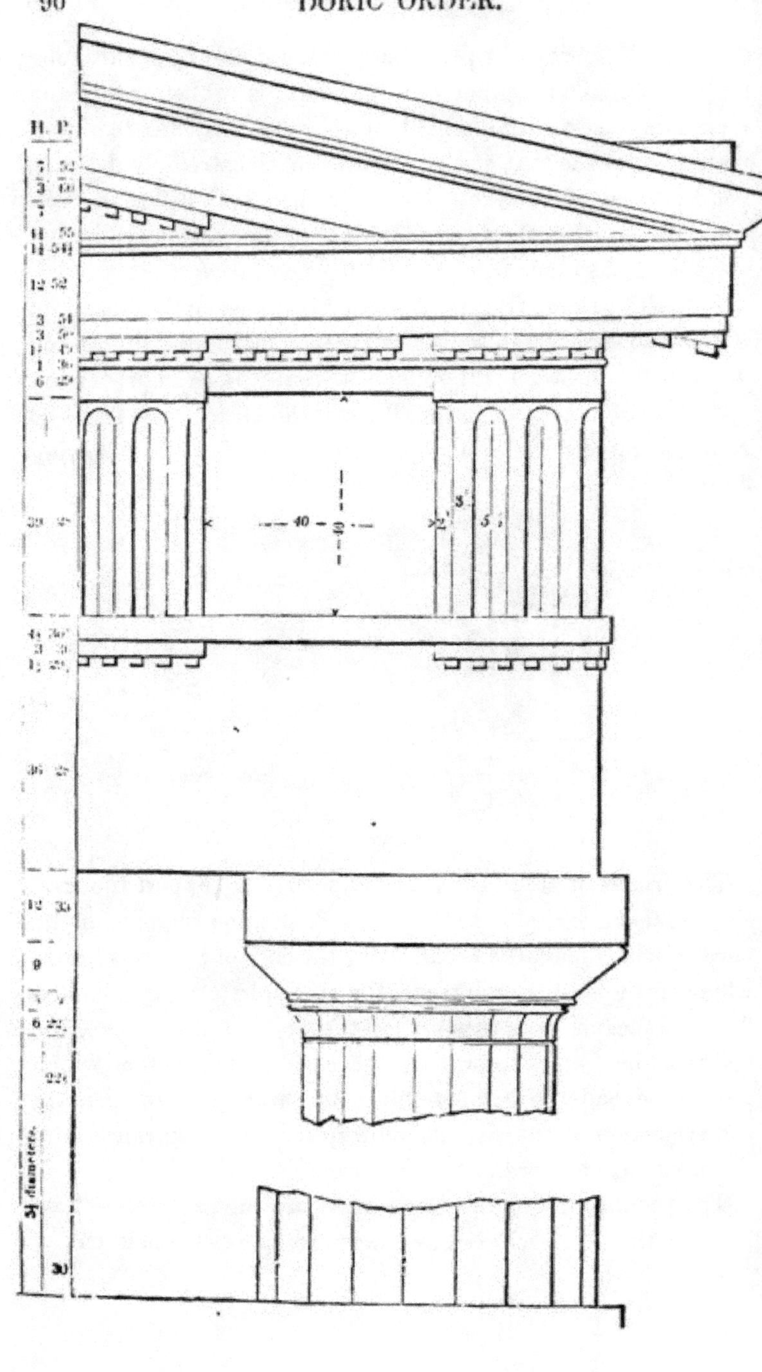
H. P.
12 52
40
12 35
9
3½ diameters.
30

their surface; the absence of a base to the column—as also of fillets between the flutings of the column, and the plainness of the capital. The triglyphs are to be so disposed that the width of the metopes—the spaces between the triglyphs—shall be equal to their height.

199.—The *intercolumniation*, or space between the columns, is regulated by placing the centres of the columns under the centres of the triglyphs—except at the angle of the building; where, as may be seen in *Fig.* 120, one edge of the triglyph must be over the centre of the column.* Where the columns are so disposed that one of them stands beneath every other triglyph, the arrangement is called, *mono-triglyph*, and is most common. When a column is placed beneath every third triglyph, the arrangement is called *diastyle;* and when beneath every fourth, *aræostyle.* This last style is the worst, and is seldom adopted.

200.—The Doric order is suitable for buildings that are destined for national purposes, for banking-houses, &c. Its appearance, though massive and grand, is nevertheless rich and graceful. The Patent Office at Washington, and the Custom-House at New York, are good specimens of this order.

---

* GRECIAN DORIC ORDER. *When the width to be occupied by the whole front is limited; to determine the diameter of the column.*

The relation between the parts may be expressed thus:

$$x = \frac{60\,a}{d\,(b + c) + (60 - c)}.$$

Where a equals the width in feet occupied by the columns and their intercolumniations taken collectively, measured at the base; b equals the width of the metope, in minutes; c equals the width of the triglyphs in minutes; d equals the number of metopes, and x equals the diameter in feet.

Example.—A front of six columns—hexastyle—61 feet wide; the frieze having one triglyph over each intercolumniation, or mono-triglyph. In this case, there being five intercolumniations and two metopes over each, therefore there are $5 \times 2 = 10$ metopes. Let the metope equal 42 minutes and the triglyph equal 28. Then $a = 61$; $b = 42$; $c = 28$; and $d = 10$; and the formula above becomes,

$$x = \frac{60 \times 61}{10\,(42 + 28) + (60 - 28)} = \frac{60 \times 61}{10 \times 70 + 32} = \frac{3660}{732} = 5 \text{ feet} = \text{the diameter required.}$$

Example.—An octastyle front, 8 columns, 184 feet wide, three metopes over each intercolumniation, 21 in all, and the metope and triglyph 42 and 28, as before. Then,

$$x = \frac{60 \times 184}{21\,(42 + 28) + (60 - 28)} = \frac{11040}{1502} = 7{\cdot}35\tfrac{3}{1502} \text{ feet} = \text{the diameter required.}$$

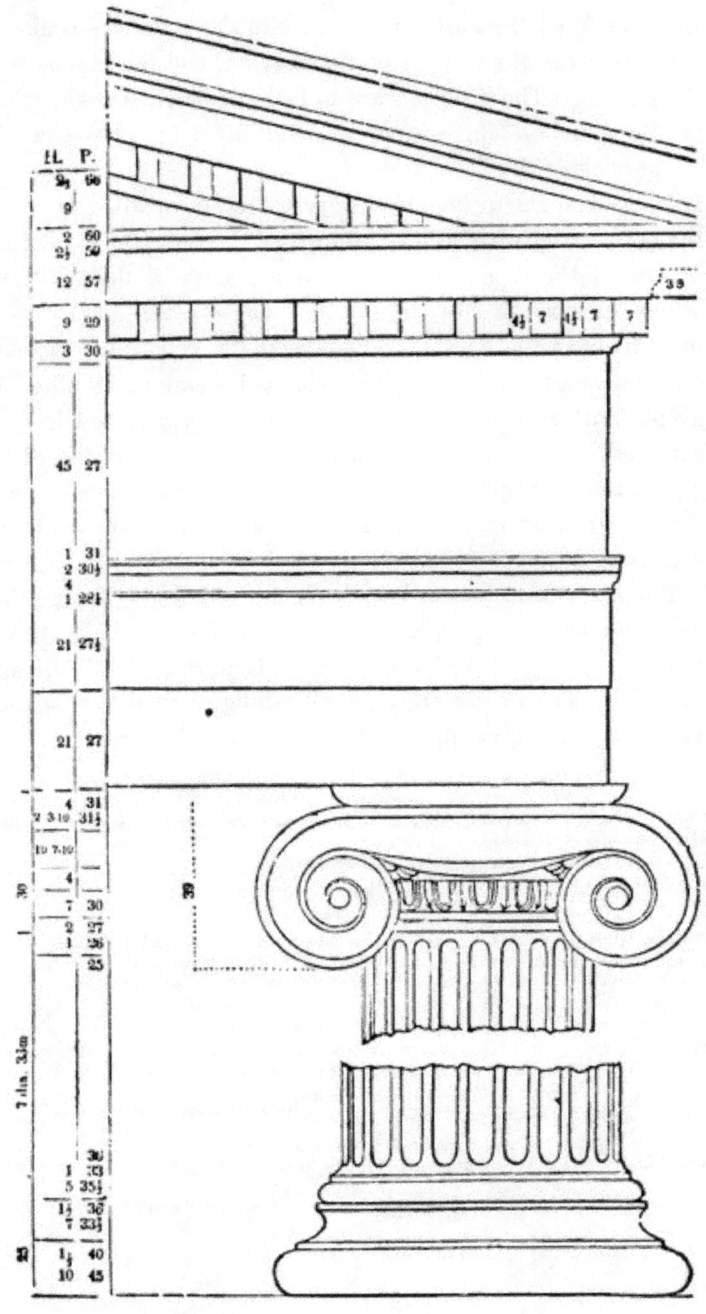

201.—The Ionic Order. (*Fig.* 121.) The Doric was for some time the only order in use among the Greeks. They gave their attention to the cultivation of it, until perfection seems to have been attained. Their temples were the principal objects upon which their skill in the art was displayed; and as the Doric order seems to have been well fitted, by its massive proportions, to represent the character of their male deities rather than the female, there seems to have been a necessity for another style which should be emblematical of feminine graces, and with which they might decorate such temples as were dedicated to the goddesses. Hence the origin of the Ionic order. This was invented, according to historians, by Hermogenes of Alabánda; and he being a native of Caria, then in the possession of the Ionians, the order was called, the Ionic.

202.—The distinguishing features of this order are the *volutes*, or spirals of the capital; and the *dentils* among the bed-mouldings of the cornice: although in some instances, dentils are wanting. The volutes are said to have been designed as a representation of curls of hair on the head of a matron, of whom the whole column is taken as a semblance.

203.—The intercolumniation of this and the other orders—both Roman and Grecian, with the exception of the Doric—are distinguished as follows. When the interval is one and a half diameters, it is called, *pycnostyle*, or columns thick-set; when two diameters, *systyle;* when two and a quarter diameters, *eustyle;* when three diameters, *diastyle;* and when more than three diameters, *aræostyle*, or columns thin-set. In all the orders, when there are four columns in one row, the arrangement is called, *tetrastyle;* when there are six in a row, *hexastyle;* and when eight, *octastyle.*

204.—The Ionic order is appropriate for churches, colleges, seminaries, libraries, all edifices dedicated to literature and the arts, and all places of peace and tranquillity. The front of the

Merchants' Exchange, New York city, is a good specimen of this order.

205.—*To describe the Ionic volute.* Draw a perpendicular from *a* to *s*, (*Fig.* 122,) and make *a s* equal to 20 min. or to $\frac{4}{7}$ of the whole height, *a c;* draw *s o*, at right angles to *s a*, and equal to $1\frac{1}{4}$ min.; upon *o*, with $2\frac{1}{2}$ min. for radius, describe the eye of the volute; about *o*, the centre of the eye, draw the square, *r t* 1 2, with sides equal to half the diameter of the eye, viz. $2\frac{1}{2}$ min., and divide it into 144 equal parts, as shown

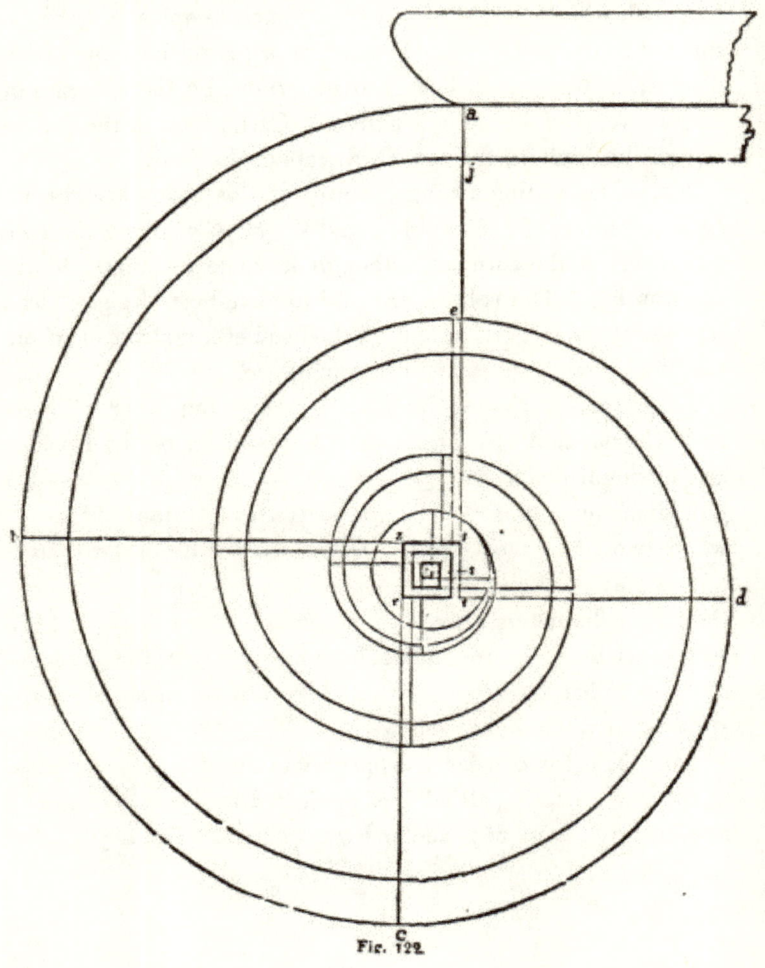

Fig. 122.

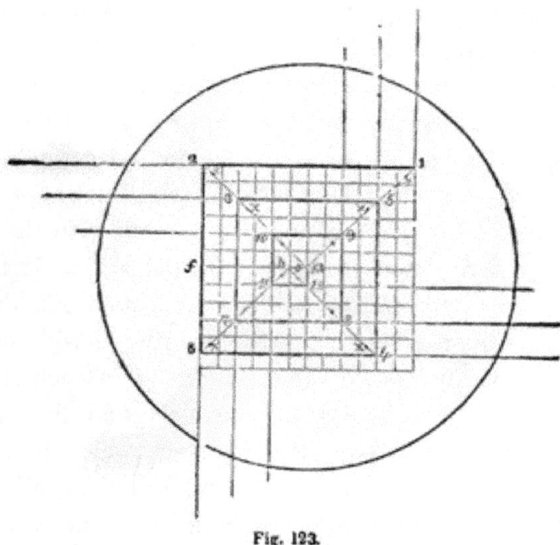

Fig. 123.

at *Fig.* 123. The several centres in rotation are at the angles formed by the heavy lines, as figured, 1, 2, 3, 4, 5, 6, &c. The position of these angles is determined by commencing at the point, 1, and making each heavy line one part less in length than the preceding one. No. 1 is the centre for the arc, *a b*, (*Fig.* 122;) 2 is the centre for the arc, *b c;* and so on to the last. The inside spiral line is to be described from the centres, *x*, *x*, *x*, &c., (*Fig.* 123,) being the centre of the first small square towards the middle of the eye from the centre for the outside arc. The breadth of the fillet at *a j*, is to be made equal to $2\frac{3}{16}$ min. This is for a spiral of *three* revolutions; but one of any number of revolutions, as 4 or 6, may be drawn, by dividing *o f*, (*Fig.* 123,) into a corresponding number of equal parts. Then divide the part nearest the centre, *o*, into two parts, as at *h;* join *o* and 1, also *o* and 2; draw *h* 3, parallel to *o* 1, and *h* 4, parallel to *o* 2; then the lines, *o* 1, *o* 2, *h* 3, *h* 4, will determine the length of the heavy lines, and the place of the centres. (See *Art.* 489.)

206.—The CORINTHIAN ORDER, (*Fig.* 125,) is in general like the Ionic, though the proportions are lighter. The Corinthian displays a more airy elegance, a richer appearance; but its distinguishing feature is its beautiful capital. This is generally supposed to have had its origin in the capitals of the columns of Egyptian temples; which, though not approaching it in elegance, have yet a similarity of form with the Corinthian. The oft-repeated story of its origin which is told by Vitruvius—an architect who flourished in Rome, in the days of Augustus Cæsar—though pretty generally considered to be fabulous, is nevertheless worthy of being again recited. It is this: a young lady of Corinth was sick, and finally died. Her nurse gathered into a deep basket, such trinkets and keepsakes as the lady had been fond of when alive, and placed them upon her grave; covering the basket with a flat stone or tile, that its contents might not be disturbed. The basket was placed accidentally upon the stem of an acanthus plant, which, shooting forth, enclosed the basket with its foliage; some of which, reaching the tile, turned gracefully over in the form of a volute.

Fig. 124.

A celebrated sculptor, Calimachus, saw the basket thus decorated, and from the hint which it suggested, conceived and constructed a capital for a column. This was called Corinthian from the fact that it was invented and first made use of at Corinth.

207.—The Corinthian being the gayest, the richest, and most lovely of all the orders, it is appropriate for edifices which are dedicated to amusement, banqueting and festivity—for all places where delicacy, gayety and splendour are desirable.

208.—In addition to the three regular orders of architecture,

CORINTHIAN ORDER.—Fig. 125.

it was sometimes customary among the Greeks—and afterwards among other nations—to employ representations of the human form, instead of columns, to support entablatures; these were called *Persians* and *Caryatides.*

209.—PERSIANS are statues of men, and are so called in commemoration of a victory gained over the Persians by Pausanias. The Persian prisoners were brought to Athens and condemned to abject slavery; and in order to represent them in the lowest state of servitude and degradation, the statues were loaded with the heaviest entablature, the Doric.

210.—CARYATIDES are statues of women dressed in long robes after the Asiatic manner. Their origin is as follows. In a war between the Greeks and the Caryans, the latter were totally vanquished, their male population extinguished, and their females carried to Athens. To perpetuate the memory of this event, statues of females, having the form and dress of the Caryans, were erected, and crowned with the Ionic or Corinthian entablature. The caryatides were generally formed of about the human size, but the persians much larger; in order to produce the greater awe and astonishment in the beholder. The entablatures were proportioned to a statue in like manner as to a column of the same height.

211.—These semblances of slavery have been in frequent use among moderns as well as ancients; and as a relief from the stateliness and formality of the regular orders, are capable of forming a thousand varieties; yet in a land of liberty such marks of human degradation ought not to be perpetuated.

212.—ROMAN STYLES. Strictly speaking, Rome had no architecture of her own—all she possessed was borrowed from other nations. Before the Romans exchanged intercourse with the Greeks, they possessed some edifices of considerable extent and merit, which were erected by architects from Etruria; but Rome was principally indebted to Greece for what she acquired of the art. Although there is no such thing as

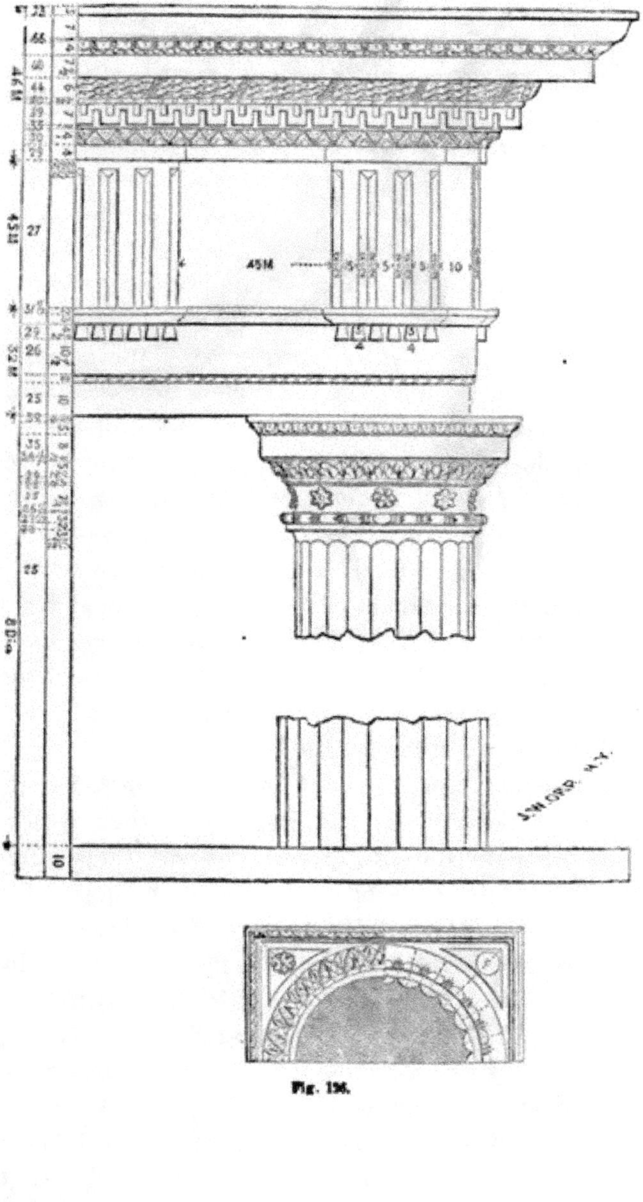

Fig. 136.

an architecture of Roman invention, yet no nation, perhaps, ever was so devoted to the cultivation of the art as the Roman. Whether we consider the number and extent of their structures, or the lavish richness and splendour with which they were adorned, we are compelled to yield to them our admiration and praise. At one time, under the consuls and emperors, Rome employed 400 architects. The public works —such as theatres, circuses, baths, aqueducts, &c.—were, in extent and grandeur, beyond any thing attempted in modern times. Aqueducts were built to convey water from a distance of 60 miles or more. In the prosecution of this work, rocks and mountains were tunnelled, and valleys bridged. Some of the latter descended 200 feet below the level of the water; and in passing them the canals were supported by an arcade, or succession of arches. Public baths are spoken of as large as cities; being fitted up with numerous conveniences for exercise and amusement. Their decorations were most splendid; indeed, the exuberance of the ornaments alone was offensive to good taste. So overloaded with enrichments were the baths of Diocletian, that on an occasion of public festivity, great quantities of sculpture fell from the ceilings and entablatures, killing many of the people.

213.—The three orders of Greece were introduced into Rome in all the richness and elegance of their perfection. But the luxurious Romans, not satisfied with the simple elegance of their refined proportions, sought to improve upon them by lavish displays of ornament. They transformed in many instances, the true elegance of the Grecian art into a gaudy splendour, better suited to their less refined taste. The Romans remodelled each of the orders: the Doric, (*Fig.* 126,) was modified by increasing the height of the column to 8 diameters; by changing the echinus of the capital for an ovolo, or quarter round, and adding an astragal and neck below it; by placing the *centre*, instead of one edge, of the first triglyph

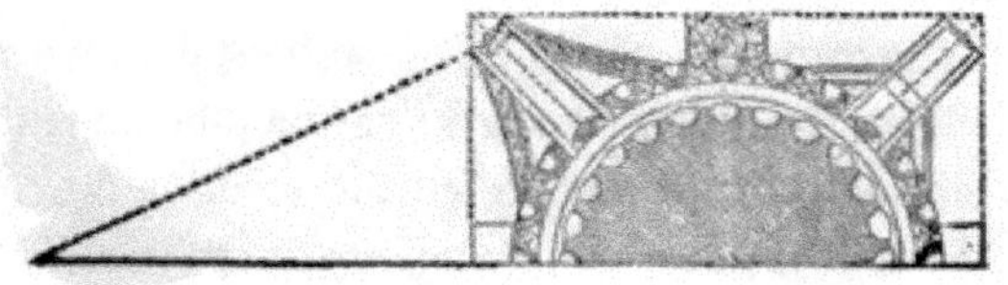

Fig. 127.

over the centre of the column; and introducing horizontal instead of inclined mutules in the cornice, and in some instances dispensing with them altogether. The Ionic was modified by diminishing the size of the volutes, and, in some specimens, introducing a new capital in which the volutes were diagonally arranged, (*Fig.* 127.) This new capital has been termed *modern* Ionic. The favorite order at Rome and her colonies was the Corinthian, (*Fig.* 128.) But this order, the Roman artists in their search for novelty, subjected to many alterations—especially in the foliage of its capital. Into the upper part of this, they introduced the modified Ionic capital; thus combining the two in one. This change was dignified with the importance of an *order*, and received the appellation, Composite, or *Roman:* the best specimen of which is found in the Arch of Titus, (*Fig.* 129.) This style was not much used among the Romans themselves, and is but slightly appreciated now.

214.—The Tuscan Order is said to have been introduced to the Romans by the Etruscan architects, and to have been the only style used in Italy before the introduction of the Grecian orders. However this may be, its similarity to the Doric order gives strong indications of its having been a rude imitation of that style: this is very probable, since history informs us that the Etruscans held intercourse with the Greeks at a remote period. The rudeness of this order prevented its extensive use in Italy. All that is known concerning it is from Vitruvius—no remains of buildings in this style being found among ancient ruins.

215.—For mills, factories, markets, barns, stables, &c., where utility and strength are of more importance than beauty, the improved modification of this order, called the *modern* Tuscan, (*Fig.* 130,) will be useful; and its simplicity recommends it where economy is desirable.

216.—Egyptian Style. The architecture of the ancient

Fig. 128.

Fig. 129.

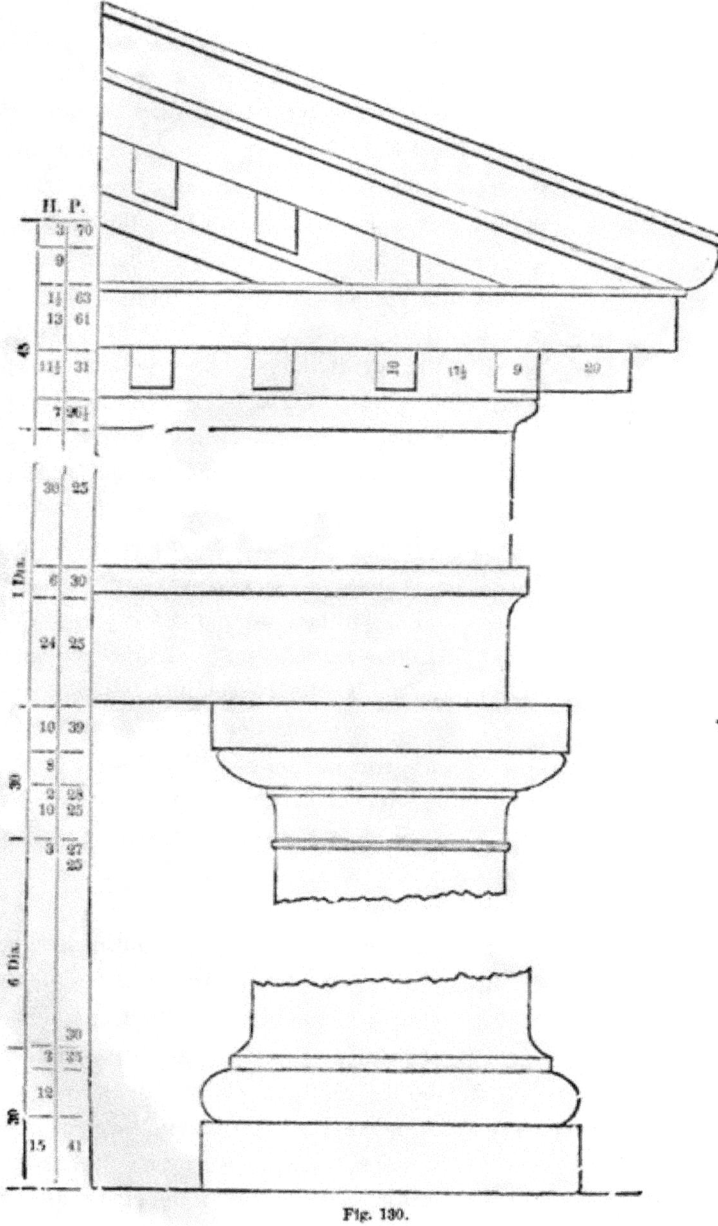
H. P.
3 70
9
1½ 63
13 61
45
11½ 31
7 26½
10
17½
9
20
30 25
6 30
1 Dia.
24 25
10 39
8
30
2 28
10 25
3 27
25
6 Dia.
30
3 35
12
30
15 41

Fig. 130.

Egyptians—to which that of the ancient Hindoos bears some resemblance—is characterized by boldness of outline, solidity and grandeur. The amazing labyrinths and extensive artificial lakes, the splendid palaces and gloomy cemeteries, the gigantic pyramids and towering obelisks, of the Egyptians, were works of immensity and durability; and their extensive remains are enduring proofs of the enlightened skill of this once-powerful, but long since extinct nation. The principal features of the Egyptian Style of architecture are—uniformity of plan, never deviating from right lines and angles; thick walls, having the outer surface slightly deviating inwardly from the perpendicular; the whole building low; roof flat, composed of stones reaching in one piece from pier to pier, these being supported by enormous columns, very stout in proportion to their height; the shaft sometimes polygonal, having no base but with a great variety of handsome capitals, the foliage of these being of the palm, lotus and other leaves; entablatures having simply an architrave, crowned with a huge cavetto ornamented with sculpture; and the intercolumniation very narrow, usually 1½ diameters and seldom exceeding 2½. In the remains of a temple, the walls were found to be 21 feet thick; and at the gates of Thebes, the walls at the foundation were 50 feet thick and perfectly solid. The immense stones of which these, as well as Egyptian walls generally, were built, had both their inside and outside surfaces faced, and the joints throughout the body of the wall as perfectly close as upon the outer surface. For this reason, as well as that the buildings generally partake of the pyramidal form, arise their great solidity and durability. The dimensions and extent of the buildings may be judged from the temple of Jupiter at Thebes, which was 1400 feet long and 300 feet wide—exclusive of the porticos, of which there was a great number.

It is estimated by Mr. Gliddon, U. S. consul in Egypt, that not less than 25,000,000 tons of hewn stone were employed in the erection of the Pyramids of Memphis alone,—or enough to construct 3,000 Bunker-Hill monuments. Some of the blocks are 40

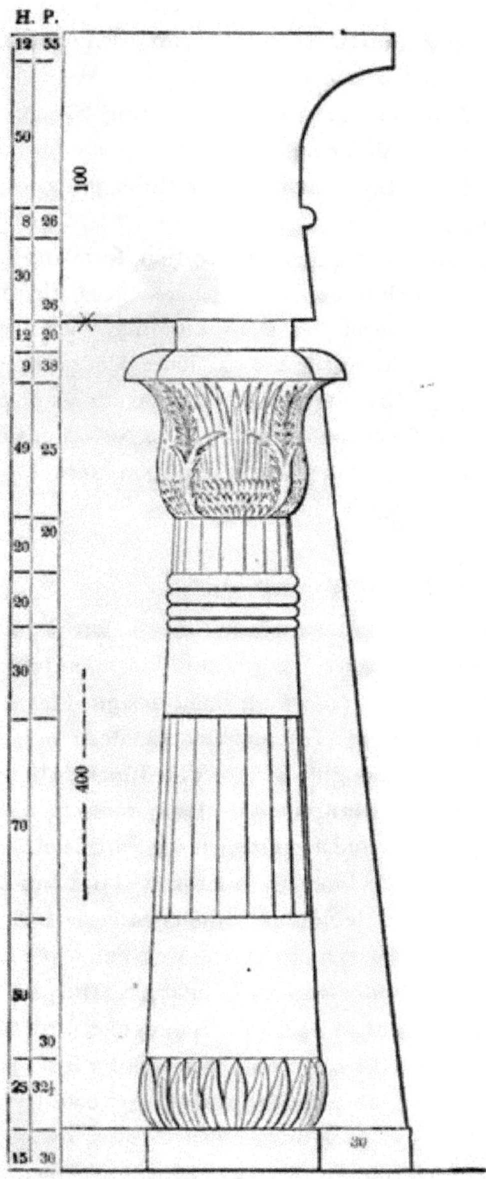
H. P.
12 55
50
100
8 26
30
26
12 20
9 38
49 25
20 20
20
30
400
70
50
30
25 32½
15 30
30

Fig. 131.

feet long, and polished with emery to a surprising degree. It is conjectured that the stone for these pyramids was brought, by rafts and canals, from a distance of 6 or 7 hundred miles.

217.—The general appearance of the Egyptian style of architecture is that of solemn grandeur—amounting sometimes to sepulchral gloom. For this reason it is appropriate for cemeteries, prisons, &c.; and being adopted for these purposes, it is gradually gaining favour.

A great dissimilarity exists in the proportion, form and general features of Egyptian columns. In some instances, there is no uniformity even in those of the same building, each differing from the others either in its shaft or capital. For practical use in this country, *Fig.* 131 may be taken as a standard of this style. The Halls of Justice in Centre-street, New-York city, is a building in general accordance with the principles of Egyptian architecture.

Buildings in General.

218.—That style of architecture is to be preferred in which utility, stability and regularity, are gracefully blended with grandeur and elegance. But as an arrangement designed for a warm country would be inappropriate for a colder climate, it would seem that the style of building ought to be modified to suit the wants of the people for whom it is designed. High roofs to resist the pressure of heavy snows, and arrangements for artificial heat, are indispensable in northern climes; while they would be regarded as entirely out of place in buildings at the equator.

219.—Among the Greeks, architecture was employed chiefly upon their temples and other large buildings; and the proportions of the orders, as determined by them, when executed to such large dimensions, have the happiest effect. But when used for small buildings, porticos, porches, &c., especially in country-places, they are rather heavy and clumsy; in such cases, more slender proportions will be found to produce a better effect. The

English cottage-style is rather more appropriate, and is becoming extensively practised for small buildings in the country.

220.—Every building should bear an expression suited to its destination. If it be intended for national purposes, it should be magnificent—grand; for a private residence, neat and modest; for a banqueting-house, gay and splendid; for a monument or cemetery, gloomy—melancholy; or, if for a church, majestic and graceful. By some it has been said—"somewhat dark and gloomy, as being favourable to a devotional state of feeling;" but such impressions can only result from a misapprehension of the nature of true devotion. "Her ways are ways of *pleasantness*, and all her paths are peace." The church should rather be a type of that brighter world to which it leads.

221.—However happily the several parts of an edifice may be disposed, and however pleasing it may appear as a whole, yet much depends upon its *site*, as also upon the character and style of the structures in its immediate vicinity, and the degree of cultivation of the adjacent country. A splendid country-seat should have the out-houses and fences in the same style with itself, the trees and shrubbery neatly trimmed, and the grounds well cultivated.

222.—Europeans express surprise that so many houses in this country are built of wood. And yet, in a new country, where wood is plenty, that this should be so is no cause for wonder. Still, the practice should not be encouraged. Buildings erected with brick or stone are far preferable to those of wood; they are more durable; not so liable to injury by fire, nor to need repairs; and will be found in the end quite as economical. A wooden house is suitable for a temporary residence only; and those who would bequeath a dwelling to their children, will endeavour to build with a more durable material. Wooden cornices and gutters, attached to brick houses, are objectionable—not only on account of their frail nature, but also because they render the building liable to destruction by fire.

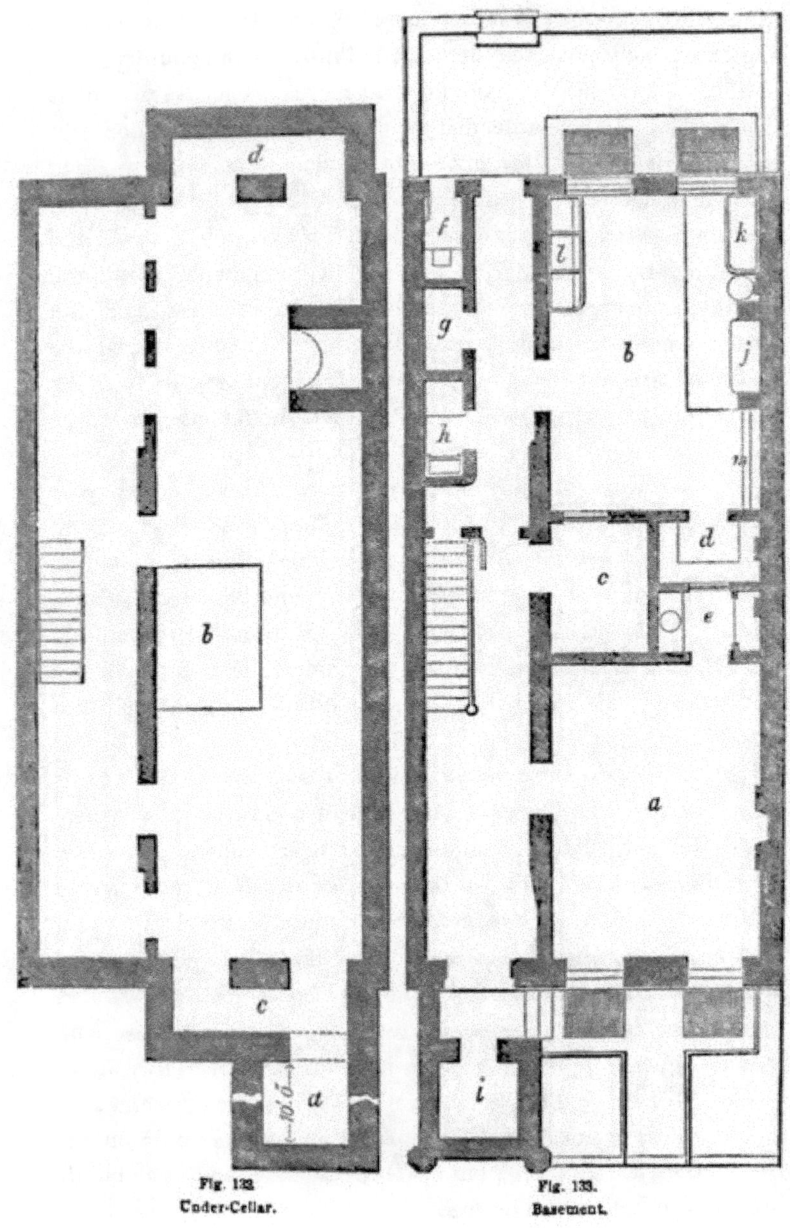

Fig. 132.
Under-Cellar.

Fig. 133.
Basement.

223.—Dwelling houses are built of various dimensions and styles, according to their destination; and to give designs and directions for their erection, it is necessary to know their situation and object. A dwelling intended for a gardener, would require very different dimensions and arrangements from one intended for a retired gentleman—with his servants, horses, &c.; nor would a house designed for the city be appropriate for the country. For city houses, arrangements that would be convenient for one family might be very inconvenient for two or more. *Fig.* 132, 133, 134, 135, 136, and 137, represent the *ichnographical projection*, or ground-plan, of the floors of an ordinary city house, designed to be occupied by one family only. *Fig.* 139 is an *elevation*, or front-view, of the same house: all these plans are drawn at the same scale—which is that at the bottom of *Fig.* 139.

Fig. 132 is a Plan of the Under-Cellar.

a, is the coal-vault, 6 by 10 feet.
b, is the furnace for heating the house.
c, *d*, are front and rear areas.

Fig. 133 is a Plan of the Basement.

a, is the library, or ordinary dining-room, 15 by 20 feet.
b, is the kitchen, 15 by 22 feet.
c, is the store-room, 6 by 9 feet.
d, is the pantry, 4 by 7 feet.
e, is the china closet, 4 by 7 feet.
f, is the servants' water-closet.
g, is a closet.
h, is a closet with a dumb-waiter to the first story above.
i, is an ash closet under the front stoop.
j, is the kitchen-range.
k, is the sink for washing and drawing water
l, are wash trays.

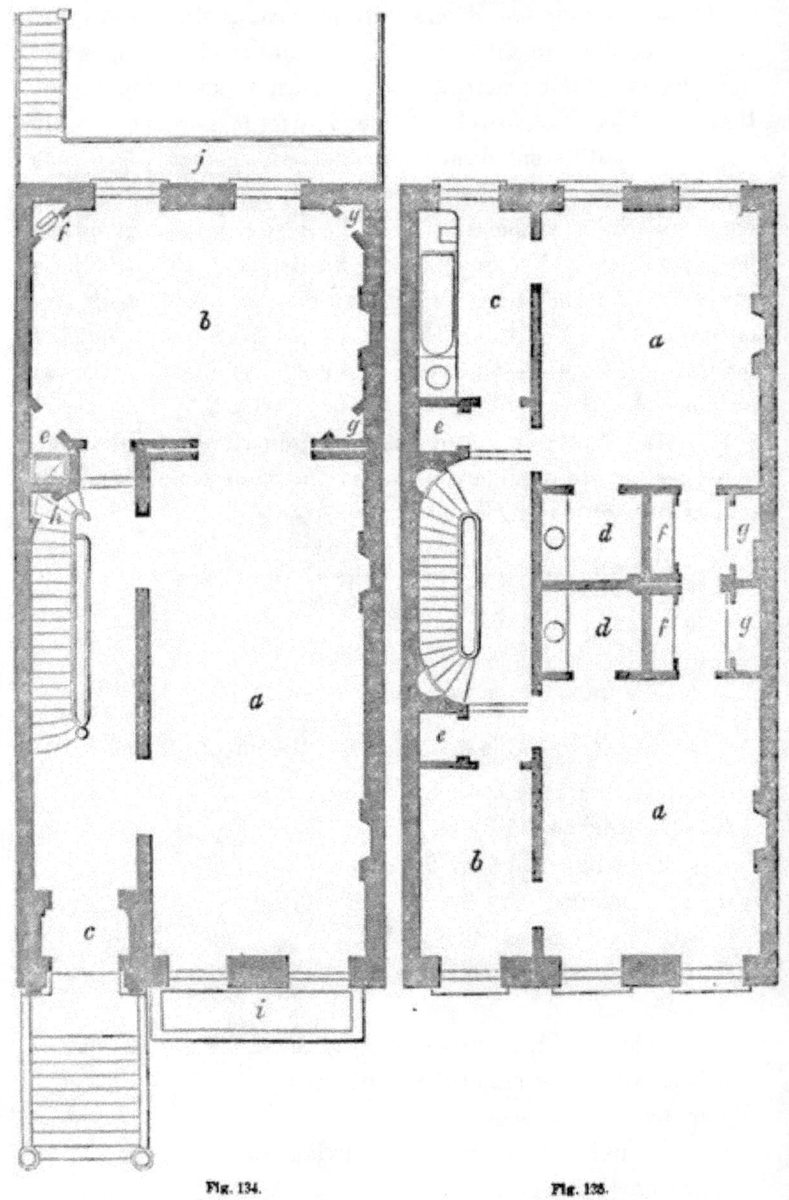

Fig. 134.
First Story

Fig. 135.
Second Story.

Fig. 134 is a Plan of the First Story.

a, is the parlor, 15 by 34 feet.

b, is the dining-room, 16 by 23 feet.

c, is the vestibule.

e, is the closet containing the dumb-waiter from the basement.

f, is the closet containing butler's sink.

g, *g*, are closets.

h, is a closet for hats and cloaks.

i, *j*, are front and rear balconies.

Fig. 135 is the Second Story.

a, *a*, are chambers, 15 by 19 feet.

b, is a bed-room, 7½ by 13 feet.

c, is the bath-room, 7½ by 13 feet.

d, *d*, are dressing-rooms, 6 by 7½ feet.

e, *e*, are closets.

f, *f*, are wardrobes.

g, *g*, are cupboards.

Fig. 136 is the Third Story.

a, *a*, are chambers, 15 by 19 feet.

b, *b*, are bed-rooms, 7½ by 13 feet.

c, *c*, are closets.

d, is a linen closet, 5 by 7 feet.

e, *e*, are dressing-closets.

f, *f*, are wardrobes.

g, *g*, are cupboards.

Fig. 137 is the Fourth Story.

a, *a*, are chambers, 14 by 17 feet.

b, *b*, are bed-rooms, 8½ by 17 feet.

c, *c*, *c*, are closets.

d, is the step-ladder to the roof.

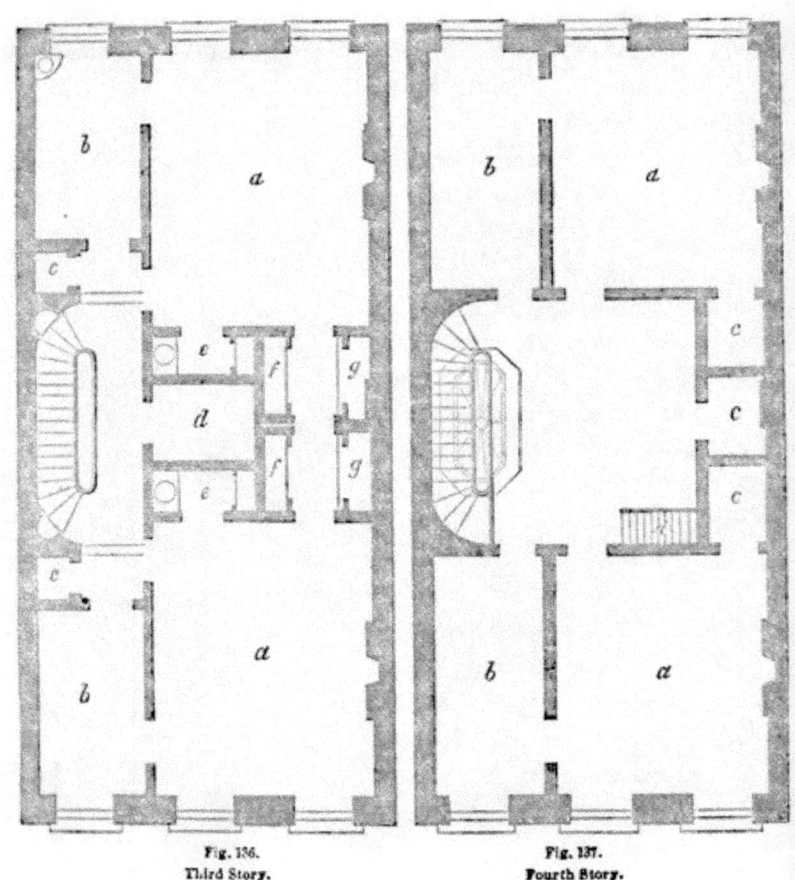

Fig. 136.
Third Story.

Fig. 137.
Fourth Story.

Fig. 138 is the Section of the House showing the heights of the several stories.

Fig. 139 is the Front Elevation.

The size of the house is 25 feet front by 55 feet deep; this is about the average depth, although some are extended to 60 and 65 feet in depth.

These are introduced to give some general ideas of the principles to be followed in designing city houses. In placing the chimneys in the parlours, set the chimney-breasts equi-distant from the ends of the room. The basement chimney-breasts may be placed nearly in the middle of the side of the room, as there is but one flue to pass through the chimney-breast above; but in the second story, as there are two flues, one from the basement and one from the parlour, the breast will have to be placed nearly perpendicular over the parlour breast, so as to receive the flues within the jambs of the fire-place. As it is desirable to have the chimney-breast as near the middle of the room as possible, it may be placed a few inches towards that point from over the breast below. So in arranging those of the stories above, always make provision for the flues from below.

224.—In placing the stairs, there should be at least as much room in the passage at the side of the stairs, as upon them; and in regard to the length of the passage in the second story, there must be room for the doors which open from each of the principal rooms into the hall, and more if the stairs require it. Having assigned a position for the stairs of the second story, now generally placed in the centre of the depth of the house, let the winders of the other stories be placed perpendicularly over and under them; and be careful to provide for head-room. To ascertain this, when it is doubtful, it is well to draw a vertical section of the whole stairs; but in ordinary cases, this is not necessary. To dispose the windows properly, the

Fig. 138.
Section.

Fig. 139.
Elevation.

middle window of each story should be exactly in the middle of the front; but the pier between the two windows which light the parlour, should be in the centre of that room; because when chandeliers or any similar ornaments, hang from the centre-pieces of the parlour ceilings, it is important, in order to give the better effect, that the pier-glasses at the front and rear, be in a range with them. If both these objects cannot be attained, an approximation to each must be attempted. The piers should in no case be less in width than the window openings, else the blinds or shutters when thrown open will interfere with one another; in general practice, it is well to make the outside piers $\frac{2}{3}$ of the width of one of the middle piers. When this is desirable, deduct the amount of the three openings from the width of the front, and the remainder will be the amount of the width of all the piers; divide this by 10, and the product will be $\frac{1}{3}$ of a middle pier; and then, if the parlour arrangements do not interfere, give twice this amount to each corner pier, and three times the same amount to each of the middle piers.

PRINCIPLES OF ARCHITECTURE.

225.—In the construction of the first habitations of men, frail and rude as they must have been, the first and principal object was, doubtless, utility—a mere shelter from sun and rain. But as successive storms shattered the poor tenement, man was taught by experience the necessity of building with an idea to durability. And when in his walks abroad, the symmetry, proportion and beauty of nature met his admiring gaze, contrasting so strangely with the misshapen and disproportioned work of his own hands, he was led to make gradual changes; till his abode was rendered not only commodious and durable, but pleasant in its appearance; and building became a fine art, having utility for its basis.

226.—In all designs for buildings of importance, utility, durability and beauty, the first great principles of architecture, should be pre-eminent. In order that the edifice be useful, commodious and comfortable, the arrangement of the apartments should be such as to fit them for their several destinations; for public assemblies, oratory, state, visitors, retiring, eating, reading, sleeping, bathing, dressing, &c.—these should each have its own peculiar form and situation. To accomplish this, and at the same time to make their relative situation agreeable and pleasant, producing regularity and harmony, require in some instances much skill and sound judgment. Convenience and regularity are very important, and each should have due attention; yet when both cannot be obtained, the latter should in most cases give place to the former. A building that is neither convenient nor regular, whatever other good qualities it may possess, will be sure of disapprobation.

227.—The utmost importance should be attached to such arrangements as are calculated to promote health: among these, *ventilation* is by no means the least. For this purpose, the ceilings of the apartments should have a respectable height; and the sky-light, or any part of the roof that can be made moveable, should be arranged with cord and pullies, so as to be easily raised and lowered. Small openings near the ceiling, that may be closed at pleasure, should be made in the partitions that separate the rooms from the passages—especially for those rooms which are used for sleeping apartments. All the apartments should be so arranged as to secure their being easily kept *dry* and *clean*. In dwellings, suitable apartments should be fitted up for *bathing* with all the necessary apparatus for conveying the water.

228.—To i sure stability in an edifice, it should be designed upon well-known geometrical principles: such as science has demonstrated to be necessary and sufficient fcr firmness and dura bility. It is well, also, that it have the *appearance* of stability as well as the *reality;* for should it seem tottering and unsafe, the sensation of fear, rather than those of admiration and pleasure, will be excited in the beholder. To secure certainty and accuracy in the application of those principles, a knowledge of the strength and other properties of the materials used, is indispensable; and in order that the whole design be so made as to be capable of execution, a practical knowledge of the requisite mechanical operations is quite important.

229.—The elegance of an architectural design, although chiefly depending upon a just proportion and harmony of the parts, will be promoted by the introduction of ornaments—provided this be judiciously performed. For enrichments should not only be of a proper character to suit the style of the building, but should also have their true position, and be bestowed in proper quantity. The most common fault, and one which is prominent in Roman architecture, is an excess of enrichment: an error which is carefully to be guarded against. But those who take the Grecian models for their standard, will not be liable to go to that extreme. In ornamenting a cornice, or any other assemblage of mouldings, at least every alternate member should be left plain; and those that are near the eye should be more finished than those which are distant. Although the characteristics of good architecture are utility and elegance, in connection with durability, yet some buildings are designed expressly for use, and others again for ornament: in the former, utility, and in the latter, beauty, should be the governing principle.

230.—The builder should be intimately acquainted with the principles upon which the essential, elementary parts of a building are founded. A scientific knowledge of these will insure certainty and security, and enable the mechanic to erect the most extensive and lofty edifices with confidence. The more important parts are the foundation, the column, the wall, the lintel, the arch, the vault, the dome and the roof. A separate description of the peculiarities of each, would seem to be necessary; and cannot perhaps be better expressed than in the following language of a modern writer on this subject.

231.—"In laying the FOUNDATION of any building, it is necessary to dig to a certain depth in the earth, to secure a solid basis, below the reach of frost and common accidents. The most solid basis is rock, or gravel which has not been moved. Next to these are clay and sand, provided no other excavations have been made in the immediate neighbourhood. From this basis a stone wall is carried up to the surface of the ground, and constitutes the foundation. Where it is intended that the superstructure shall press unequally, as at its piers, chimneys, or columns, it is sometimes of use to occupy the space between the points of pressure by an inverted arch. This distributes the pressure equally, and prevents the foundation from springing between the different points. In loose or muddy situations, it is always unsafe to build, unless we can reach the solid bottom below. In salt marshes and flats, this is done by depositing timbers, or driving wooden piles into the earth, and raising walls upon them. The preservative quality of the salt will keep these timbers unimpaired for a great length of time, and makes the foundation equally secure with one of brick or stone.

232.—The simplest member in any building, though by no means an essential one to all, is the COLUMN, or *pillar*. This is a perpendicular part, commonly of equal breadth and thickness, not intended for the purpose of enclosure, but simply for the support of some part of the superstructure. The principal force which a column has to resist, is that of perpendicular pressure. In its shape, the shaft of a column should not be exactly cylindrical, but, since the lower part must support the weight of the superior part, in addition to the weight which presses equally on the whole column, the thickness should gradually decrease from bottom to top. The outline of columns should be a little curved, so as to represent a portion of a very long spheroid, or paraboloid, rather than of a cone. This figure is the joint result of two calculations, independent of beauty of appearance. One of these is, that the form best adapted for stability of base is that of a cone; the other is, that the figure, which would be of equal strength throughout for supporting a superincumbent weight, would be generated by the revolution of two parabolas round the axis of the column, the vertices of the curves being at its extremities. The swell of the shafts of columns was called the *entasis* by the ancients. It has been lately found, that the columns of the Parthenon, at Athens, which have been commonly supposed straight, deviate about an inch from a straight line, and that their greatest swell is at about one third of their height. Columns in the antique orders are usually made to diminish one sixth or one seventh of their diameter, and sometimes even one fourth. The Gothic pillar is commonly of equal thickness throughout.

233.—The Wall, another elementary part of a building, may be considered as the lateral continuation of the column, answering the purpose both of enclosure and support. A wall must diminish as it rises, for the same reasons, and in the same proportion, as the column. It must diminish still more rapidly if it extends through several stories, supporting weights at different heights. A wall, to possess the greatest strength, must also consist of pieces, the upper and lower surfaces of which are horizontal and regular, not rounded nor oblique. The walls of most of the ancient structures which have stood to the present time, are constructed in this manner, and frequently have their stones bound together with bolts and cramps of iron. The same method is adopted in such modern structures as are intended to possess great strength and durability, and, in some cases, the stones are even dove-tailed together, as in the light-houses at Eddystone and Bell Rock. But many of our modern stone walls, for the sake of cheapness, have only one face of the stones squared, the inner half of the wall being completed with brick; so that they can, in reality, be considered only as brick walls faced with stone Such walls are said to be liable to become convex outwardly, from the difference in the shrinking of the cement. *Rubble* walls are made of rough, irregular stones, laid in mortar. The stones should be broken, if possible, so as to produce horizontal surfaces The *coffer* walls of the ancient Romans were made by enclosing successive portions of the intended wall in a box, and filling it with stones, sand, and mortar, promiscuously. This kind of structure must have been extremely insecure. The Pantheon, and various other Roman buildings, are surrounded with a double brick wall, having its vacancy filled up with loose bricks and cement. The whole has gradually consolidated into a mass of great firmness.

The *reticulated* walls of the Romans, having bricks with oblique surfaces, would, at the present day, be thought highly unphilosophical. Indeed, they could not long have stood, had it not been for the great strength of their cement. Modern brick walls are laid with great precision, and depend for firmness more upon their position than upon the strength of their cement. The bricks being laid in horizontal courses, and continually overlaying each other, or *breaking joints*, the whole mass is strongly interwoven, and bound together. Wooden walls, composed of timbers covered with boards, are a common, but more perishable kind. They require to be constantly covered with a coating of a foreign substance, as paint or plaster, to preserve them from spontaneous decomposition. In some parts of France, and elsewhere, a kind of wall is made of earth, rendered compact by ramming it in moulds or cases. This method is called building in *pisé*, and is much more durable than the nature of the material would lead us to suppose. Walls of all kinds are greatly strengthened by angles and curves, also by projections, such as pilasters, chimneys and buttresses. These projections serve to increase the breadth of the foundation, and are always to be made use of in large buildings, and in walls of considerable length.

234.—The LINTEL, or *beam*, extends in a right line over a vacant space, from one column or wall to another. The strength of the lintel will be greater in proportion as its transverse vertical diameter exceeds the horizontal, the strength being always as the square of the depth. The *floor* is the lateral continuation or connection of beams by means of a covering of boards.

235.—The ARCH is a transverse member of a building, answering the same purpose as the lintel, but vastly exceeding it in strength. The arch, unlike the lintel, may consist of any number of constituent pieces, without impairing its strength. It is, however, necessary that all the pieces should possess a uniform shape,—the shape of a portion of a wedge,—and that the joints, formed by the contact of their surfaces, should point towards a common centre. In this case, no one portion of the arch can be displaced or forced inward; and the arch cannot be broken by any force which is not sufficient to crush the materials of which it is made. In arches made of common bricks, the sides of which are parallel, any *one* of the bricks might be forced inward, were it not for the adhesion of the cement. Any *two* of the bricks, however, by the disposition of their mortar, cannot collectively be forced inward. An arch of the proper form, when complete, is rendered stronger, instead of weaker, by the pressure of a considerable weight, provided this pressure be uniform. While building, however, it requires to be supported by a centring of the shape of its internal surface, until it is complete. The upper stone of an arch is called the *key-stone*, but is not more essential than any other. In regard to the shape of the arch, its most simple form is that of the semi-circle. It is, however, very frequently a smaller arc of a circle, and, still more frequently, a portion of an ellipse. The simplest theory of an arch supporting itself only, is that of Dr. Hooke. The arch, when it has only its own weight to bear, may be considered as the inversion of a chain, suspended at each end. The chain hangs in such a form, that the weight of each link or portion is held in equilibrium by the result of two forces acting at its extremities; and these forces, or tensions, are produced, the one by the weight of the portion of the chain below the link, the other by the same weight increased by that of the link itself, both of them acting originally in a vertical direction. Now, supposing the chain inverted, so as to constitute an arch of the same form and weight, the relative situations of the forces will be the same, only they will act in contrary directions, so that they are compounded in a similar manner, and balance each other on the same conditions.

The arch thus formed is denominated a *catenary* arch. In common cases, it differs but little from a circular arch of the extent of about one third of a whole circle, and rising from the abutments with an obliquity of about 30 degrees from a perpendicular. But though the catenary arch is the best form for supporting its own weight, and also all additional weight which presses in a vertical direction, it is not the best form to resist lateral pressure, or pressure like that of fluids, acting equally in all directions. Thus the arches of bridges and similar structures, when covered with loose stones and earth, are pressed sideways, as well as vertically, in the same manner as if they supported a weight of fluid. In this case, it is necessary that the arch should arise more perpendicularly from the abutment, and that its general figure should be that of the longitudinal segment of an ellipse. In small arches, in common buildings, where the disturbing force is not great, it is of little consequence what is the shape of the curve. The outlines may even be perfectly straight, as in the tier of bricks which we frequently see over a window. This is, strictly speaking, a real arch, provided the surfaces of the bricks tend towards a common centre. It is the weakest kind of arch, and a part of it is necessarily superfluous, since no greater portion can act in supporting a weight above it, than can be included between two curved or arched lines.

Besides the arches already mentioned, various others are in use The *acute* or *lancet* arch, much used in Gothic architecture, is described usually from two centres outside the arch. It is a strong arch for supporting vertical pressure. The *rampant* arch is one in which the two ends spring from unequal heights. The *horse-shoe* or *Moorish* arch is described from one or more centres placed above the base line. In this arch, the lower parts are in danger of being forced inward. The *ogee* arch is concavo-convex, and therefore fit only for ornament. In describing arches, the upper surface is called the *extrados*, and the inner, the *intrados*. The springing lines are those where the intrados meets the abutments, or supporting walls. The *span* is the distance from one springing line to the other. The wedge-shaped stones, which form an arch, are sometimes called *voussoirs*, the uppermost being the key-stone. The part of a pier from which an arch springs is called the *impost*, and the curve formed by the upper side of the voussoirs, the *archivolt*. It is necessary that the walls, abutments and piers, on which arches are supported, should be so firm as to resist the lateral *thrust*, as well as vertical pressure, of the arch. It will at once be seen, that the lateral or sideway pressure of an arch is very considerable, when we recollect that every stone, or portion of the arch, is a wedge, a part of whose force acts to separate the abutments. For want of attention to this circumstance, important mistakes have been committed, the strength of buildings materially impaired, and their ruin accelerated. In some cases,, the want of lateral firmness in the walls is compensated by a bar of iron stretched across the span of the arch, and connecting the abutments, like the tie-beam of a roof. This is the case in the cathedral of Milan and some other Gothic buildings.

In an arcade, or continuation of arches, it is only necessary that the outer supports of the terminal arches should be strong enough to resist horizontal pressure. In the intermediate arches, the lateral force of each arch is counteracted by the opposing lateral force of the one contiguous to it. In bridges, however, where individual arches are liable to be destroyed by accident, it is desi rable that each of the piers should possess sufficient horizontal strength to resist the lateral pressure of the adjoining arches.

236.—The VAULT is the lateral continuation of an arch, serving to cover an area or passage, and bearing the same relation to the arch that the wall does to the column. A simple vault is constructed on the principles of the arch, and distributes its pressure equally along the walls or abutments. A complex or *groined* vault is made by two vaults intersecting each other, in which case the pressure is thrown upon springing points, and is greatly increased at those points. The groined vault is common in Gothic architecture.

237.—The DOME, sometimes called *cupola*, is a concave covering to a building, or part of it, and may be either a segment of a sphere, of a spheroid, or of any similar figure. When built of stone, it is a very strong kind of structure, even more so than the arch, since the tendency of each part to fall is counteracted, not only by those above and below it, but also by those on each side. It is only necessary that the constituent pieces should have a common form, and that this form should be somewhat like the frustum of a pyramid, so that, when placed in its situation, its four angles may point toward the centre, or axis, of the dome. During the erection of a dome, it is not necessary that it should be supported by a centring, until complete, as is done in the arch. Each circle of stones, when laid, is capable of supporting itself without aid from those above it. It follows that the dome may be left open at top, without a key-stone, and yet be perfectly secure in this respect, being the reverse of the arch. The dome of the Pantheon, at Rome, has been always open at top, and yet has stood unimpaired for nearly 2000 years. The upper circle of stones, though apparently the weakest, is nevertheless often made to support the additional weight of a lantern or tower above it. In several of the largest cathedrals, there are two domes, one within the other, which contribute their joint support to the lantern, which rests upon the top. In these buildings, the dome rests upon a circular wall, which is supported, in its turn, by arches upon massive pillars or piers. This construction is called building upon *pendentives*, and gives open space and room for passage beneath the dome. The remarks which have been made in regard to the abutments of the arch, apply equally to the walls immediately supporting a dome. They must be of sufficient thickness and solidity to resist the lateral pressure of the dome, which is very great. The walls of the Roman Pantheon are of great depth and solidity. In order that a dome in itself should be perfectly secure, its lower parts must not be too nearly vertical, since, in this case, they partake of the nature of perpendicular walls, and are acted upon by the spreading force of the parts above them. The dome of St. Paul's church, in London, and some others of similar construction, are bound with chains or hoops of iron, to prevent them from spreading at bottom. Domes which are made of wood depend, in part, for their strength, on their internal carpentry. The Halle du Bled, in Paris, had originally a wooden dome more than 200 feet in diameter, and only one foot in thickness. This has since been replaced by a dome of iron (See *Art.* 389.)

238.—The Roof is the most common and cheap method of covering buildings, to protect them from rain and other effects of the weather. It is sometimes flat, but more frequently oblique, in its shape. The flat or platform-roof is the least advantageous for shedding rain, and is seldom used in northern countries. The *pent* roof, consisting of two oblique sides meeting at top, is the most common form. These roofs are made steepest in cold climates, where they are liable to be loaded with snow. Where the four sides of the roof are all oblique, it is denominated a *hipped* roof, and where there are two portions to the roof, of different obliquity, it is a *curb*, or *mansard* roof. In modern times, roofs are made almost exclusively of wood, though frequently covered with incombustible materials. The internal structure or carpentry of roofs is a subject of considerable mechanical contrivance

The roof is supported by *rafters*, which abut on the walls on each side, like the extremities of an arch. If no other timbers existed, except the rafters, they would exert a strong lateral pressure on the walls, tending to separate and overthrow them. To counteract this lateral force, a *tie-beam*, as it is called, extends across, receiving the ends of the rafters, and protecting the wall from their horizontal thrust. To prevent the tie-beam from *sagging*, or bending downward with its own weight, a *king-post* is erected from this beam, to the upper angle of the rafters, serving to connect the whole, and to suspend the weight of the beam. This is called *trussing*. *Queen-posts* are sometimes added, parallel to the king-post, in large roofs; also various other connecting timbers. In Gothic buildings, where the vaults do not admit of the use of a tie-beam, the rafters are prevented from spreading, as in an arch, by the strength of the buttresses.

In comparing the lateral pressure of a high roof with that of a low one, the length of the tie-beam being the same, it will be seen that a high roof, from its containing most materials, may produce the greatest pressure, as far as weight is concerned. On the other hand, if the weight of both be equal, then the low roof will exert the greater pressure; and this will increase in proportion to the distance of the point at which perpendiculars, drawn from the end of each rafter, would meet. In roofs, as well as in wooden domes and bridges, the materials are subjected to an internal strain, to resist which, the cohesive strength of the material is relied on. On this account, beams should, when possible, be of one piece. Where this cannot be effected, two or more beams are connected together by *splicing*. Spliced beams are never so strong as whole ones, yet they may be made to approach the same strength, by affixing lateral pieces, or by making the ends overlay each other, and connecting them with bolts and straps of iron. The tendency to separate is also resisted, by letting the two pieces into each other by the process called *scarfing*. *Mortices*, intended to *truss* or suspend one piece by another, should be formed upon similar principles.

Roofs in the United States, after being boarded, receive a secondary covering of shingles. When intended to be incombustible, they are covered with slates or earthern tiles, or with sheets of lead, copper or tinned iron. Slates are preferable to tiles, being lighter, and absorbing less moisture. Metallic sheets are chiefly used for flat roofs, wooden domes, and curved and angular surfaces, which require a flexible material to cover them, or have not a sufficient pitch to shed the rain from slates or shingles. Various artificial compositions are occasionally used to cover roofs, the most common of which are mixtures of tar with lime, and sometimes with sand and gravel."—*Ency. Am.* (See *Art.* 354.)

SECTION III.—MOULDINGS, CORNICES, &c.

MOULDINGS.

239.—A moulding is so called, because of its being of the same determinate shape along its whole length, as though the whole of it had been cast in the same mould or form. The regular mouldings, as found in remains of ancient architecture, are eight in number; and are known by the following names:

Fig. 140.

Annulet, band, cincture, fillet, listel or square.

Fig. 141.

Astragal or bead.

Fig. 142.

Torus or tore.

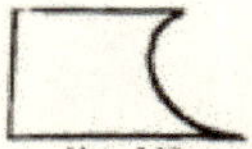

Fig. 143.

Scotia, trochilus or mouth.

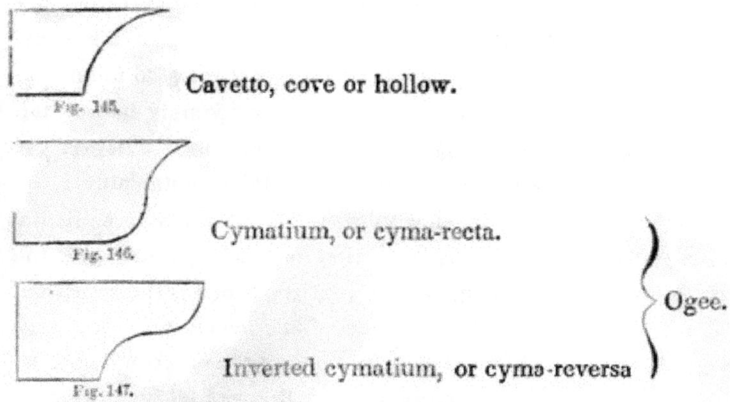

Fig. 145. Fig. 146. Fig. 147.

Some of the terms are derived thus: fillet, from the French word *fil*, thread. Astragal, from *astragalos*, a bone of the heel —or the curvature of the heel. Bead, because this moulding, when properly carved, resembles a string of beads. Torus, or tore, the Greek for *rope*, which it resembles, when on the base of a column. Scotia, from *shotia*, darkness, because of the strong shadow which its depth produces, and which is increased by the projection of the torus above it. Ovolo, from *ovum*, an egg, which this member resembles, when carved, as in the Ionic capital. Cavetto, from *cavus*, hollow. Cymatium, from *kumaton* a wave.

240.—Neither of these mouldings is peculiar to any one of the orders of architecture, but each one is common to all; and although each has its appropriate use, yet it is by no means confined to any certain position in an assemblage of mouldings The use of the fillet is to bind the parts, as also that of the astragal and torus, which resemble ropes. The ovolo and cyma-reversa are strong at their upper extremities, and are therefore used to support projecting parts above them. The cyma-recta and cavetto, being weak at their upper extremities, are not used as supporters, but are placed uppermost to cover and shelter the other parts. The scotia is introduced in the base of a column, to separate the upper and lower torus, and to produce a pleasing variety and relief. The form of the bead, and that of the torus, is the same; the reasons for giving distinct names to them are, that the torus, in every order, is always considerably larger than the bead, and is placed among the base mouldings, whereas the bead is never placed there, but on the capital or entablature; the torus, also, is seldom carved, whereas the bead is; and while the torus among the Greeks is frequently elliptical in its form, the bead retains its circular shape. While the scotia is the reverse of the torus, the cavetto is the reverse of the ovolo, and the cyma-recta and cyma-reversa are combinations of the ovolo and cavetto.

241.—The curves of mouldings, in Roman architecture, were most generally composed of parts of circles; while those of the Greeks were almost always elliptical, or of some one of the conic sections, but rarely circular, except in the case of the bead, which was always, among both Greeks and Romans, of the form of a semi-circle. Sections of the cone afford a greater variety of forms than those of the sphere; and perhaps this is one reason why the Grecian architecture so much excels the Roman. The quick turnings of the ovolo and cyma-reversa, in particular, when exposed to a bright sun, cause those narrow, well-defined streaks of light, which give life and splendour to the whole.

242.—A *profile* is an assemblage of essential parts and mouldings. That profile produces the happiest effect which is composed of but few members, varied in form and size, and arranged so that the plane and the curved surfaces succeed each other alternately.

243.—*To describe the Grecian torus and scotia.* Join the extremities, *a* and *b*, (*Fig.* 148;) and from *f*, the given projection of the moulding, draw *f o*, at right angles to the fillets; from *b*, draw *b h*, at right angles to *a b*; bisect *a b* in *c*; join *f* and *c*, and upon *c*, with the radius, *c f*, describe the arc, *f h*, cutting *b h* in *h*; through *c*, draw *d e*, parallel with the fillets; make *d c* and *c e*, each equal to *b h*; then *d e* and *a b* will be conjugate diame-

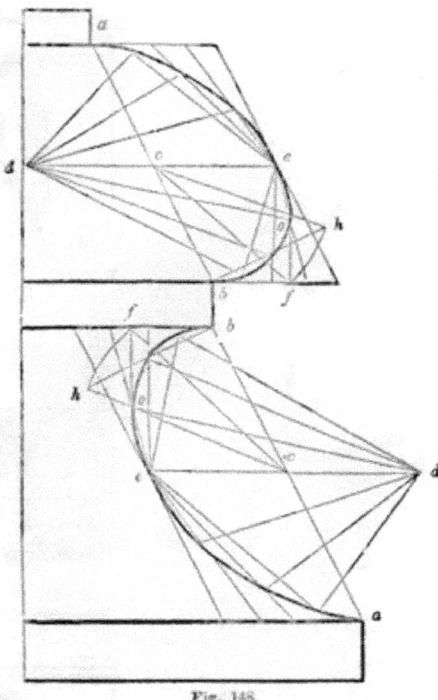

Fig. 148.

ters of the required ellipse. To describe the curve by intersection of lines, proceed as directed at *Art.* 118 and *note;* by a trammel, see *Art.* 116; and to find the foci, in order to describe it with a string, see *Art.* 115.

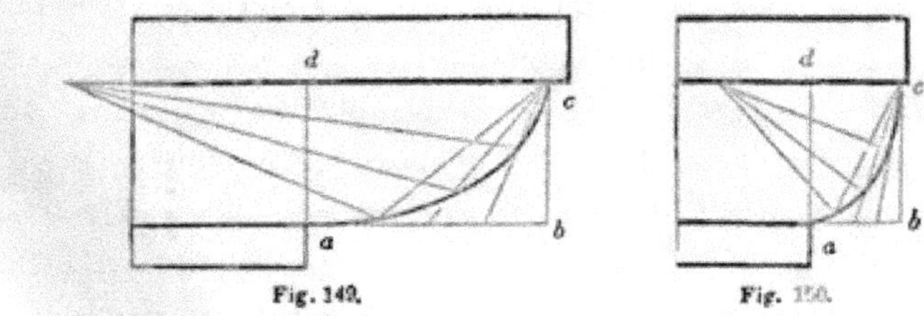

Fig. 149.

Fig. 150.

244.—*Fig.* 149 to 156 exhibit various modifications of the Grecian ovolo, sometimes called echinus. *Fig.* 149 to 153 are

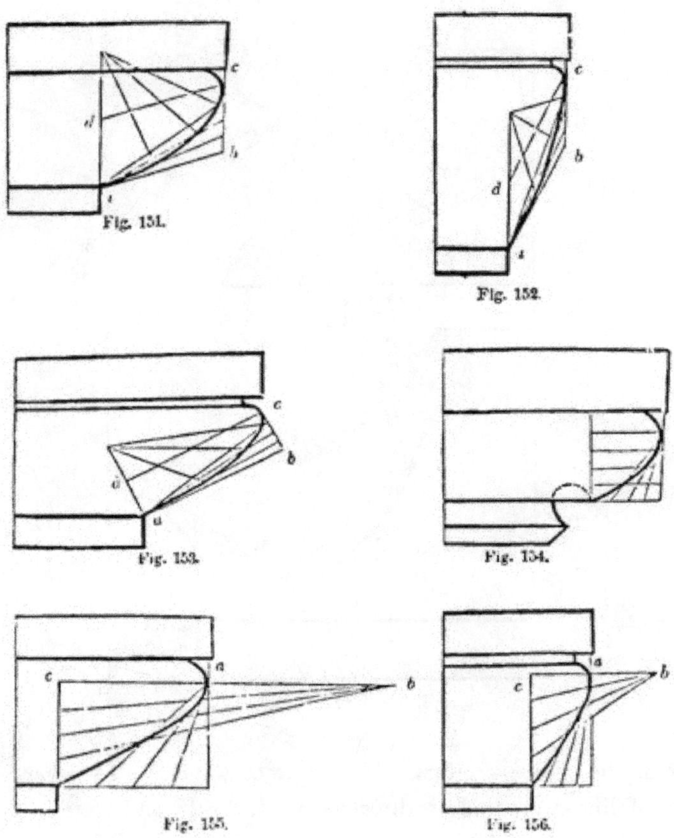

Fig. 151.

Fig. 152.

Fig. 153.

Fig. 154.

Fig. 155.

Fig. 156.

elliptical, *a b* and *b c* being given tangents to the curve; parallel to which, the semi-conjugate diameters, *a d* and *d c*, are drawn. In *Fig.* 149 and 150, the lines, *a d* and *d c*, are semi-axes, the tangents, *a b* and *b c*, being at right angles to each other. To draw the curve, see *Art.* 118. In *Fig.* 153, the curve is parabolical, and is drawn according to *Art.* 127. In *Fig.* 155 and 156, the curve is hyperbolical, being described according to *Art.* 128. The length of the transverse axis, *a b*, being taken at pleasure in order to flatten the curve, *a b* should be made short in proportion to *a c*.

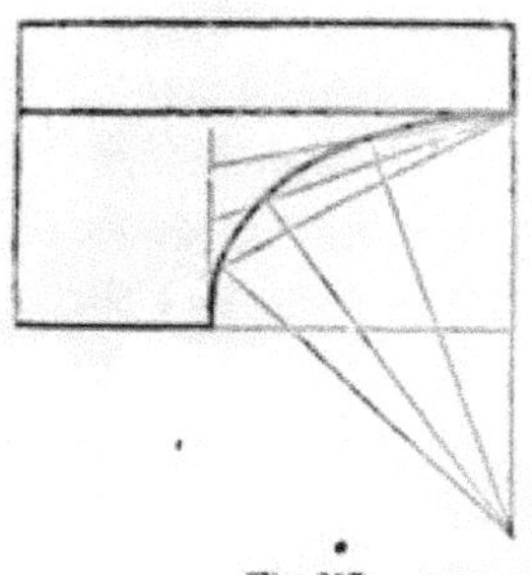
Fig. 157.

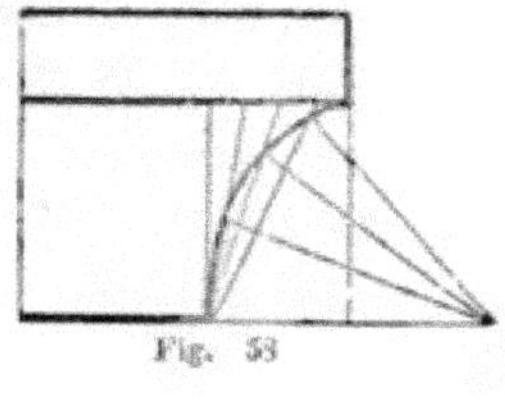
Fig. 158

245.—*To describe the Grecian cavetto,* (*Fig.* 157 and 158,) having the height and projection given, see *Art.* 118.

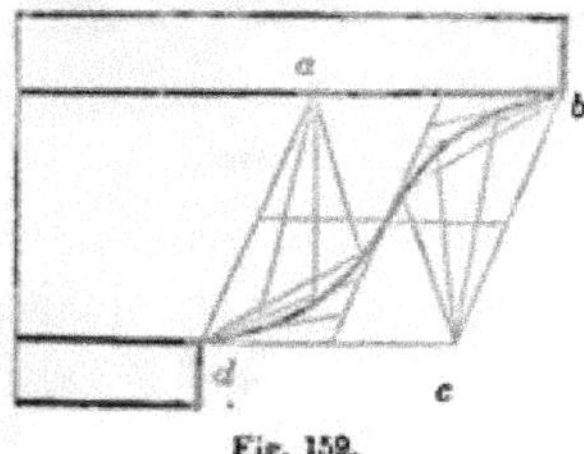

Fig. 159.

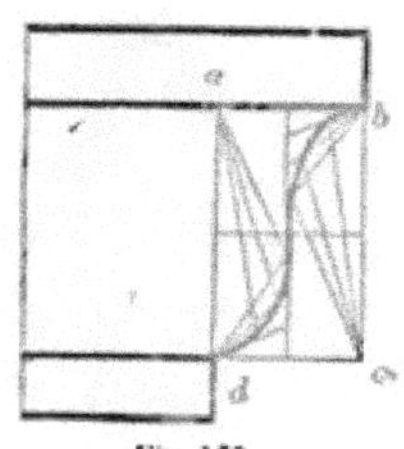

Fig. 160.

246.—*To describe the Grecian cyma-recta.* When the projection is more than the height, as at *Fig.* 159, make *a b* equal to the height, and divide *q b c d* into 4 equal parallelograms; then proceed as directed in note to *Art.* 118. When the projection is less than the height, draw *d a*, (*Fig.* 160,) at right angles to *a b;* complete the rectangle, *a b c d;* divide this into 4 equal rectangles, and proceed according to *Art.* 118.

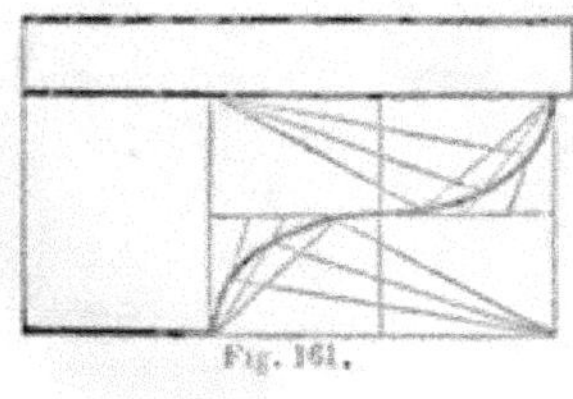
Fig. 161.

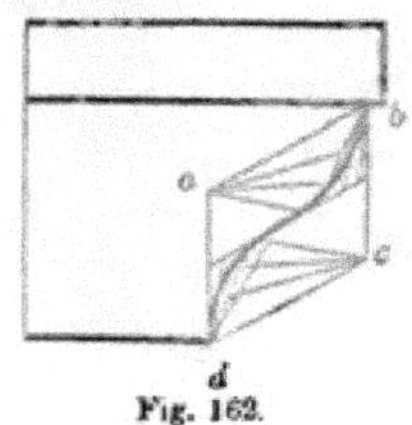

Fig. 162.

247.—*To describe the Grecian cyma-reversa.* When the

projection is more than the height, as at *Fig.* 161, proceed as directed for the last figure; the curve being the same as that, the position only being changed. When the projection is less than the height, draw *a d*, (*Fig.* 162,) at right angles to the fillet. make *a d* equal to the projection of the moulding: then proceed as directed for *Fig.* 159.

248.—Roman mouldings are composed of parts of circles, and have, therefore, less beauty of form than the Grecian. The bead and torus are of the form of the semi-circle, and the scotia, also, in some instances; but the latter is often composed of two quadrants, having different radii, as at *Fig.* 163 and 164, which resemble the elliptical curve. The ovolo and cavetto are generally a quadrant, but often less. When they are less, as at *Fig.* 167, the centre is found thus: join the extremities, *a* and *b*, and bisect *a b* in *c*; from *c*, and at right angles to *a b*, draw *c d*, cutting a level line drawn from *a* in *d*; then *d* will be the centre. This moulding projects less than its height. When the projection is more than the height, as at *Fig.* 169, extend the line from *c* until

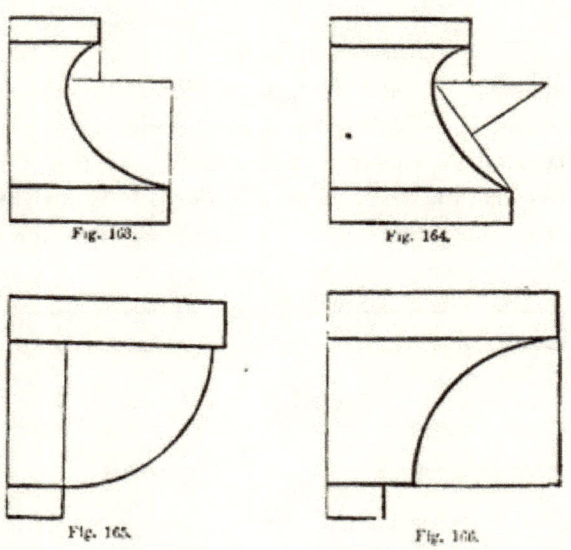

Fig. 163. Fig. 164.

Fig. 165. Fig. 166.

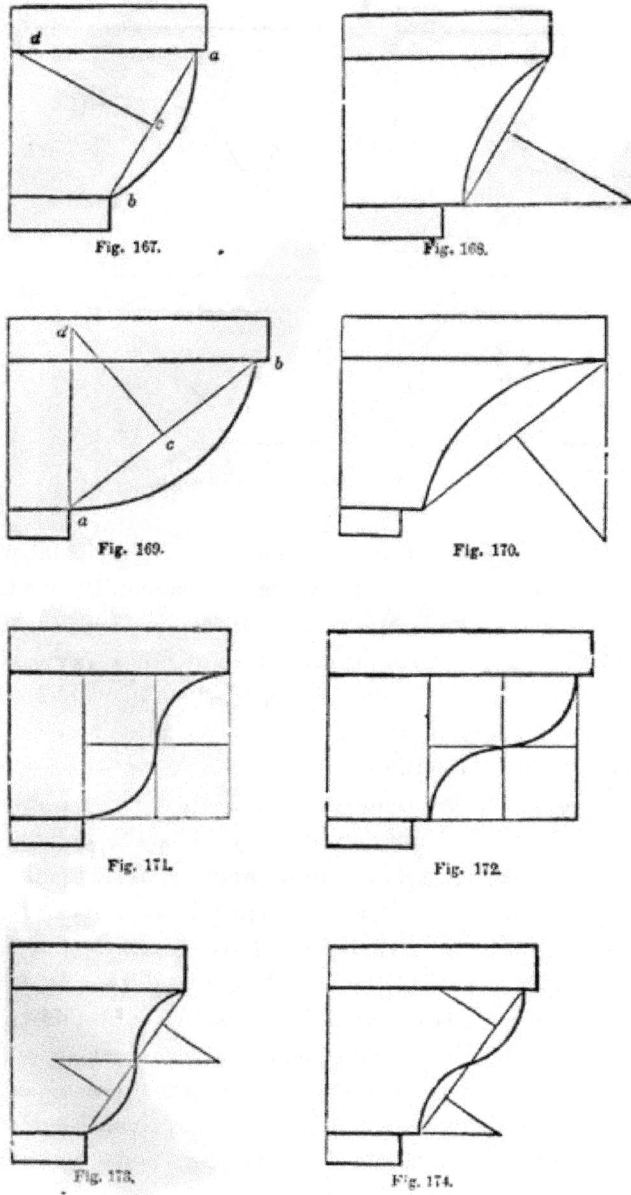

Fig. 167.

Fig. 168.

Fig. 169.

Fig. 170.

Fig. 171.

Fig. 172.

Fig. 173.

Fig. 174.

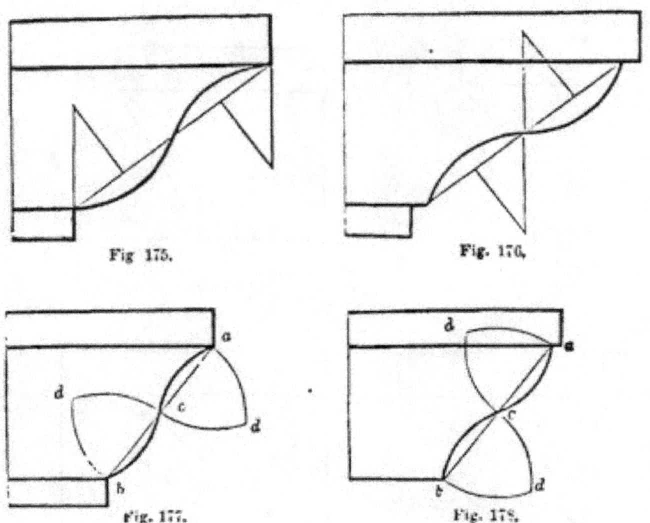

Fig 175. Fig. 176.

Fig. 177. Fig. 178.

it cuts a perpendicular drawn from *a*, as at *d*; and that will be the centre of the curve. In a similar manner, the centres are found for the mouldings at *Fig.* 164, 168, 170, 173, 174, 175, and 176. The centres for the curves at *Fig.* 177 and 178, are found thus: bisect the line, *a b*, at *c*; upon *a*, *c* and *b*, successively, with *a c* or *c b* for radius, describe arcs intersecting at *d* and *d*; then those intersections will be the centres.

249.—*Fig.* 179 to 186 represent mouldings of modern invention. They have been quite extensively and successfully used in inside finishing. *Fig.* 179 is appropriate for a bed-moulding under a low projecting shelf, and is frequently used under mantle-shelves. The tangent, *i h*, is found thus: bisect the line, *a b*, at *c*, and *b c* at *d*; from *d*, draw *d e*, at right angles to *c b*; from *b*, draw *b f*, parallel to *e d*; upon *b*, with *b d* for radius, describe the arc, *d f*; divide this arc into 7 equal parts, and set one of the parts from *s*, the limit of the projection, to *o*; make *o h* equal to *o e*; from *h*, through *c*, draw the tangent, *h i*; divide *b h*, *h c*, *c i* and *i a*, each into a like number of equal parts, and draw the in-

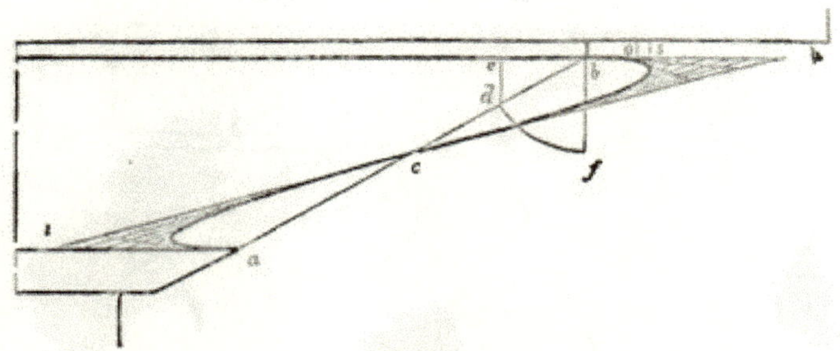

Fig. 179.

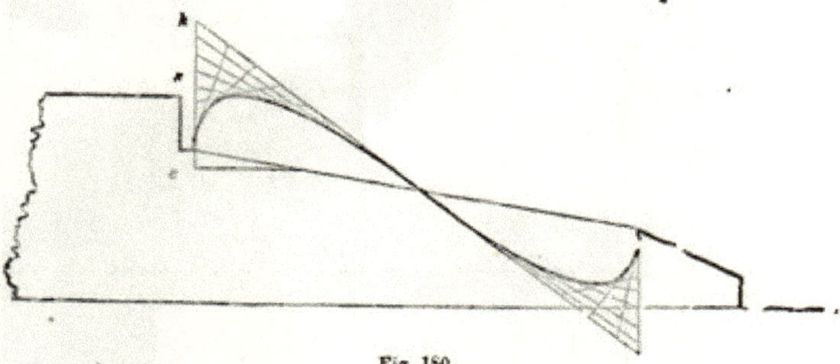

Fig 180.

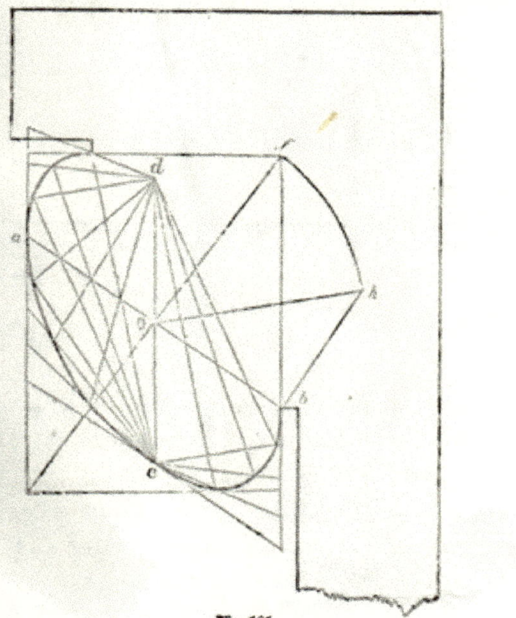

Fig. 181.

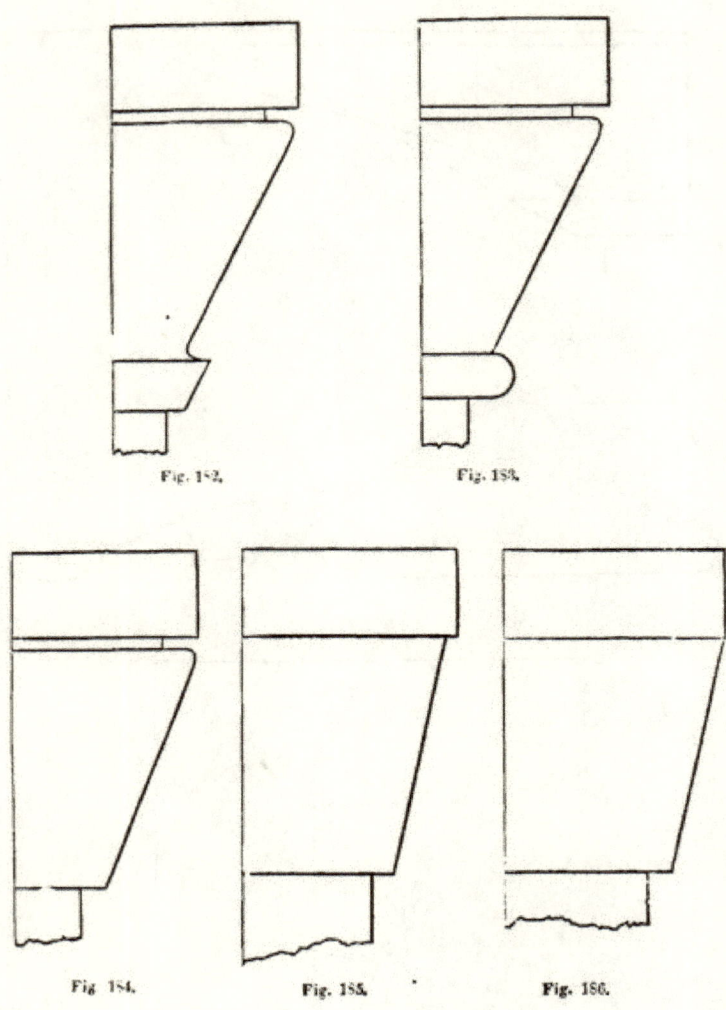

Fig. 182. Fig. 183.

Fig. 184. Fig. 185. Fig. 186.

tersecting lines as directed at *Art.* 89. If a bolder form is desired, draw the tangent, *i h*, nearer horizontal, and describe an elliptic curve as shown in *Fig.* 148 and 181. *Fig.* 180 is much used on base, or skirting of rooms, and in deep panelling. The curve is found in the same manner as that of *Fig.* 179. In this case, however, where the moulding has so little projection in comparison with its height, the point, *e*, being found as in the last figure, *h s* may be made equal to *s e*, instead of *o e* as in the last figure. *Fig.* 181 is appropriate for a crown moulding of a cornice. In this figure the height and projection are given; the direction of the diameter, *a b*, drawn through the middle of the diagonal, *e f*, is taken at pleasure; and *d c* is parallel to *a e*. To find the length of *d c*, draw *b h*, at right angles to *a b*; upon *o*, with *o f* for radius, describe the arc, *f h*, cutting *b h* in *h*; then make *o c* and *o d*, each equal to *b h*.* To draw the curve, see note to *Art.* 118. *Fig.* 182 to 186 are peculiarly distinct from ancient mouldings, being composed principally of straight lines; the few curves they possess are quite short and quick

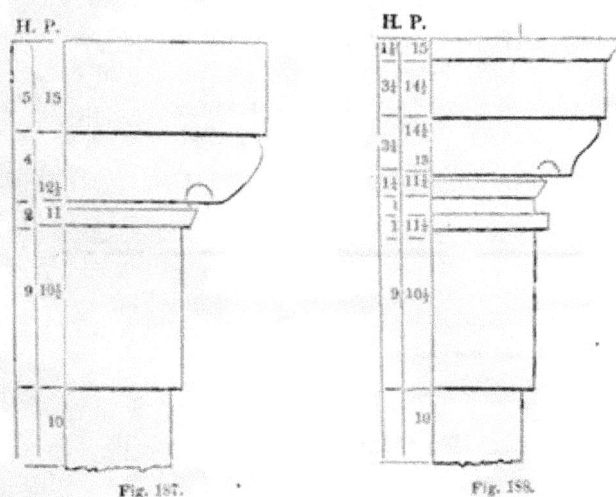

Fig. 187.

Fig. 188.

250.—*Fig.* 187 and 188 are designs for antæ caps. The

* The manner of ascertaining the length of the conjugate diameter, *d c*, in this figure, and also in *Fig.* 148, 198 and 199 is new, and is important in this application. It is founded upon well-known mathematical principles, viz: All the parallelograms that may be circumscribed about an ellipsis are equal to one another, and consequently any one is equal to the rectangle of the two axes. And again: the sum of the squares of every pair of conjugate diameters is equal to the sum of the squares of the two axes.

diameter of the antæ is divided into 20 equal parts, and the height and projection of the members are regulated in accordance with those parts, as denoted under *H* and *P*, height and projection. The projection is measured from the middle of the antæ. These will be found appropriate for porticos, doorways, mantel-pieces, door and window trimmings, &c. The height of the antæ for mantel-pieces, should be from 5 to 6 diameters, having an entablature of from 2 to $2\frac{1}{4}$ diameters. This is a good proportion, it being similar to the Doric order. But for a portico these proportions are much too heavy; an antæ, 15 diameters high, and an entablature of 3 diameters, will have a better appearance.

CORNICES.

251.—*Fig.* 189 to 197 are designs for eave cornices, and *Fig.* 198 and 199 are for stucco cornices for the inside finish of rooms. In some of these the projection of the uppermost member from the facia, is divided into twenty equal parts,

Fig 189.

and the various members are proportioned according to those parts, as figured under *H* and *P*.

Fig. 190.

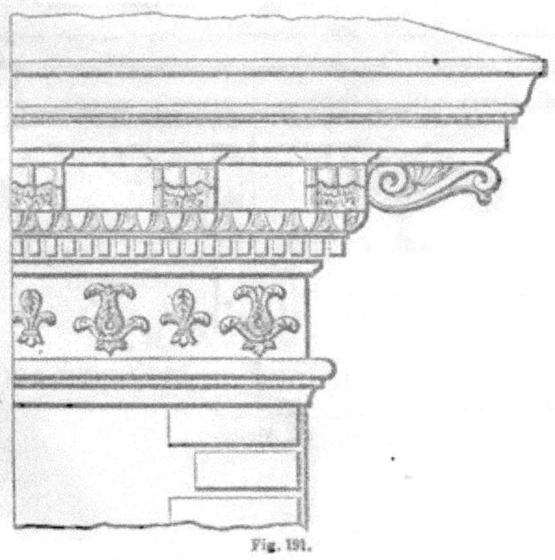

Fig. 191.

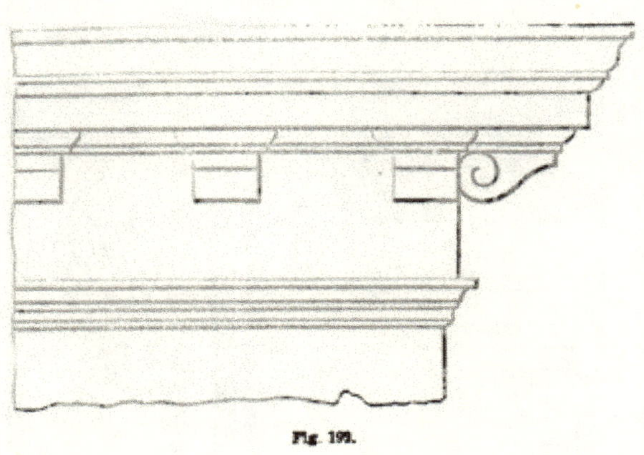

Fig. 192.

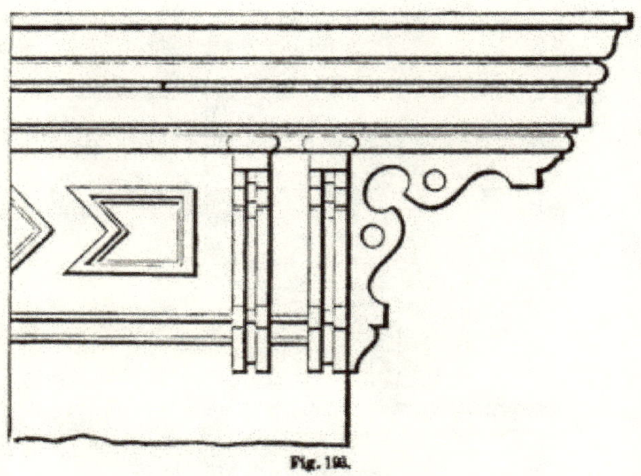

Fig. 193.

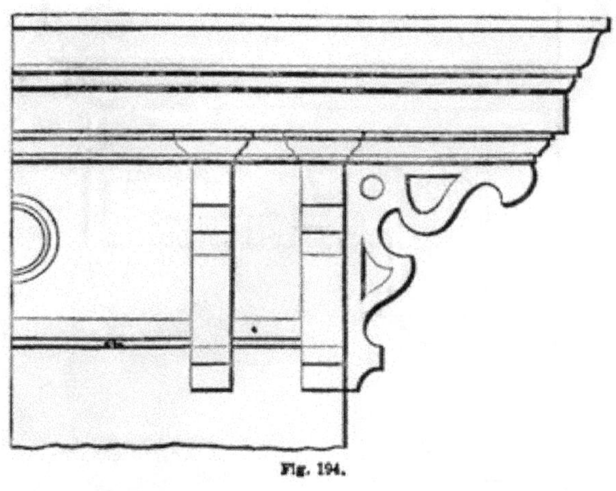

Fig. 194.

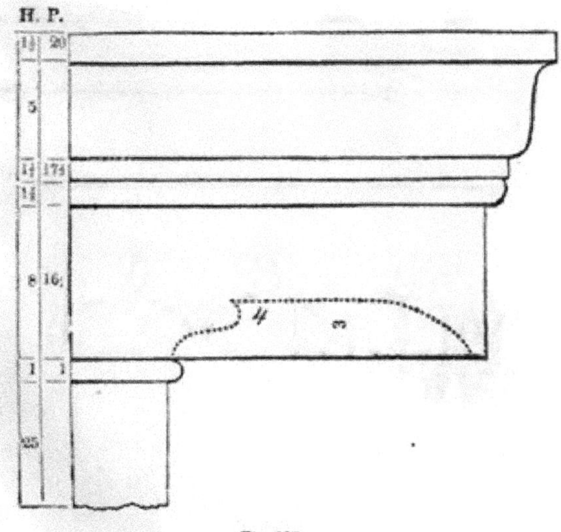

Fig. 195.

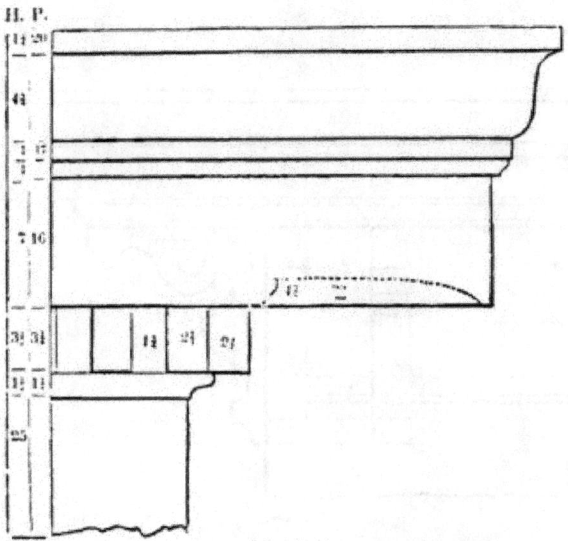

Fig. 196.

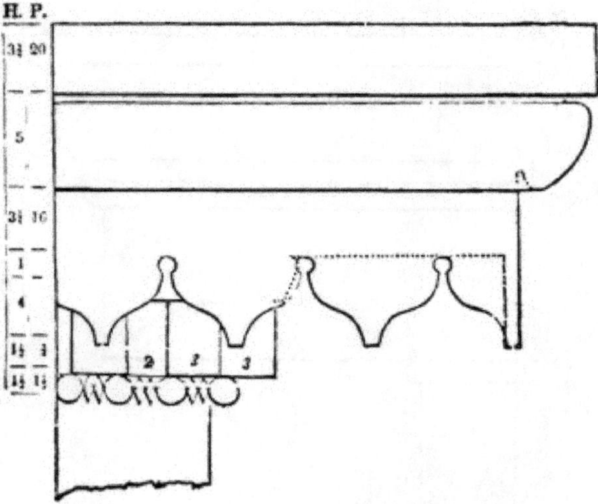

Fig. 197.

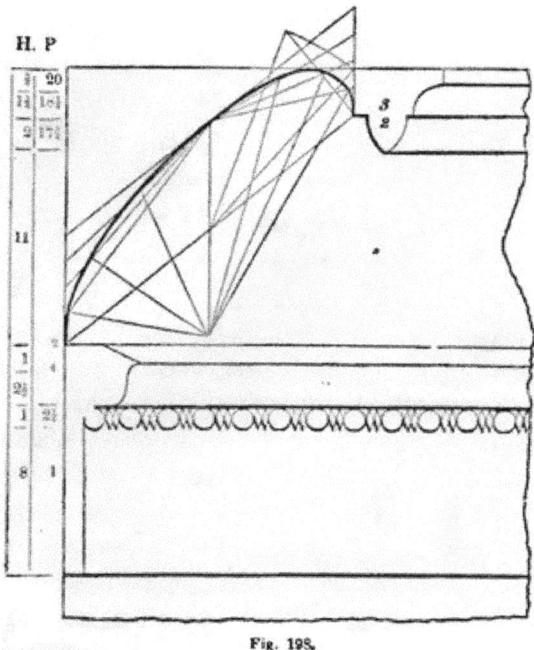

Fig. 198.

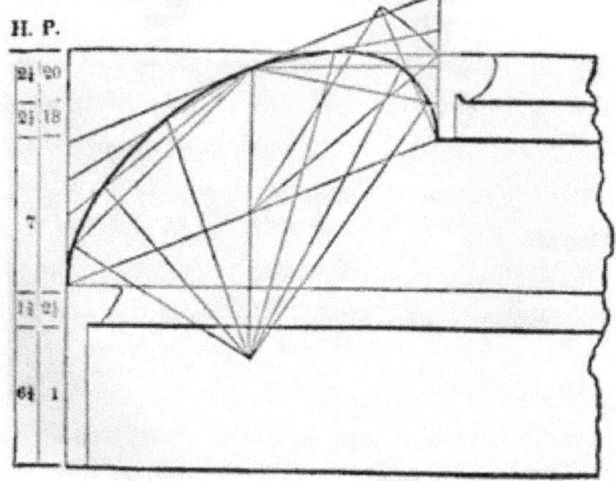

Fig. 199.

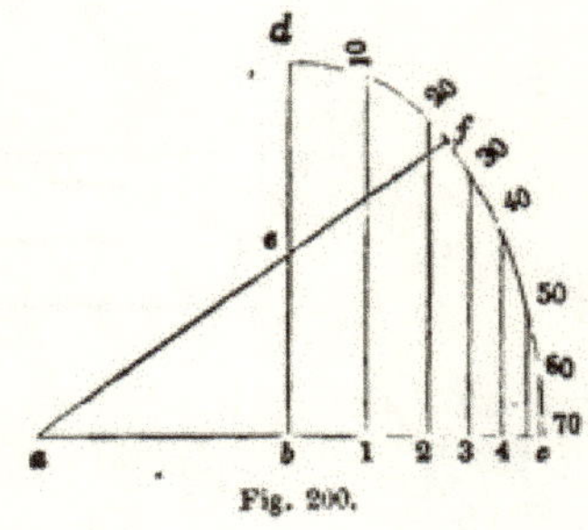

Fig. 200.

252.—*To proportion an eave cornice in accordance with the height of the building.* Draw the line, *a c*, (*Fig.* 200,) and make *b c* and *b a*, each equal to 36 inches; from *b*, draw *b d*, at right angles to *a c*, and equal in length to ⅞ of *a c;* bisect *b d* in *e*, and from *a*, through *e*, draw *a f;* upon *a*, with *a c* for radius, describe the arc, *c f*, and upon *e*, with *e f* for radius, describe the arc. *f d;* divide the curve, *d f c*, into 7 equal parts, as at 10, 20, 30, &c., and from these points of division, draw lines to *b c*, parallel to *d b;* then the distance, *b* 1, is the projection of a cornice for a building 10 feet high; *b* 2, the projection at 20 feet high; *b* 3, the projection at 30 feet, &c. If the projection of a cornice for a building 34 feet high, is required, divide the arc between 30 and 40 into 10 equal parts, and from the fourth point from 30, draw a line to the base, *b c*, parallel with *b d;* then the distance of the point, at which that line cuts the base, from *b*, will be the projection required. So proceed for a cornice of any height within 70 feet. The above is based on the supposition that 36 inches is the proper projection for a cornice 70 feet high. This, for general purposes, will be found correct; still, the length of the line, *b c*, may be varied to suit the judgment of those who think differently.

Having obtained the projection of a cornice, divide it into 20 equal parts, and apportion the several members according to its destination—as is shown at *Fig.* 195, 196, and 197.

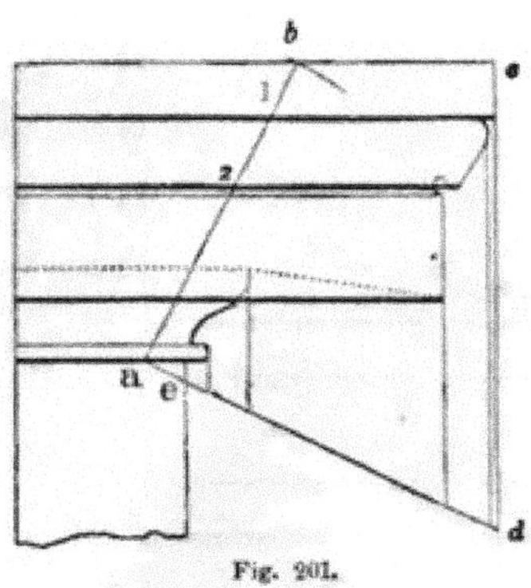

Fig. 201.

253.—*To proportion a cornice according to a smaller given one.* Let the cornice at *Fig.* 201 be the given one. Upon any point in the lowest line of the lowest member, as at *a*, with the height of the required cornice for radius, describe an intersecting arc across the uppermost line, as at *b*; join *a* and *b*; then *b* 1 will be the perpendicular height of the upper fillet for the proposed cornice, 1 2 the height of the crown moulding—and so of all the members requiring to be enlarged to the sizes indicated on this line. For the projection of the proposed cornice, draw *a d*, at right angles to *a b*, and *c d*, at right angles to *b c*; parallel with *c d*, draw lines from each projection of the given cornice to the line, *a d*; then *e d* will be the required projection for the proposed cornice, and the perpendicular lines falling upon *e d* will indicate the proper projection for the members.

254.—*To proportion a cornice according to a larger given one.* Let *A*, (*Fig.* 202,) be the given cornice. Extend *a o* to *b*, and draw *c d*, at right angles to *a b*; extend the horizontal lines of the cornice, *A*, until they touch *o d*; place the height of the proposed cornice from *o* to *e*, and join *f* and *e*; upon *o*, with the projection of the given cornice, *o a*, for radius, describe the quadrant, *a d*; from *d*, draw *d b*, parallel to *f e*; upon *o*, with *o b* for radius, describe the quadrant, *b c*; then *o c* will be the proper projection for the proposed cornice. Join *a* and *c*; draw lines from the

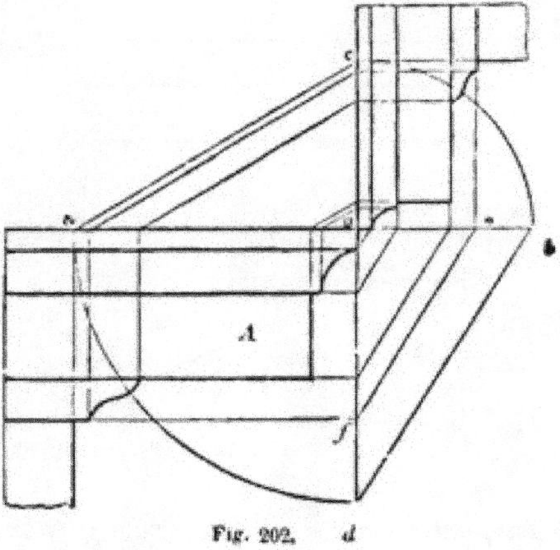

Fig. 202.

projection of the different members of the given cornice to *a o*, parallel to *o d;* from these divisions on the line, *a o*, draw lines to the line, *o c*, parallel to *a c;* from the divisions on the line, *o f*, draw lines to the line, *o e*, parallel to the line, *f e;* then the divisions on the lines, *o e* and *o c*, will indicate the proper height and projection for the different members of the proposed cornice. In this process, we have assumed the height, *o e*, of the proposed cornice to be given; but if the projection, *o c*, alone be given, we can obtain the same result by a different process. Thus: upon *o*, with *o c* for radius, describe the quadrant, *c b;* upon *o*, with *o a* for radius, describe the quadrant, *a d;* join *d* and *b;* from *f*, draw *f e*, parallel to *d b;* then *o e* will be the proper height for the proposed cornice, and the height and projection of the different members can be obtained by the above directions. By this problem, a cornice can be proportioned according to a *smaller* given one as well as to a *larger;* but the method described in the previous article is much more simple for that purpose.

255.—*To find the angle-bracket for a cornice.* Let *A*, (*Fig.* 203,) be the wall of the building, and *B* the given bracket, which, for the present purpose, is turned down horizontally. The angle-bracket, *C*, is obtained thus: through the extremity, *a*, and paral-

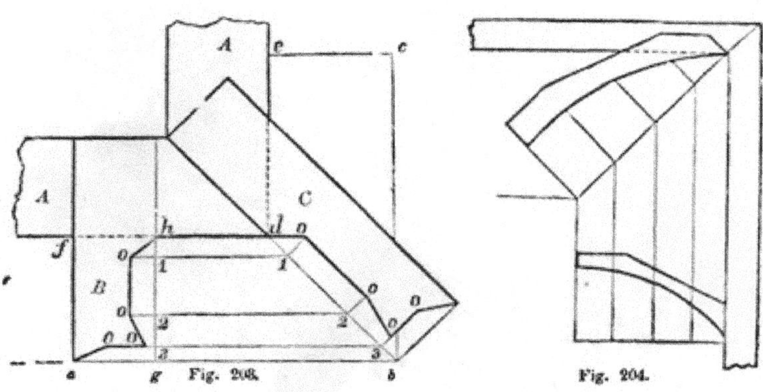

Fig. 203.

Fig. 204.

lel with the wall, *f d*, draw the line, *a b* ; make *e c* equal *a f*, and through *c*, draw *c b*, parallel with *e d* ; join *d* and *b*, and from the several angular points in *B*, draw ordinates to cut *d b* in 1, 2 and 3; at those points erect lines perpendicular to *d b* ; from *h*, draw *h g*, parallel to *f a* ; take the ordinates, 1 *o*, 2 *o*, &c., at *B*, and transfer them to *C*, and the angle-bracket, *C*, will be defined. In the same manner, the angle-bracket for an internal cornice, or the angle-rib of a coved ceiling, or of groins, as at *Fig*. 204, can be found.

256.—*A level crown moulding being given, to find the raking moulding and a level return at the top.* Let *A*, (*Fig*. 205,) be the given moulding, and *A b* the rake of the roof. Divide the curve of the given moulding into any number of parts, equal or unequal, as at 1, 2, and 3; from these points, draw horizontal lines to a perpendicular erected from *c* ; at any convenient place on the rake, as at *B*, draw *a c*, at right angles to *A b* ; also, from *b*, draw the horizontal line, *b a* ; place the thickness, *d a*, of the moulding at *A*, from *b* to *a*, and from *a*, draw the perpendicular line, *a e* ; from the points, 1, 2, 3, at *A*, draw lines to *C*, parallel to *A b* ; make *a* 1, *a* 2 and *a* 3, at *B* and at *C*, equal to *a* 1, &c., at *A* ; through the points, 1, 2 and 3, at *B*, trace the curve—this will be the proper form for the raking moulding. From 1, 2 and

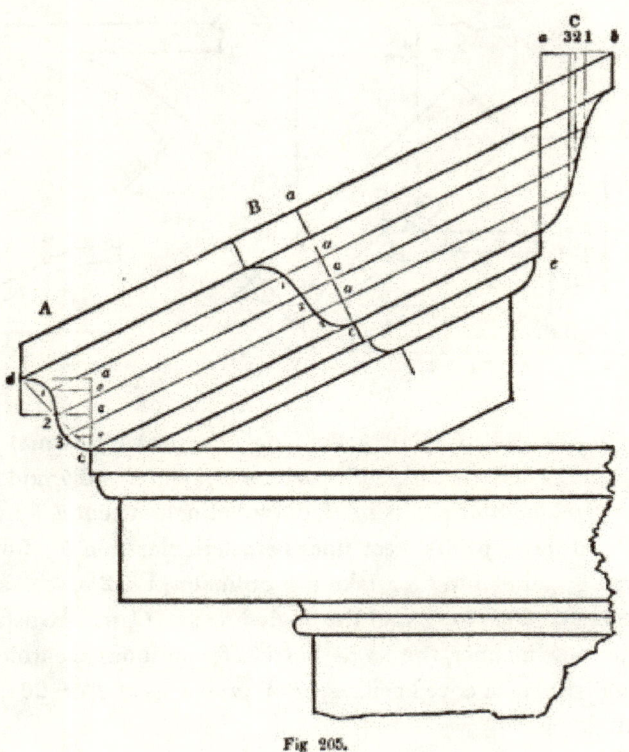

Fig 205.

3, at *C*, drop perpendiculars to the corresponding ordinates from 1, 2 and 3, at *A*; through the points of intersection, trace the curve—this will be the proper form for the *return* at the top.

SECTION IV.—FRAMING.

257.—This subject is, to the carpenter, of the highest importance; and deserves more attention and a larger place in a volume of this kind, than is generally allotted to it. Something, indeed, has been said upon the geometrical principles, by which the several lines for the joints and the lengths of timber, may be ascertained; yet, besides this, there is much to be learned. For however precise or workmanlike the joints may be made, what will it avail, should the system of framing, from an erroneous position of its timbers, &c., change its form, or become incapable of sustaining even its own weight? Hence the necessity for a knowledge of the laws of pressure and the strength of timber. These being once understood, we can with confidence determine the best position and dimensions for the several timbers which compose a floor or a roof, a partition or a bridge. As systems of framing are more or less exposed to heavy weights and strains, and, in case of failure, cause not only a loss of labour and material, but frequently that of life itself, it is very important that the materials employed be of the proper quantity and quality to serve their destination. And, on the other hand, any superfluous material is not only useless, but a positive injury, it being an unnecessary load upon the points of support. It is necessary, therefore, to know the *least* quantity of timber that will suffice for strength. The greatest fault in framing is that of using an excess of material. Economy, at least, would seem to require that this evil be abated.

Before proceeding to consider the principles upon which a system of framing should be constructed, let us attend to a few of the elementary laws in *Mechanics*, which will be found to be of great value in determining those principles.

258.—Laws of Pressure. (1.) A heavy body always exerts a pressure equal to its own weight in a vertical direction. Example: Suppose an iron ball, weighing 100 lbs., be supported upon the top of a perpendicular post, (*Fig.* 220;) then the pressure exerted upon that post will be equal to the weight of the ball; viz., 100 lbs. (2.) But if two inclined posts, (*Fig.* 206,) be substituted for the perpendicular support, the united pressures upon these posts will be more than equal to the weight, and will be in proportion to their position. The farther apart their feet are spread the greater will be the pressure, and *vice versa*. Hence tremendous strains may be exerted by a comparatively small weight. And it follows, therefore, that a piece of timber intended for a strut or post, should be so placed that its axis may coincide, as near as possible, with the direction of the pressure. The direction of the pressure of the weight, *W*, (*Fig.* 206,) is in the vertical line, *b d*; and the weight, *W*, would fall in that line, if the two posts were removed, hence the best position for a support

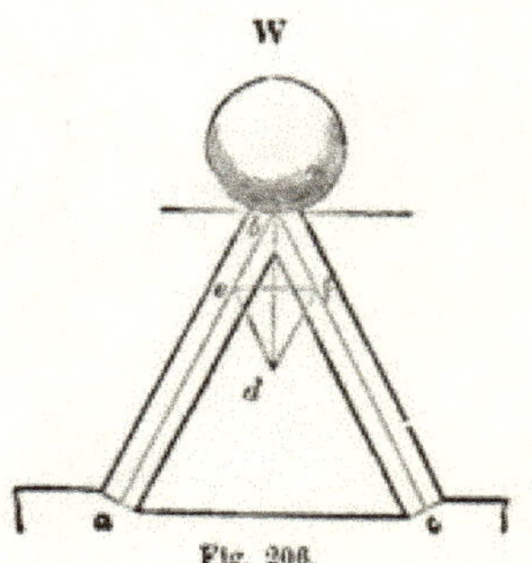

Fig. 206.

for the weight would be in that line. But, as it rarely occurs in systems of framing that weights can be supported by any single resistance, they requiring generally two or more supports, (as in the case of a roof supported by its rafters,) it becomes important, therefore, to know the exact amount of pressure any certain weight is capable of exerting upon oblique supports. Now it has been ascertained that the three lines of a triangle, drawn parallel with the direction of three concurring forces in equilibrium, are in proportion respectively to these forces. For example, in *Fig.* 206, we have a representation of three forces concurring in a point, which forces are in equilibrium and at rest; thus, the weight, *W*, is one force, and the resistance exerted by the two pieces of timber are the other two forces. The direction in which the first force acts is vertical—downwards; the direction of the two other forces is in the axis of each piece of timber respectively. These three forces all tend towards the point, *b*.

Draw the axes, *a b* and *b c*, of the two supports; make *b d* vertical, and from *d* draw *d e* and *d f* parallel with the axes, *b c* and *b a*, respectively. Then the triangle, *b d e*, has its lines parallel respectively with the direction of the three forces; thus, *b d* is in the direction of the weight, *W*, *d e* parallel with the axis of the timber *b c*, and *e b* is in the direction of the timber *a b*. In accordance with the principle above stated, the lengths of the sides of the triangle, *b d e*, are in proportion respectively to the three forces aforesaid; thus—

As the length of the line, *b d*,
Is to the number of pounds in the weight, *W*,
So is the length of the line, *b e*,
To the number of pounds' pressure resisted by the timber, *a b*.

Again—

As the length of the line, *b d*,
Is to the number of pounds in the weight, *W*,

So is the length of the line, *d e*,
To the number of pounds' pressure resisted by the timber, *b c*.

And again—

As the length of the line, *b e*,
Is to the pounds' pressure resisted by *a b*,
So is the length of the line, *d e*,
To the pounds' pressure resisted by *b c*.

These proportions are more briefly stated thus—

1*st*. $b\,d : W :: b\,e : P$,

P being used as a symbol to represent the number of pounds' pressure resisted by the timber, *a b*.

2*nd*. $b\,d : W :: d\,e : Q$,

Q representing the number of pounds' pressure resisted by the timber, *b c*.

3*d*. $b\,e : P :: d\,e : Q$.

259.—This relation between lines and pressures is important, and is of extensive application in ascertaining the pressures induced by known weights throughout any system of framing. The parallelogram, *b e d f*, is called the *Parallelogram of Forces;* the two lines, *b e* and *b f*, being called the *components*, and the line *b d* the *resultant*. Where it is required to find the *components* from a given resultant, (*Fig.* 206,) it is not needed to draw the fourth line, *d f*, for the triangle, *b d e*, gives the desired result. But when the *resultant* is to be ascertained from given components, (*Fig.* 212,) it is more convenient to draw the fourth line.

260.—*The Resolution of Forces* is the finding of two or more forces, which, acting in different directions, shall exactly balance the pressure of any given *single* force. To make a practical application of this, let it be required to ascertain the oblique pressure in *Fig.* 206. In this *Fig.* the line *b d* measures half an inch, (0·5 inch,) and the line *b e* three-tenths of an inch, (0·3 inch.) Now if the weight, *W*, be supposed to be 1200 pounds, then the first stated proportion above,

$$b\,d : W :: b\,e : P,$$

becomes

$$0{\cdot}5 : 1200 :: 0{\cdot}3 : P.$$

And since the product of the means divided by one of the extremes gives the other extreme, this proportion may be put in the form of an *equation*, thus—

$$\frac{1200 \times 0{\cdot}3}{0{\cdot}5} = P.$$

Performing the arithmetical operation here indicated, that is, multiplying together the two quantities above the line, and dividing the product by the quantity under the line, the quotient will be equal to the quantity represented by P, viz., the pressure resisted by the timber, $a\,b$. Thus—

$$\begin{array}{r} 1200 \\ 0{\cdot}3 \\ \hline 0{\cdot}5)360{\cdot}0 \\ \hline 720 = P. \end{array}$$

The strain upon the timber, $a\,b$, is, therefore, equal to 720 pounds; and the strain upon the other timber, $b\,c$, is also 720 pounds; for in this case, the two timbers being inclined equally from the vertical, the line $e\,d$ is therefore equal to the line $b\,e$.

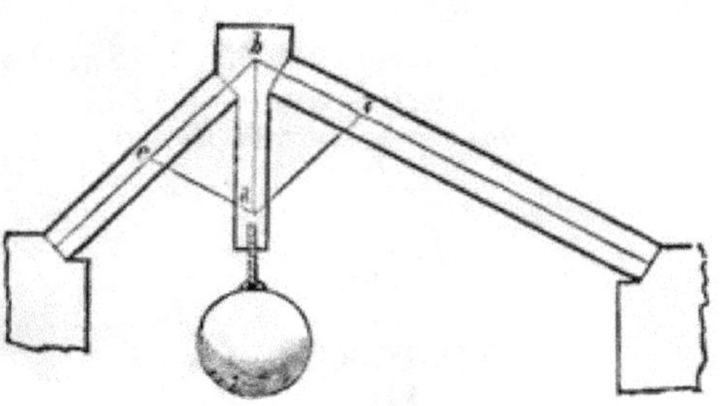

Fig. 207.

261.—In *Fig.* 207, the two supports are inclined at different angles, and the pressures are proportionately unequal. The supports are also unequal in length. The length of the supports does not alter the amount of pressure from the concentrated load supported; but generally long timbers are not so capable of resistance as shorter ones. They yield more readily laterally, as they are not so stiff, and shorten more, as the compression is in proportion to the length. To ascertain the pressures in *Fig.* 207, let the weight suspended from *b d* be equal to two and three-quarter tons, (2·75 tons.) The line *b d* measures five and a half tenths of an inch, (0·55 inch,) and the line *b e* half an inch, (0·5 inch.) Therefore, the proportion

$$b\,d : W :: b\,e : P, \text{ becomes } 0{\cdot}55 : 2{\cdot}75 :: 0{\cdot}5 : P,$$

$$\text{and } \frac{2{\cdot}75 \times 0{\cdot}5}{0{\cdot}55} = P.$$

```
       2·75
        0·5
      ------
0·55)1·375(2·5 = P.
     110
     ---
      275
      275
      ---
```

The strain upon the timber, *b e*, is, therefore, equal to two and a half tons.

Again, the line *e d* measures four-tenths of an inch, (0·4 inch;) therefore, the proportion

$$b\,d : W :: e\,d : Q, \text{ becomes } 0{\cdot}55 : 2{\cdot}75 :: 0{\cdot}4 : Q,$$

$$\text{and } \frac{2{\cdot}75 \times 0{\cdot}4}{0{\cdot}55} = Q.$$

```
       2·75
        0·4
      ------
0·55)1·100(2 = Q.
     110
     ---
```

The strain upon the timber, *b f*, is, therefore, equal to two tons.

262.—Thus it is seen that the united pressures exerted by a weight upon two inclined supports always exceed the weight. In the last case 2¼ tons exerts a pressure of 2½ and two tons, equal together to 4½ tons; and in the former case, 1200 pounds exerts a pressure of twice 720 pounds, equal to 1440 pounds. The smaller the angle of inclination to the horizontal, the greater will be the pressure upon the supports. So, in the frame of a roof, the strain upon the rafters decreases gradually with the increase of the angle of inclination to the horizon, the length of the rafter remaining the same.

263.—This is true in comparing systems of framing with each other; but in a system where the concentrated weight to be supported is not in the middle, (see *Fig.* 207,) and, in consequence, the supports are not inclined equally, the strain will be *greatest* upon the support that has the greatest inclination to the horizon.

264.—In ordinary cases, in roofs for example, the load is not concentrated but is that of the framing itself. Here the *amount* of the load will be in proportion to the length of the rafter, and the rafter increases in length with the increase of the angle of inclination, the span remaining the same. So it is seen that in enlarging the angle of inclination to the horizon in order to lessen the oblique thrust, the load is increased in consequence of the elongation of the rafter, thus increasing the oblique thrust. Hence there is a limit to the angle of inclination. A rafter will have the least oblique thrust when its angle of inclination to the horizon is 35° 16′ nearly. This angle is attained very nearly when the rafter rises 8½ inches per foot; or, when the height, *B C*, (*Fig.* 216,) is to the base, *A C*, as 8½ is to 12, or as 0·7071 is to 1·0.

265.—Correct ideas of the comparative pressures exerted upon timbers, according to their position, will be readily formed by drawing various designs of framing, and estimating the several strains in accordance with the parallelogram of forces, always drawing the triangle, *b d e*, so that the three lines shall be parallel with the three forces, or pressures, respectively. The *length* of the lines forming this triangle is unimportant, but it will be found more convenient if the line drawn parallel with the *known* force is made to contain as many inches as the known force contains pounds, or as many tenths of an inch as pounds, or as many inches as tons, or tenths of an inch as tons: or, in general, as many divisions of any convenient scale as there are units of weight or pressure in the known force. If drawn in this manner, then the number of divisions of the same scale found in the other two lines of the triangle will equal the units of pressure or weight of the other two forces respectively, and the pressures sought will be ascertained simply by applying the scale to the lines of the triangle.

For example, in *Fig.* 207, the vertical line, *b d*, of the triangle, measures fifty-five hundredths of an inch, (0·55 inch;) the line, *b e*, fifty-hundredths, (0·50 inch;) and the line, *e d*, forty, (0·40 inch.) Now, if it be supposed that the vertical pressure, or the weight suspended below *b d*, is equal to 55 pounds, then the pressure on *b e* will equal 50 pounds, and that on *e d* will equal 40 pounds; for, by the proportion above stated,

$$b\,d : W :: b\,e : P,$$
$$55 : 55 :: 50 : 50;$$

and so of the other pressure.

266.—If a scale cannot be had of equal proportions with the forces, the arithmetical process will be shortened somewhat by making the line of the triangle that represents the *known* weight equal to unity of a decimally divided scale, then the other lines will be measured in tenths or hundredths; and in the numerical statement of the proportions between the lines and forces, the first term being unity, the fourth term will be ascertained simply by multiplying the second and third terms together.

For example, if the three lines are 1, 0·7 and 1·3, and the known weight is 6 tons, then

$$b\,d : W :: b\,e : P, \text{ becomes}$$

$$1 : 6 :: 0·7 : P = 4·2,$$

equals four and two-tenths tons. Again—

$$b\,d : W :: e\,d : Q, \text{ becomes}$$

$$1 : 6 :: 1·3 : Q = 7·8,$$

equals seven and eight-tenths tons.

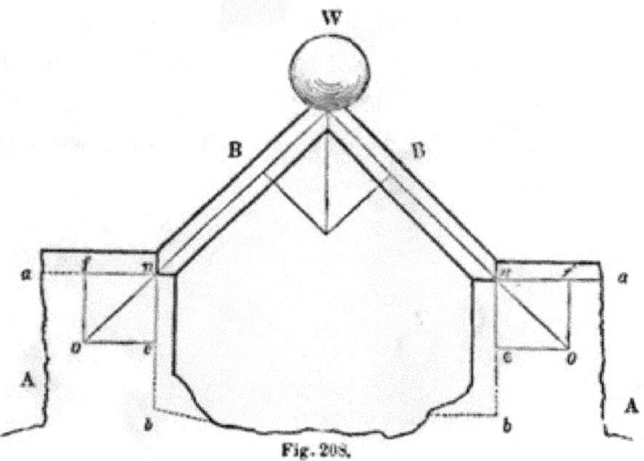

Fig. 208.

267.—In *Fig.* 208 the weight, W, exerts a pressure on the struts in the direction of their length; their feet, $n\ n$, have, therefore, a tendency to move in the direction $n\ o$, and would so move, were they not opposed by a sufficient resistance from the blocks, A and A. If a piece of each block be cut off at the horizontal line, $a\ n$, the feet of the struts would slide away from each other along that line, in the direction, $n\ a$; but if, instead of these, two pieces were cut off at the vertical line, $n\ b$, then the struts would descend vertically. To estimate the horizontal and the vertical pressures exerted by the struts, let $n\ o$ be made equal (upon any scale of equal parts) to the number of tons with which the strut is pressed; construct the parallelogram of forces by drawing *o e* parallel to *a n*, and *o f* parallel to *b n*; then *n f*, (by the same scale,) shows the number of tons pressure that is exerted by the strut in the direction *n a*, and *n e* shows the amount exerted in the direction *n b*. By constructing designs similar to this, giving various and dissimilar positions to the struts, and then estimating the pressures, it will be found in every case that the horizontal pressure of one strut is exactly equal to that of the other, however much one strut may be inclined more than the other; and also, that the united vertical pressure of the two struts is exactly equal to the weight, *W*. (In this calculation the weight of the timbers has not been taken into consideration, simply to avoid complication to the learner. In practice it is requisite to include the weight of the framing with the load upon the framing.)

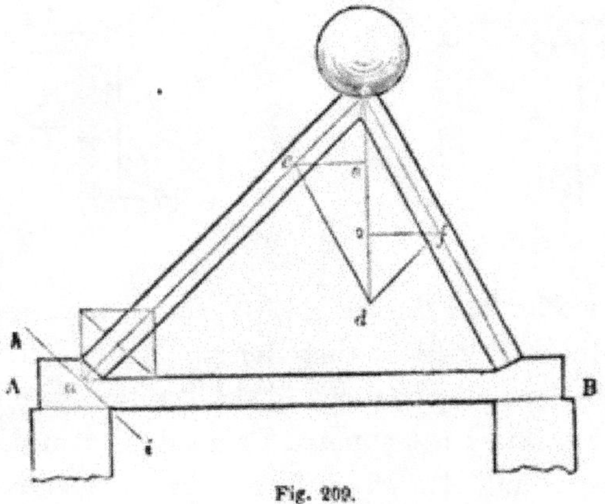

Fig. 209.

268.—Suppose that the two struts, *B* and *B*, (*Fig.* 208,) were rafters of a roof, and that instead of the blocks, *A* and *A*, the walls of a building were the supports: then, to prevent the walls from being thrown over by the thrust of *B* and *B*, it would be desirable to remove the horizontal pressure. This may be done by uniting the feet of the rafters with a rope, iron rod, or piece of timber, as in *Fig.* 209. This figure is similar to the truss of a roof. The horizontal strains on the tie-beam, tending to pull it asunder in the direction of its length, may be measured at the foot of the rafter, as was shown at *Fig.* 208; but it can be more readily and as accurately measured, by drawing from *f* and *e* horizontal lines to the vertical line, *b d*, meeting it in *o* and *o*; then *f o* will be the horizontal thrust at *B*, and *e o* at *A*; these will be found to equal one another. When the rafters of a roof are thus connected, all tendency to thrust the walls horizontally is removed, the only pressure on them is in a vertical direction, being equal to the weight of the roof and whatever it has to support. This pressure is beneficial rather than otherwise, as a roof having trusses thus formed, and the trusses well braced to each other, tends to steady the walls.

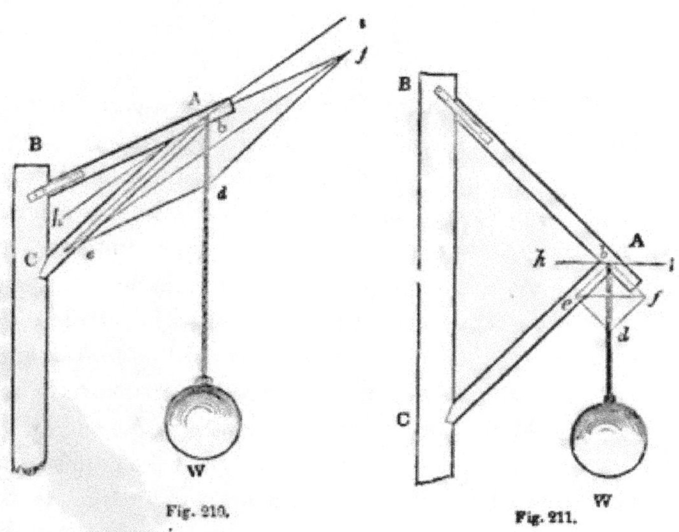

Fig. 210. Fig. 211.

269.—*Fig.* 210 and 211 exhibit methods of framing for supporting the equal weights, *W* and *W*. Suppose it be required to measure and compare the strains produced on the pieces, *A B* and *A C*. Construct the parallelogram of forces, *e b f d*, according to *Art*. 258. Then *b f* will show the strain on *A B*, and *b e* the strain on *A C*. By comparing the figures, *b d* being equal in each, it will be seen that the strains in *Fig*. 210 are about three times as great as those in *Fig*. 211: the position of the pieces, *A B* and *A C*, in *Fig*. 211, is therefore far preferable.

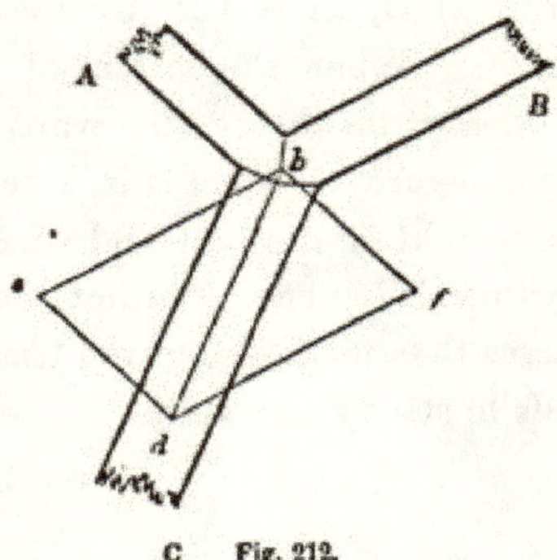

Fig. 212.

270.—The *Composition of Forces* consists in ascertaining the direction and amount of *one* force, which shall be just capable of balancing *two or more* given forces, acting in different directions. This is only the reverse of the resolution of forces, and the two are founded on one and the same principle, and may be solved in the same manner. For example, let *A* and *B*, (*Fig*. 212,) be two pieces of timber, pressed in the direction of their length towards *b*—*A* by a force equal to 6 tons weight, and *B* equal to 9. To find the *direction* and *amount* of pressure they would unitedly exert, draw the lines, *b e* and *b f*, in a line with the axes of the timbers, and make *b e* equal to the pressure exerted by *B*, viz., 9; also make *b f* equal to the pressure on *A*, viz., 6, and complete the parallelogram of forces, *e b f d;* then *b d*, the diagonal of the parallelogram, will be the *direction*, and its length, 9·25, will be the *amount*, of the united pressures of *A* and of *B*. The line, *b d*, is termed the *resultant* of the two forces, *b f* and *b e*. If *A* and *B* are to be supported by one post, *C*, the best position for that post will be in the direction of the diagonal, *b d*; and it will require to be sufficiently strong to support the united pressures of *A* and of *B*, which are equal to 9·25 or $9\frac{1}{4}$ tons.

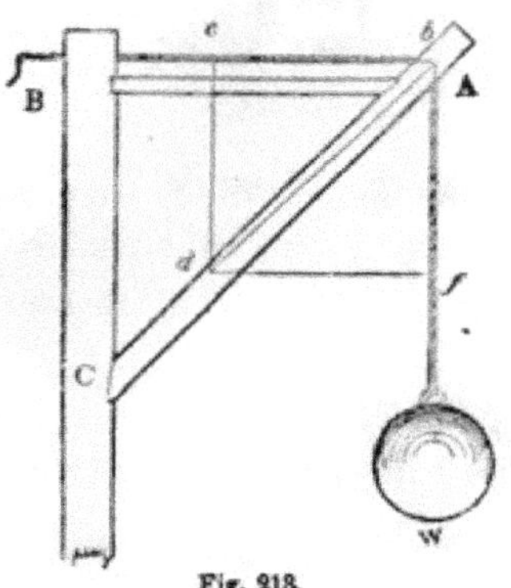

Fig. 213.

271.—Another example: let *Fig.* 213 represent a piece of framing commonly called a crane, which is used for hoisting heavy weights by means of the rope, *B b f*, which passes over a pulley at *b*. This is similar to *Fig.* 210 and 211, yet it is materially different. In those figures, the strain is in one direction only, viz., from *b* to *d*; but in this there are two strains, from *A* to *B* and from *A* to *W*. The strain in the direction *A B* is evidently equal to that in the direction *A W*. To ascertain the best position for the strut, *A C*, make *b e* equal to *b f*, and complete the parallelogram of forces, *e b f d*; then draw the diagonal, *b d*, and it will be the position required. Should the foot, *C*, of the strut be placed either higher or lower, the strain on *A C* would be increased. In constructing cranes, it is advisable, in order that the piece, *B A*, may be under a gentle pressure, to place the foot of the strut a trifle lower than where the diagonal, *b d*, would indi cate, but never higher.

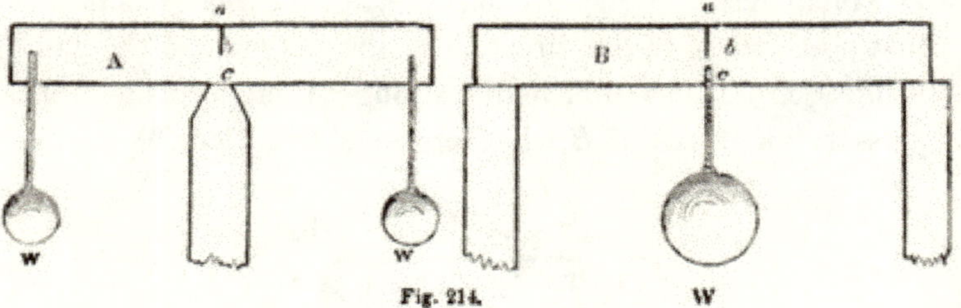

Fig. 214.

272.—*Ties and Struts.* Timbers in a state of tension are called *ties*, while such as are in a state of compression are termed *struts*. This subject can be illustrated in the following manner:

Let *A* and *B*, (*Fig.* 214,) represent beams of timber supporting the weights, *W*, *W* and *W*; *A* having but one support, which is in the middle of its length, and *B* two, one at each end. To show the nature of the strains, let each beam be sawed in the middle from *a* to *b*. The effects are obvious: the cut in the beam, *A*, will open, whereas that in *B* will close. If the weights are heavy enough, the beam, *A*, will break at *b*; while the cut in *B* will be closed perfectly tight at *a*, and the beam be very little injured by it. But if, on the other hand, the cuts be made in the bottom edge of the timbers, from *c* to *b*, *B* will be seriously injured, while *A* will scarcely be affected. By this it appears evident that, in a piece of timber subject to a pressure across the direction of its length, the fibres are exposed to contrary strains. If the timber is supported at both ends, as at *B*, those from the top edge down to the middle are compressed in the direction of their length, while those from the middle to the bottom edge are in a state of tension; but if the beam is supported as at *A*, the contrary effect is produced; while the fibres at the middle of either beam are not at all strained. The strains in a framed truss are of the same nature as those in a single beam. The truss for a roof, being supported at each end, has its tie-beam in a state of tension, while its rafters are compressed in the direction of their length. By this, it appears highly important that pieces in a state of tension should be distinguished from such as are compressed, in order that the former may be preserved continuous. A strut may be constructed of two or more pieces; yet, where there are many joints, it will not resist compression so well.

273.—*To distinguish ties from struts.* This may be done by the following rule. In *Fig.* 206, the timbers, *a b* and *b c*, are the *sustaining* forces, and the weight, *W*, is the *straining* force; and, if the support be removed, the straining force would move from the point of support, *b*, towards *d*. Let it be required to ascertain whether the sustaining forces are *stretched* or *pressed* by the straining force. *Rule:* upon the direction of the straining force, *b d*, as a diagonal, construct a parallelogram, *e b f d*, whose sides shall be parallel with the direction of the sustaining forces, *a b* and *c d;* through the point, *b*, draw a line, parallel to the diagonal, *e f;* this may then be called the dividing line between ties and struts. Because all those supports which are on that side of the dividing line, which the straining force would occupy if unresisted, are compressed, while those on the other side of the dividing line are stretched.

In *Fig.* 206, the supports are both compressed, being on that side of the dividing line which the straining force would occupy if unresisted. In *Fig.* 210 and 211, in which *A B* and *A C* are the sustaining forces, *A C* is compressed, whereas *A B* is in a state of tension; *A C* being on that side of the line, *h i*, which the straining force would occupy if unresisted, and *A B* on the opposite side. The place of the latter might be supplied by a chain or rope. In *Fig.* 209, the foot of the rafter at *A* is sustained by two forces, the wall and the tie beam, one perpendicular and the other horizontal: the direction of the straining force is indicated by the line, *b a*. The dividing line, *h i*, ascertained by the rule, shows that the wall is pressed and the tie-beam stretched.

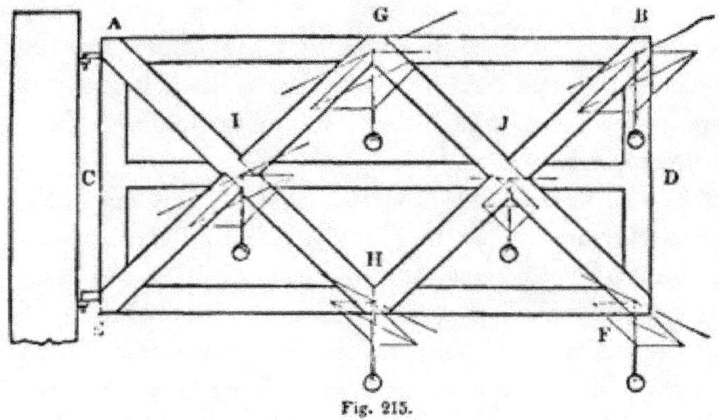

Fig. 215.

274.—Another example: let *E A B F*, (*Fig.* 215,) represent a gate, supported by hinges at *A* and *E*. In this case, the *straining* force is the weight of the materials, and the direction of course *vertical*. Ascertain the dividing line at the several points, *G*, *B*, *I*, *J*, *H* and *F*. It will then appear that the force at *G* is sustained by *A G* and *G E*, and the dividing line shows that the former is stretched and the latter compressed. The force at *H* is supported by *A H* and *H E*—the former stretched and the latter compressed. The force at *B* is opposed by *H B* and *A B*, one pressed, the other stretched. The force at *F* is sustained by *G F* and *F E*, *G F* being stretched and *F E* pressed. By this it appears that *A B* is in a state of tension, and *E F*, of compression; also, that *A H* and *G F* are stretched, while *B H* and *G E* are compressed: which shows the necessity of having *A H* and *G F*, each in one whole length, while *B H* and *G E* may be, as they are shown, each in two pieces. The force at *J* is sustained by *G J* and *J H*, the former stretched and the latter compressed.

The piece, $C\,D$, is neither stretched nor pressed, and could be dispensed with if the joinings at J and I could be made as effectually without it. In case $A\,B$ should fail, then $C\,D$ would be in a state of tension.

275.—*The centre of gravity.* The centre of gravity of a uniform prism or cylinder, is in its axis, at the middle of its length; that of a triangle, is in a line drawn from one angle to the middle of the opposite side and at one-third of the length of the line from that side; that of a right-angled triangle, at a point distant from the perpendicular equal to one-third of the base, and distant from the base equal to one-third of the perpendicular; that of a pyramid or cone, in the axis and at one-quarter of the height from the base.

276.—The centre of gravity of a trapezoid, (a four-sided figure having only two of its sides parallel,) is in a line joining the centres of the two parallel sides, and at a distance from the longest of the parallel sides equal to the product of the length into the sum of twice the shorter added to the longer of the parallel sides, divided by three times the sum of the two parallel sides. Algebraically thus—

$$d = \frac{l\,(2\,a + b)}{3\,(a + b)}$$

where d equals the distance from the longest of the parallel sides, l the length of the line joining the two parallel sides, and a the shorter and b the longer of the parallel sides.

Example.—A rafter, 25 feet long, has the larger end 14 inches wide, and the smaller end 10 inches wide, how far from the larger end is the centre of gravity located?

Here, $l = 25$, $a = \frac{10}{12}$, and $b = \frac{14}{12}$,

hence $d = \frac{l\,(2\,a + b)}{3\,(a + b)} = \frac{25\,(2 \times \frac{10}{12} + \frac{14}{12})}{3\,(\frac{10}{12} + \frac{14}{12})} = \frac{25 \times \frac{34}{12}}{3 \times \frac{24}{12}} =$

$\frac{25 \times 34}{3 \times 24} = \frac{850}{72} = 11{\cdot}8 = 11$ feet $9\frac{5}{8}$ inches nearly.

In irregular bodies with plain sides, the centre of gravity

may be found by balancing them upon the edge of a prism—upon the edge of a table—in two positions, making a line each time upon the body in a line with the edge of the prism, and the intersection of those lines will indicate the point required. Or suspend the article by a cord or thread attached to one corner or edge; also, from the same point of suspension, hang a plumb-line, and mark its position on the face of the article; again, suspend the article from another corner or side, (nearly at right angles to its former position,) and mark the position of the plumb-line upon its face; then the intersection of the two lines will be the centre of gravity.

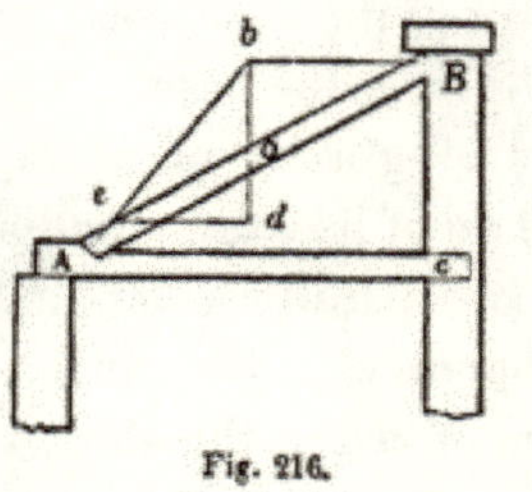

Fig. 216.

277.—*The effect of the weight of inclined beams.* An inclined post or strut, supporting some heavy pressure applied at its upper end, as at *Fig.* 209, exerts a pressure at its foot in the direction of its length, or nearly so. But when such a beam is loaded uniformly over its whole length, as the rafter of a roof, the pressure at its foot varies considerably from the direction of its length. For example, let *A B*, (*Fig.* 216,) be a beam leaning against the wall, *B c*, and supported at its foot by the abutment, *A*, in the beam, *A c*, and let *o* be the centre of gravity of the beam. Through *o*, draw the vertical line, *b d*, and from *B*, draw the horizontal line, *B b*, cutting *b d* in *b*; join *b* and *A*, and *b A* will be the *direction* of the thrust. To prevent the beam from loosing its footing, the joint at *A* should be made at right angles to *b A*. The *amount* of pressure will be found thus: let *b d*, (by any scale of equal parts,) equal the number of tons upon the beam, $A\ B$; draw $d\ e$, parallel to $B\ b$; then $b\ e$, (by the same scale,) equals the pressure in the direction, $b\ A$; and $e\ d$, the pressure against the wall at B—and also the horizontal thrust at A, as these are always equal in a construction of this kind.

278.—The horizontal thrust of an inclined beam, (*Fig.* 216,) —the effect of its own weight—may be calculated thus:

Rule.—Multiply the weight of the beam in pounds by its base, $A\ C$, in feet, and by the distance in feet of its centre of gravity, o, (see *Art.* 275 and 276,) from the lower end, at A; and divide this product by the product of the length, $A\ B$, into the height, $B\ C$, and the quotient will be the horizontal thrust in pounds. This may be stated thus: $H = \frac{d\ b\ w}{h\ l}$, where d equals the distance of the centre of gravity, o, from the lower end; b equals the base, $A\ C$; w equals the weight of the beam; h equals the height, $B\ C$; l equals the length of the beam; and H equals the horizontal thrust.

Example.—A beam, 20 feet long, weighs 300 pounds; its centre of gravity is at 9 feet from its lower end; it is so inclined that its base is 16 feet and its height 12 feet; what is the horizontal thrust?

Here $\frac{d\ b\ w}{h\ l}$ becomes $\frac{9 \times 16 \times 300}{12 \times 20} = \frac{9 \times 4 \times 25}{5} = 9 \times 4 \times 5$ $= 180 = H =$ the horizontal thrust.

This rule is for cases where the centre of gravity does not occur at the middle of the length of the beam, although it is applicable when it *does* occur at the middle; yet a shorter rule will suffice in this case,—and it is thus:—

Rule.—Multiply the weight of the rafter in pounds by the base, $A\ C$, (*Fig.* 216,) in feet, and divide the product by twice the height, $B\ C$, in feet; and the quotient will be the horizontal thrust, when the centre of gravity occurs at the middle of the beam.

If the inclined beam is loaded with an equally distributed load, add this load to the weight of the beam, and use this *total* weight in the rule instead of the weight of the beam. And generally, if the centre of gravity of the combined weights of the beam and load does not occur at the centre of the length of the beam then the former rule is to be used.

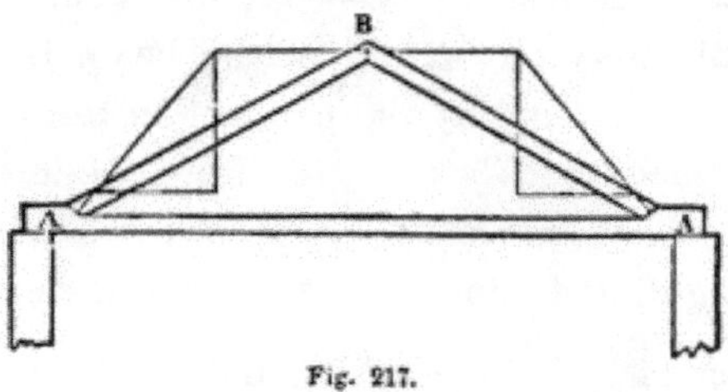

Fig. 217.

279.—In *Fig.* 217, two equal beams are supported at their feet by the abutments in the tie-beam. This case is similar to the last; for it is obvious that each beam is in precisely the position of the beam in *Fig.* 216. The horizontal pressures at *B*, being equal and opposite, balance one another; and their horizontal thrusts at the tie-beam are also equal. (See *Art.* 268—*Fig.* 209.) When the height of a roof, (*Fig.* 217,) is one-fourth of the span, or of a shed, (*Fig.* 216,) is one-half the span, the horizontal thrust of a rafter, whose centre of gravity is at the middle of its length, is exactly equal to the weight distributed uniformly over its surface.

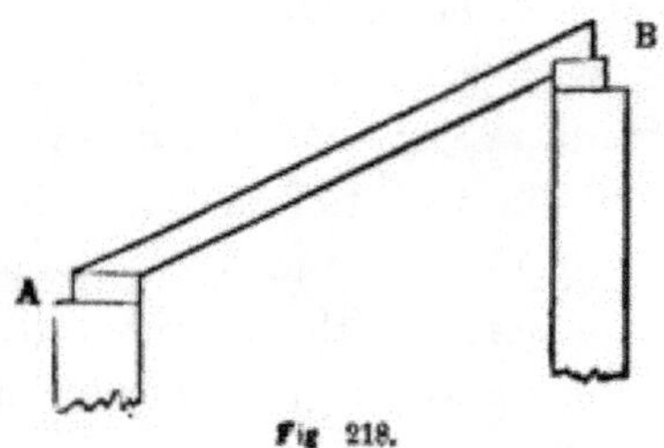

Fig 218.

280.—In shed, or *lean-to* roofs, as *Fig.* 216, the horizontal pressure will be entirely removed, if the bearings of the rafters, as *A B*, (*Fig.* 218,) are made horizontal—provided, however, that the rafters and other framing do not bend between the points of support. If a beam or rafter have a natural curve, the convex or rounding edge should be laid uppermost.

281.—A beam laid horizontally, supported at each end and uniformly loaded, is subject to the greatest strain at the middle of its length. Hence mortices, large knots and other defects, should be kept as far as possible from that point; and, in resting a load upon a beam, as a partition upon a floor beam, the weight should be so adjusted, if possible, that it will bear at or near the ends.

Twice the weight that will break a beam, acting at the centre of its length, is required to break it when equally distributed over its length; and precisely the same deflection or *sag* will be produced on a beam by a load equally distributed, that five-eighths of the load will produce if acting at the centre of its length.

282.—When a beam, supported at each end on horizontal bearings, (the beam itself being either horizontal or inclined,) has its load equally distributed, the amount of pressure caused by the load on each point of support is equal to one half the load; and this is also the case, when the load is concentrated at the middle of the beam, or has its centre of gravity at the middle of the beam; but, when the load is unequally distributed or concentrated, so that its centre of gravity occurs at some other point than the middle of the beam, then the amount of pressure caused by the load on one of the points of support is unequal to that on the other. The precise amount on each may be ascertained by the following rule.

Rule.—Multiply the weight *w*, (*Fig.* 219,) by its distance, *C B*, from its nearest point of support, *B*, and divide the product by the length, *A B*, of the beam, and the quotient will be the

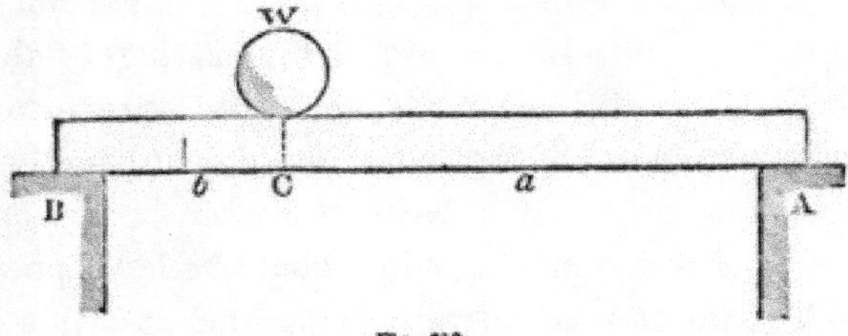

Fig. 219.

amount of pressure on the *remote* point of support, A. Again, deduct this amount from the weight, w, and the remainder will be the amount of pressure on the near point of support, B; or, multiply the weight, w, by its distance, $A\ C$, from the remote point of support, A, and divide the product by the length, $A\ B$, and the quotient will be the amount of pressure on the *near* point of support, B.

When l equals the length, $A\ B$; $a = A\ C$; $b = C\ B$, and $w =$ the load, then

$$\frac{w\ b}{l} = A = \text{the amount of pressure at } A, \text{ and}$$

$$\frac{w\ a}{l} = B = \text{the amount of pressure at } B.$$

Example.—A beam, 20 feet long between the bearings, has a load of 100 pounds concentrated at 3 feet from one of the bearings, what is the portion of this weight sustained by each bearing?

Here $w = 100$; a, 17; b, 3; and l, 20.

$$\text{Hence } A = \frac{w\ b}{l} = \frac{100 \times 3}{20} = 15.$$

$$\text{And } B = \frac{w\ a}{l} = \frac{100 \times 17}{20} = 85.$$

Load on A = 15 pounds.
Load on B = 85 pounds.
Total weight = 1(0 pounds.

RESISTANCE OF MATERIALS.

283.—Before a roof truss, or other piece of framing, can be properly designed, two things are required to be known. The one is, the effect of gravity acting upon the various parts of the intended structure; the other, the power of resistance possessed by the materials of which the framing is to be constructed. In the preceding pages, the former subject having been treated of, it remains now to call attention to the latter.

284.—Materials used in construction are constituted in their structure either of fibres (threads) or of grains, and are termed, the former fibrous, the latter granular. All woods and wrought metals are fibrous, while cast iron, stone, glass, &c., are granular. The strength of a granular material lies in the power of attraction, acting among the grains of matter of which the material is composed, by which it resists any attempt to separate its grains or particles of matter. A fibre of wood or of wrought metal has a strength by which it resists being compressed or shortened, and finally crushed; also a strength by which it resists being extended or made longer, and finally sundered. There is another kind of strength in a fibrous material; it is the adhesion of one fibre to another along their sides, or the lateral adhesion of the fibres.

285.—In the strain applied to a piece of timber, as a post supporting a weight imposed upon it, (*Fig.* 220,) we have an instance of an attempt to shorten the fibres of which the timber is composed. The strength of the timber in this case is termed the *resistance to compression*. In the strain on a piece of timber like a king-post or suspending piece, (*A*, *Fig.* 221,) we have an instance of an attempt to extend or lengthen the fibres of the material. The strength here exhibited is termed the *resistance to tension*. When a piece of timber is strained like a floor beam, or any horizontal piece carrying a load, (*Fig.* 222,) we have an instance in which the two strains of

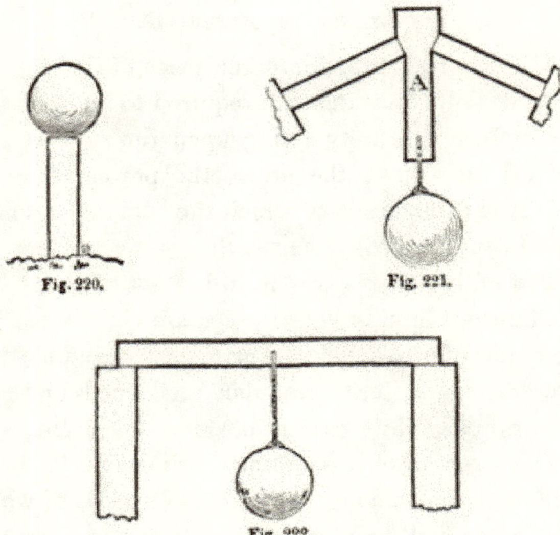

Fig. 220. Fig. 221.

Fig. 222.

compression and tension are brought into action; the fibres of the upper portion of the beam being compressed, and those of the under part being stretched. This kind of strength of timber is termed *resistance to cross strains.* In each of these three kinds of strain to which timber is subjected, the power of resistance is in a measure due to the *lateral* adhesion of the fibres, not so much perhaps in the simple tensile strain, yet to a considerable degree in the compressive and cross strains. But the power of timber, by which it resists a pressure acting compressively in the direction of the length of the fibres, tending to separate the timber by splitting off a part, as in the case of the end of a tie beam, against which the foot of the rafter presses—is wholly due to the lateral adhesion of the fibres.

286.—The *strength* of materials is that power by which they resist *fracture*, while the *stiffness* of materials is that quality which enables them to resist *deflection* or sagging. A knowledge of their *strength* is useful, in order to determine their limits of size to sustain given weights safely; but a knowledge of their *stiffness* is more important, as in almost all constructions it is desirable not only that the load be safely sustained, but that no appearance of weakness be manifested by any sensible deflection or sagging.

I.—RESISTANCE TO COMPRESSION.

287.—The resistance of materials to the force of compression may be considered in four several ways, viz.:

1st. When the pressure is applied to the fibres longitudinally, and on short pieces.

2d. When the pressure is applied to the fibres longitudinally, and on long pieces.

3d. When the pressure is applied to the fibres longitudinally, and so as to split off the part pressed against, causing the fibres to separate by sliding.

4th. When the pressure is applied to the fibres transversely.

Posts having their height less than ten times their least side will crush before bending; these belong to the first case: while posts, whose height is ten times their least side, or more than ten times, will bend before crushing; these belong to the second case.

288.—In the above first and fourth cases of compression, experiment has shown that the resistance is in proportion to the number of fibres pressed, that is, in proportion to the area. For example, if 5,000 pounds is required to crush a prism with a base 1 inch square, it will require 20,000 pounds to crush a prism having a base of 2 by 2 inches, equal to 4 inches area; because 4 times 5,000 equals 20,000. Experiment has also shown that, in the third case, the resistance is in proportion to the area of the surface separated without regard to the form of the surface.

289.—In the second case of compression, the resistance is in proportion to the area of the cross section of the piece, multiplied by the square of its thickness, and inversely in proportion to the square of the length, multiplied by the weight. When the piece is square, it will bend and break in the direction of its diagonal; here, the resistance is in proportion to the square of the diagonal multiplied by the square of the diagonal, and inversely proportional to the square of the length multiplied by the weight. If the piece is round or cylindrical, its resistance will be in accordance with the square of the diameter multiplied by the square of the diameter, and inversely proportional to the square of the length, multiplied by the weight.

290.—These relations between the dimensions of the piece strained and its resistance, have resulted from the discussion of the subject by various authors, and rules based upon these relations are in general use, yet their accuracy is not fully established. Some experiments, especially those by Prof. Hodgkinson, have shown that the resistance is in proportion to a less power of the diameter, and inversely to a less power of the height; yet the variance is not great, and inasmuch as the material is restricted in the rules to a strain decidedly within its limits of resistance, no serious error can be made in the use of rules based on the aforesaid relations.

291.—*Experiments.* In the investigation of the laws applicable to the resistance of materials, only such of the relations of the parts have been considered as apply alike to wood and metal, stone and glass, or other material, leaving to experiment the task of ascertaining the compactness and cohesion of particles, and the tenacity and adhesion of fibres; those qualities upon which depend the superiority of one kind of material over another, and which is represented in the rules by a *constant* number, each specific kind of material having its own special *constant*, obtained by experimenting on specimens of that peculiar material.

292.—The following table exhibits the results of experiments on such woods as are in most common use in this country for the purpose of construction. The resistance of timber of the

TABLE I.—COMPRESSION.

Kind of Material.	Specific Gravity.	To crush Fibres longitudinally.	Value of C in the Rules.	Pressure applied longitudinally to separate Fibres.	Value of H in the Rules.	To crush Fibres transversely $\frac{1}{20}$ inch deep.	Value of P in the Rules. Sensible Impression.
		Pounds per in.		Pounds per in.		Pounds per in.	
White wood.	·397	2432	600			600	300
Mahogany (Baywood),	·439	3527	880			1300	650
Ash,	·517	4175	1040			2300	1150
Spruce,	·369	4199	1050	470	160	500	250
Chestnut,	·491	4791	1200	690	230	950	475
White pine,	·388	4806	1200	490	160	600	300
Ohio pine,	·586	4809	1200	388	130	1250	625
Oak,	·612	5316	1330	780	260	1900	950
Hemlock,	·423	5400	1350	540	180	600	300
Black walnut,	·421	5594	1400			1600	800
Maple,	·574	6061	1515			2050	1025
Cherry,	·494	6477	1620			1900	950
White oak,	·774	6660	1665			2000	1000
Georgia pine,	·613	6767	1700	510	170	1700	850
Locust,	·762	7652	1910	1180	400	2100	1050
Live oak,	·916	7936	1980			5100	2550
Mahogany (St. Domingo),	·837	8280	2070			4300	2150
Lignum vitæ,	1·282	8650	2160			5800	2900
Hickory,	·877	9817	2450			3100	1550

same name varies much; depending as it obviously must on the soil in which it grew, on its age before and after cutting, on the time of year when cut, and on the manner in which it has been kept since it was cut. And of wood from the same tree, much depends upon its location, whether at the butt or towards the limbs, and whether at the heart or at the sap, or at a point midway from the centre to the circumference of the tree. The pieces submitted to experiment were of ordinary good quality, such as would be deemed proper to be used in framing. The prisms crushed were 2 inches long, and from 1 inch to 1½ inches square; some were wider one way than the other, but all containing in area of cross section from 1 to 2 inches. There were generally three specimens of each kind. The weight given in the table is the average crushing weight per superficial inch.

In the preceding table the first column contains the specific gravity of the several kinds of wood, showing their comparative density. The weight in pounds of a cubic foot of any kind of wood or other material, is equal to its specific gravity multiplied by 62·5; this number being the weight in pounds of a cubic foot of water. The second column contains the weight in pounds required to crush a prism having a base of one inch square; the pressure applied to the fibres longitudinally. The third column contains the value of C in the rules; C being equal to one-fourth of the crushing weight in the preceding column. The fourth column contains the weight in pounds, which, applied to the fibres longitudinally, is required to force off a part of the piece, causing the fibres to separate by sliding, the surface separated being one inch square. The fifth column contains the value of H in the rules, H being equal to one third of the weight in the preceding column. The sixth column contains the weight in pounds required to crush the piece when the pressure is applied to the fibres transversely, the piece being one inch thick, and the surface crushed being one inch square, and depressed one twentieth of an inch deep. The seventh column contains the value of P in the rules; P being the weight in pounds applied to the fibres transversely, which is required to make a sensible impression one inch square on the side of the piece, this being the greatest weight that would be proper for a post to be loaded with per inch surface of bearing, resting on the side of the kind of wood set opposite in the table. A greater weight would, in proportion to the excess, crush the side of the wood under the post, and proportionably derange the framing, if not cause a total failure. It will be observed that the measure of this resistance is useful in limiting the load on a post according to the kind of material contained, not in the *post*, but in the *timber upon which* the post presses.

293.—In Table II. are the results of experiments made to test the resistance of materials to flexure: first, the flexure produced by compression, the force acting on the ends of the fibres longitudinally; secondly, the flexure arising from the effects of a cross strain, the force acting on the side of the fibres transversely, the beams being laid on chairs or rests. Of white oak, No. 1, there were eight specimens, of 2 by 4 inches, and 3½ feet long, seasoned more than a year after they were prepared for experiment. Of the other kinds of wood there were from three to five specimens of each, of 1¼ by 2¼ inches, and from 1½ to 2¾ feet long. Of the cast iron there were six specimens, of 1 inch square and 1 foot long; and of the wrought iron there were five specimens of American, three of ¾ by 2 inches, and two of 1⅛ inches square, and three specimens of common English, ½ by 2 inches; the eight specimens being each 19 inches long, clear bearing. In each case the result is the average of the stiffness of the several specimens. The numbers contained in the second column are the weights producing the first degree of flexure in a post or strut, where the post or strut is one foot long and one inch square; so, likewise, the numbers in the fifth column, and which are represented in the rules by E, are the weights required to deflect a beam one inch, where the beam is one foot long, clear bearing, and one inch square.—(See remarks upon this, Art. (321.) The numbers in the third column are equal to one-half of those in the second. The numbers contained in the fourth column, and represented by n in the rules, show the greatest *rate* of deflection that the material may be subjected to without injury. This rate multiplied by the length in feet, equals the total deflection within the limits of elasticity.

TABLE II.—FLEXURE.

Kind of Material.	Specific Gravity.	Under Compression.		Under Cross Strain.	
		Pounds producing the first degree of flexure.	Value of B in the Rules.	Value of π in the Rules.	Value of E in the Rules
Hemlock, ...	0·402	2640	1320	0·08794	1240
Spruce, ...	·432	4190	2095	0·09197	1550
White pine, ...	·407	2350	1175	0·1022	1750
Ohio yellow pine, ...	·586	6000	3000	0·049	1970
Chestnut, ...	·52	7720	3860	0·07541	2330
White oak, No. 1, ...	·82			0·09152	2520
White oak, No. 2, ...	·805	6950	3475	0·0567	2590
Georgia pine, ...	·755	9660	4830	0·07723	2970
Locust, ...	·863	10920	5460	0·06615	3280
Cast iron, ...	7·042			0·0148	30500
Wrought iron, common English,	7·576			0·03717	45500
Wrought iron, American, . .	7·576			0·04038	51400

PRACTICAL RULES FOR COMPRESSION.

First Case.

294.—To find the weight that can be safely sustained by a post, when the height of the post is less than ten times the diameter if round, or ten times the thickness if rectangular, and the direction of the pressure coinciding with the axis.

Rule I.—Multiply the area of the cross-section of the post, in inches, by the value of C in Table I., the product will be the required weight in pounds.

$$A\ C = w. \qquad (1.)$$

Example.—A Georgia pine post is 6 feet high, and in cross-section, 8 × 12 inches, what weight will it safely sustain? The area = 8 × 12 = 96 inches; this multiplied by 1700, the value of C, in the table, set opposite Georgia pine, the result, 163,200, is the weight in pounds required. It will be observed that the weight would be the same for a Georgia pine post of any height less than 10 times 8 inches = 80 inches = 6 feet 8 inches, provided its breadth and thickness remain the same, 12 and 8 inches.

295.—To find the area of the cross-section of a post to sustain a given weight safely, the height of the post being less than ten times the diameter if round, or ten times the least side if rectangular; the pressure coinciding with the axis.

Rule II.—Divide the given weight in pounds by the value of C, in Table I., and the product will be the required area in inches

$$\frac{w}{C} = A. \qquad (2.)$$

Example.—A weight of 38,400 pounds is to be sustained by a white pine post 4 feet high, what must be its area of section in order to sustain the weight safely? Here, 38,400 divided by 1200, the value of C, in Table I., set opposite white pine, gives a quotient of 32; this, therefore, is the required area, and such a post may be 5 × 6·4 inches. To find the least side, so that it shall not be less than one-tenth of the height, divide the height, reduced to inches, by 10, and make the least side to exceed this quotient. The area, divided by the least side so determined, will give the wide side. If, however, by this process, the first side found should prove to be the greatest, then the size of the post is to be found by Rule VII., VIII., or IX.

296.—If the post is to be round, by reference to the Table of Circles in the Appendix, the diameter will be found in the column of diameters, set opposite to the area of the post found in the column of areas, or opposite to the next nearest area. For example, suppose the required area, as just found by the example under Rule II., is 32; by reference to the column of areas, 33·183 is the nearest to 32, and the diameter set opposite is 6·5. The post may, therefore, be 6½ inches diameter.

Second Case.

297.—To ascertain the weight that can be sustained safely

by a post whose height is, at least, ten times its least side if rectangular, or ten times its diameter if round, the direction of the pressure coinciding with the axis.

Rule III.—*When the post is round* the weight may be found by this rule: Multiply the square of the diameter in inches by the square of the diameter in inches, and multiply the product by 0·589 times the value of B, in Table II., divide this product by the square of the height in feet, and the quotient will be the required weight in pounds.

$$w = \frac{0{\cdot}589\, B\, D^2\, D^2}{h^2} = \frac{0{\cdot}589\, B\, D^4}{h^2} \qquad (3.)$$

Example.—What weight will a Georgia pine post sustain safely, whose diameter is 10 inches and height 10 feet? The square of the diameter is 100; 100 × 100 = 10,000. And 10,000 by 0·589 times 4830, the value of B, Table II., set opposite Georgia pine, = 28,448,700, and this divided by 100, the square of the height, equals 284,487, the weight required, in pounds.

Rule IV.—*If the post be rectangular* the weight is found by this rule: Multiply the area of the cross-section of the post by the square of the thickness, both in inches, and by the value of B, Table II. Divide the product by the square of the height in feet, and the quotient will be the required weight in pounds.

$$w = \frac{A\, t^2\, B}{h^2} = \frac{b\, t^3\, B}{h^2} \qquad (4.)$$

Example.—What weight will a white pine post sustain safely, whose height is 12 feet, and sides 8 and 12 inches respectively? The area = 8 × 12 = 96 inches; the square of the thickness, 8, = 64. The area by the square of the thickness, 96 × 64, = 6144; and this by 1175, the value of B, for white pine, equals 7,219,200. This, divided by 144, the square of the height = 50,133⅓, the required weight in pounds.

Rule V.—*If the post be square*, the weight is found by this rule: Multiply the value of B, Table II., by the square of the area of the post in inches, and divide the product by the square of the height in feet, and the quotient will be the required weight in pounds.

$$w = \frac{A^2 B}{h^2} = \frac{D^4 B}{h^2}. \qquad (5.)$$

Example.—What weight will a white oak post sustain safely, whose height is 9 feet, and sides each 6 inches? The value of B, set opposite white oak, is 3475; this, by (36 × 36 =) 1296, the square of the area, equals 4,503,600. This product, divided by 81, the square of the height, gives for quotient, 55,600, the required weight in pounds.

298.—To ascertain the size of a post to sustain safely a given weight when the height of the post is at least ten times the least side or diameter.

Rule VI.—*When the post is to be round* or cylindrical, the size may be obtained by this rule: Divide the weight in pounds by 0·589 times the value of B, Table II., and extract the square root of the product; multiply the square root by the height in feet, and the square root of this product will be the diameter of the post in inches.

$$D = \sqrt{h\sqrt{\frac{w}{\cdot 589\,B}}} = \sqrt[4]{\frac{h^2\,w}{\cdot 589\,B}}. \qquad (6.)$$

Example.—What must be the diameter of a locust post, 10 feet high, to sustain safely 40,000 pounds? Here 0·589 times 5,460, the value of B for locust, Table II., equals 3215·9. The weight, 40,000, divided by 3215·9, equals 12·438. The square root of this, 3·5268, multiplied by 10, the height, equals 35·268, and the square root of this is 5·9386 or $5\frac{15}{16}$ inches, the required diameter of the post.

Rule VII.—*If the post is to be rectangular*, the size may be obtained by this rule: Multiply the square of the height in feet by the weight in pounds, and divide the product by the value of B, Table II. Now, if the breadth is known, divide the quotient by the breadth in inches, and the cube root of this quotient will be the thickness in inches. But if the thickness is known, and the breadth desired, divide, instead, by the cube of the thickness in inches, and the quotient will be the breadth in inches.

$$t = \sqrt[3]{\frac{h^2 w}{B b}} \qquad (7.)$$

$$b = \frac{h^2 w}{B t^3} \qquad (8.)$$

Example.—What thickness must a hemlock post have, whose breadth is 4 inches and height 12 feet, to sustain safely 1,000 pounds? The square of the height equals 144; this, by 1,000, the weight, equals 144,000. This, divided by 1,320, the value of B for hemlock, Table II., equals 109·091. This, divided by 4, the breadth, equals 27·273, and the cube root of this is 3·01, a trifle over 3 inches, and this is the thickness required.

Another Example.—What breadth must a spruce post have, whose thickness is 4 inches and height 10 feet, to sustain safely 10,000 pounds? The square of the height, 100, by 10,000, the weight, equals 1,000,000. This, divided by 2095, the value of B, Table II., for spruce, equals 477·09; and this, divided by 64, the cube of the thickness, equals 7·45, nearly 7½ inches, the breadth required.

Rule VIII.—*If the post is to be square,* the size may be obtained by this rule. Divide the weight in pounds by the value of B, Table II., and multiply the square root of the product by the height in feet, and the square root of this product will be the dimension of a side of the post in inches.

$$t = \sqrt{h\sqrt{\frac{w}{B}}} = \sqrt[4]{\frac{h^2 w}{B}} \qquad (9.)$$

Example.—What dimension must the side of a square post

have, whose height is 15 feet, the post being of Georgia pine, to sustain safely 50,000 pounds? The weight 50,000, divided by 4830, the value of *B*, Table II., for Georgia pine, equals 10·352. The square root of this, 3·2175, multiplied by 15, the height, equals 48·362, and the square root of this is 6·9472, nearly 7 inches, the size of a side of the required post.

299.—A square post is not the stiffest that can be made from a given amount of material. The stiffest rectangular post is that whose sides are in proportion as 6 is to 10. When this proportion is desired it may be obtained by the following rule.

Rule IX.—Divide six-tenths of the weight in pounds by the value of *B*, Table II., and extract the square root of the quotient; multiply the square root by the height in feet, and then the square root of this product will be the thickness in inches. The breadth is equal to the thickness divided by 0·6.

$$t = \sqrt{h\sqrt{\frac{0{\cdot}6\,w}{B}}} = \sqrt[4]{\frac{0{\cdot}6\,h^2\,w}{B}} \qquad (10.)$$

$$b = \frac{t}{0{\cdot}6} \qquad (11.)$$

Example.—What must be the breadth and thickness of a white pine post, 10 feet high, to sustain safely 25,000 pounds. Here $\frac{6}{10}$ of 25,000, the weight, divided by 1175, the value of *B*, Table II., for white pine, equals 12·766. The square root of this, 3·5729, multiplied by 10, the height, equals 35·729, and the square root of this is 5·977, nearly 6 inches, the thickness required. Th's, divided by 0·6, equals 10, equals the breadth in inches required.

300.—The sides of a post may be obtained in any desirable proportion by Rule IX., simply by changing the decimal 0·6 to such decimal as will be in proportion to unity as one side is to be to the other. For example, if it be desired to have the sides in proportion as 10 is to 9, then 0·9 is the required decimal; if as 10 is to 8, then 0·8 is the decimal; if as 10 is to 7, then 0·7 is the decimal to be used in place of 0·6 in the rule. And generally let b equal the broad side and t the narrow side, or let these letters represent respectively the numbers that the sides are to be in proportion to; then, where x equals the decimal sought, $b : t :: 1 : x = \frac{t}{b} =$ the required decimal, or *fraction*. For a fraction may be used in place of the decimal, where it would be more convenient, as is the case when the sides are desired to be in proportion as 3 to 2. Here $3 : 2 :: 1 : x = \frac{2}{3}$. This fraction should be used in the rule in place of the decimal 0·6—rather than its equivalent decimal; simply because the decimal contains many figures, and therefore would not be convenient. The decimal equivalent to $\frac{2}{3}$ is 0·666666 +.

Third Case.

301.—To ascertain what weight may be sustained safely by the resistance of a given area of surface, when the weight tends to split off the part pressed against by causing one surface to slide on the other, in case of fracture.

Rule X.—Multiply the area of the surface by the value of H, in Table I., and the product will be the weight required in pounds.

$$A\,H = w. \qquad (12.)$$

Example.—The foot of a rafter is framed into the end of its tie-beam, so that the uncut substance of the tie-beam is 15 inches long from the end of the tie-beam to the joint of the rafter; the tie-beam is of white pine, and is six inches thick; what amount of horizontal thrust will this end of the tie-beam sustain, without danger of having the end of the tie-beam split off? Here the area of surface that sustains the pressure is 6 by 15 inches, equal to 90 inches. This, multiplied by 160, the value of H, set opposite to white pine, Table I., gives a product of 14,400, and this is the required weight in pounds.

302.—To ascertain the area of surface that is required to sustain a given weight safely, when the weight tends to split off the part pressed against, by causing one surface to slide on the other, in case of fracture.

Rule XI.—Divide the given weight in pounds by the value of H, Table I., and the quotient will be the required area in inches.

$$A = \frac{w}{H}. \qquad (13.)$$

Example.—The load on a rafter causes a horizontal thrust at its foot of 40,000 pounds, tending to split off the end of the tie-beam, what must be the length of the tie-beam beyond the line, where the foot of the rafter is framed into it, the tie-beam being of Georgia pine, and nine inches thick? The weight, or horizontal thrust, 40,000, divided by 170, the value of H, Table I., set opposite Georgia pine, gives a quotient of 235·3. This, the area of surface in inches, divided by 9, the breadth of the surface strained, (equal to the thickness of the tie-beam,) the quotient, 26·1, is the length in inches from the end of the tie-beam to the rafter joint, say 26 inches.

303.—A knowledge of this kind of resistance of materials is useful, also, in ascertaining the length of framed tenons, so as to prevent the pin, or key, with which they are fastened from tearing out; and, also, in cases where tie-beams, or other timber under a tensile strain, are spliced, this rule gives the length of the joggle on each end of the splice.

Fourth Case.

304.—To ascertain what weight a post may be loaded with, so as not to crush the surface against which it presses.

Rule XII.—Multiply the area of the post in inches by the value of P, Table I., and the product is the weight required in pounds,

$$w = A\,P. \qquad (14.)$$

Example.—A post, 8 by 10 inches, stands upon a white pine girder; the area equals $8 \times 10 = 80$ inches. This, by 300, the value of P, Table I., set opposite white pine, the product, 24,000, is the required weight in pounds.

305.—To ascertain what area a post must have in order to prevent the post, loaded with a given weight, from crushing the surface against which it presses.

Rule XIII.—Divide the given weight in pounds by the value of P, Table I., and the quotient will be the area required in inches.

$$A = \frac{w}{P}. \qquad (15.)$$

Example.—A post standing on a Georgia pine girder is loaded with 100,000 pounds, what must be its area? The weight, 100,000, divided by 850, the value of P, Table I., set opposite Georgia pine, the quotient, 117·65, is the required area in inches. The post may be 10 by 11¾, or 10×12 inches, or, if square, each side will be 10·84 inches, or 12¼ inches diameter, if round.

II.—RESISTANCE TO TENSION.

306.—The resistance of materials to the force of stretching, as exemplified in the case of a rope from which a weight is suspended, is termed the *resistance to tension.* In fibrous materials, this force will be different in the same specimen, in accordance with the *direction* in which the force acts, whether in the direction of the length of the fibres, or at right angles to the direction of their length. It has been found that, in hard woods, the resistance in the former direction is about 8 to 10 times what it is in the latter; and in soft woods, straight grained, such as white pine, the resistance is from 16 to 20 times. A knowledge of the resistance in the direction of the fibres is the most useful in practice.

307.—In the following table, the experiments recorded were

to test this resistance in such woods, also iron, as are in common use. Each specimen was turned cylindrical, and about 2 inches diameter, and then the middle part for 10 inches in length reduced to ⅝ths of an inch diameter, at the middle of the reduced part, and gradually increased toward each end, where it was about an eighth of an inch larger at its junction with the enlarged end.

TABLE III.—TENSION.

Kind of Material.	Specific Gravity.	Weight producing fracture per square inch.	Value of T in the Rules.
		Pounds.	
Hickory, ...	0·751	20,700	3,450
Locust, ...	·794	15,900	2,650
Maple, ...	·694	15,400	2,567
White pine, ...	·458	14,200	2,367
Ash, ...	·608	11,700	1,950
Oak, ...	·728	10,000	1,667
White oak, ...	·774	17,000	2,833
Georgia pine, ...	·650	17,000	2,833
Cast iron, from	7·200	17,000	2,833
Cast iron, to	7·600	30,000	5,000
American wrought iron, 2 in. diam.,	7·000	30,000	5,000
Do. do. ⅝ and ¼ do.,	7·800	55,000	9,166
Do. do. wire, No. 3, . .		102,000	17,000
Do. do. do. No. 0, . .		74,500	12,416
Do. annealed do. No. 0, . .		53,000	8,833

308.—The value of T in the rules, as contained in the last column of the above table, is one-sixth of the weight producing fracture per square inch of cross section, as recorded in the preceding column. This proportion of the breaking weight is deemed the proper one, from the fact that in practice, through defects in workmanship, the attachments *may* be so made as to cause the strain to act along *one side* of the piece, instead of through its axis; and as in this case it has been found, that fracture will be produced with ⅓ of the strain that can be sustained through the axis, therefore one half of this reduced strain, (equal to ⅙ of the strain through the axis), is the largest that a due regard to security will permit to be used. And in some cases it may be deemed advisable to load the material with even a still smaller strain.

309.—To ascertain the weight or pressure that may be safely applied to a beam as a tensile strain.

Rule XIV.—Multiply the area of the cross section of the beam in inches by the value of *T*, Table III., and the product will be the required weight in pounds. The cross section here intended is that taken at the smallest part of the beam or rod. A beam is usually cut with mortices in framing; the area will probably be smallest at the severest cutting: the area used in the rule must be only of the uncut fibres.

$$A\ T = w. \qquad (16.)$$

Example.—A tie-beam of a roof truss is of white pine, and 6 × 10 inches; the cutting for the foot of the rafter reduces the uncut area to 40 inches: what amount of horizontal thrust from the foot of the rafter will this tie-beam safely sustain? Here 40 times 2,367, the value of *T*, equals 94,680, the required weight in pounds.

310.—To ascertain the sectional area of a beam or rod that will sustain a given weight safely, when applied as a tensile strain.

Rule XV.—Divide the given weight in pounds by the value of *T*, Table III., and the quotient will be the area required in inches: this will be the smallest area of uncut fibres. If the piece is to be cut for mortices, or for any other purpose, then a sufficient addition is to be made to the result found by the rule.

$$\frac{w}{T} = A. \qquad (17.)$$

Example.—A rafter produces a thrust horizontally of 80,000 pounds; the tie-beam is to be of oak: what must the area of the cross section of the tie-beam be, in order to sustain the rafter safely? The given weight, 80,000, divided by 1,667, the value of *T*, the quotient, 48, is the area of uncut fibres. This should have usually one-half of its amount added to it as an allowance for cutting; therefore $48 + 24 = 72$. The tie-beam may be 6×12 inches.

311.—In these rules nothing has been said of an allowance for the weight of the beam itself, in cases where the beam is placed vertically, and the weight suspended from the end. Usually, in timber, this is small in comparison with the load, and may be neglected; although in very long timbers, and where accuracy is decidedly essential, it may form a part of the rule.

312.—Taking the effect of the weight of the beam into account, the relation existing between the weights and parts of the beam, may be stated algebraically thus:—

$$A\,T = w + k \qquad (18.)$$

Where A equals the area of the section of uncut fibres, T equals the tabular constant in the rules, which is equal to the load that may be safely trusted on a rod of like material with the beam and one inch square; w equals the load, and k equals the weight of the beam. Now, the weight of the beam equals its cubical contents in feet, multiplied by the weight of a cubic foot of like material; and a cubic foot of the material equals 62·5 times its specific gravity, while the cubical contents of the beam in feet equals $\frac{R}{144}\,l$, where R equals the sectional area in inches, and l equals the length in feet. Hence—

$$k = 62{\cdot}5\,f\,\frac{R}{144}\,l, \qquad (19.)$$

where f equals the specific gravity. It will be observed that A equals the sectional area of the uncut fibres, while R equals the sectional area of the entire beam; and, where the excess of R over A may be stated as a proportional part of A, or when $A + n\,A = R$, (n being a decimal in proportion to unity, as the excess of R over A is to A,) or

$$\frac{R - A}{A} = n.$$ Then, [from (18.)]—

$$A\,T = w + k.$$
$$= w + 62{\cdot}5\,\frac{A + n\,A}{144}\,f\,l$$
$$= w + \frac{62{\cdot}5}{144}\,A\,(1 + n)\,f\,l,$$
$$= w + 0{\cdot}434\,(n + 1)\,A\,f\,l;$$
$$\text{and } w = A\,T - 0{\cdot}434\,(n + 1)\,A\,f\,l,$$
$$w = A\,(T - 0{\cdot}434\,(n + 1)\,f\,l;) \qquad (20.)$$
$$\text{and } A = \frac{w}{T - 0{\cdot}434\,(n + 1)\,f\,l.} \qquad (21.)$$

When A is found, to find R, we have from

$$R = A + n\,A,$$
$$R = A\,(n + 1.) \qquad (22.)$$

As the excess of R over A decreases, n also decreases, until finally, when $R = A$, n becomes zero. For—

$$n = \frac{R - A}{A},$$

and when $A = R$, then

$$n = \frac{R - R}{R} = \frac{0}{R} = 0.$$

When n equals zero, it disappears from the rules, and (20) becomes

$$w = A\,(T - 0{\cdot}434\,f\,l) \qquad (23.)$$

and (21) becomes

$$A = \frac{w}{T - 0{\cdot}434\,f\,l,} \qquad (24.)$$

and (22) becomes

$$R = A, \qquad (25.)$$

313.—These rules stated in words at length are as follows:—

To ascertain the weight that may be suspended safely from a vertical beam, when the weight of the beam itself is to be taken into account, and when a portion of the fibres are cut in framing.

Rule XVI.—From the sectional area of the beam, deduct

the sectional area of uncut fibres, and divide the remainder by the sectional area of the uncut fibres, and to the quotient add unity; multiply this sum by 0·434 times the specific gravity of the beam, and by its length in feet; substract this product from the value of T, Table III., and the remainder, multiplied by the sectional area of the uncut fibres, will be the required weight in pounds.

$$w = A\,(T - 0{\cdot}434\,(n + 1) f\,l.) \qquad (20.)$$

Example.—A white pine beam, set vertically, 5 × 9 inches and 30 feet long, is so cut by mortices as to have remaining only 5 × 6 inches sectional area of uncut fibres: what weight will such a beam sustain safely, as a tensile strain? The uncut fibres, 5 × 6 = 30, deducted from the area of the beam, 5 × 9 = 45, there remains 15. This remainder, divided by 30, the area of the uncut fibres, the quotient is 0·5. This added to unity, the sum is 1·5. This, by 0·434 times 0·458, the specific gravity set opposite white pine in Table III., and by 30, the length of the beam in feet, the product is 8·95. This product, deducted from 2,367, the value of T set opposite white pine in Table III., the remainder is 2,358·05. This remainder multiplied by 30, the sectional area of the uncut fibres, the product, 70,741·5, is the required weight in pounds.

314.—When the beam is uncut for mortices or other purposes, the former part of the rule is not needed; the weight will then be found by the following rule.

Rule XVII.—Deduct 0·434 times the specific gravity of the beam, multiplied by its length in feet, from the value of T, Table III.; the remainder, multiplied by the sectional area of the beam in inches, will be the required weight in pounds.

$$w = A\,(T - 0{\cdot}434\,f\,l). \qquad (23.)$$

Example.—A Georgia pine beam, set vertically, is 25 feet long and 7 × 9 inches in sectional area: what weight will it sustain safely, as a tensile strain? By the rule, 0·434 times

0·65, the specific gravity of Georgia pine, as in Table III., multiplied by 25, the length in feet, the product is 7·05. This product, deducted from 2,833, the value of T, Table III., set opposite Georgia pine, and the remainder, 2,825·95, multiplied by 63, the sectional area, the product, 178,034·85, is the required weight in pounds.

315.—To ascertain the sectional area of a vertical beam that will safely sustain a given tensile strain, where the weight of the beam itself is to be considered.

Rule XVIII.—Where the beam is cut for mortices or other purposes, let the relative proportion of the uncut fibres to those that are cut, be as 1 is to n, (n being a decimal to be fixed on at pleasure.) Then to the value of n add unity, and multiplying the sum by 0·434 times the specific gravity in Table III., and by the length in feet. Deduct this product from the value of T, Table III., divide the given weight in pounds by this remainder, and the quotient will be the area of the uncut fibres in inches. Add unity to the value of n, as above, and multiply the sum by the area of the uncut fibres; the product will be the required area of the beam in inches.

$$A = \frac{w}{T - 0·434\,(n + 1) f\, l}, \qquad (21.)$$

$$R = A\,(n + 1), \qquad (22.)$$

Example.—A vertical beam of white oak, 30 feet long, is required to resist effectually a tensile strain of 80,000 pounds: what must be its sectional area? The relative proportion of the uncut fibres is to be to those that are cut as 1 is to 0·4. To 0·4, the value of n, add 1·; the sum is 1·4. This, by 0·434 times ·774, the specific gravity of white oak in Table III., and by 30, the length, the product is 14·109. This, deducted from 2,833, the value of T for white oak in Table III., the remainder is 2,818·891. The given weight, 80,000, divided by 2,818·891, the remainder, as above, the quotient, 28·38, is the area of the uncut fibres. This multiplied by the sum of 0·4 and 1·, (or the value of n and unity = 1·4,) the product, 39·732, is the required area of the beam in inches.

316.—When the fibres are uncut, then their sectional area equals the area of the beam, and may be found by the following rule.

Rule XIX.—Deduct 0·434 times the specific gravity in Table III., multiplied by the length in feet, from the value of T, Table III., and divide the weight in pounds by the remainder. The quotient will be the required area in inches.

$$A = \frac{w}{T - 0{\cdot}434\, f\, l}. \qquad (24.)$$

Example.—A vertical beam of locust, 15 feet long, fibres all uncut, is required to sustain a tensile strain equal to 25,000 pounds: what must be its area? Here 0·434 times ·794, the specific gravity for locust in Table III., multiplied by 15, the length in feet, is 5·17. This, from 2,650, the value of T for locust, Table III., the remainder is 2,644·83. The given weight, 25,000, divided by 2,644·83, the remainder, as above, the quotient, 9·45, will be the required area in inches.

III.—RESISTANCE TO CROSS-STRAINS.

317.—A load placed upon a beam, laid horizontal or inclined, tends to bend it, and if the weight be proportionally large, to break it. The power in the material that resists this bending or breaking, is termed the *resistance to cross-strains*, or transverse strains. While in posts or struts the material is compressed or shortened, and in ties and suspending-pieces it is extended or lengthened; in beams subjected to cross-strains the material is both compressed and extended. (See *Art.* 254.) When the beam is bent, the fibres on the concave side are compressed, while those on the convex side are extended. The line where these two portions of the beam meet—that is, the portion compressed and the portion extended—the horizontal line of juncture, is termed the *neutral* line or plane. It is so called because at this line the fibres are neither compressed nor extended, and hence are under no strain whatever. The location of this line or plane is not far from the middle of the depth of the beam, when the strain is not sufficient to injure the elasticity of the material; but it removes towards the concave or convex side of the beam as the strain is increased, until, at the period of rupture, its distance from the top of the beam is in proportion to its distance from the bottom of the beam as the tensile strength of the material is to its compressive strength.

318.—In order that the strength of a beam be injured as little as possible by the cutting required in framing, all mortices should be located at or near the middle of the depth. There is a prevalent idea among some, who are aware that the upper fibres of a beam are compressed when subject to cross-strains, that it is not injurious to cut these top fibres, provided that the cutting be for the insertion of another piece of timber—as in the case of *gaining* the ends of beams into the side of a girder. They suppose that the piece filled in will as effectually resist the compression as the part removed would have done, had it not been taken out. Now, besides the effect of shrinkage, which of itself is quite sufficient to prevent the proper resistance to the strain, there is the mechanical difficulty of fitting the joints perfectly throughout; and, also, a great loss in the power of resistance, as the material is so much less capable of resistance when pressed at right angles to the direction of the fibres, than when directly with them, as the results of the experiments in the tables show.

319.—In treating upon the resistance to cross-strains, the subject is divided naturally into two parts, viz. *stiffness* and *strength:* the former being the power to resist deflection or bending, and the latter the resistance to rupture.

320.—*Resistance to Deflection.* When a load is placed upon a beam supported at each end, the beam bends more or less; the distance that the beam descends under the operation of the load, measured at the middle of its length, is termed its *deflection*. In an investigation of the laws of deflection it has been demonstrated, and experiments have confirmed it, that while the elasticity of the material remains uninjured by the pressure, or is injured in but a small degree, the amount of deflection is directly in proportion to the weight producing it, and is as the cube of the length; and, in pieces of rectangular sections, it is inversely proportional to the breadth and the cube of the depth: or, inversely proportional to the fourth power of the *side* of a square beam or of the *diameter* of a cylindrical one. Or, when l equals the length between the supports, w the weight or pressure, b the breadth, d the depth, and p the deflection; then—

$$\frac{l^3 w}{b d^3 p} = E, \qquad (26.)$$

equals a *constant quantity* for beams of all dimensions made from a *like material*. Also,

$$\frac{l^3 w}{s^4 p} = E, \qquad (27.)$$

where s equals a side of a square beam; and

$$\frac{l^3 w}{0{\cdot}589\, D^4 p} = E, \qquad (28.)$$

where D equals the diameter of a cylindrical beam. The constant here is less than in the case of the square and of the rectangular beams. It is as much less as the circular beam is less stiff than a square beam whose side is equal to the diameter of the cylindrical one. The constant, E, is therefore multiplied by the decimal 0·589.

321.—It may be observed that E in (26) and (27) would be equal to w, in case the dimensions of the beam and the amount of deflection were each made equal to unity; and in (28) equal to w divided by 0·589. That is, when in (26) the length is 1, the breadth 1, and the depth 1, then E would be equal to the weight that would depress the beam from its original line equal to 1. Thus—

$$E = \frac{l^3 w}{b d^3 p} = \frac{1^3 \times w}{1 \times 1^3 \times 1} = w,$$

the dimensions all taken in inches except the length, and this taken in feet. This is an extreme state of the case, for in most kinds of material this amount of depression would exceed the limits of elasticity; and hence the rule would here fail to give the correct relation among the dimensions and pressure. For the law of deflection as above stated, (the deflection being equal for equal weights,) is true only while the depressions are small in comparison with the length. Nothing useful is, therefore, derived from this position of the question, except to give an idea of the nature of the quantity represented by the constant, E; it being in reality a measure of the stiffness of the kind of material used in comparing one material with another. Whatever may be the dimensions of the beam, E, calculated by (26,) will always be the same quantity for the same material; but when various materials are used, E will vary according to the flexibility or stiffness of each particular material. For example, E will be much greater for iron than for wood; and again, among the various kinds of wood, it will be larger for the stiff woods than for those that are flexible.

322.—If the amount of deflection that would be proper in beams used in framing generally, (such as floor beams, girders and rafters,) were agreed upon, the rules would be shortened, and the labor of calculation abridged. Tredgold proposed to make the deflection in proportion to the length of the beam, and the amount at the rate of one-fortieth of an inch (= 0·025 inch) for every foot of length. He was undoubtedly right in the manner and probably so in the rate; yet, as this is a matter of opinion, it were better perhaps to leave the rate of deflection open for the decision of those who use the rules, and then it may be varied to suit the peculiarities of each case that may arise. Any deflection within the limits of the elasticity of the material, may be given to beams used for some purposes, while others require to be restricted to that amount of deflection that shall not be perceptible to a casual observer. Let n represent, in the decimal of an inch, the rate of deflection per foot of the length of the beam; then the product of n, multiplied by the number of feet contained in the length of the beam, will equal the total deflection, $= n\,l$. Now, if $n\,l$ be substituted for p in the formulas, (26,) (27) and (28,) they will be rendered more available for general use. For example, let this substitution be made in (26,) and there results—

$$E = \frac{l^3 w}{b\,d^3\,n\,l} = \frac{l^2 w}{b\,d^3\,n}, \qquad (29.)$$

where l is in feet, and b, d and n in inches; and for (27)—

$$E = \frac{l^3 w}{s^4\,n\,l} = \frac{l^2 w}{s^4\,n}; \qquad (30.)$$

also for (28)—

$$E = \frac{l^3 w}{0{\cdot}589\,D^4\,n\,l} = \frac{l^2 w}{0{\cdot}589\,D^4\,n}, \qquad (31.)$$

where the notation is as before, with also s and D in inches. In these formulas, w represents the weight in pounds *concentrated* at the middle of the length of the beam. If the weight, instead thereof, is *equally distributed* over the length of the beam, then, since $\frac{5}{8}$ of it concentrated at the middle will deflect a beam to the same depth that the whole does when equally distributed, (*Art.* 281,) therefore—

$$E = \frac{\frac{5}{8}\,w\,l^2}{b\,d^3\,n}, \qquad (32.)$$

$$E = \frac{\frac{5}{8}\,w\,l^2}{s^4\,n}, \qquad (33.)$$

$$E = \frac{\frac{5}{8}\,w\,l^2}{0{\cdot}589\,D^4\,n}, \qquad (34.)$$

where w equals the whole of the equally distributed load. Again, if the load is borne by more beams than one, laid parallel to each other—as, for example, a series or tier of floor beams—and the load is equally distributed over the supported surface or floor; then, if f represents the number of pounds of the load contained on each square foot of the floor, or the pounds' weight per foot superficial, and c represents the distance in feet between each two beams, or rather the distance from their centres, and l the length of the beam in feet, in the clear, between the supports at the ends; then $c\,l$ will equal the area of surface supported by one of the beams, and $f\,c\,l$ will represent the load borne by it, equally distributed over its length. Now, if this representation of the load be substituted for w in (32,) (33) and (34) there results—

$$E = \frac{\frac{5}{8} f\,c\,l\,l^3}{b\,d^3\,n} = \frac{\frac{5}{8} f\,c\,l^4}{b\,d^3\,n}, \qquad (35.)$$

$$E = \frac{\frac{5}{8} f\,c\,l\,l^3}{s^4\,n} = \frac{\frac{5}{8} f\,c\,l^4}{s^4\,n}, \qquad (36.)$$

$$E = \frac{\frac{5}{8} f\,c\,l\,l^3}{0{\cdot}589\,D^4\,n} = \frac{\frac{5}{8} f\,c\,l^4}{0{\cdot}589\,D^4\,n} \qquad (37.)$$

Practical Rules and Examples.

323.—To ascertain the weight, placed upon the middle of a beam, that will cause a given deflection.

Rule XX.—Multiply the area of the cross-section of the beam by the square of the depth and by the rate of the deflection, all in inches; multiply the product by the value of E, Table II., and divide this product by the square of the length in feet, and the quotient will be the weight in pounds required.

Example.—What weight can be supported upon the middle of a Georgia pine girder, ten feet long, eight inches broad, and ten inches deep, the deflection limited to three-tenths of an inch, or at the rate of 0·03 of an inch per foot of the length: Here the area equals $8 \times 10 = 80$; the square of the depth equals $10 \times 10 = 100$: $80 \times 100 = 8{,}000$; this by 0·03, the rate of deflection, the product is 240; and this by 2970, the value of E for Georgia pine, Table II., equals 712,800. This product, divided by 100, the square of the length, the quotient, 7,128, is the weight required in pounds.

Rule XXI.—Where the beam is square the weight may be found by the preceding rule or by this:—Multiply the square of the area of the cross-section by the rate of deflection, both in inches, and the product by the value of *E*, Table II., and divide this product by the square of the length in feet, and the quotient will be the weight required in pounds.

Example.—What weight placed on the middle of a spruce beam will deflect it seven-tenths of an inch, the beam being 20 feet long, 6 inches broad, and 6 inches deep? Here the area is $6 \times 6 = 36$, and its square is $36 \times 36 = 1296$; the rate of deflection is equal to the total deflection divided by the length, $= \frac{0 \cdot 7}{20} = 0 \cdot 035$; therefore, $1296 \times 0 \cdot 035 = 45 \cdot 36$, and this by 1550, the value of *E* for spruce, Table II., equals 70,308. This, divided by 400, the square of the length, equals 175·77, the required weight in pounds.

Rule XXII.—When the beam is round find the weight by this rule:—Multiply the square of the diameter of the cross-section by the square of the diameter, and the product by the rate of deflection, all in inches, and this product by 0·589 times the value of *E*, Table II. This last product, divided by the square of the length in feet, will give the required weight in pounds.

Example.—What weight on the middle of a round white pine beam will cause a deflection of 0·028 of an inch per foot, the beam being 10 inches diameter and 20 feet long? The square of the diameter equals $10 \times 10 = 100$; $100 \times 100 = 10{,}000$; this by the rate, 0·028, $= 280$, and this by $0 \cdot 589 \times 1750$, the value of *E*, Table II., for white pine, equals 288,610. This last product, divided by 400, the square of the length, equals 721·5, the required weight in pounds.

324.—To ascertain the weight that will produce a given de

flection, when the weight is equally distributed over the length of the beam.

Rule XXIII.—The rules for this are the same as the three preceding rules, with this modification, viz., instead of the square of the length, divide by five-eighths of the square of the length.

325.—In a series or tier of beams, to ascertain the weight per foot, equally distributed over the supported surface, that will cause a given deflection in the beam.

Rule XXIV.—The rules for this are the same as Rules XX., XXI., and XXII., with this modification, viz., instead of the square of the length, divide by the product of the distance apart in feet between each two beams, (measured from the centres of their breadths,) multiplied by five-eighths of the cube of the length, and the quotient will be the required weight in pounds that may be placed upon each superficial foot of the floor or other surface supported by the beams. In this and all the other rules, the weight of the material composing the beams, floor, and other parts of the constructions is understood to be a part of the load. Therefore from the ascertained weight deduct the weight of the framing, floor, plastering, or other parts of the construction, and the remainder will be the neat load required.

Example.—In a tier of white pine beams, 4 × 12 inches, 20 feet long, placed 16 inches or 1⅓ feet from centres, what weight per foot superficial may be equally distributed over the floor covering said beams—the rate of deflection to be not more than 0·025 of an inch per foot of the length of the beams. Proceeding by Rule XX. as above modified, the area of the cross-section, 4 × 12, equals 48; this by 144, the square of the depth, equals 6912, and this by 0·025, the rate of deflection, equals 172·8. Then this product, multiplied by 1750, the value of *E*, Table II., for white pine, equals 302,400. The distance between the centres of the beams is 1⅓ feet, the cube of the length is 8,000, and $\frac{1}{3}$ by $\frac{5}{2}$ of 8,000 equals $6{,}666\frac{2}{3}$. The above 302,400, divided by $6{,}666\frac{2}{3}$, the quotient, 45·36, equals the required weight in pounds per foot superficial. The weight of beams, floor plank, cross-furring, and plastering occurring under every square foot of the surface of the floor, is now to be ascertained. Of the timber in every 16 inches by 12 inches, there occurs 4 × 12 inches, one foot long; this equals one-third of a cubic foot. Now, by proportion, if 16 inches in width contains $\frac{1}{3}$ of a cubic foot, what will 12 inches in width contain? $\frac{\frac{1}{3} \times 12}{16} = \frac{12}{3 \times 16} = \frac{3}{12} = \frac{1}{4}$ of a cubic foot. The floor plank (Georgia pine) is 12 × 12 inches, and $1\frac{1}{4}$ inches thick, equal to $\frac{1\frac{1}{4}}{12}$ of a cubic foot, equals $\frac{\frac{5}{4}}{12}$, equals $\frac{5}{48}$. Of the furring strips, 1 × 2 inches, placed 12 inches from centres, there will occur one of a foot long in every superficial foot. Now, since in a cubic foot there is 144 rods, one inch square and one foot long, therefore, this furring strip, 1 × 2 × 12 inches, equals $\frac{2}{144} = \frac{1}{72}$ of a cubic foot. The weight of the timber and furring strips, being of white pine, may be estimated together: $\frac{1}{4} + \frac{1}{72} = \frac{18}{72} + \frac{1}{72} = \frac{19}{72}$ of a cubic foot. White pine varies from 23 to 30 pounds. If it be taken at 30 pounds, the beam and furring together will weigh $30 \times \frac{19}{72}$ pounds, equals 7·92 pounds. Georgia pine may be taken at 50 pounds per cubic foot;* the weight of the floor plank, then, is $50 \times \frac{5}{48} = 5{\cdot}21$ pounds. A superficial foot of lath and plastering will weigh about 10 lbs. Thus, the white pine, 7·92, Georgia pine, 5·21, and the plastering, 10, together equal 23·13 pounds; this from 45·36, as before ascertained, leaves 22·23, say $22\frac{1}{4}$ pounds, the neat weight per foot superficial that may be equally distributed over the floor as its load.

* To get the weight of wood or any other material, multiply its specific gravity by 62·5. For Specific Gravities see Tables I., II., and III. and the Appendix for Weight of Materials.

326.—To ascertain the weight when the beam is laid not horizontal, but *inclined.*

Rule XXV.—In each of the foregoing rules, multiply the result there obtained by the length in feet, and divide the product by the horizontal distance between the supports in feet, and the quotient will be the required weight in pounds.

The foregoing Rules, stated algebraically, are placed in the following table :—

TABLE IV.—STIFFNESS OF BEAMS; WEIGHT.

When the beam is laid	When the weight is	When the beam is		
		Rectangular.	Square.	Round.
Horizontal	Concentrated at middle, *w*, in pounds, equals	(38.) $\frac{E n b d^3}{l^2}$	(39.) $\frac{E n s^4}{l^2}$	(40.) $\frac{\cdot589\, E n D^4}{l^2}$
Horizontal	Equally distributed, *w*, in pounds, equals	(41.) $\frac{E n b d^3}{\frac{5}{8} l^2}$	(42.) $\frac{E n s^4}{\frac{5}{8} l^2}$	(43.) $\frac{\cdot9424\, E n D^4}{l^2}$
Horizontal	By the foot superficial, *f*, in pounds, equals	(44.) $\frac{E n b d^3}{\frac{5}{8} c l^3}$	(45.) $\frac{E n s^4}{\frac{5}{8} c l^3}$	(46.) $\frac{\cdot9424\, E n D^4}{c l^3}$
Inclining	Concentrated at middle, *w*, in pounds, equals	(47.) $\frac{E n b d^3}{l h}$	(48.) $\frac{E n s^4}{l h}$	(49.) $\frac{\cdot589\, E n D^4}{l h}$
Inclining	Equally distributed, *w*, in pounds, equals	(50.) $\frac{E n b d^3}{\frac{5}{8} l h}$	(51.) $\frac{E n s^4}{\frac{5}{8} l h}$	(52.) $\frac{\cdot9424\, E n D^4}{l h}$
Inclining	By the foot superficial, *f*, in pounds, equals	(53.) $\frac{E n b d^3}{\frac{5}{8} c l^2 h}$	(54.) $\frac{E n s^4}{\frac{5}{8} c l^2 h}$	(55.) $\frac{\cdot9424\, E n D^4}{c l^2 h}$

In the above table, *b* equals the breadth, and *d* equals the depth of cross-section of beam; *s* equals the breadth of a side of a square beam, and *D* equals the diameter of a round beam; *n* equals the rate of deflection per foot of the length;

D, s, b, d and n, all in inches; l equals the length, c equals the distance between two parallel beams measured from the centres of their breadth; h equals the horizontal distance between the supports of an inclined beam; l, c and h in feet; w equals the weight in pounds on the beam; f equals the weight upon each superficial foot of a floor or roof supported by two or more beams laid parallel and at equal distances apart; E is a constant, the value of which is found in Table II.; r is any decimal, chosen at pleasure, in proportion to unity, as b is to d, from which proportion b equals $d\,r$.

327.—To ascertain the dimensions of the cross-section of a beam to support the required weight with a given deflection.

Rule XXVI.—Preliminary. *When the weight is concentrated at the middle of the length.* Multiply the weight in pounds by the square of the length in feet, and divide the product by the product of the rate of deflection multiplied by the value of E, Table II., and the quotient equals a quantity which may be represented by M—referred to in succeeding rules.

$$\frac{w\,l^2}{E\,n} = M. \qquad (56.)$$

Rule XXVII.—Preliminary. *When the weight is equally distributed over the length.* Multiply five-eighths of the weight in pounds by the square of the length in feet, and divide the product by the rate of deflection multiplied by the value of E, Table II., and the quotient equals a quantity which may be represented by N—referred to in succeeding rules.

$$\frac{\frac{5}{8}\,w\,l^2}{E\,n} = N. \qquad (57.)$$

Rule XXVIII.—Preliminary. *When the weight is given per foot superficial and supported by two or more beams.* Multiply the distance apart between two of the beams, (measured from the centres of their breadth,) by the cube of the length, both in feet, and multiply the product by five-eighths of the weight per foot superficial; divide this product by the product of the rate of deflection, multiplied by the value of E Table II., and the quotient equals a quantity which may be represented by U—referred to in succeeding rules.

$$\frac{\frac{5}{8} f c l^3}{E n} = U. \qquad (58.)$$

Rule XXIX.—Preliminary. *When the beam is laid not horizontal, but inclining.* In Rules XXVI. and XXVII., instead of the *square of the length* multiply by the length, and by the *horizontal distance* between the supports, in feet. And in Rule XXVIII., instead of the *cube of the length*, multiply by the *square of the length*, and by the *horizontal distance* between the supports, in feet.

From (56)

$$\frac{w l h}{E n} = M_{,}. \qquad (59.)$$

From (57)

$$\frac{\frac{5}{8} w l h}{E n} = N_{,}. \qquad (60.)$$

From (58)

$$\frac{\frac{5}{8} f c l^2 h}{E n} = U_{,}. \qquad (61.)$$

Rule XXX.—*When the beam is rectangular to find the dimensions of the cross-section.* Divide the quantity represented by M, N or U, (in preceding preliminary rules,) by the breadth in inches, and the cube root of the quotient will equal the required depth in inches. Or, divide the quantity represented by M, N or U, by the cube of the depth in inches, and the quotient will equal the required breadth in inches. Or, again, if it be desired to have the breadth and depth in proportion, as r is to unity, (where r equals any required decimal.) divide the quantity represented by M, N or U, by the value of r, and extract the square root of the quotient: and the square root extracted the second time, will equal the depth in inches. Multiply the depth thus found by the value of r, and the product will equal the breadth in inches.

Example.—To find the depth. A beam of spruce, laid on supports with a clear bearing of 20 feet, is required to support a load of 1674 pounds at the middle, and the deflection not to exceed 0·05 of an inch per foot; what must be the depth when the breadth is 5 inches. By Rule XXVI. for load at middle: the product of 1674, the weight, by 400, the square of the length, equals 669,600. The product of 0·05, the rate of deflection, multiplied by 1550, the value of *E*, from Table II., set opposite spruce, is 77·5. The aforesaid product, 669,600, divided by 77·5, equals 8640, the value of *M*. Then by Rule XXX., 8640, the value of *M*, divided by 5, the breadth, the quotient is 1728, and 12, the cube root of this, found in the table of the Appendix, equals the required depth in inches.

Example.—To find the breadth. Suppose that in the last example it were required to have the depth 13 inches; in that case what must be the breadth? The value of *M*, 8640, as just found, divided by 2197, the cube of the depth, equals 3·9326, the required breadth—nearly 4 inches.

Example.—To find both breadth and depth, and in a certain proportion. Suppose, in the above example, that neither the breadth nor the depth are given, but that they are desired to be in proportion as 0·5 is to 1·0. Now, having ascertained the value of *M*, by Rule XXVI., to be 8640, as above, then, by Rule XXX., 8640, divided by 0·5, the ratio, gives for quotient 17,280. The square root of this (by the table in the Appendix,) is 131·45, and the square root of this square root is 11·465, the required depth. The breadth equals 11·465 × 0·5, which equals 5·7325. The depth and breadth may be 11½ by 5¾ inches. In cases where the load is *equally distributed* over the length of the beam, the process is precisely the same as set forth in the three preceding examples, except that *five-eighths* of the weight is to be used in place of the *whole* weight; and hence it would be a useless repetition to give examples to illustrate such cases.

Example.—When the weight is per foot superficial to find the depth. A floor is to be constructed to support 500 pounds on every superficial foot of its surface. The beams to be ot white pine, 16 feet long in the clear of the supports or walls, placed 16 inches apart, from centres, to be 4 inches thick, and the *amount* of deflection not objectionable provided it be within the limits of elasticity. Proceeding by Rule XXVIII., the product of 1⅓ feet, (equal to 16 inches,) multiplied by 4096, the cube of the length, equals 5461⅓. This, multiplied by 312·5, (equal to ⅝ of the weight,) equals 1,706,666. The largest rate of deflection within the limits of the elasticity of white pine is 0·1022, as per Table II. This, multiplied by 1750, the value of *E* for white pine, Table II., equals 178·85. The former product, 1,706,666, divided by the latter, 178·85, equals 9,542·5, the value of *U*. Now, by Rule XXX., this value of *U*, 9,542·5, divided by 4, the breadth, equals 2385·6, the cube root of which, 13·362, is the required depth—nearly 13⅜ inches.

Example.—To find the breadth. Suppose, in the last example, that the depth is known but not the breadth, and that the depth is to be 13 inches. Having found the value of *U*, as before, to be 9542·5, then by Rule XXX., dividing 9542·5, the value of *U*, by 2197, the cube of the depth, gives a quotient of 4·3434 and this equals the breadth—nearly 4⅜ inches.

Example.—To find the depth and breadth in a given proportion. Suppose, in the above example, that the breadth and depth are both unknown, and that it is desired to have them in proportion as 0·7 is to 1·0. Having found the value of *U*, as before, to be 9542·5, then by Rule XXX., dividing 9542·5, the value of *U*, by 0·7, the quotient is 13,632, the square root of which is 116·75, and the square root of this is 10·805, the depth in inches. Then 10·805, multiplied by 0·7, the product, 7·5635, is the breadth in inches. The size may be 7$\frac{9}{16}$ by 10$\frac{13}{16}$ inches.

328.—*Example.—In the case of inclined beams to find the*

depth. A beam of white pine, 10 feet long in the clear of the bearing, and laid at such an inclination that the *horizontal* distance between the supports is 9 feet, is required to support 12,000 pounds at the centre of its length, with the greatest allowable deflection within the limits of elasticity; what must be its depth when its breadth is fixed at 6 inches? By reference to Table II. it is seen that the greatest value of *n*, within the limits of elasticity, is 0·1022. By Rule XXVI., for concentrated load, and Rule XXIX., for inclined beams, 12,000, the weight, multiplied by 10, the length, and by 9, the horizontal distance, equals 1,080,000. The product of 0·1022, the greatest rate of deflection, by 1750, the value of *E*, Table II., for white pine, equals 178·85. Dividing 1,080,000 by 178·85, the quotient is 6038·58, the value of *M*. Now, by Rule XXX., for rectangular beams, 6038·58, the value of *M*, divided by 6, the breadth, the quotient is 1006·43. The cube root of this, 10·02, a trifle over 10 inches, is the depth required.

Example.—In case of inclined beams to find the breadth. In the last example suppose the depth fixed at 12 inches; then by Rule XXX., 6038·58, the value of *M*, as above found, divided by 1728, the cube of the depth, equals 3·4945, or nearly 3½ inches—the breadth required.

Example.—Again, *in case the breadth and depth are to be in a certain proportion;* as, for example, as 0·4 is to unity. Then by Rule XXX., 6038·58, the value of *M*, found as above, divided by 0·4, equals 15,096·45, the square root of which is 122·87, and the square root of this square root is 11·0843, a trifle over 11 inches—the depth required. Again, 11 multiplied by the decimal 0·4, (as above,) equals 4·4, a little over 4⅜ inches—the breadth required.

In the three preceding examples, the weight is understood to be concentrated at the middle. If, however, the weight had been equally distributed, the same process would have been used to obtain the dimensions of the cross-section, with only one exception: viz. $\frac{5}{8}$ of the weight instead of the whole weight would have been used. (See Rule XXVII.)

Example.—In case of inclined beams; the weight per foot superficial, and borne by two or more beams. A tier of spruce beams, laid with a clear bearing of 10 feet, and at 20 inches apart from centres, and laid so inclining that the horizontal distance between bearings is 8 feet, are required to sustain 40 pounds per superficial foot, with a deflection not to exceed 0·02 inch per foot of the length; what must be the depth when the breadth is 3 inches? Proceeding by Rule XXIX. for inclined beams, and by Rule XXVIII., $1\frac{2}{3}$, (= 20 inches,) the distance from centres, multiplied by 100, the square of the length, and by 8, the horizontal distance between bearings, equals $1{,}333\frac{1}{3}$; this, by $\frac{5}{8} \times 40$, five-eighths of the weight, equals $33{,}333\frac{1}{3}$. This, divided by $0{\cdot}02 \times 1550$, the rate of deflection, by the value of *E*, Table II., for spruce, equal to 31, equals 1075·27, the value of *U*. Now by Rule XXX. for rectangular beams, 1075·27, divided by 3, the breadth, equals 358·42, the cube root of which, 7·1, is the required depth in inches.

Example.—The same as the preceding; but to find the breadth, when the depth is fixed at 8 *inches.* By Rule XXX., 1075·27, the value of *U*, divided by 512, the cube of the depth, equals 2·1—the breadth required in inches.

Example.—The same as the next but one preceding; but to find the breadth and depth in the proportion of 0·3 *to* 1·0, *or* 3 *to* 10. By Rule XXX., 1075·27, the value of *U*, divided by 0·3, the value of *r*, equals 3584·23. The square root of this is 59·869, and the square root of this square root is 7·737—the depth required in inches. This 7·737, multiplied by 0·3, the value of *r*, equals 2·3211—the required breadth in inches. The dimensions may, therefore, be $2\frac{5}{16}$ by $7\frac{3}{4}$ inches.

Rule XXXI.—*When the beam is square to find the side.* Extract the square root of the quantity represented by *M*, *N* or U, in preliminary Rules XXVI., XXVII. and XXVIII., and the square root of this square root will equal the side required.

Example.—A beam of chestnut, having a clear bearing of 8 feet, is required to sustain at the middle a load of 1500 pounds; what must be the size of its sides in order that the deflection shall not exceed 0·03 inch per foot of its length? By Rule XXVI., 1500, the load, multiplied by 64, the square of the length, equals 96,000. This product divided by 0·03 times 2330, the value of E, Table II., for chestnut, gives a quotient of 1373·4, the quantity represented by M. Now by Rule XXXI., the square root of 1373·4 is 37·05, and the square root of this is 6·087. The beam must, therefore, be 6 inches square. In this example, had the load, instead of being concentrated at the middle, been equally distributed over its length, the side would have been equal to the side just found, multiplied by the fourth root of $\frac{5}{8}$ or of 0·625, equal to 6·087 × 0·889 = 5·4 inches. (See Rules XXVII. and XXXI.)

Example.—In the case where the weight is per foot superficial and borne by two or more beams. A floor, the beams of which are of oak, and placed 20 inches or $1\frac{2}{3}$ feet apart from centres, and which have a clear bearing of 20 feet, is required to sustain 200 pounds per superficial foot, the deflection not to exceed 0·025 inch per foot of the length, and the beam to be square. By Rule XXVIII., $1\frac{2}{3}$, the distance from centres, multiplied by 8000, the cube of the length, equals $13,333\frac{1}{3}$; and this by 125, (being $\frac{5}{8}$ of 200 pounds,) equals $1,666,666\frac{2}{3}$. Dividing this by 0·025 times 2520, the value of E, Table II., for oak, the quotient is 26,455—a number represented by U. Now by Rule XXXI., the square root of this number is 162·65. and the square root of this square root is 12·753—the required side. The beam may be $12\frac{3}{4}$ inches square.

Example.—Inclined square beams, load at middle. A bar of cast-iron, 6 feet long in the clear of bearings, and laid inclining so that the horizontal distance between the bearings is 5 feet, is required to sustain at the middle 3000 pounds, and the deflection not to exceed 0·01 inch per foot of its length; what must be the size of its sides?

By Rule XXVI. for load at middle, modified by Rule XXIX. for inclined beams; 3000, the weight, multiplied by 6, the length, and by 5, the horizontal distance between bearings, equals 90,000. The rate of deflection, 0·01, by 30,500, the value of E, Table II., for cast-iron, equals 305; and 9000 divided by 305, equals 295·082, the value of M. Now by Rule XXXI. for square beams, the square root of 295·082 is 17·18, the square root of which is 4·145—the size of the side required; a trifle over 4⅛, the bar may, therefore, be 4⅛ inches square.

Example.—Same as preceding, but the weight equally distributed. By Rule XXVII. ⅝ of the weight is to be used instead of the weight; therefore 295·082, the value of M, as above, multiplied by ⅝, will equal 184·426, the value of N. By Rule XXXI. the square root of 184·426 is 13·58, the square root of which is 3·685—the size of the side required; equal to nearly $3\frac{11}{16}$ inches square.

Example.—Same as preceding case, but the weight per foot superficial, and sustained by 2 or more bars, placed 2 feet from centres, the load being 250 pounds per foot superficial. By Rule XXVIII., modified by Rule XXIX., the distance from centres, 2, multiplied by 36, the square of the length, and by 5, the horizontal distance, equals 360. This by 156·25, five-eighths of the weight, equals 56,250. The rate of deflection, 0·01, by 30,500, the value of E, Table II., for cast-iron, equals 305. The above 56,250, divided by 305, equals 184·426, the value of U. Now by Rule XXXI. the square root of 184·426, the value of U, is 13·58, the square root of which is 3·685—the size of the side required. It will be observed that this result is precisely like that in the last example. This is as it should be, for each beam has to sustain the weight on 2 × 6

= 12 superficial feet, equal to 12 × 250, equal 3000 pounds; and all the other conditions are parallel.

Rule XXXII.—*When the beam is round to find the diameter.* Divide the value of *M*, *N* or *U*, found by Rules XXVI., XXVII. or XXVIII., by the decimal 0·589, and extract the square root: and the square root of this square root will be the diameter required.

Example.—In the case of a concentrated load at middle. A round bar of American iron, of 5 feet clear bearing, is required to sustain 800 pounds at the middle, with a deflection not to exceed 0·02 inch per foot; what must be its diameter? By Rule XXVI. for load at middle, 800, the weight, multiplied by 25, the square of the length, equals 20,000. The rate of deflection, 0·02, by 51,400, the value of *E*, Table II., for American wrought iron, equals 1028. The above 20,000, divided by 1028, equals 19·4552, the value of *M*. Now, by Rule XXXII., 19·4552, the value of *M*, divided by 0·589 equals 33·03, the square root of which is 5·747, and the square root of this is 2·397, nearly 2·4, the diameter required in inches, equal to 2⅜ large.

Example.—Same case as the preceding, but the load equally distributed. By Rule XXVII., five-eighths of the weight is to be used instead of the whole weight; therefore the above 33·03, multiplied by ⅝, equals 20·64375, the square root of which is 4·544, and the square root of this square root is 2·132, the diameter required in inches, 2⅛ inches large.

Example.—When the weight is per foot superficial, and sustained by two or more bars or beams. The conditions being the same as in the preceding examples, but the weight, 100 pounds per foot, is to be sustained on a series of round rods, placed 18 inches apart from centres, equal 1·5 feet. By Rule XXVIII., for weight per foot superficial, 1·5, the distance from centres, multiplied by 125, the cube of the length, and by 62·5, five-eighths of the weight, equals 11,718·75. This divided by 1028, the product of the rate of deflection by the value of *E*, as found in the preceding example, equals 11·4, the value of *U*. Now by Rule XXVII., 11·4, the value of *U*, divided by 0·589, equals 19·42, the square root of which is 4·407, and the square root of this square root is 2·099, the diameter required—very nearly $2\frac{1}{10}$ inches.

Example.—When the beam is round and laid inclining, the weight concentrated at the middle. A round beam of white pine, 20 feet long between bearings, and laid inclining so that the horizontal distance between bearings is 18 feet, is required to support 1250 pounds at the middle, with a deflection not to exceed 0·05 inch per foot; what must be its diameter? By Rule XXVI. for load at middle, modified by Rule XXIX. for inclined beams, 1250, the weight, multiplied by 20, the length, and by 18, the horizontal distance, equals 450,000. The rate of deflection, 0·05, multiplied by 1750, the value of *E*, Table II., for white pine, equals 87·5. The above 450,000 divided by 87·5, equals 5142·86, the value of *M*. Now by Rule XXXII. for round beams, 5142·86, the value of *M*, divided by 0·589, equals 8731·5, the square root of which is 93·44, and the square root of this square root is 9·667, the diameter required—equal to $9\frac{2}{3}$ inches.

Example.—Same as in preceding example, but the weight equally distributed. By Rule XXVII., five-eighths of the weight is to be used instead of the whole weight, therefore 8731·5, the result in the last example just previous to taking the square root, multiplied by $\frac{5}{8}$, equals 5457·2, the square root of which is 73·87, and the square root of this square root is 8·59, the diameter required—nearly $8\frac{5}{8}$ inches.

Example.—Same as in the next but one preceding example, but the weight per foot superficial, and supported by two or more beams. A series of round hemlock poles or beams, 10 feet long clear bearing, laid inclining so as that the horizontal distance between the supports equals 7 feet, and laid 2 feet and 6 inches apart from centres, are required to support 20 pounds per superficial foot without regard to the amount of deflection, provided that the elasticity of the material be not injured; what must be their diameter? By Rule XXVIII. for weight per foot superficial, modified by Rule XXIX. for inclined beams, 2·5, the distance from centres, multiplied by 100, the square of the length, and by 7, the horizontal distance between bearings, and by five-eighths of the weight, 12·5, equals 21,875. The greatest value of *n*, Table II., for hemlock, 0·08794, multiplied by 1240, the value of *E*, Table II., for hemlock, equals 109·0456. The above 21,875, divided by 109·0456, equals 200·6, the value of *U*. Now by Rule XXXII., the above 200·6, divided by 0·589, equals 340·6, the square root of which is 18·46, and the square root of this square root is 4·296, the diameter required—equal to $4\frac{5}{16}$ inches nearly.

329.—The greater the depth of a beam in proportion to the thickness, the greater the strength. But when the difference between the depth and the breadth is great, the beam must be stayed, (as at *Fig.* 228,) to prevent its falling over and breaking sideways. Their shrinking is another objection to deep beams; but where these evils can be remedied, the advantage of increasing the depth is considerable. The following rule is, *to find the strongest form for a beam out of a given quantity of timber.*

Rule.—Multiply the length in feet by the decimal, 0·6, and divide the given area in inches by the product; and the square of the quotient will give the depth in inches.

Example.—What is the strongest form for a beam whose given area of section is 48 inches, and length of bearing 20 feet? The length in feet, 20, multiplied by the decimal, 0·6, gives 12; the given area in inches, 48, divided by 12, gives a quotient of 4, the square of which is 16—this is the depth in inches; and the breadth must be 3 inches. A beam 16 inches by 3 would bear twice as much as a square beam of the same area of section; which shows how important it is to make beams deep and thin. In many old buildings, and even in new ones, in country places, the very reverse of this has been practised; the principal beams being oftener laid on the broad side than on the narrower one.

The foregoing rules, stated algebraically, are placed in the following table.

TABLE V.—STIFFNESS OF BEAMS; DIMENSIONS.

When the beam is laid	When the weight is	Rectangular.			Square.	Round.
		Value of depth.	Value of breadth.	When $b = dr$, value of d.	Value of a side.	Value of the diameter
Horizontal	Concentrated at middle	(62.) $\sqrt[3]{\frac{w l^3}{E n b}}$	(63.) $\frac{w l^3}{E n d^3}$	(64.) $\sqrt[4]{\frac{w l^3}{E n r}}$	(65.) $\sqrt[4]{\frac{w l^3}{E n}}$	(66.) $\sqrt[4]{\frac{w l^3}{\cdot 589\, E n}}$
	Equally distributed	(67.) $\sqrt[3]{\frac{\frac{5}{8} w l^3}{E n b}}$	(68.) $\frac{\frac{5}{8} w l^3}{E n d^3}$	(69.) $\sqrt[4]{\frac{\frac{5}{8} w l^3}{E n r}}$	(70.) $\sqrt[4]{\frac{\frac{5}{8} w l^3}{E n}}$	(71.) $\sqrt[4]{\frac{w l^3}{\cdot 9424\, E n}}$
	By the foot superficial	(72.) $\sqrt[3]{\frac{\frac{5}{8} f c l^4}{E n b}}$	(73.) $\frac{\frac{5}{8} f c l^4}{E n d^3}$	(74.) $\sqrt[4]{\frac{\frac{5}{8} f c l^4}{E n r}}$	(75.) $\sqrt[4]{\frac{\frac{5}{8} f c l^4}{E n}}$	(76.) $\sqrt[4]{\frac{f c l^4}{\cdot 9424\, E n}}$
Inclining	Concentrated at middle	(77.) $\sqrt[3]{\frac{w l h}{E n b}}$	(78.) $\frac{w l h}{E n d^3}$	(79.) $\sqrt[4]{\frac{w l h}{E n r}}$	(80.) $\sqrt[4]{\frac{w l h}{E n}}$	(81.) $\sqrt[4]{\frac{w l h}{\cdot 589\, E n}}$
	Equally distributed	(82.) $\sqrt[3]{\frac{\frac{5}{8} w l h}{E n b}}$	(83.) $\frac{\frac{5}{8} w l h}{E n d^3}$	(84.) $\sqrt[4]{\frac{\frac{5}{8} w l h}{E n r}}$	(85.) $\sqrt[4]{\frac{\frac{5}{8} w l h}{E n}}$	(86.) $\sqrt[4]{\frac{w l h}{\cdot 9424\, E n}}$
	By the foot superficial	(87.) $\sqrt[3]{\frac{\frac{5}{8} f c l^3 h}{E n b}}$	(88.) $\frac{\frac{5}{8} f c l^3 h}{E n d^3}$	(89.) $\sqrt[4]{\frac{\frac{5}{8} f c l^3 h}{E n r}}$	(90.) $\sqrt[4]{\frac{f c l^3 h}{E n}}$	(91.) $\sqrt[4]{\frac{f c l^3 h}{\cdot 9424\, E n}}$

In the above table, b equals the breadth, and d the depth of cross-section of beam; n equals the rate of deflection per foot of the length; b, d and n, all in inches. Also, l equals the length, c the distance between two parallel beams measured from the centres of their breadth, and h equals the horizontal distance between the supports of an inclined beam; l, c and h, all in feet. Again, w equals the weight on the beam, f equals the weight upon each superficial foot of a floor or roof, supported by two or more beams laid parallel and at equal distances apart; w and f in pounds. And r is any decimal, chosen at pleasure, in proportion to unity, as b is to d—from which proportion $b = d\,r$. E is a constant the value of which is found in Table II.

330.—*To ascertain the scantling of the stiffest beam that can be cut from a cylinder.* Let $d\ a\ c\ b$, (*Fig.* 223,) be the section, and e the centre, of a given cylinder. Draw the diameter, $a\ b$; upon a and b, with the radius of the section, describe the arcs, $d\ e$ and $e\ c$; join d and a, a and c, c and b, and b and d; then the rectangle, $d\ a\ c\ b$, will be a section of the beam required.

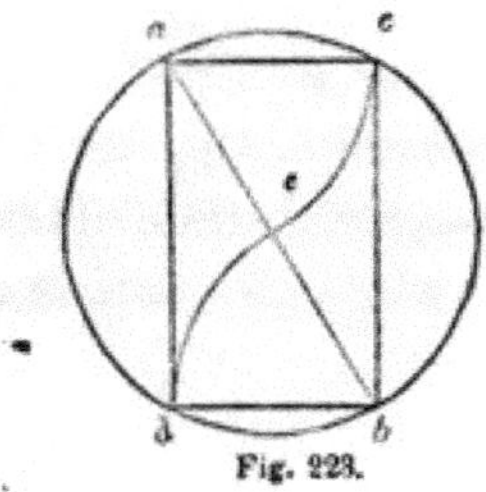

Fig. 223.

331.—*Resistance to Rupture.*—The resistance to *deflection* having been treated of in the preceding articles, it now remains to speak of the other branch of resistance to cross strains, namely, the resistance to *rupture.* When a beam is laid horizontally and supported at each end, its strength to resist a cross strain, caused by a weight or vertical pressure at the middle of its length, is directly as the breadth and square of the depth and inversely as the length. If the beam is square, or the depth equal to the breadth, then the strength is directly as the cube of a side of the beam and inversely as the length, and if the beam is round the strength is directly as the cube of the diameter and inversely as the length.

When the weight is concentrated at any point in the length, the strength of the beam is directly as the length, breadth, and square of the depth, and inversely as the product of the two parts into which the length is divided by the point at which the weight is located.

When the beam is laid not horizontal but inclining, the strength is the same as in each case above stated, and also in proportion, inversely as the cosine of the angle of inclination with the horizon, or, which is the same thing, directly as the length and inversely as the horizontal distance between the points of support.

When the weight is equally diffused over the length of a beam, it will sustain just twice the weight that could be sustained at the middle of its length.

A beam secured at one end only, will sustain at the other end just one-quarter of the weight that could be sustained at its middle were the beam supported at each end.

These relations between the strain and the strength exist in all materials. For any particular kind of material,

$$\frac{w\,l}{b\,d^2} = S; \qquad (92.)$$

S, representing a constant quantity for all materials of like strength. The superior strength of one kind of material over another is ascertained by experiment; the value of S being ascertained by a substitution of the dimensions of the piece tried for the symbols in the above formula. Having thus obtained the value of S, the formula, by proper inversion, becomes useful in ascertaining the dimensions of a beam that will require a certain weight to break it; or to ascertain the weight that will be required to break a certain beam. It will be observed in the preceding formula, that if each of the dimensions of the beam equal unity, then $w = S$. Hence, S is equal to the weight required to break a beam one inch square and one foot long. The values of S, for various materials, have been ascertained from experiment, and are here recorded :—

TABLE VI.—STRENGTH.

Materials.	Value of S.	Number of Experiments.
Green plate-glass	178	4
Spruce	345	5
Hemlock	363	7
Soft white pine	390	9
Hard white pine	449	1
Ohio yellow pine	454	2
Chestnut	503	2
Georgia pine	510	7
Oak	574	2
Locust	742	2
Cast-iron (from 1550 to 2280)	1926	29

The specimens broken were of various dimensions, from one foot long to three feet, and from one inch square to one by three inches. The cast-iron specimens were of the various kinds of iron used in this country in the mechanic arts. S may be taken at 2,000 for a good quality of cast-iron. It is usual in determining the dimensions of a beam to suppose it capable of sustaining safely one-third of the breaking weight, and yet Tredgold asserts that one-fifth of the breaking weight will in time injure the beam so as to give it a permanent set or bend, and Hodgkinson says that cast-iron is injured by any weight however small, or, in other words, that it has no elastic power. However this may be, experience has proved cast-iron quite reliable in sustaining safely immense weights for a long period. Practice has shown that beams will sustain safely from one-third to one-sixth of their breaking weight. If the load is laid on quietly, and is to remain where laid, at rest, beams may be trusted with one-third of their breaking weight, but if the load is moveable, or subject to vibration, one-quarter, one-fifth, or even, in some cases, one-sixth is quite a sufficient proportion of the breaking load.

332.—The dimensions of beams should be ascertained only by means of the rules for the *stiffness* of materials, (*Arts.* 320, 323, *et seq.*,) as these rules show more accurately the amount of pressure the material is capable of sustaining without injury. Yet owing to the fact that the rules for the *strength* of materials are somewhat shorter, they are more frequently used than those for the *stiffness* of materials. In order that the *proportion* of the breaking weight may be adjusted to suit circumstances it is well to introduce into the formula a symbol to represent it. The proportion represented by the symbol may then be varied at discretion. Let this symbol be a; a decimal in proportion to unity as the safe load is to the breaking load, then $S\,a$ will equal the safe load. Hence,

$$w = \frac{S\,a\,b\,d^2}{l} \qquad (93.)$$

for a safe load at middle on a horizontal beam supported at both ends; and

$$w = \frac{2\,S\,a\,b\,d^2}{l} \qquad (94.)$$

for a safe load equally diffused over the length of the beam; and

$$f = \frac{2\,S\,a\,b\,d^2}{c\,l^2} \qquad (95.)$$

for the load, per superficial foot, that can be sustained safely upon a floor supported by two or more beams, c being the distance in feet from centres between each two beams, and f the load in pounds per superficial foot of the floor. Generally, in (93,) (94.) and (95,) w equals the load in pounds; S, a constant, the value of which is found in Table VI.; a a decimal, in proportion to unity as the safe load is to the breaking load; l the length in feet between the bearings; and b and d the breadth and depth in inches.

TO FIND THE WEIGHT.

333.—The formulas for ascertaining the weight in the several cases are arranged in the following table, where c, f, w, S, a, l, b and d represent as above; and also s equals a side of a square beam; D equals the diameter of a cylindrical beam; m and n equal respectively the two parts into which the length is divided by the point at which the weight is located; and h equals the horizontal distance between the supports of an inclined beam.

TABLE VII.—STRENGTH OF BEAMS; SAFE WEIGHT.

When the beam is laid	When the weight is	When the beam is		
		Rectangular.	Square.	Round.
Horizontal	Concentrated at middle, w, in pounds, equals	(96.) $\frac{S\,a\,b\,d^2}{l}$	(97.) $\frac{S\,a\,s^3}{l}$	(98.) $\frac{\cdot 589\,D^3\,S\,a}{l}$
	Equally distributed, w, in pounds, equals	(99.) $\frac{2\,S\,a\,b\,d^2}{l}$	(100.) $\frac{2\,S\,a\,s^3}{l}$	(101.) $\frac{1\cdot 178\,D^3\,S\,a}{l}$
	By the foot superficial, f, in pounds, equals	(102.) $\frac{2\,S\,a\,b\,d^2}{c\,l^2}$	(103.) $\frac{2\,S\,a\,s^3}{c\,l^2}$	(104.) $\frac{1\cdot 178\,D^3\,S\,a}{c\,l^2}$
	Concentrated at any point in the length, w, in pounds, equals	(105.) $\frac{S\,a\,b\,d^2\,l}{4\,m\,n}$	(106.) $\frac{S\,a\,l\,s^3}{4\,m\,n}$	(107.) $\frac{\cdot 147\,D^3\,S\,a\,l}{m\,n}$
Inclining	Concentrated at middle, w, in pounds, equals	(108.) $\frac{S\,a\,b\,d^2}{h}$	(109.) $\frac{S\,a\,s^3}{h}$	(110.) $\frac{\cdot 589\,D^3\,S\,a}{h}$
	Equally distributed, w, in pounds, equals	(111.) $\frac{2\,S\,a\,b\,d^2}{h}$	(112.) $\frac{2\,S\,a\,s^3}{h}$	(113.) $\frac{1\cdot 178\,D^3\,S\,a}{h}$
	By the foot superficial, f, in pounds, equals	(114.) $\frac{2\,S\,a\,b\,d^2}{c\,h\,l}$	(115.) $\frac{2\,S\,a\,s^3}{c\,h\,l}$	(116.) $\frac{1\cdot 178\,D^3\,S\,a}{c\,h\,l}$
	Concentrated at any point in the length, w, in pounds, equals	(117.) $\frac{S\,a\,b\,d^2\,l^2}{4\,h\,m\,n}$	(118.) $\frac{S\,a\,l^2\,s^3}{4\,h\,m\,n}$	(119.) $\frac{\cdot 147\,D^3\,S\,a\,l^2}{h\,m\,n}$

Practical Rules and Examples.

Rule XXXIII.—To find the weight that may be *supported safely* at the *middle* of a beam laid *horizontally*. Multiply the value of *S*, Table VI., by a decimal that is in proportion to unity as the safe weight is to the breaking weight, and divide the product by the length in feet. Then, if the beam is rectangular, multiply this quotient by the breadth and by the square of the depth, and the product will be the required weight in pounds; or, if the beam is square, multiply the said quotient, instead, by the cube of a side of the beam and the product will be the required weight in pounds; but, if the beam is round, multiply the aforesaid quotient, instead, by ·589 times the cube of the diameter, and the product will be the required weight in pounds.

Example.—What weight will a *rectangular* white pine beam, 20 feet long, and 3 by 10 inches, sustain safely at the middle, the portion of the breaking weight allowable being 0·3? By the above rule, 390, the value of *S* for white pine, Table VI., multiplied by 0·3, the decimal referred to, equals 117, and this divided by 20, the length, the quotient is 5·85. Now the beam being rectangular, this quotient multiplied by 3 and by 100, the breadth and the square of the depth, the product, 1755, is the desired weight in pounds.

Example.—If the above beam had been *square*, and 6 by 6 inches, then the quotient, 5·85, multiplied by 216, the cube of 6, a side, the product, 1263·6, is the weight required in pounds.

Example.—If the above beam had been *round*, and 6 inches diameter, then the above quotient, 5·85, multiplied by ·589 times 216, the cube of the diameter, the product, 744·26, would be the required weight in pounds.

Rule XXXIV.—To find the weight that may be *supported safely* when *equally distributed* over the length of a beam, laid *horizontally*. Multiply the result obtained, by Rule

XXXIII., by 2, and the product will be the required weight in pounds.

Example.—In the example, under Rule XXXIII., the safe weight at middle of *rectangular* beam is found to be 1755 pounds. This multiplied by 2, the product, 3510, is the weight the beam will bear safely if equally distributed over its length.

Example.—So in the case of the *square* beam, 2527·2 pounds is the weight, equally distributed, that may be safely sustained.

Example.—And for the *round* beam 1488·52 is the required weight.

Rule XXXV.—To ascertain the weight *per superficial foot* that may be *safely sustained* on a floor resting on two or more beams laid *horizontally* and parallel. Multiply twice the value of *S*, Table VI., by the decimal that is in proportion to unity, as the safe weight is to the breaking weight, and divide the product by the square of the length, in feet, multiplied by the distance apart, in feet, between the beams measured from their centres. Now, if the beams are *rectangular*, multiply this quotient by the breadth and by the square of the depth, both in inches, and the product will be the required weight in pounds; or if the beams are *square*, multiply said quotient, instead, by the cube of a side of a beam and the product will be the required weight in pounds. But if the beams are *round*, multiply the aforesaid quotient, instead, by ·589 times the cube of the diameter, and the product will be the weight required in pounds.

Example.—What weight may be safely sustained on each foot superficial of a floor resting on spruce beams, 10 feet long, 3 by 9 inches, placed 16 inches, or 1⅓ feet, from centres: the portion of the breaking weight allowable being 0·25? By the Rule, 690, twice the value of spruce, Table VI., multiplied by 0·25, the decimal aforesaid, equals 172·5. This product divided by 100, the square of the length, multiplied by 1⅓, the distance from centres, equals 1·294. Now this quotient multiplied by 3, the breadth, and by 81, the square of the depth, the product 314·44 is the required weight in pounds.

Had these beams been *square*, and 6 by 6 inches, the required weight would be 279·5 pounds.

Or, if *round*, and 6 inches diameter, 164·63 pounds.

Rule XXXVI.—To ascertain the weight that may be *sustained safely* on a beam when *concentrated* at *any point* of its length. Multiply the value of *S*, Table VI., by the decimal in proportion to unity, as the safe weight is to the breaking weight, and by the length in feet, and divide the product by four times the product of the two parts, in feet, into which the length is divided, by the point at which the weight is concentrated. Then, if the beam is *rectangular*, multiply this quotient by the breadth and by the cube of the depth, both in inches, and the product will be the required weight in pounds. Or, if the beam is *square*, multiply the said quotient, instead, by the cube of a side of the beam, and the product will be the required weight in pounds. But if the beam is *round*, multiply the aforesaid quotient by ·589 times the cube of the diameter, and the product will be the weight required.

Example.—What weight may be safely supported on a Georgia pine beam, 5 by 12 inches, and 20 feet long; the weight placed at 5 feet from one end, and the proportion of the breaking weight allowable being 0·2? By the rule, 510, the value of *S* for Georgia pine, Table VI., multiplied by 0·2, the decimal referred to, equals 102; this by 20, the length, equals 2040; this divided by 300, (= 4 × 5 × 15,) or 4 times the product of the two parts into which the length is divided by the point at which the weight is located, equals 6·8. The beam being *rectangular*, this quotient multiplied by 5, the breadth, and by 144, the square of the depth, equals 4896, the required weight.

A beam, 8 inches *square*, other conditions being the same as in the preceding case, would sustain safely 3481·6 pounds.

And a *round* beam, 8 inches diameter, will sustain safely, under like conditions, 2050·66 pounds.

Rule XXXVII.—To find the weight that may be *safely sustained* on *inclined beams*. Multiply the result found for horizontal beams in preceding rules, by the length, in feet, and divide the product by the horizontal distance between the supports, in feet, and the quotient will be the required weight.

Example.—What weight may be safely sustained at the middle of an oak beam, 6 × 10 inches, and 10 feet long, (set inclining, so that the horizontal distance between the supports is 8 feet,) the portion of the breaking weight allowable being 0·3? The result for a horizontal beam, by Rule XXXIII., is 10,332 pounds. This, multiplied by 10, the length, and divided by 8, the horizontal distance, equals 12,915 pounds, the required weight.

TO FIND THE DIMENSIONS.

334.—The following table exhibits, algebraically, rules for ascertaining the dimensions of beams required to support given weights; where b equals the breadth, and d the depth of a rectangular beam, in inches; l the length between supports; h the horizontal distance between the supports of an inclined beam, and c the distance apart of two parallel beams, measured from the centres of their breadth, l, h, and c, in feet; w equals the weight on a beam; f the weight on each superficial foot of a floor resting on two or more parallel beams; R equals a load on a beam, and m and n the distances, respectively, at which R is located from the two supports; also P is a weight, and g and k the distances, respectively, at which P is located from the two supports; also $m + n = l = g + k$; w, f, R, and P, all in pounds; m, n, g, and k, in feet. S is a constant, the value of which is found in Table VI.; a is a decimal in proportion to unity as the safe load is to the break ing load; r is a decimal in proportion to unity as b is to d; from which $b = d\,r$; a and r to be chosen at discretion.

TABLE VIII.—STRENGTH

When the beam is laid	When the weight is	Rectangular.	
		Value of depth.	Value of breadth.
Horizontal.	Concentrated at middle	(120.) $\sqrt{\frac{w\,l}{S\,a\,b}}$	(121.) $\frac{w\,l}{S\,a\,d^2}$
	Equally distributed	(125.) $\sqrt{\frac{w\,l}{2\,S\,a\,b}}$	(126.) $\frac{w\,l}{2\,S\,a\,d^2}$
	By the foot superficial	(130.) $\sqrt{\frac{f\,c\,l^2}{2\,S\,a\,b}}$	(131.) $\frac{f\,c\,l^2}{2\,S\,a\,d^2}$
	Concentrated at any point in the length	(135.) $\sqrt{\frac{4\,w\,m\,n}{S\,a\,b\,l}}$	(136.) $\frac{4\,w\,m\,n}{S\,a\,d^2\,l}$
	At two or more points in the length	(140.) $\sqrt{\frac{4\,(R\,m\,n + P\,g\,k + \&c.)}{S\,a\,b\,l}}$	(141.) $\frac{4\,(R\,m\,n + P\,g\,k + \&c.)}{S\,a\,d^2\,l}$
Inclining	Concentrated at middle	(145.) $\sqrt{\frac{w\,h}{S\,a\,b}}$	(146.) $\frac{w\,h}{S\,a\,d^2}$
	Equally distributed	(150.) $\sqrt{\frac{w\,h}{2\,S\,a\,b}}$	(151.) $\frac{w\,h}{2\,S\,a\,d^2}$
	By the foot superficial	(155.) $\sqrt{\frac{f\,c\,h\,l}{2\,S\,a\,b}}$	(156.) $\frac{f\,c\,h\,l}{2\,S\,a\,d^2}$
	Concentrated at any point in the length	(160.) $\sqrt{\frac{4\,h\,w\,m\,n}{S\,a\,b\,l^2}}$	(161.) $\frac{4\,h\,w\,m\,n}{S\,a\,d^2\,l^2}$
	At two or more points in the length	(165.) $\sqrt{\frac{4\,h\,(R\,m\,n + P\,g\,k + \&c.)}{S\,a\,b\,l^2}}$	(166.) $\frac{4\,h\,(R\,m\,n + P\,g\,k + \&c.)}{S\,a\,d^2\,l^2}$

OF BEAMS; DIMENSIONS.

	Square.	Round.
When $b = d\,r$, value of d.	Value of a side.	Value of the diameter.
(122.) $\sqrt[3]{\frac{w\,l}{S\,a\,r}}$	(123.) $\sqrt[3]{\frac{w\,l}{S\,a}}$	(124.) $\sqrt[3]{\frac{w\,l}{\cdot589\,S\,a}}$
(127.) $\sqrt[3]{\frac{w\,l}{2\,S\,a\,r}}$	(128.) $\sqrt[3]{\frac{w\,l}{2\,S\,a}}$	(129.) $\sqrt[3]{\frac{w\,l}{1\cdot178\,S\,a}}$
(132.) $\sqrt[3]{\frac{f\,c\,l^2}{2\,S\,a\,r}}$	(133.) $\sqrt[3]{\frac{f\,c\,l^2}{2\,S\,a}}$	(134.) $\sqrt[3]{\frac{f\,c\,l^2}{1\cdot178\,S\,a}}$
(137.) $\sqrt[3]{\frac{4\,w\,m\,n}{S\,a\,r\,l}}$	(138.) $\sqrt[3]{\frac{4\,w\,m\,n}{S\,a\,l}}$	(139.) $\sqrt[3]{\frac{w\,m\,n}{\cdot147\,S\,a\,l}}$
(142.) $\sqrt[3]{\frac{4\,(R\,m\,n + P\,g\,k + \&c.)}{S\,a\,r\,l}}$	(143.) $\sqrt[3]{\frac{4\,(R\,m\,n + P\,g\,k + \&c.)}{S\,a\,l}}$	(144.) $\sqrt[3]{\frac{(R\,m\,n + P\,g\,k + \&c.)}{\cdot147\,S\,a\,l}}$
(147.) $\sqrt[3]{\frac{w\,h}{S\,a\,r}}$	(148.) $\sqrt[3]{\frac{w\,h}{S\,a}}$	(149.) $\sqrt[3]{\frac{w\,h}{\cdot589\,S\,a}}$
(152.) $\sqrt[3]{\frac{w\,h}{2\,S\,a\,r}}$	(153.) $\sqrt[3]{\frac{w\,h}{2\,S\,a}}$	(154.) $\sqrt[3]{\frac{w\,h}{1\cdot178\,S\,a}}$
(157.) $\sqrt[3]{\frac{f\,c\,h\,l}{2\,S\,a\,r}}$	(158.) $\sqrt[3]{\frac{f\,c\,h\,l}{2\,S\,a}}$	(159.) $\sqrt[3]{\frac{f\,c\,h\,l}{1\cdot178\,S\,a}}$
(162.) $\sqrt[3]{\frac{4\,h\,w\,m\,n}{S\,a\,r\,l^2}}$	(163.) $\sqrt[3]{\frac{4\,h\,w\,m\,n}{S\,a\,l^2}}$	(164.) $\sqrt[3]{\frac{h\,w\,m\,n}{\cdot147\,S\,a\,l^2}}$
(167.) $\sqrt[3]{\frac{4\,h\,(R\,m\,n + P\,g\,k + \&c.)}{S\,a\,r\,l^2}}$	(168.) $\sqrt[3]{\frac{4\,h\,(R\,m\,n + P\,g\,k + \&c.)}{S\,a\,l^2}}$	(169.) $\sqrt[3]{\frac{h\,(R\,m\,n + P\,g\,k + \&c.)}{\cdot147\,S\,a\,l^2}}$

Practical Rules and Examples.

Rule XXXVIII.—Preliminary. *When the weight is concentrated at the middle.* Multiply the weight, in pounds, by the length, in feet, and divide the product by the value of S, Table VI., multiplied by a decimal that is in proportion to unity as the safe weight is to the breaking weight, and the quotient is a quantity which may be represented by J, referred to in succeeding rules.

$$\frac{w\,l}{S\,a} = J \qquad (170.)$$

Rule XXXIX.—Preliminary. *When the weight is equally distributed.* One-half of the quotient obtained by the preceding rule is a quantity which may be represented by K, referred to in succeeding rules.

$$\frac{w\,l}{2\,S\,a} = K \qquad (171.)$$

Rule XL.—Preliminary. *When the weight is per foot superficial.* Multiply the weight per foot superficial, in pounds, by the square of the length, in feet, and by the distance apart from centres between two parallel beams, and divide the product by twice the value of S, Table VI., multiplied by a decimal in proportion to unity as the safe weight is to the breaking weight, and the quotient is a quantity which may be represented by L, referred to in succeeding rules.

$$\frac{f\,c\,l^2}{2\,S\,a} = L \qquad (172.)$$

Rule XLI.—Preliminary. *When the weight is concentrated at any point in the length.* Multiply the distance, in feet, from the loaded point to one support, by the distance, in feet, from the same point to the other support, and by four times the weight in pounds, and divide the product by the value of S, Table VI., multiplied by a decimal in proportion to unity as the safe weight is to the breaking weight, and by the length, in feet; and the quotient is a quantity which may be represented by Q, referred to in the rules.

$$\frac{4\,w\,m\,n}{S\,a\,l} = Q \qquad (173.)$$

Rule XLII.—Preliminary. ***When two or more weights are concentrated at any points in the length of the beams.*** Multiply each weight by each of the two parts, in feet, into which the length is divided by the point at which the weight is located, and divide four times the sum of these products by the value of S, Table VI., multiplied by a decimal in proportion to unity as the safe weight is to the breaking weight, and by the length, in feet, and the quotient is a quantity which may be represented by V, referred to in the rules.

$$\frac{4\,(R\,m\,n + P\,g\,k + \&c.)}{S\,a\,l} = V \qquad (174.)$$

Rule XLIII.—Preliminary. ***When the beam is not laid horizontal, but inclining.*** In the five preceding preliminary rules, multiply the result there obtained by the horizontal distance between the supports, in feet, and divide the product by the length, in feet, and the quotient in each case is to be used for beams when inclined, as referred to in succeeding rules.

TO FIND THE DIMENSIONS.

Rule XLIV.—When the beam is *rectangular.* To find the *depth.* Divide the quantity represented in preceding rules by J, K, L, Q, or V, by the breadth, in inches, and the square root of the quotient will be the depth required in inches.

To find the *breadth.* Divide the quantity represented by J, K, L, Q, or V, by the square of the depth, and the quotient will be the required breadth, in inches.

To find both *breadth* and *depth*, when they are to be in a given proportion. Divide the quantity represented by J, K, L, Q, or V, by a decimal in proportion to unity as the breadth is to be to the depth, and the cube root of the quotient will be the *depth* in inches. Multiply the depth by the aforesaid decimal and the quotient will be the *breadth* in inches.

Example.—A locust beam, 10 feet long in the clear of the supports, is required to sustain safely 3,000 pounds at the middle of its length, the portion of the breaking weight allowable being 0·3; what is the required breadth and depth? Proceeding by the rule for weight concentrated at middle, (Rule XXXVIII.,) 3,000, the weight, by 10, the length, equals 30,000. The value of *S*, Table VI., for locust, is 742: this by 0·3 the decimal, as above, equals 222·6; the 30,000 aforesaid divided by this 222·6 equals 134·77, equals the quantity represented by *J*. Now to find the depth when the breadth is 4 inches, 134·77 divided by 4, the breadth, as above required, the quotient is 33·69, and the square root of this, 5·8, is the required depth in inches. But to find the breadth, when the depth is known, let the depth be 6 inches, then 134·77 divided by 36, the square of the depth, equals 3·74, the breadth required in inches. Again, to find both breadth and depth in a given proportion, say, as 0·6 is to 1·0. Here 134·77 divided by 0·6 equals 224·617, the cube root of which is 6·08, the required depth in inches, and 6·08 by 0·6 equals 3·648, the required breadth in inches.

Thus it is seen, in this example, that a piece of locust timber, 10 feet long, having 3,000 pounds *concentrated* at the middle of its length, as $\frac{3}{10}$ of its breaking load, is required to be 4 by $5\frac{13}{16}$ inches, or $3\frac{3}{4}$ by 6 inches, or $3\frac{5}{8}$ by $6\frac{1}{8}$ inches. If this load were *equally diffused* over the length, the dimensions required would be found to be 4 by 4·1, or 1·87 by 6, or 2·895 by 4·825 inches, in the three cases respectively.

Example.—A tier of chestnut beams, 20 feet long, placed one foot apart from centres, is required to sustain 100 pounds per superficial foot upon the floor laid upon them: this load to be 0·2 of the breaking weight; what is the required dimensions of the cross-section? By Rule XL., the rule for a load per foot superficial, 100 by 20 × 20 and by 1 equals 40,000. Twice 503, the value of S for chestnut, Table VI., by 0·2 equals 201·2. The above 40,000 divided by 201·2 equals 198·8, the value of L. Now if the breadth is known, and is 3 inches, 198·8 divided by 3 equals 66·27, the square root of which is 8·14, the required depth. But if the depth is known and is 9 inches, 198·8 divided by (9 × 9 =) 81 equals 2·454 inches, the required breadth. Again, when the breadth and depth are required in the proportion of 0·25 to 1·0, then 198·8 divided by 0·25 equals 795·2, the cube root of which is 9·265, the required depth in inches, and 9·265 by 0·25 equals 2·316, the required breadth in inches.

Example.—A cast-iron bar, 10 feet long, is required to sustain safely 5,000 pounds placed at 3 feet from one end, and consequently at 7 feet from the other end, the portion of the breaking load allowable being 0·3; what must be the size of the cross section? By Rule XLI., the rule for a concentrated load at any point in the length of the beam, 3 × 7 × 4 × 5000 = 420,000. And 1926, the value of S for cast-iron, Table VI., by 0·3 and by 10 equals 5778. The aforesaid 420,000 divided by this, 5778, equals 72·689, the value of Q. Now if the breadth is fixed at 1·5 then 72·689 divided by 1·5 equals 48·459, the square root of which is 6·96, the required depth in inches. But if the depth is fixed at 6 inches, then 72·689, the value of Q, divided by 36, the square of 6, equals 2·019, the required breadth in inches. Again, if the breadth and depth are required in the proportion 0·2 to 1·0; then Q, 72·689, divided by 0·2, equals 363·445, the cube root of which is 7·136, the required depth in inches; and 7·136 by 0·2 equals 1·427 the required breadth in inches.

Rule XLV.—When the beam is *square* to find the breadth of a side. The cube root of the quantity represented by J, K, L, Q, or V, in preceding rules, is the breadth of the side required

Example.—A Georgia pine beam, 10 feet long, is required to sustain, as 0·3 of the breaking load, a weight of 30,000 pounds equally distributed over its length, and the beam to be square, what must be the breadth of the side of such a beam? By the rule for an equally distributed load, (Rule XXXIX.,) 30,000 × 10 = 300,000, and 510 (the value of *S*, for Georgia pine, Table VI.) × 0·3 = 153. 300,000 divided by 153 equals 1960·784, and one-half of this equals 980·392, the value of *K*. Now the cube root of this is 9·934 inches, or 9$\frac{15}{16}$, the required side. Had the weight been concentrated at the middle 1960·784 would be the value of *J*, and the cube root of this 12·515, or 12½ inches, would be the size of a side of the beam.

Example.—A square oak beam, 20 feet long, is required to sustain, as 0·25 of the breaking strength, three loads, one of 8,000 pounds at 5 feet from one end, one of 7,000 pounds at 14 feet, and one of 5,000 pounds at 8 feet from one end, what must be the breadth of a side of the beam? The value of *S*, for oak, Table VI., is 574. By the rule for this case, (Rule XLII.,) 8000 × 5 × 15 equals 600,000; and 7000 × 14 × 6 equals 588,000; and 5000 × 8 × 12 equals 480,000. The sum of these products is 1,668,000; this by 4 equals 6,672,000. Now 574 × 0·25 × 20 equals 2870, and the 6,672,000 divided by the 2870 equals 2325, the number represented by *V*; the cube root of which is 13·25, the required size of a side of the beam, 13¼ inches. This is for a horizontal beam. Now if this beam be laid inclining, so that the horizontal distance between the bearings is 15 feet, then to find the size by the rule for this case, viz. XLIII., the above number *V*, equal to 2325, multiplied by 15, the horizontal distance, equals 34,875, and this divided by 20, the length, equals 1743·75. Now by Rule XLV., the cube root of this is 12·04, the required size of a side—12 inches full.

Rule XLVI.—When the beam is *round.* Divide the quantity represented by J, K, L, Q, or V by the decimal 0·589, and the cube root of the quotient will be the required diameter.

Example.—A white pine beam or pole, 10 feet long, is required to sustain, as the 0·2 of the breaking strength, a load of 5,000 pounds concentrated at the middle, what must be the diameter? The value of S, for white pine, Table VI., is 390. Now by the rule for load at middle, (XXXVIII.,) $5000 \times 10 = 50{,}000$; and $390 \times 0·2 = 78$; and $50{,}000 \div 78 = 641 = J$. By this rule, $641 \div 0·589 = 1088·28$, the cube root of which, 10·28, is the required diameter. If this beam be inclined, so that the horizontal distance between the supports is 7 feet, then to find the diameter, by Rule XLIII., value of J as above, 641 multiplied by 7 and divided by 10 equals 448·7. Now by this rule, $448·7 \div 0·589 = 761·796$, the cube root of which, 9·133, is the required diameter.

Example.—A spruce pole, 10 feet long, is required to sustain, as the 0·33⅓ or ⅓ of the breaking weight, a load of 1,000 pounds at 3 feet from one end, what must be the diameter? The value of S for spruce, Table VI., is 345. By the rule for this case, (Rule XLI.,) $3 \times 7 \times 4 \times 1000 = 84{,}000$; and $345 \times \frac{1}{3} \times 10 = 1150$; and $84{,}000 \div 1150 = 93·04$, the value of Q. Now by this rule, $73·04 \div ·589 = 124$, the cube root of which, 4·9866, is the required diameter in inches.

335.—*Systems of Framing.* In the various parts of framing known as floors, partitions, roofs, bridges, &c., each has a specific object; and, in all designs for such constructions, this object should be kept clearly in view; the various parts being so disposed as to serve the design with the least quantity of material. The simplest form is the best, not only because it is the most economical, but for many other reasons. The great number of joints, in a complex design, render the construction liable to derangement by multiplied compressions, shrinkage, and, in consequence, highly increased oblique strains; by which its stability and durability are greatly lessened.

FLOORS.

336.—Floors are most generally constructed *single*, that is, simply a series of parallel beams, each spanning the width of

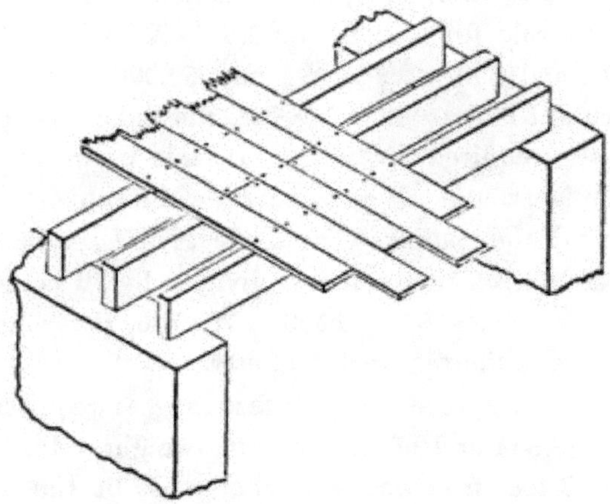

Fig 224.

the floor, as seen at *Fig.* 224. Occasionally floors are con

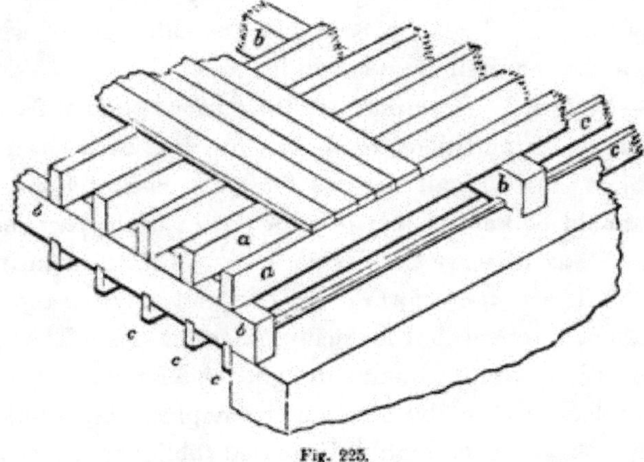

Fig. 225.

structed *double*, as at *Fig.* 225; and sometimes *framed*, as at

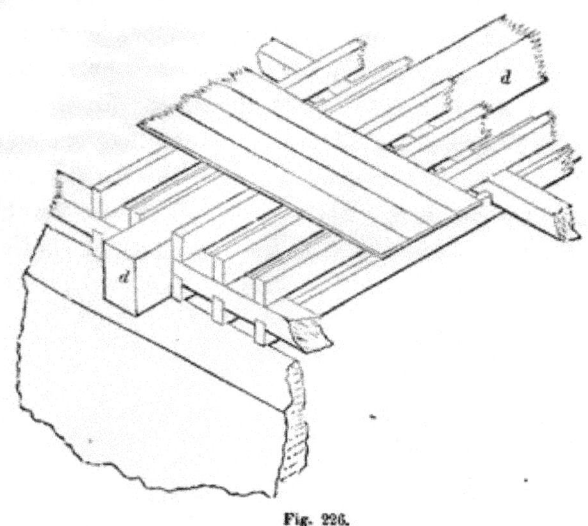

Fig. 226.

Fig. 226; but these methods are seldom practised, inasmuch as either of these require more timber than the single floor. Where lathing and plastering is attached to the floor-beams to form a ceiling below, the springing of the beams, by customary use, is liable to crack the plastering. To obviate this in good dwellings, the double and framed floors have been resorted to, but more in former times than now, as the *cross-furring* (a series of narrow strips of board or plank, nailed transversely to the underside of the beams to receive the lathing for the plastering,) serves a like purpose very nearly as well.

337.—In single floors the dimensions of the beams are to be ascertained by the preceding rules for the *stiffness* of materials. These rules give the required dimensions for the various kinds of material in common use. The rules may be somewhat abridged for ordinary use, if some of the quantities represented in the formula be made constant within certain limits. For example, if the load per foot superficial, and the rate of deflection, be fixed, then these, together with the $\frac{5}{8}$, and the constant represented by E, may be reduced to one constant. For dwellings, the load per foot may be taken at 66 pounds, as this is the weight, that has been ascertained by experiment, to arise from a crowd of people on their feet. To this add 20 for the weight of the material of which the floor is composed, and the sum, 86, is the value of f, or the weight per foot superficial for dwellings. The rate of deflection allowable for this load may be fixed at 0·03 inch per foot of the length. Then (44) transposed,

$$\frac{5fcl^3}{8En} = bd^3$$

becomes

$$\frac{5 \times 86}{8 \times 0{\cdot}03} \times \frac{cl^3}{E} = bd^3$$

which, reduced, is

$$\frac{1800}{E} \times cl^3 = bd^3 \qquad (175.)$$

Reducing $\frac{1800}{E}$ for five of the most common woods, and there results, rejecting small decimals, and putting $\frac{1800}{E} = x$, x equal, for

Georgia pine	0·6
Oak	0·7
White pine	1·0
Spruce	1·15
Hemlock	1·45

Therefore, the rule is reduced to $xcl^3 = bd^3$. And for *white pine*, the wood most used for floor beams, $x = 1{\cdot}0$, and therefore disappears from the formula, rendering it still more simple, thus,

$$cl^3 = bd^3 \qquad (176.)$$

The dimensions of beams for stores, for all ordinary business, may also be calculated by this modified rule, (175,) for it will require about 3½ times the weight used in this rule, or about

300 pounds, to increase the deflection to the limit of elasticity in white pine, and nearly that in the other woods. But for warehouses, taking the rate of deflection at its limit, and fixing the weight per foot at 500 pounds, including the weight of the material of which the floor is constructed, and letting y represent the constant, then

$$y\,c\,l^3 = b\,d^3 \qquad (177.)$$

and y equals, for

Georgia pine	1·85
Oak	1·35
White pine	1·75
Spruce	2·2
Hemlock	2·85

338.—Hence to find the dimensions of floor beams for *dwellings* when the rate of deflection is 0·03 inch per foot, or for *ordinary stores* when the load is about 300 pounds per foot, and the deflection caused by this weight is within the limits of the elasticity of the material, we have the following rule:

Rule XLVII.—Multiply the cube of the length by the distance apart between the beams, (from centres,) both in feet, and multiply the product by the value of x, (*Art.* 337.) Now to find the breadth, divide this product by the cube of the depth in inches, and the quotient will be the breadth in inches. But if the depth is sought, divide the said product by the breadth in inches, and the cube root of the quotient will be the depth in inches; or, if the breadth and depth are to be in proportion, as r is to unity, r representing any required decimal, then divide the aforesaid product by the value of r, and extract the square root of the quotient, and the square root of this square root will be the depth required in inches, and the depth multiplied by the value of r will be the breadth in inches.

Example.—To find the breadth. In a dwelling or ordinary store what must be the breadth of the beams, when placed 15 inches from centres, to support a floor covering a span of 16 feet, the depth being 11 inches, the beams of oak? By the rule, 4096, the cube of the length, by 1¼, the distance from centres, and by 0·7, the value of *x*, for oak, equals 3584. This divided by 1331, the cube of the depth, equals 2·69 inches, or $2\frac{11}{16}$ inches, the required breadth.

Example.—To find the depth. The conditions being the same as in the last example, what must be the depth when the breadth is 3 inches. The product, 3584, as above, divided by 3, the breadth, equals $1194\frac{2}{3}$; the cube root of this is 10·61, or $10\frac{5}{8}$ inches nearly.

Example.—To find the breadth and depth in proportion, say, as 0·3 to 1·0. The aforesaid product, 3584, divided by 0·3, the value of *r*, equals $11,946\frac{2}{3}$, the square root of which is 109·3, and the square root of this is 10·45, the required depth. This multiplied by 0·3, the value of *r*, equals 3·135, the required breadth, the beam is therefore to be $3\frac{1}{8}$ by $10\frac{1}{2}$ inches.

339.—And to find the breadth and depth of the beams for a floor of a *warehouse* sufficient to sustain 500 pounds per foot superficial, (including weight of the material in the floor,) with a deflection not exceeding the limits of the elasticity of the material, we have the following rule:

Rule XLVIII.—The same as XLVII., with the exception that the value of *y* (*Art.* 337) is to be used instead of the value of *x*.

Example.—To find the breadth. The beams of a warehouse floor are to be of Georgia pine, with a clear bearing between the walls of 15 feet, and placed 14 inches from centres, what must be the breadth when the depth is 11 inches? By the rule, 3375, the cube of the length, and $1\frac{1}{6}$, the distance from centres, and 1·35, the value of *y*, for Georgia pine, all multiplied together, equals 5315·625; and this product divided by 1331, the cube of the depth, equals 3·994, the required depth, or 4 inches.

Example.—*To find the depth.* The conditions remaining, as in last example, what must be the depth when the breadth is 3 inches? 5315·625, the said product, divided by 3, the breadth, equals 1771·875, and the cube root of this, 12·1, or 12 inches, is the depth required.

Example.—*To find the breadth and depth in a given proportion*, say, 0·35 to 1·0. 5315·625 aforesaid, divided by 0·35, the value of r, equals 15187·5, the square root of which is 121·8, and the square root of this square root is 11·04, or 11 inches, the required depth. And 11·04 multiplied by 0·35, the value of r, equals 3·864, the required breadth—$3\frac{7}{8}$ inches.

340.—It is sometimes desirable, when the breadth and depth of the beams are fixed, or when the beams have been sawed and are now ready for use, to know the distance from centres at which such beams should be placed, in order that the floor be sufficiently stiff. In this case, (175,) transposed, and putting $x = \frac{1800}{E}$, there results

$$c = \frac{b\,d^3}{x\,l^3} \qquad (178.)$$

This in words, at length, is, as follows:

Rule XLIX.—Multiply the cube of the depth by the breadth, both in inches, and divide the product by the cube of the length, in feet, multiplied by the value of x, for dwellings, and for ordinary stores, or by y for warehouses; and the quotient will be the distance apart from centres in feet.

Example.—A span of 17 feet, in a dwelling, is to be covered by white pine beams, 3 × 12 inches, at what distance apart from centres must they be placed? By the rule, 1728, the cube of the depth, multiplied by 3, the breadth, equals 5184. The cube of 17 is 4913, this by 1·0, the value of x, for white pine, equals 4913. The aforesaid 5184, divided by this, 4913, equals 1·055 feet, or 1 foot and $\frac{2}{3}$ of an inch.

341.—Where chimneys, flues, stairs, etc., occur to interrupt

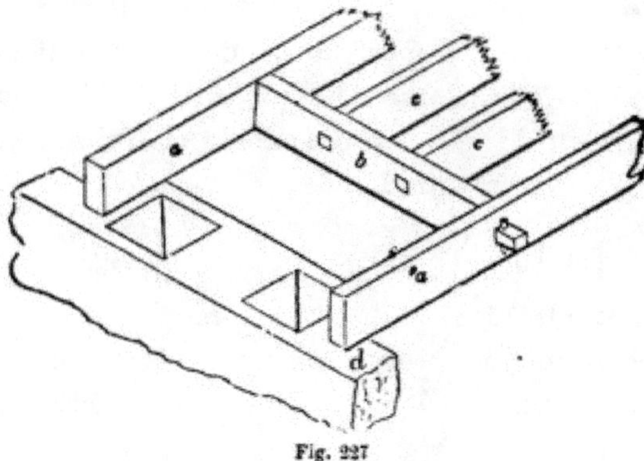

Fig. 227

the bearing, the beams are framed into a piece, *b*, (*Fig.* 227,) called a *header*. The beams, *a a*, into which the header is framed, are called *trimmers* or *carriage-beams*. These framed beams require to be made thicker than the common beams. The header must be strong enough to sustain one-half of the weight that is sustained upon the *tail* beams, *c c*, (the wall at the opposite end or another header there sustaining the other half,) and the trimmers must each sustain one-half of the weight sustained by the header in *addition* to the weight it supports as a common beam. It is usual in practice to make these framed beams one inch thicker than the common beams for dwellings, and two inches thicker for heavy stores. This practice in ordinary cases answers very well, but in extreme cases these dimensions are not proper. Rules applicable generally must be deduced from the conditions of the case—the load to be sustained and the strength of the material.

342.—For the header, formula (68,) Table V., is applicable. The weight, represented by *w*, is equal to the superficial area of the floor supported by the header, multiplied by the load on every superficial foot of the floor. This is equal to the length of the header multiplied by half the length of the tail beams, and by 86 pounds for dwellings and ordinary stores, or by 500 pounds for warehouses. Calling the length of the tail beams, in feet, g, formula (68,) becomes

$$b = \tfrac{1}{72} \frac{f g l^3}{E d^3 n}.$$

Then if f equals 86, and n equals 0·03, there results

$$b = \frac{900 \, g \, l^3}{E d^3} \qquad (179.)$$

This in words, is, as follows:

Rule L.—Multiply 900 times the length of the tail beams by the cube of the length of the header, both in feet. The product, divided by the cube of the depth, multiplied by the value of E, Table II., will equal the breadth, in inches, for *dwellings* or *ordinary stores*.

Example.—A header of white pine, for a dwelling, is 10 feet long, and sustains tail beams 20 feet long, its depth is 12 inches, what must be its breadth? By the rule, $900 \times 20 \times 10^3 = 18,000,000$. This, divided by $(12^3 \times 1750 =) 3,024,000$, equals 5·95, say 6 inches, the required breadth.

For heavy *warehouses* the rule is the same as the above, only using 1550 in the place of the 900. This constant may be varied, at discretion, to anything between 900 and 5000, in accordance with the use to which the floor is to be put.

343.—In regard to the trimmer or carriage beam, formula (136,) Table VIII., is applicable. The load thrown upon the trimmer, in *addition* to its load as a common beam, is equal to one-half of the load on the header, and therefore, as has been seen in last article, is equal to one-half of the superficial area of the floor, supported by the tail beams, multiplied by the weight per superficial foot of the load upon the floor; therefore, when the length of the header, in feet, is represented by j, and the length of the tail beam by n, w equals $\frac{j}{2} \times \frac{n}{2} \times f$, equals $\frac{1}{4} f j n$, and therefore (136,) of Table VIII., becomes

$$b = \frac{f j m n^2}{S a d^2 l}$$

equals the *additional* thickness to be given to a common beam when used as a trimmer, and for *dwellings* when f equals 86 and a equals 0·3, this part of the formula reduces to $286\frac{2}{3}$, or, for simplicity, call it 300, which would be the same as fixing f at 90 instead of 86. Then we have

$$b = \frac{300\, j\, m\, n^2}{S\, d^2\, l} \qquad (180.)$$

This, in words, is as follows:

Rule LI.—*For dwellings.* Multiply 300 times the length of the header by the square of the length of the tail beams, and by the difference in length of the trimmer and tail beams, all in feet. Divide this product by the square of the depth in inches, multiplied by the length of the trimmer in feet, and by the value of S, Table VI., and the quotient added to the thickness of a common beam of the floor, will equal the required thickness of the trimmer beam.

Example.—A tier of 3 × 12 inch beams of white pine, having a clear bearing of 20 feet, has a framed well-hole at one side, of 5 by 12 feet, the header being 12 feet long, what must be the thickness of the trimmer beams? By the rule, 300 × 12 × 15^2 × 5, divided by the product of 12^2 × 20 × 390, equals 3·6, and this added to 3, the thickness of one of the common beams, equals 6·6, the breadth required, $6\frac{1}{2}$ inches.

For *stores* and *warehouses* the rule is the same as the above, only the constant, 300, must be enlarged in proportion to the load intended for the floor, making it as high as 1600 for heavy warehouses.

344.—When a framed opening occurs at any point removed from the wall, requiring two headers, then the load from the headers rest at two points on the carriage beam, and here formula (141.) Table VIII., is applicable. In this special case this formula reduces to

$$b = \frac{300\, j\, (m^2\, n + k^2\, g)}{S\, d^2\, l} \qquad (181.)$$

where b equals the *additional* thickness, in inches, to be given to the carriage beam over the thickness of the common beams; j, the length of the header, in feet; m and k the length, in feet, respectively, of the two sets of tail beams, and $m + n = k + g = l$.

The constant in the above, (181,) is for *dwellings;* if the floor is to be loaded more than dwelling floors, then it must be increased in proportion to the increase of load up to as high as 1600 for *warehouses.*

Rule LII.—*Trimmer beams for framed openings occurring so as to require two headers.* Multiply the square of the length of each tail beam by the difference of length of the tail beam and trimmer, all in feet, and add the products; multiply their sum by 300 times the length, in feet, of the header, and divide this product by the product of the square of the depth, in inches, by the length, in feet, and by the value of S, Table VI.; and the quotient, added to the thickness of a common beam of the tier, will equal the thickness of the trimmer beams.

Example.—A tier of white pine beams, 4 × 14 inches, 20 feet long, is to have an opening of 5 × 10 feet, framed so that the length of one series of tail beams is 7 feet, the other 8 feet, what must be the breadth of the trimmers? Here, $(7^2 \times 13) + (8^2 \times 12)$ equals 1405. This by 300 × 10 equals 4,215,000. This divided by 1,528,800 $(= 14^2 + 20 \times 390)$ equals 2·75, and this added to 4, the breadth, equals 6·75, or $6\frac{3}{4}$, the breadth required, in inches.

345.—Additional stiffness is given to a floor by the insertion of *bridging* strips, or struts, as at *a a*, (*Fig.* 228.) These prevent the turning or twisting of the beams, and when a weight is placed upon the floor, concentrated over one beam, they prevent this beam from descending below the adjoining beams to the injury of the plastering upon the underside. It is usual to insert a course of bridging at every 5 to 8 feet of the length

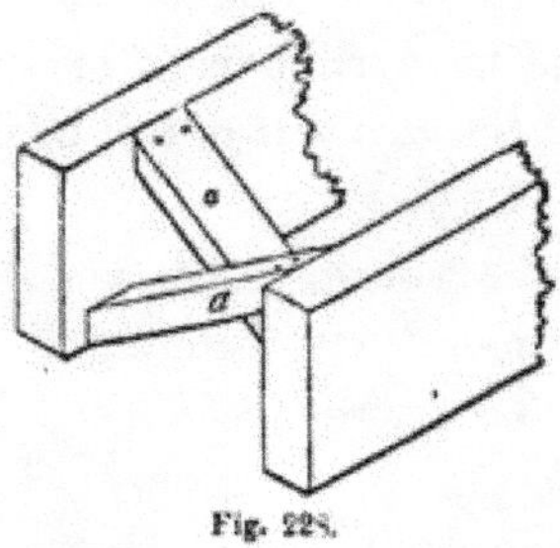

Fig. 228.

of the beam. Strips of board or plank nailed to the underside of the floor beams to receive the lathing, are termed *cross-furring*, and should not be over 2 inches wide, and placed 12 inches from centres. It is desirable that all furring be narrow, in order that the *clinch* of the mortar be interrupted but little. When it is desirable to prevent the passage of sound, the openings between the beams, at about 3 inches from the upper edge, are closed by short pieces of boards, which rest on cleets, nailed to the beam along its whole length. This forms a floor, on which mortar is laid from 1 to 2 inches deep. This is called *deafening*.

346.—When the distance between the walls of a building is great, it becomes requisite to introduce girders, as an additional support, beneath the beams. The dimensions of girders may be ascertained by the general rules for stiffness. Formulas (72,) (73.) and (74,) Table V., are applicable, taking f, at 86, for dwellings and ordinary stores, and increased in proportion to the load, up to 500, for heavy warehouses. When but one girder occurs, in the length of the beam, the distance from centres, c, is equal to one-half the length of the beam.

347.—When the breadth of a girder is more than about 12 inches, it is recommended to divide it by sawing from end to end, vertically through the middle, and then to bolt it to gether with the sawn sides outwards. This is not to strengthen the girder, as some have supposed, but to reduce the size of the timber, in order that it may dry sooner. The operation affords also an opportunity to examine the heart of the stick—a necessary precaution; as large trees are frequently in a state of decay at the heart, although outwardly they are seemingly sound. When the halves are bolted together, thin slips of wood should be inserted between them at the several points at which they are bolted, in order to leave sufficient space for the air to circulate between. This tends to prevent decay; which will be found first at such parts as are not exactly tight, nor yet far enough apart to permit the escape of moisture.

348.—When girders are required for a long bearing, it is usual to truss them; that is, to insert between the halves two pieces of oak which are inclined towards each other, and which meet at the centre of the length of the girder, like the rafters of a roof-truss, though nearly if not quite concealed within the girder. This, and many similar methods, though extensively practised, are generally worse than useless; since it has been ascertained that, in nearly all such cases, the operation has positively *weakened* the girder.

A girder may be strengthened by mechanical contrivance, when its depth is required to be greater than any one piece of

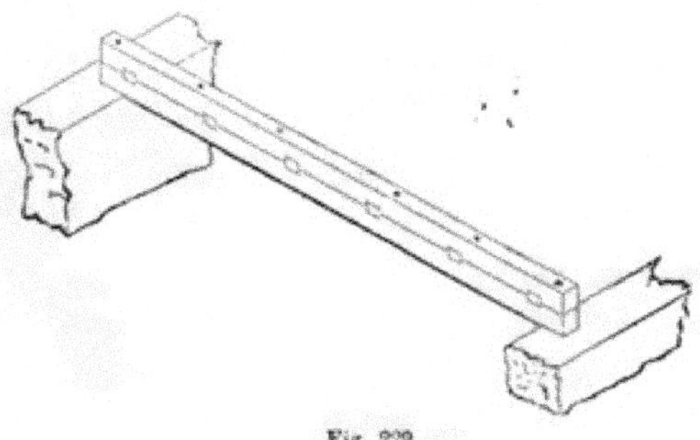

Fig. 229.

timber will allow. *Fig.* 229 shows a very simple yet invaluable method of doing this. The two pieces of which the girder is composed are bolted, or pinned together, having keys inserted between to prevent the pieces from sliding. The keys should be of hard wood, well seasoned. The two pieces should be about equal in depth, in order that the joint between them may be in the neutral line. (See *Art.* 317.) The thickness of the keys should be about half their breadth, and the amount of their united thicknesses should be equal to a trifle over the depth and one-third of the depth of the girder. Instead of bolts or pins, iron hoops are sometimes used; and when they can be procured, they are far preferable. In this case, the girder is diminished at the ends, and the hoops driven from each end towards the middle.

349.—Beams may be spliced, if none of a sufficient length can be obtained, though not at or near the middle, if it can be avoided. (See *Art.* 281.) Girders should rest from 9 to 12 inches on the wall, and a space should be left for the air to circulate around the ends, that the dampness may evaporate. Floor-timbers are supported at their ends by walls of considerable height. They should not be permitted to rest upon intervening partitions, which are not likely to settle as much as the walls; otherwise the unequal settlements will derange the level of the floor. As all floors, however well-constructed, settle in some degree, it is advisable to frame the beams a little higher at the middle of the room than at its sides,—as also the ceiling-joists and cross-furring, when either are used. In single floors, for the same reason, the rounded edge of the stick, if it have one, should be placed uppermost.

If the floor-plank are laid down temporarily at first, and left to season a few months before they are finally driven together and secured, the joints will remain much closer. But if the edges of the plank are planed after the first laying, they will shrink again; as it is the nature of wood to shrink after *every* planing however dry it may have been before.

PARTITIONS.

350.—Too little attention has been given to the construction of this part of the frame-work of a house. The settling of floors and the cracking of ceilings and walls, which disfigure to so great an extent the apartments of even our most costly houses, may be attributed almost solely to this negligence. A square of partitioning weighs nearly a ton, a greater weight, when added to its customary load, such as furniture, storage, &c., than any ordinary floor is calculated to sustain. Hence the timbers bend, the ceilings and cornices crack, and the whole interior part of the house settles; showing the necessity for providing adequate supports independent of the floor-timbers. A partition should, if practicable, be supported by the walls with which it is connected, in order, if the walls settle, that it may settle with them. This would prevent the separation of the plastering at the angles of rooms. For the same reason, a firm connection with the ceiling is an important object in the construction of a partition.

351.—The joists in a partition should be so placed as to discharge the weight upon the points of support. All oblique pieces in a partition, that tend not to this object, are much better omitted. *Fig.* 230 represents a partition having a door in the middle. Its construction is simple but effective. *Fig.* 231 shows the manner of constructing a partition having doors near the ends. The truss is formed above the door-heads, and the lower parts are suspended from it. The posts, *a* and *b*, are halved, and nailed to the tie, *c d*, and the sill, *e f*. The braces in a trussed partition should be placed so as to form, as near as possible, an angle of 40 degrees with the horizon. In partitions that are intended to support only their own weight,

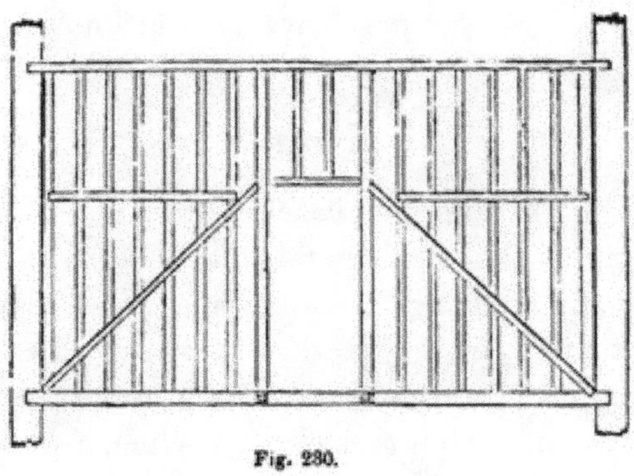

Fig. 230.

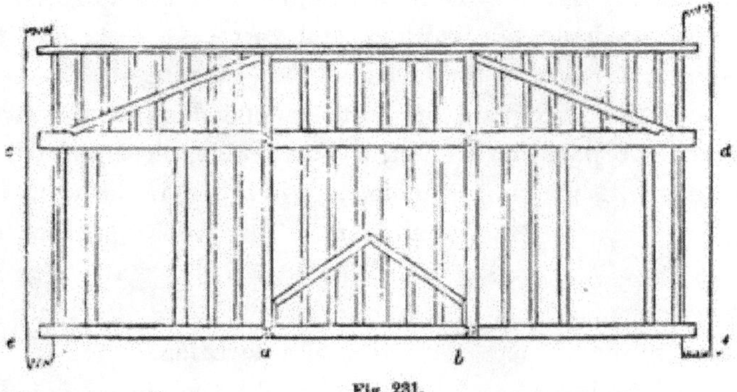

Fig. 231.

the principal timbers may be 3 × 4 inches for a 20 feet span, $3\frac{1}{2} \times 5$ for 30 feet, and 4 × 6 for 40. The thickness of the filling-in stuff may be regulated according to what is said at *Art.* 345, in regard to the width of furring for plastering. The filling-in pieces should be stiffened at about every three feet by short struts between.

All superfluous timber, besides being an unnecessary load upon the points of support, tends to injure the stability of the plastering; for, as the strength of the plastering depends, in a great measure, upon its clinch, formed by pressing the mortar through the space between the laths, the narrower the surface, therefore, upon which the laths are nailed, the less will be the quantity of plastering unclinched, and hence its greater security from fractures. For this reason, the principal timbers of the partition should have their edges reduced, by chamfering off the corners.

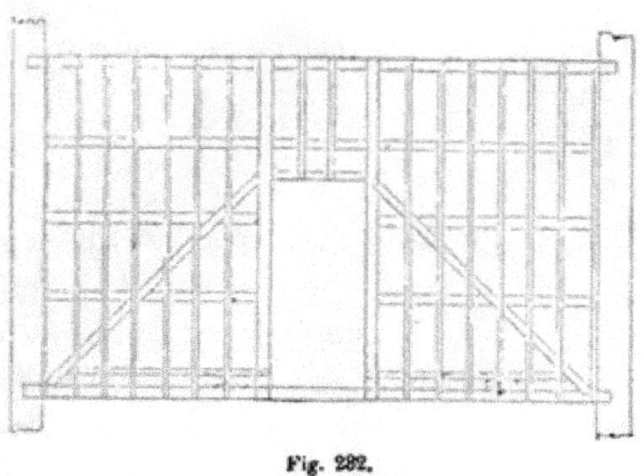

Fig. 232.

352.—When the principal timbers of a partition require to be large for the purpose of greater strength, it is a good plan to omit the upright filling-in pieces, and in their stead, to place a few horizontal pieces; in order, upon these and the principal timbers, to nail upright battens at the proper distances for lathing, as in *Fig.* 232. A partition thus constructed requires a little more space than others; but it has the advantage of insuring greater stability to the plastering, and also of preventing to a good degree the conversation of one room from being heard in the other. When a partition is required to support, in addition to its own weight, that of a floor or some other burden resting upon it, the dimensions of the timbers may be ascertained, by applying the principles which regulate the laws of pressure and those of the resistance of timber, as explained at the first part of this section. The following data, however, may assist in calculating the amount of pressure upon partitions:

White pine timber weighs from 22 to 32 pounds per cubic foot, varying in accordance with the amount of seasoning it has had. Assuming it to weigh 30 pounds, the weight of the beams and floor plank in every superficial foot of the flooring will be, when the beams are

3 × 8	inches,	and	placed	20	inches	from	centres,	6	pounds.
3 × 10	"	"	"	18	"	"	"	$7\frac{1}{2}$	"
3 × 12	"	"	"	16	"	"	"	9	"
3 × 12	"	"	"	12	"	"	"	11	"
4 × 12	"	"	"	12	"	"	"	13	"
4 × 14	"	"	"	14	"	"	"	13	"

In addition to the beams and plank, there is generally the *plastering* of the ceiling of the apartments beneath, and sometimes the *deafening*. Plastering may be assumed to weigh 9 pounds per superficial foot, and deafening 11 pounds.

Hemlock weighs about the same as white pine. A partition of 3 × 4 joists of hemlock, set 12 inches from centres, therefore, will weigh about $2\frac{1}{2}$ pounds per foot superficial, and when plastered on both sides, $20\frac{1}{2}$ pounds.

353.—When floor beams are supported at the extremities, and by a partition or girder at any point between the extremities, one-half of the weight of the whole floor will then be supported by the partition or girder. As the settling of partitions and floors, which is so disastrous to plastering, is frequently owing to the shrinking of the timber and to ill-made joints, it is very important that the timber be seasoned and the work well executed. Where practicable, the joists of a partition ought to extend down between the floor beams to the plate of the partition beneath, to avoid the settlement consequent upon the shrinkage of the floor beams.

ROOFS.*

354.—In ancient buildings, the Norman and the Gothic, the

* See also *Art.* 238.

walls and buttresses were erected so massive and firm, that it was customary to construct their roofs without a tie-beam; the walls being abundantly capable of resisting the lateral pressure exerted by the rafters. But in modern buildings, the walls are so slightly built as to be incapable of resisting scarcely any oblique pressure; and hence the necessity of constructing the roof so that all oblique and lateral strains may be removed; as, also, that instead of having a tendency to separate the walls, the roof may contribute to bind and steady them.

355.—In estimating the pressures upon any certain roof, for the purpose of ascertaining the proper sizes for the timbers, calculation must be made for the pressure exerted by the wind, and, if in a cold climate, for the weight of snow, in addition to the weight of the materials of which the roof is composed. The weight of snow will be of course according to the depth it acquires. Snow weighs 8 lbs. per cubic foot, and more when saturated with water. In a severe climate, roofs ought to be constructed steeper than in a milder one, in order that the snow may have a tendency to slide off before it becomes of sufficient weight to endanger the safety of the roof. The inclination should be regulated in accordance with the qualities of the material with which the roof is to be covered. The following table may be useful in determining the smallest inclination, and in estimating the weight of the various kinds of covering:

Material.	Inclination.	Weight upon a square foot.
Tin	Rise 1 inch to a foot	5/8 to 1 1/4 lbs.
Copper	" 1 " " "	1 to 1 1/4 "
Lead	" 2 inches " "	4 to 7 "
Zinc	" 3 " " "	1 1/2 to 2 "
Short pine shingles	" 5 " " "	1 1/2 to 2 "
Long cypress shingles	" 6 " " "	2 to 3 "
Slate	" 6 " " "	5 to 9 "

The weight of the covering, as above estimated, is that of the material only, with the weight of whatever is used to fix it to the roof, such as nails, &c. What the material is laid on, such as plank, boards or lath, is not included. The weight of plank is about 3 pounds per foot superficial; of boards, 2 pounds; and lath, about a half pound.

356.—The weights and pressures on a roof arise from the roofing, the truss, the ceiling, wind and snow, and may be stated as follows:

First, the Roofing.—On each foot superficial of the inclined surface,

Slating	will weigh about 7 lbs.
Roof plank, 1½ inches thick . .	" " " 2·7 "
Roof beams or jack rafters . .	" " " 2·3 "
	Total, 12 lbs.

This is the weight per foot on the *inclined* surface; but it is desirable to know how much per foot, measured *horizontally*, this is equal to. The horizontal measure of one foot of the inclined surface is equal to the cosine of the angle of inclination. Therefore,

$$\cos. : 1 :: p : w = \frac{p}{\cos.};$$

where p represents the pressure on a foot of the inclined surface, and w the weight of the roof per foot, measured horizontally. The cosine of an angle is equal to the base of the right-angled triangle divided by the hypothenuse, which in this case would be half the span divided by the length of the rafter, or $\frac{s}{2l}$, where s is the span, and l the length of the rafter. Hence,

$$\frac{p}{\cos.} = \frac{p}{\frac{s}{2l}} = \frac{2lp}{s};$$

or, twice the pressure per foot of *inclined* surface, multiplied by the length of the rafter, and divided by the span, will give the weight per foot measured horizontally; or,

$$24\frac{l}{s} = w \qquad (182.)$$

equals the weight per foot, measured horizontally, of the roof beams, plank, and covering for a slate roof.

Second, the ***Truss.***—The weight of the framed truss is nearly in proportion to the length of the truss, and to the distance apart at which the trusses are placed.

$$w = 5{\cdot}2\,c\,s \qquad (183.)$$

equals the weight, in pounds, of a white pine truss with iron suspension rods and a horizontal tie beam, near enough for the requirements of our present purpose; where s equals the length or span of the truss, and c the distance apart at which the trusses are placed, both in feet. It is desirable to know how much this is equal to per foot of the area over which the truss is to sustain a covering. This is found by dividing the weight of the truss by the span, and by the distance apart from centres at which the trusses are placed; or,

$$\frac{5{\cdot}2\,c\,s}{c\,s} = 5{\cdot}2 = w \qquad (184.)$$

equals the weight in pounds per foot to be allowed for the truss.

Third, the Ceiling.—The weight supported by the tie beams, viz.: that of the ceiling beams, furring and plastering, is about 9 pounds per superficial foot.

Fourth, the Wind.—The force of wind has been known as high as 50 pounds per superficial foot against a vertical surface. The effect of a horizontal force on an inclined surface is in proportion to the sine of the angle of inclination, the effect produced being in the direction at right angles to the inclined surface. The force thus acting may be resolved into forces acting in two directions—the one horizontal, the other vertical; the former tending, in the case of a roof, to thrust aside the walls on which the roof rests, and the latter acting directly on the materials of which the roof is constructed—this latter force being in proportion to the sine of the angle of inclination multiplied by the cosine. This is the *vertical* effect of the wind upon a roof, without regard to the *surface* it acts upon. The wind, acting horizontally through one foot superficial of vertical section, acts on an area of inclined surface equal to the reciprocal of the sine of inclination, and the horizontal measurement of this inclined surface is equal to the cosine of the angle of inclination divided by the sine. This is the horizontal measurement of the inclined surface, and the vertical force acting on this surface is, as above stated, in proportion to the sine multiplied by the cosine. Combining these, it is found that the vertical power of the wind is in proportion to the square of the sine of the angle of inclination. Therefore, if the power of wind against a vertical surface be taken at 50 pounds per superficial foot, then the vertical effect on a roof is equal to

$$w = 50 \sin.^2 = 50 \frac{h^2}{l^2} \qquad (185.)$$

for each piece of the inclined surface, the horizontal measurement of which equals one foot; where l equals the length of the rafter, and h the height of the roof.

Fifth, Snow.—The weight of snow will be in proportion to the depth it acquires, and this will be in proportion to the rigour of the climate of the place at which the building is to be erected. Upon roofs of most of the usual inclinations, snow, if deposited in the absence of wind, will not slide off. When it has acquired some depth, and not till then, it will have a tendency, in proportion to the angle of inclination, to slide off in a body. The weight of snow may be taken, therefore, at its weight per cubic foot, 8 pounds, multiplied by the depth it is usual for it to acquire. This, in the latitude of New York, may be stated at about 2½ feet. Its weight would, therefore, be 20 pounds per foot superficial, measured horizontally.

357.—There is one other cause of strain upon a roof; namely, the load that may be deposited in the roof when used as a room for storage, or for dormitories. But this seldom occurs. When a case of this kind does occur, allowance is to be made for it as shown in the article on floors. But in the *general* rule, now under consideration, it may be omitted.

358.—The following, therefore, comprehends all the pressures or weights that occur on roofs generally, per foot superficial;

For roof beams, plank, and slate (182)	$24\frac{l}{s}$ lbs.
" the truss (184)	5·2 "
" ceiling	9 "
" wind (185)	$50\frac{h^2}{l^2}$ "
" snow, latitude of New York	20 "

Having found the weight per foot, the total weight for any part of the roof is found by multiplying the weight per foot by the area of that part. This process will give the weight supported by braces and suspension rods, and also that supported by the rafters and tie beam. But in these last two, only *half* of the pressure of the *wind* is to be taken, for the wind will act only on one side of the roof at the same time.

The vertical pressure on the head of a brace, then, equals

$$W = 4\,c\,n\left(6\frac{l}{s} + 8{\cdot}55 + 12{\cdot}5\frac{h^2}{l^2}\right) \qquad (186.)$$

And $W = c\,p\,n$, where p equals $4\left(6\frac{l}{s} + 8{\cdot}55 + 12{\cdot}5\frac{h^2}{l^2}\right)$, equals the weight per foot.

And the aggregate load of the roof on each truss equals

$$W = 4\,c\,s\left(6\frac{l}{s} + 8{\cdot}55 + 6{\cdot}25\frac{h^2}{l^2}\right) \qquad (187.)$$

And $W = c\,q\,s$, where $q = 4\left(6\frac{l}{s} + 8{\cdot}55 + 6{\cdot}25\frac{h^2}{l^2}\right)$, equals the weight per foot; where c equals the distance apart from centres at which the trusses are placed; n the distance horizontally between the heads of the braces, or, if these are not located at equal distances, then n is the distance horizontally from a point half-way to the next brace on one side to a point half-way to the next brace on the other side; l the length of the rafter; s the span, and h the height—all in feet.

359.—By the parallelogram of forces, the weight of the roof is in proportion to the oblique thrust or pressure in the axis of the rafter, as twice the height of the roof is to the length of the rafter; or,

$$W : R :: 2\,h : l, \text{ or}$$

$$2h : l :: W : R = \frac{W\,l}{2\,h}, \qquad (188.)$$

where R equals the pressure in the axis of the rafter. And the weight of the roof is in proportion to the horizontal thrust in the tie beam, as twice the height of the roof is to half the span; or,

$$W : H :: 2\,h : \frac{s}{2}, \text{ or}$$

$$2h : \frac{s}{2} :: W : H = \frac{W\,s}{4\,h}, \qquad (189.)$$

where H equals the horizontal thrust in the tie beam; the value of W in (188) and (189) being shown at (187), and (187) being compounded as explained in *Art.* 356. The weight is that for a slate roof. If other material is used for covering, or should there be other conditions modifying the weight in any particular case, an examination of *Art.* 356 will show how to modify the formula accordingly.

360.—The pressures may be obtained geometrically, as shown in *Fig.* 233, where $A\,B$ represents the axis of the tie beam, $A\,C$ the axis of the rafter, $D\,E$ and $F\,B$ the axes of the braces, and $D\,G$, $F\,E$, and $C\,B$, the axes of the suspension rods. In this design for a truss, the distance $A\,B$ is divided into three equal parts, and the rods located at the two points of division, *G* and *E*. By this arrangement the rafter *AC* is supported at equi-distant points, *D* and *F*. The point *D* supports the rafter for a distance extending half-way to *A* and half-way to *F*, and the point *F* sustains half-way to *D* and half-way to *C*. Also, the point *C* sustains half-way to *F* and, on the other rafter, half-way to the corresponding point to *F*. And because these points of support are located at equal distances apart, therefore the load on each is the same in amount. On *DG* make *Da* equal to 100 of any decimally divided scale, and let *Da* represent the load on *D*, and draw the parallelogram *abDc*. Then, by the same scale, *Db* represents (*Art.* 258) the pressure in the axis of the rafter by the load at *D*; also, *Dc* the pressure in the brace *DE*. Draw *cd* horizontal; then *Dd* is the vertical pressure exerted by the brace *DE* at *E*. The point *F* sustains, besides the common load represented by 100 of the scale, also the vertical pressure exerted by the brace *DE*; therefore, since *Da* represents the common load on *D*, *F*, or *C*, make *Fe* equal to the sum of *Da* and *Dd*, and draw the parallelogram *Fgef*. Then *Fg*, measured by the scale, is the pressure in the axis of the rafter caused by the load at *F*, and *Ff* is the load in the axis of the brace *FB*. Draw *fh* horizontal; then *Fh* is the vertical pressure exerted by the brace *FB* at *B*. The point *C*, besides the common load represented by *Da*, sustains the vertical pressure *Fh* caused by the brace *FB*, and a like amount from the corresponding brace on the opposite side. Therefore, make *Cj* equal to the sum of *Da* and twice *Fh*, and draw *jk* parallel to the opposite rafter. Then *Ck* is the pressure in the axis of the rafter at *C*. This is not the only pressure in the rafter, although it is the total pressure at its head *C*. At the point *F*, besides the pressure *Ck*, there is *Fg*. At the point *D*, besides these two pressures, there is the pressure *Db*. At the foot, at *A*, there is still an additional pressure: while the point *D* sustains the load half-way to F and half-way to A, the point A sustains the load half-way to D. This load is, in this case, just half the load at D. Therefore draw $A\,m$ vertical, and equal to 50 of the scale, or half of $D\,a$. Extend $C\,A$ to l; draw $m\,l$ horizontal. Then $A\,l$ is the pressure in the rafter at A caused by the weight of the roof from A half-way to D. Now the total of the pressures in the rafter is equal to the sum of $A\,l + D\,b + F\,g$ added to $C\,k$. Therefore make $k\,n$ equal to the sum of $A\,l + D\,b + F\,g$, and draw $n\,o$ parallel with the opposite rafter, and $n\,j$ horizontal. Then $C\,o$, measured by the same scale, will be found equal to the total weight of the roof on both sides of $C\,B$. If $D\,a = 100$ represent the portion of the weight borne by the point D, then $C\,o$, representing the whole weight of the roof, should equal 600, (as it does by the scale,) for D supports just one-sixth of the whole load. As $C\,n$ is the total oblique thrust in the axis of the rafter at its foot, therefore $n\,j$ is the horizontal thrust in the tie beam.

361.—In stating the amount of pressures in the above as being equal to certain lines, it was so stated with the understanding that the lines were simply in proportion to the weights. To obtain the weight represented by a line, multiply its length (measured by the scale used) by the load resting at D, (or at F or C, as these are all equal in this example,) and divide the product by 100, and the quotient will be the weight required. For, as 100 of the scale is to the load it represents, so is any other dimension on the same scale to the load *it* represents.

362.—*Example.* Let $A\,B$ (*Fig.* 233) equal 26 feet, $C\,B$ 13 feet, and $A\,C$ 29 feet, and $A\,G$, $G\,E$, and $E\,B$, each $8\frac{2}{3}$ feet. Let the trusses be placed 10 feet apart. Then the weight on D, for the use of the braces and rods, is, per (186), equal to

$$4\,c\,n\left(6\frac{l}{s} + 8{\cdot}55 + 12\tfrac{1}{2}\frac{h^2}{l^2}\right)$$

$$= 4 \times 10 \times 8\tfrac{2}{3}\left(6 \times \frac{29}{52} + 8{\cdot}55 + 12\tfrac{1}{2} \times \frac{13^2}{29^2}\right)$$

$= 346\frac{2}{3} \times 14{\cdot}398$
$= 4991{\cdot}3.$

This is the common load at the points D, F, and C, and each of the lines denoting pressures multiplied by it and divided by 100, or multiplied by the quotient of $\frac{4991{\cdot}3}{100} = 49{\cdot}913$, the product will be the weight required. 49·913 may be called 50, for simplicity; therefore the pressure in the brace DE equals $112 \times 50 = 5600$ pounds, and in the brace FB, $140 \times 50 = 7000$ pounds, and in like manner for any other strain. For the rafters and tie beam the total weight, as per (187), equals

$$4\,c\,s\left(6\frac{l}{s} + 8{\cdot}55 + 6\tfrac{1}{4}\frac{h^2}{l^2}\right)$$

$$= 4 \times 10 \times 52\left(6 \times \frac{29}{52} + 8{\cdot}55 + 6\tfrac{1}{4} \times \frac{13^2}{29^2}\right)$$

$$= 2080 \times 13{\cdot}148$$

$$= 27343{\cdot}68 \text{ pounds.}$$

This is the total weight of the roof supported by one truss. The oblique thrust in the rafter $A\,C$ is, per (188), equal to

$$\frac{l\,W}{2\,h} = \frac{29 \times 27343{\cdot}68}{2 \times 13}$$

$$= 30498{\cdot}72 \text{ pounds.}$$

To obtain this oblique thrust geometrically: $C\,o$ (*Fig.* 233) represents the weight of the roof, and measures 600 by the scale; and the line Cn, representing the oblique thrust, measures 670. By the proportion, 600 : 670 :: 27343·68 : 30533·8, = the oblique thrust. The result here found is a few pounds more than the other. This is owing to the fact that the line $C\,n$ is not *exactly* 670, nor is the length of the rafter precisely 29 feet. Were the exact dimensions used in each case the results would be identical; but the result in either case is near enough for the purpose.

The horizontal strain is, per (189), equal to

$$\frac{W s}{4 h} = \frac{27343{\cdot}68 \times 52}{4 \times 13}$$

$$= 27343{\cdot}68 \text{ pounds.}$$

The result gives the horizontal thrust precisely equal to the weight. This is as it should be in all cases where the height of the roof is equal to one-fourth of the span, but not otherwise; for the result depends (189) upon this relation of the height to the span. Geometrically, the result is the same, for $C o$ and $n j$ (*Fig.* 233,) representing the weight and horizontal thrust, are precisely equal by measurement.

363.—The weight at the head of a brace is sustained partly at the foot of the brace and partly at the foot of the rafter. The *sum* of the vertical effects at these two points is just equal to the weight at the head of the brace. The portion of the weight sustained at either point is in proportion, inversely, to the horizontal distance of that point from the weight; therefore,

$$V = W \frac{g}{a}, \qquad (190.)$$

where V equals the vertical effect at the foot of the brace; W, the weight at the head of the brace; g, the horizontal distance from the foot of the rafter to the *head* of the brace; and a, the distance from the same point to the *foot* of the brace.

364.—For the oblique thrust in the brace: from the triangle $F f h$ (*Fig.* 233,)

$$F h : F f :: \text{sin.} : \text{rad.}$$

$$\text{sin.} : \text{rad.} :: V : T;$$

therefore,

$$T = \frac{V}{\text{sin.}} = V \frac{l}{h}, \qquad (191.)$$

where T equals the oblique thrust in the brace; V, the vertical pressure caused by T at the foot of the brace (190); a l and h the length and height respectively of the brace.

365.—*Example.* Brace $D E$, *Fig.* 233. In this case, equals the product of the weight per superficial foot, m plied by the area supported at the point D, equals 5000 pounds, (*Art.* 362.) The length g equals $8\frac{2}{3}$ feet, and a equals $17\frac{1}{3}$ feet. Therefore (190),

$$V = W\frac{g}{a} = 5000 \times \frac{8\frac{2}{3}}{17\frac{1}{3}} = 2500 \text{ pounds}$$

equals the vertical pressure at E caused by the brace $D\,E$. Ther for the oblique thrust, l equals 9·6 feet, and h equals 4·3 feet. Therefore, from (191),

$$T = V\frac{l}{h} = 2500 \times \frac{9{\cdot}6}{4{\cdot}3} = 5581{\cdot}4 \text{ pounds}$$

equals the oblique thrust in the brace $D\,E$. In *Art.* 362 it was found to be 5600. The discrepancy is owing to like causes of want of accuracy in the case of the rafter, as explained in *Art.* 362.

Another Example.—Brace $F\,B$, *Fig.* 233. In this case, W equals the product of the weight per superficial foot, multiplied by the area supported by the point F, added to the vertical strain caused by the brace $D\,E$. From *Art.* 362 the weight of roof on F equals 5000 pounds, and the vertical strain from brace $D\,E$ is, as just ascertained, = 2500, total 7500, equals W. The length, g, equals two-thirds of 26 feet, equals $17\frac{1}{3}$, and a equals 26 feet. Therefore, from (190),

$$V = W\frac{g}{a} = 7500 \times \frac{17\frac{1}{3}}{26} = 5000$$

equals the vertical effect at B caused by the brace $F\,B$. Then, for the oblique thrust in the brace, l equals 12·2, and h equals $8\frac{2}{3}$. Therefore, from (191)

$$T = V\frac{l}{h} = 5000 \times \frac{12{\cdot}2}{8\frac{2}{3}} = 7038{\cdot}5$$

equals the oblique thrust or strain in the axis of the brace. It was 7000 by the geometrical process, (*Art.* 362.)

366.—The strain upon the first rod, $D\,G$, equals simply the weight of the ceiling supported by it, added to the part of the tie beam it sustains. The weight of the tie beam will equal

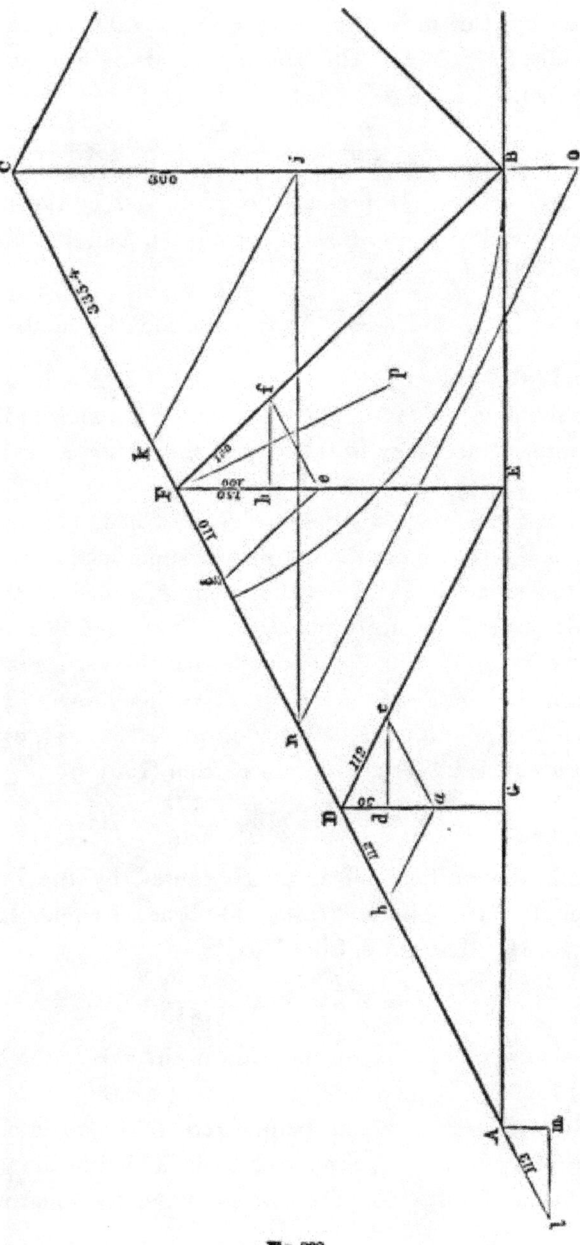

Fig. 233.

about one pound per superficial foot of the ceiling. The weight per foot for the ceiling is stated (see *Art.* 356 *third*, and 358,) at 9 pounds. To this add 1 pound for the tie beam, and the sum is 10. Then

$$N = 10\,c\,n. \qquad (192.)$$

The strain on the second tie rod equals the weight of ceiling supported, $= N$, added to the vertical effect of the strain in the brace it sustains, [see (190)] or equal to

$$O = 10\,c\,n + V. \qquad (193.)$$

The strain on the third rod is equal to N, added to the vertical effect of the strain in the brace it sustains, and this is the strain on any rod. The first rod has no brace to sustain, and the middle rod sustains two braces. In this case the strain equals

$$U = 10\,c\,n + 2\,V. \qquad (194.)$$

It may be observed that V represents the vertical strain caused by that brace that is sustained by the rod under consideration; and, as the vertical strain caused by any one brace is more than that caused by any other brace nearer the foot of the rafter, therefore the V of (193) is not equal to the V of (194). Hence a necessity for care lest the two be confounded and thus cause error.

367.—*Examples.* The rod $D\,G$ (*Fig.* 233) has a strain which equals (192)

$$N = 10\,c\,n = 10 \times 10 \times 8\tfrac{2}{3} = 867 \text{ pounds.}$$

The strain on rod $F\,E$ equals (193)

$$O = 10\,c\,n + V = 867 + 2500 = 3367 \text{ pounds.}$$

The strain on rod $C\,B$, the middle rod, equals (194)

$$U = 10\,c\,n + 2\,V = 867 + \overline{2 \times 5000} = 10867 \text{ pounds.}$$

368.—The load, and the strains caused thereby, having been discussed, it remains to speak of the resistance of the materials.

First, of the Rafter.—Generally this piece of timber is so pinioned by the roof beams or purlins as to prevent any lateral movement, and the braces keep it from deflection; therefore it is not liable to yield by flexure. Hence the manner of its yielding, when overloaded, will be by crushing at the ends, or it will crush the tie beam against which it presses. The fibres of timber yield much more readily when pressed together by a force acting at *right angles* to the direction of their length, than when it acts in a *line* with their length.

The value of timber subjected to pressure in these two ways is shown in *Art.* 292, Table I., the value per square inch of the first stated resistance being expressed by P, and that of the other by C. Timber pressed in an oblique direction yields with a force exceeding that expressed by P, and less than that by C. When the angle of inclination at which the force acts is just 45°, then the force will be an average between P and C. And for any angle of inclination, the force will vary inversely as the angle; approaching P as the angle is enlarged, and approaching C as the angle is diminished. It will be equal to C when the angle becomes zero, and equal P when the angle becomes 90°. The resistance of timber per square inch to an oblique force is therefore expressed by

$$M = P + \frac{A^\circ}{90}(C - P), \qquad (195.)$$

where A° equals the complement of the angle of inclination. In a roof, A° is the acute angle formed by the rafter with a vertical line. If no convenient instrument be at hand to measure the angle, describe an arc upon the plan of the truss—thus: with CB (*Fig.* 233) for radius, describe the arc $B\,g$, and get the length of this arc by stepping it off with a pair of dividers. Then

$$\frac{A^\circ}{90} = 0{\cdot}63\tfrac{2}{3}\,\frac{a}{h},$$

where a equals the length of the arc, and h equals $B\,C$, the height of the roof. Therefore,

$$M = P + 0{\cdot}63\tfrac{3}{8}\frac{a}{h}(C - P) \qquad (196.)$$

equals the value of timber per square inch in a tie beam, C and P being obtained from Table I., *Art.* 292. When C for the kind of wood in the tie beam exceeds C set opposite the kind of wood in the rafter, then the latter is to be used in the rules instead of the former.

369.—Having obtained the strain to which the material is subjected in a roof, and the capability of the material to resist that strain, it only remains now to state the rules for determining the dimensions of the material.

370.—To obtain the dimensions of the rafter:—It has been shown that the strain in the axis of the rafter equals (188),

$$R = W\frac{l}{2\,h}.$$

This is the strain in pounds. Timber is capable of resisting effectually, in every square inch of the surface pressed (196),

$$P + 0{\cdot}63\tfrac{3}{8}\frac{a}{h}(C - P) \text{ pounds.}$$

And when the strain and resistance are equal,

$$R = b\,d\,[P + 0{\cdot}63\tfrac{3}{8}\frac{a}{h}(C - P)],$$

where b and d are respectively the breadth and depth of the rafter. Hence

$$b\,d = \frac{R}{P + 0{\cdot}63\tfrac{3}{8}\frac{a}{h}(C - P)}. \qquad (197.)$$

Example.—(*Fig.* 233.) The strain in the axis of the rafter in this example, ascertained in *Art.* 362, is 30498·72 pounds. If the timber used be white pine, then $P = 300$ and $C = 1200$. The length of the arc $B\,g$ is 14¼ feet, and $h = 13$. Therefore

$$b\,d = \frac{30498{\cdot}72}{300 + (0{\cdot}63\tfrac{3}{8} \times \frac{14\frac{1}{4}}{13} \times 900)} = 32{\cdot}8.$$

This is the area of the abutting surface at the tie beam—say 6 by 5½ inches. At least half this amount should be added to allow for the shoulder, and for cutting at the joints fo. braces, &c. The rafter may therefore be 6 by 9 inches.

The above method is based upon the supposition that the rafter is effectually secured from flexure by the braces and roof beams. Should this not be the case, then the dimensions of the rafter are to be obtained by rules in *Art.* 298, for posts. Nevertheless, the abutting surface in the joint is to be determined by the above formula (197).

371.—To obtain the dimensions of the braces:—Usually, braces are so slender as to require their dimensions to be obtained by rules in *Art.* 298; the strain in the axis of the brace having been obtained by formula (191), or geometrically as in *Art.* 360.

The abutting surface of the joint of the brace is to be obtained, as in the case of the rafter, by formula (195); $A°$ being the number of degrees contained in the acute angle formed by the brace and a vertical line, for the joint at the tie beam; but for the joint at the rafter, $A°$ is the number of degrees contained in the acute angle formed by the brace and a line perpendicular to the rafter, or it is 90, diminished by the number of degrees contained in the acute angle formed by the rafter and brace.

Example.—Fig. 233, Brace $D\,E$, of white pine. In this brace the strain was found (*Art.* 362) to be 5600 pounds, the length of the brace is 9·6 feet. By *Art.* 298, the brace is therefore required to be 4·18 × 6 inches. For the abutting surface at the joints, for white pine, P equals 300 and, C 1200. The angle $D\,E\,F$ equals 63° 26′. By (197) and (195),

$$b\,d = \frac{T}{P + \frac{A°}{90}(C - P)} = \frac{5600}{300 + [\frac{36°\ 26}{90} \times (1200 - 300)]}$$

$$= \frac{5600}{934·5} = 6 \text{ inches.}$$

This is the area of the abutting surface of the joint at the tie beam. To obtain the joint at the rafter, the angle $F D E$ equals 53° 8′, and hence

$$b\,d = \frac{T}{P + \frac{A^\circ}{90}(C - P)} = \frac{5600}{300 + [\frac{90^\circ - 53^\circ\ 8'}{90} \times (1200 - 300)]}$$

$$= \frac{5600}{300 + (\frac{36^\circ\ 52'}{90} \times 900)} = 8{\cdot}375 \text{ inches.}$$

This is the area of the abutting surface of the joint at the rafter.

Another Example.—Brace $F B$, *Fig.* 233, of white pine, 12·2 feet long. The strain in its axis is (*Art.* 362) 7000 pounds. By *Art.* 298, the brace is required to be $5\frac{1}{4} \times 6$ inches. For the abutting surface of the joints, P equals 300, C equals 1200, and the angle $F B C$ equals 45°; therefore,

$$b\,d = \frac{7000}{300 + [\frac{45}{90} \times (1200 - 300)]} = 9\tfrac{1}{3} \text{ inches.}$$

This is the area of the abutting surface at the tie beam. For the surface at the rafter, the angle $C F B$ equals 71°, and $90 - 71 = 19$, equals the angle to be used in the formula; therefore,

$$b\,d = \frac{7000}{300 + [\frac{19}{90} \times (1200 - 300)]} = 14{\cdot}3 \text{ inches, nearly.}$$

This is the area of the abutting surface of the joint at the rafter.

372.—To obtain the dimensions of the tie beam:—A tie beam must be of such dimensions as will enable it to resist effectually the tensile strain caused by the horizontal thrust of the rafter and the cross strains arising from the weight of the ceiling, and from any load that may be placed upon it in the roof. From (17), *Art.* 310,

$$A = \frac{w}{T} = \frac{H}{T}$$

where H equals the horizontal thrust, and from (189),

$$H = \frac{W s}{4 h};$$

therefore,

$$A = \frac{H}{T} = \frac{W s}{4 h T},$$

where W equals the weight of the roof in pounds, as shown at (187); s, the span; h, the height, both in feet; and T, a constant set opposite the kind of material, in Table III.; and A equals the area of uncut fibres in the tie beam. About one-half of this should be added to allow for the requisite cutting at the joints; or, the area of the cross section of the tie beam should be equal to at least $\frac{3}{2}$ of the area of uncut fibres; or, when $b\,d$ equals the area of the tie beam, then

$$b\,d = \tfrac{3}{8} \cdot \frac{W s}{h T}. \qquad (198).$$

Example.—The weight on the truss at *Fig.* 233 is shown to be (*Art.* 362) 27343·68 pounds, say 27500 pounds; the span is 52 feet, the height 13, and the value of T for white pine is (Table III.) 2367, therefore

$$b\,d = \tfrac{3}{8} \frac{W s}{h T} = \tfrac{3}{8} \times \frac{27500 \times 52}{13 \times 2367} = 17·4 \text{ inches}$$

equals the area of cross section of the tie beam requisite to resist the tensile strain. This is smaller, as will be shown, than what is required to resist the cross strains, and this will be found to be the case generally. The weight of the ceiling is 9 pounds per superficial foot; the length of the longest unsupported part of the tie beam is $8\frac{2}{3}$ feet; then, if the deflection per lineal foot be allowed at 0·015 inch, the depth of the tie beam will be required ((72), Table V.) to be 6·14 inches. But in order effectually to resist the strains tie beams are subjected to at the hands of the workmen, in the process of framing and elevating, the area of cross section in inches should be at least equal to the length in feet. Were it possible to guard against this cause of strain, the size ascertained by the rule, 6 by 6·14, would be sufficient; but to resist this strain, the size should be 6 by 9.

There is yet one other dimension for the tie beam required, and that is, the distance at which the joint for the rafter must be located from the end of the tie beam, in order that the thrust of the rafter may not split off the part against which it presses. This may be ascertained by Rule XI., *Art.* 302, for all cases where no iron strap or bolt is used to secure the joint; but where these fastenings are used the abutment may be of any convenient length. And in using irons here, care should be exercised to have the surface of pressure against the iron of sufficient area to prevent indentation.

373.—To obtain the dimensions of the iron suspension rods. By *Art.* 310, (17),

$$A = \frac{w}{T},$$

and T varies (Table III.) from 5000 to 17000, according to the diameter inversely; for the smaller rods are stronger in proportion than the larger ones.

Example.—Taking T equal 5000, then the area of the rod $D\,G$ (*Fig.* 233) requires (*Art.* 367) to be equal to

$$A = \frac{867}{5000} = 0{\cdot}173 \text{ inch},$$

corresponding to 0·469 inch diameter. This rod may be half inch diameter.

Another Example.—The rod $F\,E$ (*Fig.* 233) is loaded with (*Art.* 367) 3367 pounds, therefore

$$A = \frac{3367}{5000} = 0{\cdot}673 \text{ inch}$$

equals the area of the rod, the corresponding diameter of which is 0·925. This rod may be one inch diameter.

Again, a third example; the rod $C\,B$. This rod is loaded with (*Art.* 367) 10867 pounds, therefore

$$A = \frac{10867}{5000} = 2{\cdot}173 \text{ inch}$$

equals the required area of the rod, the diameter corresponding to which is 1·66. This rod may therefore be $1\frac{3}{4}$ inches diameter.

374.—While discussing the principles of strains in roofs and deducing rules therefrom, the truss indicated in *Fig.* 233 has been examined throughout. The result is as follows: rafter, 6 × 9; tie beam, (6 × 6, or) 6 × 9; the first brace from the wall, $4\frac{1}{4}$ × 6 inches, with an abutting surface at the lower end of 6 inches, and at the upper end of $8\frac{3}{8}$ inches; the other brace, $5\frac{1}{4}$ × 6 inches, with an abutting surface at the lower end of $9\frac{1}{3}$ inches, and at the upper end of $14\frac{3}{16}$ inches; the shortest rod, $\frac{1}{2}$ inch diameter; the next, 1 inch diameter; and the middle rod, $1\frac{3}{4}$ inches diameter.

PRACTICAL RULES AND EXAMPLES.

For Roofs Loaded as per Art. 356.

375.—*Rule* LIII. To obtain the dimensions of the rafter Multiply the value of R (Table IX., *Art.* 376) by the span of the roof, by the length of the rafter, and by the distance apart from centres at which the roof trusses are placed, all in feet, and divide the product by the sum of twice the height of the roof multiplied by the value of P, Table I., set opposite the kind of wood used in the tie beam, added to the difference of the values of C and P in said table multiplied by $1\frac{1}{4}$ times the length of the arc that measures the acute angle formed between the rafter and a vertical line, the arc having the height of the roof for radius (see arc $B\ G$, *Fig.* 233), and the quotient will be the area of the abutting surface of the joint at the foot of the rafter. To the abutting surface add its half, and the sum will be the area of the cross section of the rafter

This rule is upon the presumption that the rafter is secured from flexure by the roof beams and by braces and ties at short intervals, as in *Fig.* 233. In roofs where the rafter does not extend up to the ridge of the roof but abuts against a horizontal straining beam (*c*, *Fig.* 237), in the rule for rafters, take for the length of the rafter the distance from the foot of the rafter to the ridge of the roof; or, a distance equal to what the rafter would be in the absence of a straining beam. The area of cross section of the straining beam should be made equal to that of the rafter, as found by the rule so modified.

Example.—Find the dimensions of a rafter for a roof truss whose span is 52 feet, and height 13; the length of the rafter being 29 feet, the trusses placed 10 feet apart from centres, and the arc measuring the angle at the head of the rafter (having the height of the roof for radius) being 14¼ feet, white pine being used in the tie beam. The height of this roof being in proportion to the span as 1 to 4, the value of R in Table IX. is 52·6; multiplying this, in accordance with the rule, by 52, the span of the roof, and by 29, the length of the rafter, and by 10, the distance between the roof trusses, the product is 793208. The value of P for white pine in Table I. is 300; multiplying this by $2 \times 13 = 26$, twice the height of the roof, the product is 7800. The value of C for white pine, (Table I.) is 1200, hence the difference of the values of C and P is $1200 - 300 = 900$; this multiplied by 1¼, and by 14¼, the length of the arc, the product is 16031; this added to the 7800 aforesaid, the sum is 23831. The aforesaid product of 793208, divided by this 23831, the quotient, 33·3, equals the area in inches of the abutting surface of the joint at the tie beam. To this add 16·7, its half, and the sum, 50, equals the area of cross section of the rafter. This divided by the thickness of the rafter, say 6 inches, the quotient, 8⅓, is the breadth. The rafter is therefore to be 6 × 8⅓ inches. It may be made 6 × 9, avoiding the fractions.

376.—The following table, calculated upon data in *Art.* 358, presents the weight per foot for roofs of various inclinations, and covered with slate.

TABLE IX.

When height of roof is to span as	The vertical strain per foot of surface supported, measured horizontally,	
	on rafters — *R* —	on braces — *Q* —
1 to 8	48 pounds	49·5 pounds.
1 " 7	48·6 "	50·5 "
1 " 6	49·4 "	51·9 "
1 " 5	50·6 "	54· "
1 " 4	52·6 "	57·6 "
1 " 3	56·3 "	64· "
1 " 2	63·7 "	76·2 "
1 " 1	81· "	101· "

To get the proportion that the height bears to the span, divide the span by the height; then unity will be to the quotient as the height is to the span. In case the quotient is not a whole number, the required value of *R* or *Q* will not be found in the above table, but may be obtained thus: multiply the decimal part of the quotient by the difference of the values of *R* set opposite the two proportions, between which the given proportion occurs as an intermediate, and subtract the product from the larger of the two said values of *R*; the remainder will be the value of *R* required. The process is the same for the values of *Q*.

Example.—A roof whose span is 60 feet, has a height of 25 feet. Then 60 divided by 25 equals 2·4. The proportion, therefore, between the height and span is 1 to 2·4. This proportion is an intermediate between 1 to 2 and 1 to 3. The values of *R*, opposite these two, are 63·7 and 56·3. The difference between these values is 7·4; this multiplied by 0·4, the decimal portion of the quotient, equals 2·96; this subtracted from 63·7, the larger value of *R*, the remainder, 60·74, is the required value of *R*.

The values of R and Q are those for a roof covered with slate weighing 7 pounds per superficial foot of the roof surface. When the roof covering is either lighter or heavier, subtract from or add to the table values, the difference of weight between 7 pounds and the weight of the covering used, and the remainder, or sum, will be the value of R or Q required.

377.—*Rule* LIV. To obtain the dimensions of braces. Multiply the value of Q (Table IX., *Art.* 376) by the distance apart in feet at which the roof trusses are placed, and by the *horizontal* distance in feet from a point half-way to the next point of support of the rafter on one side of the brace, to a corresponding point on the other side. The product will be the weight in pounds sustained at the head of the brace. To this add the vertical strain (*Art.* 360) on the suspension rod located at the head of the brace, and make a vertical line dropped from the head of the brace, as *Fe*, *Fig.* 233, equal, by any convenient scale, to this sum, and draw the parallelogram *Ffeg*. Then *Ff*, measured by the same scale, equals the pressure in the axis of the brace *FB*. Multiply this pressure in pounds by the square of the length of the brace in feet, and divide the product by the breadth of the brace in inches multiplied by the value of B (Table II., *Art.* 293). The cube root of the quotient will be the thickness of the brace in inches. If this cube root should exceed the *breadth* of the brace, the result is not correct, and the calculation will have to be made anew, taking a larger dimension for the breadth.

Example.—The brace *FB* (*Fig.* 233) is of white pine, and is required to sustain a pressure in its axis of 7000 pounds (*Art.* 362). The length of the brace is 12 feet and its breadth 6 inches, what must be its thickness? Here 7000, the pressure, multiplied by 144, the square of the length, equals 1008000. The value of B is 1175; this by 6, the breadth of the brace, equals 7050. The product 1008000 divided by the product 7050 equals 143, the cube root of which, 5·23, is the required thickness of the brace in inches. The brace will therefore be 5·23 by 6 inches, or 5¼ by 6.

378.—*Rule* LV. To obtain the area of the abutting surface of the ends of braces. Divide the number of degrees contained in the complement of the angle of inclination by 90, and multiply the quotient by the difference of the values of *C* and *P*, set opposite the kind of wood in the tie beam or rafter, in Table I., *Art.* 292; and to the product add the said value of *P*, and by the sum divide the pressure in the axis of the brace, and the quotient will be the area of the abutting surface.

The complement of the angle of inclination referred to is, for the foot of the brace, the acute angle contained between the brace and a vertical line; and for the head of the brace, the acute angle contained between the brace and a line perpendicular to the rafter.

Example.—To find the abutting surface of the ends of the brace *F B* (*Fig.* 233). The complement of the angle of inclination, for the *foot* of the brace, is that contained between the lines *F B* and *F E*, and measures by the protractor, 45°. The tie beam is of white pine, and the values of *P* and *C* for this wood are 300 and 1200 respectively, and the pressure in the axis of the brace is 7000 pounds. Now by the rule, 45 divided by 90 equals 0·5, this by the 900, the difference of the values of *C* and *P*, equals 450; to this add 300, the value of *P*, and the sum is 750. The pressure in the axis of the brace, 7000, divided by this 750, equals 9⅓, the required area of the abutting surface at the foot of the rafter. The complement of the angle of inclination for the *head* of the brace is that contained between the lines *B F* and *F p*, and measures by the protractor 19°. The rafter being of white pine, the values of *P* and *C* are as before. By the rule, 19 divided by 90 equals 0·2⅑, and this multiplied by 900, the difference of the values of *P* and *C*, equals 190; to this add 300, the value of *P*, and the sum is 490. The pressure, 7000, divided by this 490, equals 14·3 inches, the required area of the abutting surface at the head of the brace.

379.—To obtain the dimensions of the tie beam. Tie beams are subjected to two kinds of strain—tensile and transverse.

Rule LVI.—To guard against the tensile strain, multiply the value of *R* (Table IX., *Art.* 376) by three times the distance apart at which the trusses are placed, and by the square of the span of the truss, both in feet. Divide this product by the value of *T*, (Table III., *Art.* 308) set opposite the kind of wood in the tie beam, multiplied by 8 times the height of the roof in feet, and by the breadth of the tie beam in inches. The quotient will be the required depth in inches.

The result thus obtained is usually smaller than that required to resist the cross strain to which the tie beam is subjected. The dimensions required to resist this strain, where there is simply the weight of the ceiling to support, may be obtained by this rule:

Rule LVII.—Multiply the cube of the longest unsupported part of the tie beam by 400 times the distance apart at which the trusses are placed, both in feet; and divide the product by the breadth of the tie beam in inches, multiplied by the value of *E*, (Table II., *Art.* 293) set opposite the kind of wood in the tie beam, and the cube root of the quotient will be the required depth of the tie beam in inches.

The result thus obtained may not be sufficient, in some cases, to resist the strains to which the tie beam is subjected in the hands of the workmen during the process of framing.

Rule LVIII.—To resist these strains the area of cross section in inches should be at least equal to the length in feet.

Example.—The tie beam in *Fig.* 233. For this case we have the value of *R* 52·6, the trusses placed 10 feet from centres, the span 52 feet, the height 13 feet, the breadth 6 inches, and the value of *T* 2367. Then by the rule, $52{\cdot}6 \times 3 \times 10 \times 52^2$

= 4266912, and 2367 × 8 × 13 × 6 = 1477008; the former product divided by the latter, the quotient equals 2·9, equals the required depth of the tie beam in inches. The other strains will require the depth to be more. To resist the cross strains, we have the longest unsupported part of the tie beam $8\frac{3}{8}$ feet, (this dimension is frequently greater than this,) distance from centres 10 feet, and breadth 6 inches. Then, by the rule, $8\frac{3}{8}^3 \times 400 \times 10 = 2603852$, and 6 × 1750 = 10500; the former product divided by the latter, the quotient is 248, the cube root of which, 6·28, equals the required depth in inches. The tie beam therefore is to be 6 by 6·28 inches, or 6 × 7 inches. But if not guarded against severe accidental strains from careless handling this size would be too small. It would, in this case, require to be 52 inches area of cross section, say 6 × 9 inches.

380.—To obtain the diameter of the suspension rods, when made of good wrought iron.

Rule LIX.—Divide the weight or vertical strain, in pounds, by 4000. The square root of the quotient will be the required diameter of the rod in inches.

Example.—A suspension rod is required to sustain 16000 pounds, what must be its diameter? Dividing by 4000, the quotient is 4; the square root of which, 2, is the required diameter.

The vertical strain on any rod is equal to the weight of so much of the ceiling as is supported by the rod, added to the vertical strain caused by each brace that is footed in the tie beam at the rod. The weight of the ceiling supported by a rod, is equal to ten times the distance apart in feet at which the trusses are placed, multiplied by half the distance in feet between the two next points of support, one on either side of the rod. The vertical strain caused by the braces can be ascertained geometrically, as in *Art.* 360.

381.—When the suspension rods are located as in *Fig.* 233, dividing the span into equal parts, the diameter of the rods may be obtained without the preliminary calculation of the strain, as follows:

Rule LX.—For the first rod from the wall. Multiply the distance apart at which the trusses are placed by the distance apart between the suspension rods, and divide the product by 400. The square root of the quotient will be the required diameter of the rod.

Example.—Rod *D G*, *Fig.* 233. In this figure the rods are located at $8\frac{2}{3}$ feet apart, and the distance between the trusses is 10 feet. Therefore, $10 \times 8\frac{2}{3} = 86\frac{2}{3}$; this divided by 400, the quotient is 0·2167, the square root of which, 0·465, is the required diameter. The diameter may be half an inch.

Rule LXI.—For the second rod from the wall. To the value of *Q* (Table IX., *Art.* 376) add 20, and multiply the sum by the distance apart at which the trusses are placed and by the distance between the rods, both in feet, and divide the product by 8000. The square root of the quotient will be the required diameter.

Example.—Rod *F E*, *Fig.* 233. The distances apart in this case are as stated in last example. The value of *Q* is 57·6, and when added to 20 equals 77·6. Therefore, $77{\cdot}6 \times 10 \times 8\frac{2}{3} = 6673\frac{1}{3}$; this divided by 8000, the quotient is 0·8341, the square root of which, 0·91, is the required diameter. This rod may be one inch diameter.

Rule LXII.—For the centre rod. To the value of *Q* (Table IX., *Art.* 376) add 5, and multiply the sum by the distance apart at which the trusses are placed and by the distance apart between the rods, both in feet, and divide the product by 2000. The square root of the quotient will equal the required diameter.

Example.—Rod *C B*, *Fig.* 233. The distances apart as before, and the value of *Q* the same. To *Q* add 5, and the sum is 62·6. Then $62{\cdot}6 \times 10 \times 8\frac{2}{3} = 5425\frac{1}{3}$; this divided by 2000, the quotient is 2·7126, the square root of which, 1·647, equals the required diameter. This rod may be $1\frac{3}{4}$ inches diameter.

382.—For all wrought iron straps and bolts the dimensions may be found by this rule.

Rule LXIII.—Divide the tensile strain on the piece, in pounds, by 5000, and the quotient will be the area of cross section of the required bar or bolt, in inches.

383.—Roof-beams, jack-rafters, and purlins. All pieces of timber subject to cross strains will sustain safely much greater strains when extended in one piece over two, three, or more distances between bearings; therefore roof-beams, jack-rafters, and purlins should, if possible, be made in as long lengths as practicable; the roof-beams and purlins laid on, not framed into, the principal rafters, and extended over at least two spaces, the joints alternating on the trusses; and likewise the jack-rafters laid on the purlins in long lengths. The dimensions of these several pieces may be obtained by the following rule:

Rule LXIV.—From the value of Q (Table IX., *Art.* 376) deduct 10, and multiply the remainder by 33 times the distance from centres in feet at which the pieces are placed, and by the cube of the distance between bearings in feet; divide the product by the value of E (Table II., *Art.* 293) for the kind of wood used and extract the square root of the quotient. The square root of this square root will be the required depth in inches. Multiply the depth thus obtained by the decimal 0·6, and the product will be the required breadth in inches.

Example.—Roof-beams of white pine placed 2 feet from centres, resting on trusses placed 10 feet from centres, the height and the span of the roof being in proportion as 1 to 4. In this case the value of Q is 57·6. By the rule, $57·6 - 10 = 47·6$, and $47·6 \times 33 \times 2 \times 10^3 = 3141600$. This divided by 1750, the quotient is 1795·2, the square root of which is 42·37, and the square root of 42·37 is 6·5, the required depth. This multiplied by 0·6 equals 3·9, the required breadth. These roof beams may therefore be 4 by 6½ inches.

384.—Five examples of roofs are shown at *Figs.* 234, 235, 236, 237, and 238. In *Fig.* 234, *a* is an iron suspension rod, *b b* are braces. In *Fig.* 235, *a*, *a*, and *b* are iron rods, and *d d*, *c c*, are braces. In *Fig.* 236, *a b* are iron rods, *d d* braces, and *c* the straining beam. In *Fig.* 237, *a a*, *b b*,

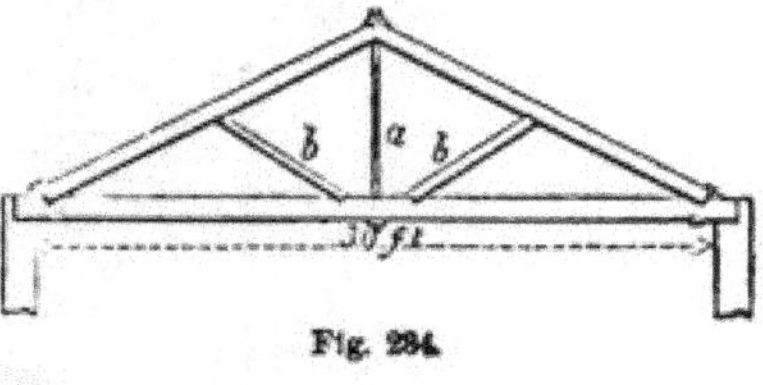

Fig. 234.

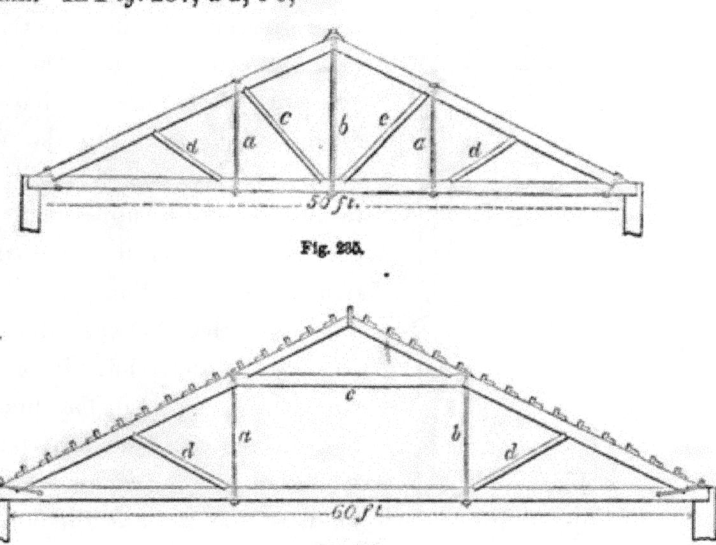

Fig. 235.

Fig. 236.

are iron rods, *e e*, *d d*, are braces, and *c* is a straining beam. In *Fig.* 238, purlins are located at *P P*, &c.; the inclined beam that lies upon them is the jack-rafter; the post at the ridge is the king post, the others are queen posts. In this design the tie beam is increased in height along the middle by a strengthening piece (*Art.* 348), for the purpose of sustaining additional weight placed in the room formed in the truss.

385.—*Fig.* 239 shows a method of constructing a truss having a *built-rib* in the place of principal rafters. The proper form for the curve is that of a parabola, (*Art.* 127.) This curve, when as flat as is described in the figure, approximates so near to that of the circle, that the latter may be used in its stead. The height, *a b*, is just half of *a c*, the curve to pass through the middle of the rib. The rib is composed of two series of abutting pieces, bolted together. These pieces should be as long as the dimensions of the timber will admit, in order that there may be but few joints. The suspending pieces are in halves, notched and bolted to the tie-beam and rib, and a purlin is framed upon the upper end of each. A truss of this construction needs, for ordinary roofs, no diagonal braces between the suspending pieces, but if extra strength is required the braces may be added. The best place for the suspending pieces is at the joints of the rib. A rib of this kind will be sufficiently strong, if the area of its section contain about one-fourth more timber, than is required for that of a rafter for a roof of the same size. The proportion of the depth to the thickness should be about as 10 is to 7.

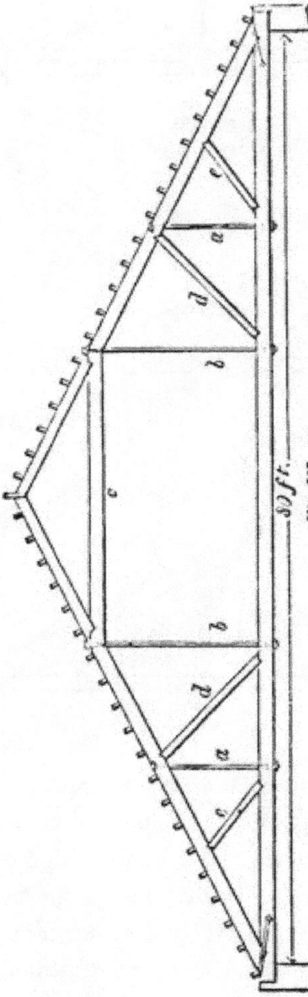

Fig. 237.

386—Some writers have given designs for roofs similar to *Fig.* 240, having the tie-beam omitted for the accommodation of an arch in the ceiling. This and all similar designs are seriously objectionable, and should always be avoided; as the small height gained by the omission of the tie-beam can never

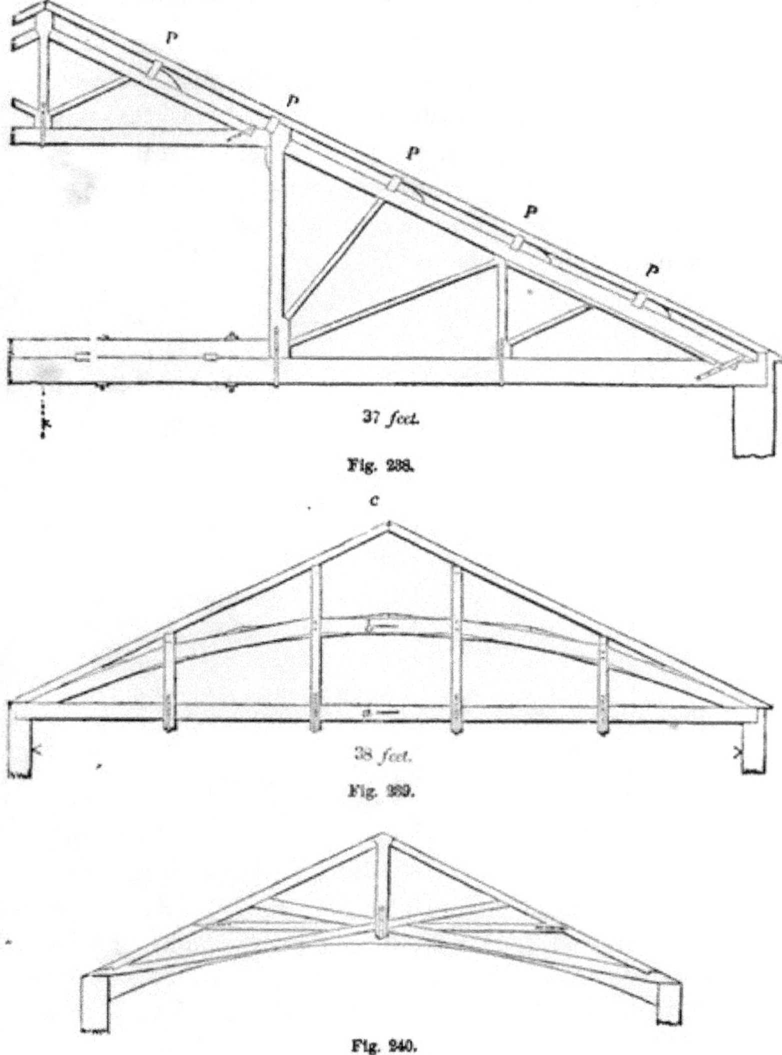

Fig. 238.

Fig. 239.

Fig. 240.

compensate for the powerful lateral strains, which are exerted by the oblique position of the supports, tending to separate the walls. Where an arch is required in the ceiling, the best plan is to carry up the walls as high as the top of the arch. Then, by using a horizontal tie-beam, the oblique strains will be entirely removed. Many a public building, by my own observation, has been all but ruined by the settling of the roof, consequent upon a defective plan in the formation of the truss in this respect. It is very necessary, therefore, that the horizontal tie-beam be used, except where the walls are made so strong and firm by buttresses, or other support, as to prevent a possibility of their separating.

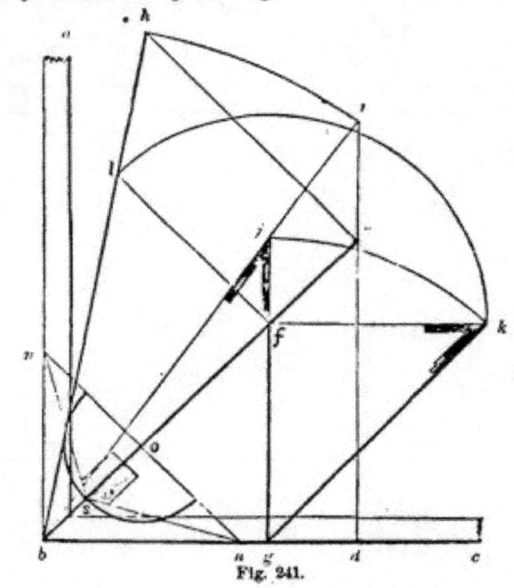

Fig. 241.

387.—*Fig.* 241 is a method of obtaining the proper lengths and bevils for rafters in a hip-roof: *a b* and *b c* are walls at the angle of the building; *b e* is the seat of the hip-rafter and *g f* of a jack or cripple rafter. Draw *e h*, at right angles to *b e*, and make it equal to the rise of the roof; join *b* and *h*, and *h b* will be the length of the hip-rafter. Through *e*, draw *d i*, at right angles to *b c*; upon *b*, with the radius, *b h*, describe the arc, *h i*, cutting *d i* in *i*; join *b* and *i*, and extend *g f* to meet *b i* in *j*; then *g j* will be the length of the jack-rafter. The length of each jack-rafter is found in the same manner—by extending its seat to cut the line, *b i*. From *f*, draw *f k*, at right angles to *f g*, also *f l*, at right angles to *b e*; make *f k* equal to *f l* by the arc, *l k*, or make *g k* equal to *g j* by the arc, *j k*; then the angle at *j* will be the *top-bevil* of the jack-rafters, and the one at *k* will be the *down-bevil*.*

388.—*To find the backing of the hip-rafter.* At any convenient place in *b e*, (*Fig.* 241,) as *o*, draw *m n*, at right angles to *b e*; from *o*, tangical to *b h*, describe a semi-circle, cutting *b e* in *s*; join *m* and *s* and *n* and *s*; then these lines will form at *s* the proper angle for beviling the top of the hip-rafter.

DOMES.†

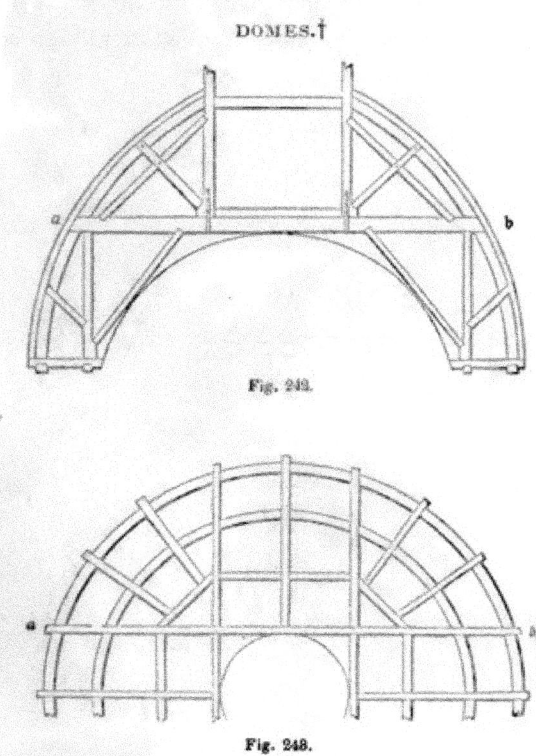

Fig. 242.

Fig. 243.

* The lengths and bevils of rafters for *roof-valleys* can also be found by the above process

† See also *Art.* 237.

389.—The most usual form for domes is that of the sphere, the base being circular. When the interior dome does not rise too high, a horizontal tie may be thrown across, by which any degree of strength required may be obtained. *Fig.* 242 shows a section, and *Fig.* 243 the plan, of a dome of this kind, *a b* being the tie-beam in both. Two trusses of this kind, (*Fig.* 242,) parallel to each other, are to be placed one on each side of the opening in the top of the dome. Upon these the whole framework is to depend for support, and their strength must be calculated accordingly. (See the first part of this section, and *Art.* 356.) If the dome is large and of importance, two other trusses may be introduced at right angles to the foregoing, the tie-beams being preserved in one continuous length by framing them high enough to pass over the others.

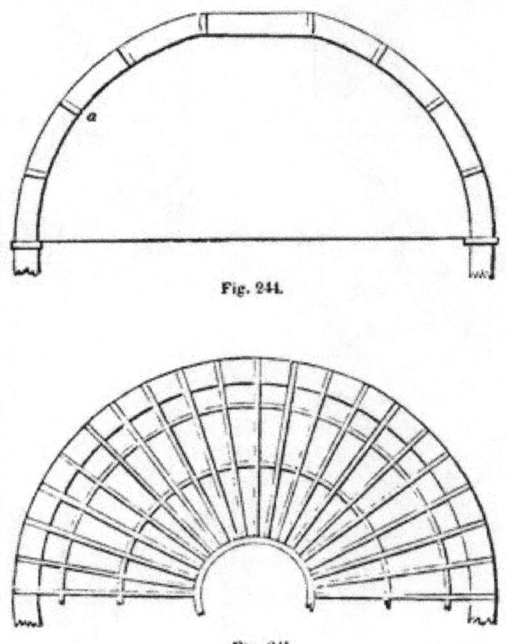

Fig. 244.

Fig. 245.

390.—When the interior dome rises too high to admit of a leve.

tie-beam, the framing may be composed of a succession of ribs standing upon a continuous circular curb of timber, as seen at *Fig.* 244 and 245,—the latter being a plan and the former a section. This curb must be well secured, as it serves in the place of a tie-beam to resist the lateral thrust of the ribs. In small domes, these ribs may be easily cut from wide plank; but, where an extensive structure is required, they must be built in two thicknesses so as to *break joints*, in the same manner as is described for a roof at *Art.* 385. They should be placed at about two feet apart at the base, and strutted as at *a* in *Fig.* 244.

391.—The scantling of each thickness of the rib may be as follows:

For domes of		24	feet diameter,	1×8	inches.
"	"	36	"	1½×10	"
"	"	60	"	2×13	"
"	"	90	"	2½×13	"
"	"	108	"	3×13	"

392.—Although the outer and the inner surfaces of a dome may be finished to any curve that may be desired, yet the framing should be constructed of such a form, as to insure that the *curve of equilibrium* will pass through the middle of the depth of the framing. The nature of this curve is such that, if an arch or dome be constructed in accordance with it, no one part of the structure will be less capable than another of resisting the strains and pressures to which the whole fabric may be exposed. The curve of equilibrium for an arched vault or a roof, where the load is equally diffused over the whole surface, is that of a parabola, (*Art.* 127;) for a dome, having no *lantern*, tower or cupola above it, a *cubic parabola*, (*Fig.* 246;) and for one having a tower, &c., above it, a curve approaching that of an hyperbola must be adopted, as the greatest strength is required at its upper parts. If the curve of a dome be circular, (as in the vertical section, *Fig.* 244,) the pressure will have a tendency to burst the dome outwards at about one-third of its height. Therefore, when this form is used in the construction of an extensive dome, an iron band should be placed around the framework at that height; and whatever may be the form of the curve, a band or tie of some kind is necessary around or across the base.

If the framing be of a form less convex than the curve of equilibrium, the weight will have a tendency to crush the ribs inwards, but this pressure may be effectually overcome by strutting between the ribs; and hence it is important that the struts be so placed as to form continuous horizontal circles.

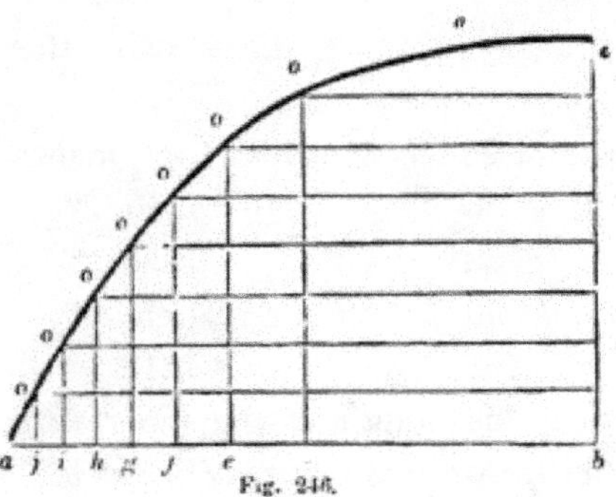

Fig. 246.

393.—*To describe a cubic parabola.* Let *a b*, (*Fig.* 246,) be the base and *b c* the height. Bisect *a b* at *d*, and divide *a d* into 100 equal parts; of these give *d e* 26, *e f* 18¼, *f g* 14½, *g h* 12¼, *h i* 10¾, *i j* 9½, and the balance, 8¾, to *j a*; divide *b c* into 8 equal parts, and, from the points of division, draw lines parallel to *a b*, to meet perpendiculars from the several points of division in *a b*, at the points, *o*, *o*, *o*, &c. Then a curve traced through these points will be the one required.

394.—Small domes to light stairways, &c., are frequently made elliptical in both plan and section; and as no two of the ribs in one quarter of the dome are alike in form, a method for obtaining the curves is necessary.

395.—*To find the curves for the ribs of an elliptical dome.* Let *a b c d*, (*Fig.* 247,) be the plan of a dome, and *e f* the seat

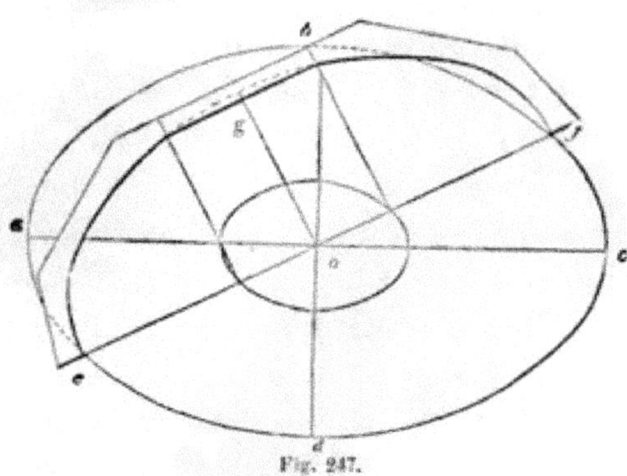

Fig. 247.

of one of the ribs. Then take *e f* for the transverse *axis* and twice the rise, *o g*, of the dome for the conjugate, and describe (according to *Art.* 115, 116, &c.,) the semi-ellipse, *e g f*, which will be the curve required for the rib, *e g f*. The other ribs are found in the same manner.

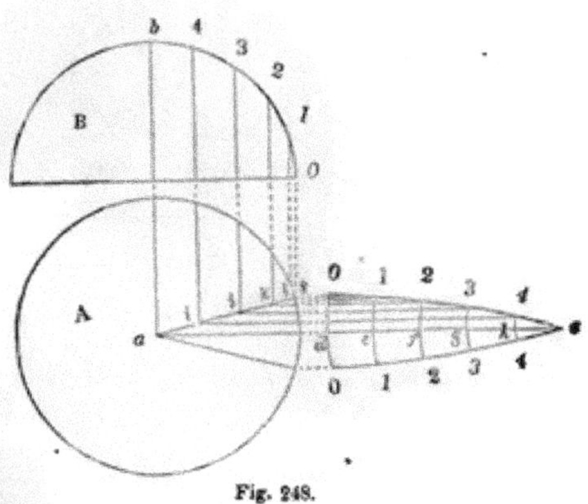

Fig. 248.

396.—*To find the shape of the covering for a spherical dome.* Let *A*, (*Fig.* 248,) be the plan and *B* the section of a given dome. From *a*, draw *a c*, at right angles to *a b*; find the stretch-out, (*Art.* 92,) of *o b*, and make *d c* equal to it; divide the arc, *o b*, and the line, *d c*, each into a like number of equal parts, as 5, (a large number will insure greater accuracy than a small one;) upon *c*, through the several points of division in *c d*, describe the arcs, *o d o*, 1 *e* 1, 2 *f* 2, &c.; make *d o* equal to half the width of one of the boards, and draw *o s*, parallel to *a c*; join *s* and *a*, and from the points of division in the arc, *o b*, drop perpendiculars, meeting *a s* in *i j k l*; from these points, draw *i* 4, *j* 3, &c., parallel to *a c*; make *d o*, *e* 1, &c., on the lower side of *a c*, equal to *d o*, *e* 1, &c., on the upper side; trace a curve through the points, *o*, 1, 2, 3, 4, *c*, on each side of *d c*; then *o c o* will be the proper shape for the board. By dividing the circumference of the base, *A*, into equal parts, and making the bottom, *o d o*, of the board of a size equal to one of those parts, every board may be made of the same size. In the same manner as the above, the shape of the covering for sections of another form may be found, such as an ogee, cove, &c.

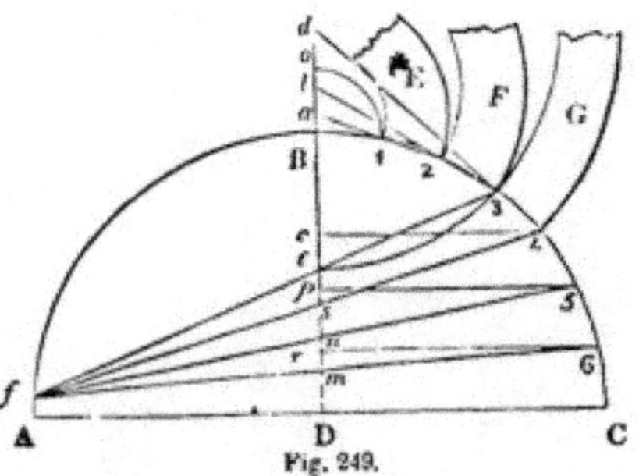

Fig. 249.

397.—*To find the curve of the boards when laid in horizontal courses.* Let *A B C*, (*Fig.* 249,) be the section of a given dome, and *D B* its axis. Divide *B C* into as many parts as there are to be courses of boards, in the points, 1, 2, 3, &c.; through 1 and 2, draw a line to meet the axis extended at *a*; then *a* will be the centre for describing the edges of the board, *E*. Through 3 and 2, draw 3 *b*; then *b* will be the centre for describing *F*. Through 4 and 3, draw 4 *d*; then *d* will be the centre for *G*. *B* is the centre for the arc, 1 *o*. If this method is taken to find the centres for the boards at the base of the dome, they would occur so distant as to make it impracticable: the following metl od is preferable for this purpose. *G* being the last board obtained by the above method, extend the curve of its inner edge until it meets the axis, *D B*, in *e*; from 3, through *e*, draw 3 *f*, meeting the arc, *A B*, in *f*; join *f* and 4, *f* and 5 and *f* and 6, cutting the axis, *D B*, in *s*, *n* and *m*; from 4, 5 and 6, draw lines parallel to *A C* and cutting the axis in *c*, *p* and *r*; make *c* 4, (*Fig.* 250,)

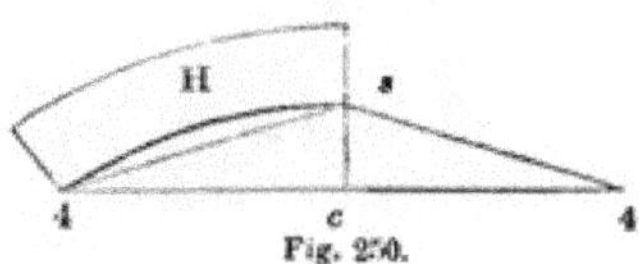

Fig. 250.

equal to *c* 4 in the previous figure, and *c s* equal to *c s* also in the previous figure; then describe the inner edge of the board, *H*, according to *Art.* 87: the outer edge can be obtained by gauging from the inner edge. In like manner proceed to obtain the next board—taking *p* 5 for half the chord and *p n* for the height of the segment. Should the segment be too large to be described easily, reduce it by finding intermediate points in the curve, as at *Art.* 86.

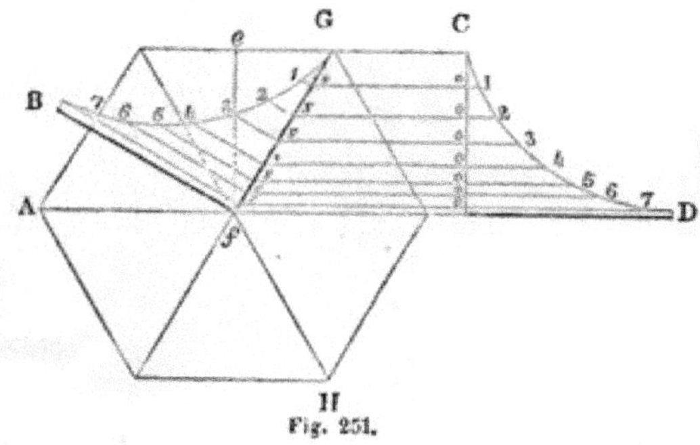

Fig. 251.

398.—*To find the shape of the angle-rib for a polygonal dome.* Let *A G H*, (*Fig.* 251,) be the plan of a given dome, and

C D a vertical section taken at the line, *e f*. From 1, 2, 3, &c. in the arc, *C D*, draw ordinates, parallel to *A D*, to meet *f G*, from the points of intersection on *f G*, draw ordinates at right-angles to *f G*; make *s* 1 equal to *o* 1, *s* 2 equal to *o* 2, &c.; then *G f B*, obtained in this way, will be the angle-rib required. The best position for the sheathing-boards for a dome of this kind is horizontal, but if they are required to be bent from the base to the vertex, their shape may be found in a similar manner to that shown at *Fig.* 248.

BRIDGES.

399.—Various plans have been adopted for the construction of bridges, of which perhaps the following are the most useful. *Fig.* 252 shows a method of constructing wooden bridges, where the banks of the river are high enough to permit the use of the tie-beam, *a b*. The upright pieces, *c d*, are notched and bolted on in pairs, for the support of the tie-beam. A bridge of this construction exerts no lateral pressure upon the abutments. This method may be employed even where the banks of the river are low, by letting the timbers for the roadway rest immediately upon the tie-beam. In this case, the framework above will serve the purpose of a railing.

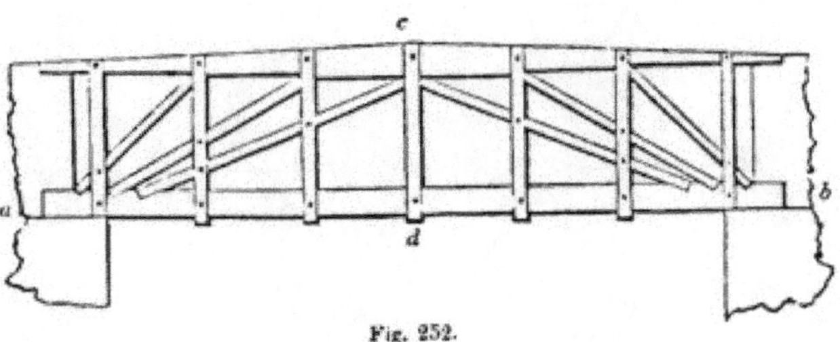

Fig. 252.

400.—*Fig.* 253 exhibits a wooden bridge without a tie-beam. Where staunch buttresses can be obtained, this method may be recommended; but if there is any doubt of their stability, it

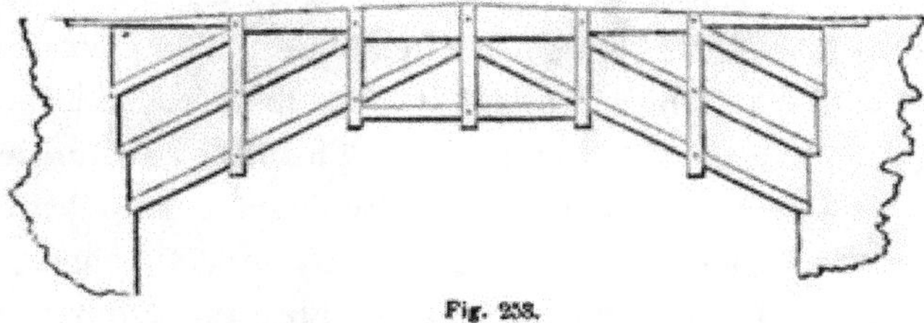

Fig. 253.

should not be attempted, as it is evident that such a system of framing is capable of a tremendous lateral thrust.

Fig. 254.

401.—*Fig.* 254 represents a wooden bridge in which a *built-rib*, (see *Art.* 385,) is introduced as a chief support. The curve of equilibrium will not differ much from that of a parabola: this, therefore, may be used—especially if the rib is made gradually a little stronger as it approaches the buttresses. As it is desirable that a bridge be kept low, the following table is given to show the least rise that may be given to the rib.

Span in feet.	Least rise in feet.	Span in feet.	Least rise in feet.	Span in feet.	Least rise in feet.
30	0·5	120	7	280	24
40	0·8	140	8	300	28
50	1·4	160	10	320	32
60	2	180	11	350	39
70	2½	200	12	380	47
80	3	220	14	400	53
90	4	240	17		
100	5	260	20		

The rise should never be made less than this, but in all cases greater if practicable; as a small rise requires a greater quantity of timber to make the bridge equally strong. The greatest uniform weight with which a bridge is likely to be loaded is, probably, that of a dense crowd of people. This may be estimated at 66 pounds per square foot, and the framing and gravelled roadway at 234 pounds more; which amounts to 300 pounds on a square foot. The following rule, based upon this estimate, may be useful in determining the area of the ribs. *Rule* LXV.—Multiply the width of the bridge by the square of half the span, both in feet; and divide this product by the rise in feet, multiplied by the number of ribs; the quotient, multiplied by the decimal, 0·0011, will give the area of each rib in feet. When the roadway is only planked, use the decimal, 0·0007, instead of 0·0011. *Example.*—What should be the area of the ribs for a bridge of 200 feet span, to rise 15 feet, and be 30 feet wide, with 3 curved ribs? The half of the span is 100 and its square is 10,000; this, multiplied by 30, gives 300,000, and 15, multiplied by 3, gives 45; then 300,000, divided by 45, gives 6666⅔, which, multiplied by 0·0011, gives 7·333 feet, or 1056 inches for the area of each rib. Such a rib may be 24 inches thick by 44 inches deep, and composed of 6 pieces, 2 in width and 3 in depth.

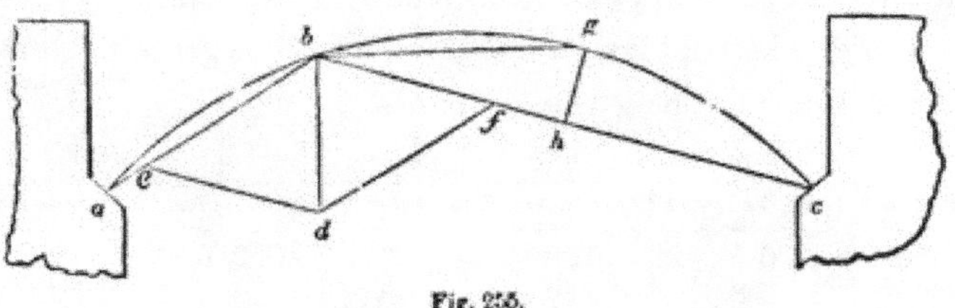

Fig. 255.

402.—The above rule gives the area of a rib, that would be requisite to support the greatest possible *uniform* load. But in large bridges, a *variable* load, such as a heavy wagon, is capable of exerting much greater strains; in such cases, therefore, the rib should be made larger. The greatest concentrated load a bridge will be likely to encounter, may be estimated at from about 20 to 50 thousand pounds, according to the size of the bridge. This is capable of exerting the greatest strain, when placed at about one-third of the span from one of the abutments, as at *b* (*Fig.* 255.) The weakest point of the segment, *b g c*, is at *g*, the most distant point from the chord line. The pressure exerted at *b* by the above weight, may be considered to be in the direction of the chord lines, *b a* and *b c*; then, by constructing the parallelogram of forces, *c b f d*, according to *Art.* 258, *b f* will show the pressure in the direction, *b c*. Then the scantling for the rib may be found by the following rule.

Rule LXVI.—Multiply the pressure in pounds in the direction *b c*, by the distance *g h*, and by the square of the distance *b c*, both in feet; and divide the product by the united breadth in inches of the several ribs, multiplied by the value of *B* (Table II., *Art.* 293) for the kind of wood used; and the cube root of the quotient will be the required depth of the rib in inches.

Example.—A bridge is to have three white pine ribs each 20 inches wide; the pressure in the direction *b c*, (*Fig.* 255) is equal to 60,000 pounds, the distance *b c* equals 60 feet, and the distance *g h* equals 10 feet. What must be the depth of the ribs, the value of *B* (Table II.) being for white pine 1175? Here, by the rule, $60,000 \times 10 \times 60^2 = 2,160,000,000$. Then $1175 \times 3 \times 20 = 70,500$. The former product divided by the latter equals 30,638, the cube root of which, 31·29, equals the required depth in inches. The ribs are, therefore, to be 20 by 31½ inches.

403.—In constructing these ribs, if the span be not over 50 feet, each rib may be made in two or three thicknesses of timber, (three thicknesses is preferable,) of convenient lengths bolted together; but, in larger spans, where the rib will be such as to render it difficult to procure timber of sufficient breadth, they may be constructed by bending the pieces to the proper curve and bolting them together. In this case, where timber of sufficient length to span the opening cannot be obtained and scarfing is necessary, such joints must be made as will resist both tension and compression, (see *Fig.* 264) To ascertain the greatest depth for the pieces which compose the rib, so that the process of bending may not injure their elasticity, multiply the radius of curvature in feet by the decimal, 0·05, and the product will be the depth in inches. *Example.*—Suppose the curve of the rib to be described with a radius of 100 feet, then what should be the depth ? The radius in feet, 100, multiplied by 0·05, gives a product of 5 inches. White pine or oak timber, 5 inches thick, would freely bend to the above curve ; and, if the required depth of such a rib be 20 inches, it would have to be composed of at least 4 pieces. Pitch pine is not quite so elastic as white pine or oak—its thickness may be found by using the decimal, 0·046, instead of 0·05.

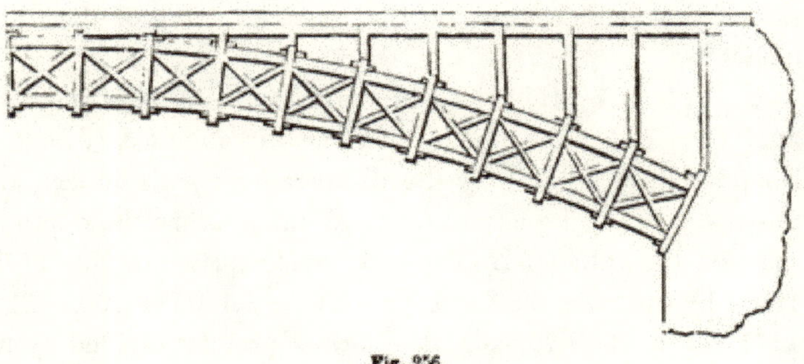

Fig. 256.

404.—When the span is over 250 feet, a *framed* rib, formed as in *Fig.* 256, would be preferable to the foregoing. Of this, the upper and the lower edges are formed as just described, by bending the timber to the proper curve. The pieces that tend to the centre of the curve, called *radials,* are notched and bolted on in pairs, and the cross-braces are halved together in the middle, and abut end to end between the radials. The distance between the ribs of a bridge should not exceed about 8 feet. The roadway should be supported by vertical standards bolted to the ribs at about every 10 to 15 feet. At the place where they rest on the ribs, a double, horizontal tie should be notched and bolted on the back of the ribs, and also another on the under side; and diagonal braces should be framed between the standards, over the space between the ribs, to prevent lateral motion. The timbers for the roadway may be as light as their situation will admit, as all useless timber is only an unnecessary load upon the arch.

405.—It is found that if a roadway be 18 feet wide, two carriages can pass one another without inconvenience. Its width, therefore, should be either 9, 18, 27 or 36 feet, according to the amount of travel. The width of the foot-path should be 2 feet for every person. When a stream of water has a rapid current, as few piers as practicable should be allowed to obstruct its course; otherwise the bridge will be liable to be swept away by freshets. When the span is not over 300 feet, and the banks of the river are of sufficient height to admit of it, only one arch should be employed. The rise of the arch is limited by the form of the roadway, and by the height of the banks of the river (See *Art.* 401.) The rise of the roadway should not exceed one in 24 feet, but, as the framing settles about one in 72, the roadway should be framed to rise one in 18, that it may be one in 24 after settling. The commencement of the arch at the abutments—the *spring*, as it is termed, should not be below high-water mark: and the bridge should be placed at right angles with the course of the current.

406.—The best material for the abutments and piers of a bridge, is stone; and, if possible, stone should be procured for the purpose. The following rule is to determine the extent of the abutments, they being rectangular, and built with stone weighing 120 lbs. to a cubic-foot. *Rule* LXVII.—Multiply the square of the height of the abutment by 160, and divide this product by the weight of a square foot of the arch, and by the rise of the arch; add unity to the quotient, and extract the square-root. Diminish the square-root by unity, and multiply the root, so diminished, by half the span of the arch, and by the weight of a square-foot of the arch. Divide the last product by 120 times the height of the abutment, and the quotient will be the thickness of the abutment. *Example.*—Let the height of the abutment from the base to the springing of the arch be 20 feet, half the span 100 feet, the weight of a square foot of the arch, including the greatest possible load upon it, 300 pounds, and the rise of the arch 18 feet—what should be its thickness? The square of the height of the abutment, 400, multiplied by 160, gives 64,000, and 300 by 18, gives 5400; 64,000, divided by 5400, gives a quotient of 11·852, one added to this makes 12·852, the square-root of which is 3·6; this, less one, is 2·6; this, multiplied by 100, gives 260, and this again by 300, gives 78,000; this, divided by 120 times the height of the abutment, 2400, gives 32 feet 6 inches, the thickness required.

The dimensions of a pier will be found by the same rule. For, although the thrust of an arch may be balanced by an adjoining arch, when the bridge is finished, and while it remains uninjured; yet, during the erection, and in the event of one arch being destroyed, the pier should be capable of sustaining the entire thrust of the other.

407.—Piers are sometimes constructed of timber, their principal strength depending on piles driven into the earth, but such piers should never be adopted where it is possible to avoid them: for, being alternately wet and dry, they decay much sooner than the upper parts of the bridge. Spruce and elm are considered good for piles. Where the height from the bottom of the river to the roadway is great, it is a good plan to cut them off at a little below low-water mark, cap them with a horizontal tie. and upon this erect the posts for the support of the roadway. This method cuts off the part that is continually wet from that which is only occasionally so, and thus affords an opportunity for replacing the upper part. The pieces which are immersed will last a great length of time, especially when of elm; for it is a well-established fact, that timber is less durable when subject to alternate dryness and moisture, than when it is either continually wet or continually dry. It has been ascertained that the piles under London bridge, after having been driven about 600 years, were not materially decayed. These piles are chiefly of elm, and wholly immersed.

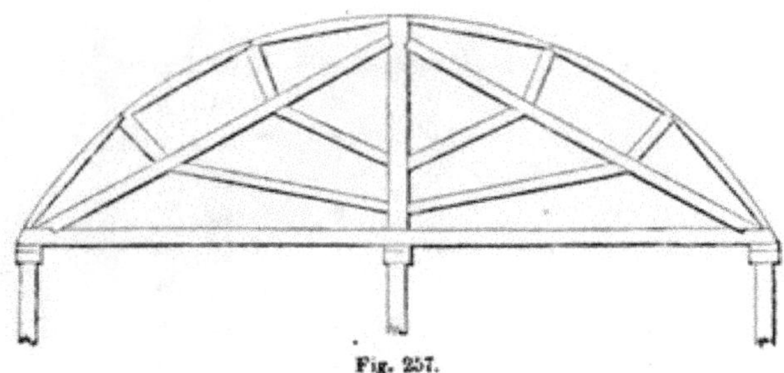

Fig. 257.

408.—*Centres for stone bridges.* *Fig.* 257 is a design for a centre for a stone bridge where intermediate supports, as piles driven into the bed of the river, are practicable. Its timbers are so distributed as to sustain the weight of the arch-stones as they are being laid, without destroying the original form of the centre; and also to prevent its destruction or settlement, should any of the piles be swept away. The most usual error in badly-constructed centres is, that the timbers are disposed so as to cause the framing to rise at the crown, during the laying of the arch-stones up the sides. To remedy this evil, some have loaded the crown with heavy stones; but a centre properly constructed will need no such precaution.

Experiments have shown that an arch-stone does not press upon the centring, until its bed is inclined to the horizon at an angle of from 30 to 45 degrees, according to the hardness of the stone, and whether it is laid in mortar or not. For general purposes, the point at which the pressure commences, may be considered to be at that joint which forms an angle of 32 degrees with the horizon. At this point, the pressure is inconsiderable, but gradually increases towards the crown. The following table gives the *portion* of the weight of the arch stones that presses upon the framing at the various angles of inclination formed by the bed of the stone with the horizon. The pressure perpendicular to the curve is equal to the weight of the arch stone multiplied by the decimal

·0,	when	the	angle	of	inclination	is	32	degrees.
·04	"	"	"	"	"	"	34	"
·08	"	"	"	"	"	"	36	"
·12	"	"	"	"	"	"	38	"
·17	"	"	"	"	"	"	40	"
·21	"	"	"	"	"	"	42	"
·25	"	"	"	"	"	"	44	"
·29	"	"	"	"	"	"	46	"
·33	"	"	"	"	"	"	48	"
·37	"	"	"	"	"	"	50	"
·4	"	"	"	"	"	"	52	"
·44	"	"	"	"	"	"	54	"
·48	"	"	"	"	"	"	56	"
·52	"	"	"	"	"	"	58	"
·54	"	"	"	"	"	"	60	"

From this it is seen that at the inclination of 44 degrees the pressure equals one-quarter the weight of the stone; at 57 degrees, half the weight; and when a vertical line, as *a b*, (*Fig.* 258,) passing through the centre of gravity of the arch-stone, does not fall within its bed, *c d*, the pressure may be considered equal to the whole weight of the stone. This will be the case at about 60 degrees, when the depth of the stone is double its breadth. The direction of these pressures is considered in a line with the radius of the curve. The weight upon a centre being known, the pressure may be estimated and the timber calculated accordingly. But it must be remembered that the whole weight is never placed upon the framing at once—as seems to have

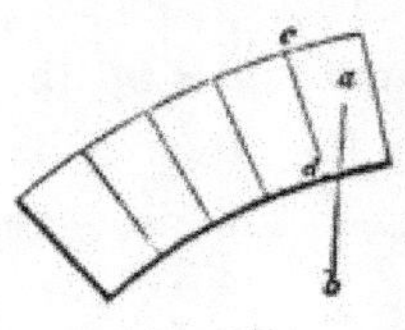

Fig. 258.

been the idea had in view by the designers of some centres. In building the arch, it should be commenced at each buttress at the same time, (as is generally the case,) and each side should progress equally towards the crown. In designing the framing, the effect produced by each successive layer of stone should be considered. The pressure of the stones upon one side should, by the arrangement of the struts, be counterpoised by that of the stones upon the other side.

409.—Over a river whose stream is rapid, or where it is necessary to preserve an uninterrupted passage for the purposes of navigation, the centre must be constructed without intermediate supports, and without a continued horizontal tie at the

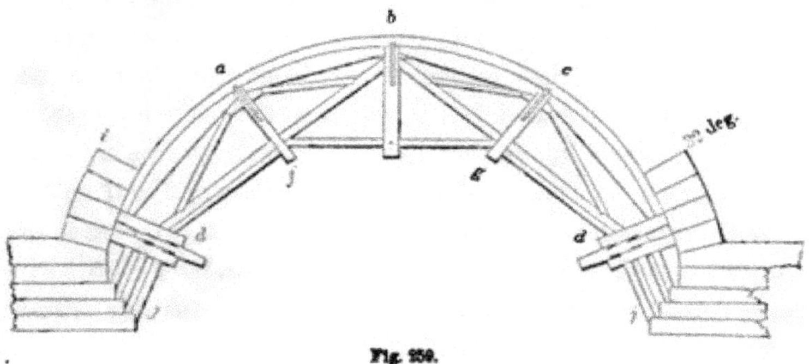

Fig. 259.

base; such a centre is shown at *Fig.* 259. In laying the stones from the base up to *a* and *c*, the pieces, *b d* and *b d*, act as ties to prevent any rising at *b*. After this, while the stones are being laid from *a* and from *c* to *b*, they act as struts: the piece, *f g*, is added for additional security. Upon this plan, with some variation to suit circumstances, centres may be constructed for any span usual in stone-bridge building.

410.—In bridge centres, the principal timbers should abut, and not be intercepted by a suspension or radial piece between. These should be in halves, notched on each side and bolted. The timbers should intersect as little as possible, for the more joints the greater is the settling; and halving them together is a bad practice, as it destroys nearly one-half the strength of the timber. Ties should be introduced across, especially where many timbers meet; and as the centre is to serve but a temporary purpose, the whole should be designed with a view to employ the timber afterwards for other uses. For this reason, all unnecessary cutting should be avoided.

411.—Centres should be sufficiently strong to preserve a staunch and steady form during the whole process of building; for any shaking or trembling will have a tendency to prevent the mortar or cement from *setting*. For this purpose, also, the centre should be lowered a trifle immediately after the key-stone is laid, in order that the stones may take their bearing before the mortar is set; otherwise the joints will open on the under side. The trusses, in centring, are placed at the distance of from 4 to 6 feet apart, according to their strength and the weight of the arch. Between every two trusses, diagonal braces should be introduced to prevent lateral motion.

412.—In order that the centre may be easily lowered, the frames, or trusses, should be placed upon wedge-formed sills; as is shown at *d*, (*Fig.* 259.) These are contrived so as to admit of the settling of the frame by driving the wedge, *d*, with a maul, or, in large centres, with a piece of timber mounted as a battering-ram. The operation of lowering a centre should be very slowly performed, in order that the parts of the arch may take their bearing uniformly. The wedge pieces, instead of being placed parallel with the truss, are sometimes made sufficiently long and laid through the arch, in a direction at right angles to that shown at *Fig.* 259. This method obviates the necessity of stationing men beneath the arch during the process of lowering; and was originally adopted with success soon after the occurrence of an accident, in lowering a centre, by which nine men were killed.

413.—To give some idea of the manner of estimating the pres

sures, in order to select timber of the proper scantling, calculate (*Art.* 408) the pressure of the arch-stones from *i* to *b*, (*Fig.* 259,) and suppose half this pressure concentrated at *a*, and acting in the direction *a f*. Then, by the parallelogram of forces, (*Art.* 258,) the strain in the several pieces composing the frame, *b d a*, may be computed. Again, calculate the pressure of that portion of the arch included between *a* and *c*, and consider half of it collected at *b*, and acting in a vertical direction; then, by the parallelogram of forces, the pressure on the beams, *b d* and *b d*, may be found. Add the pressure of that portion of the arch which is included between *i* and *b* to half the weight of the centre, and consider this amount concentrated at *d*, and acting in a vertical direction; then, by constructing the parallelogram of forces, the pressure upon *d j* may be ascertained.

414.—The strains having been obtained, the dimensions of the several pieces in the frames *b a d* and *b c d*, may be found by computation, as directed in the case of roof trusses, from *Arts.* 375 to 380. The tie-beams *b d*, *b d*, if made of sufficient size to resist the compressive strain acting upon them from the load at *b*, will be more than large enough to resist the tensile strain upon them during the laying of the first part of the arch-stones below *a* and *c*.

415.—In the construction of arches, the *voussoirs*, or arch-stones, are so shaped that the joints between them are perpendicular to the curve of the arch, or to its tangent at the point at which the joint intersects the curve. In a circular arch, the joints tend toward the centre of the circle: in an elliptical arch, the joints may be found by the following process:

Fig. 260.

416.—*To find the direction of the joints for an elliptical arch.* A joint being wanted at *a*, (*Fig.* 260,) draw lines from that point to the foci, *f* and *f*; bisect the angle, *f a f*, with the line, *a b*; then *a b* will be the direction of the joint.

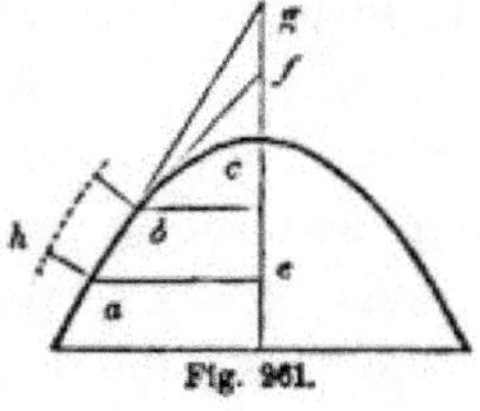

Fig. 261.

417.—*To find the direction of the joints for a parabolic arch* A joint being wanted at *a*, (*Fig.* 261,) draw *a e*, at right angles to the axis, *e g*; make *c g* equal to *c e*, and join *a* and *g*; draw *a h*, at right angles to *a g*; then *a h* will be the direction of the joint. The direction of the joint from *b* is found in the same manner. The lines, *a g* and *b f*, are tangents to the curve at those points respectively; and any number of joints in the curve may be obtained, by first ascertaining the tangents, and then drawing lines at right angles to them.

JOINTS.

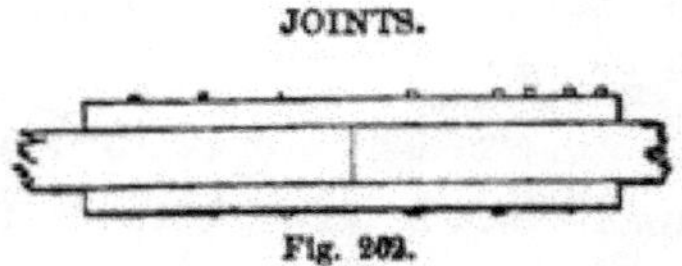
Fig. 262.

418.—*Fig.* 262 shows a simple and quite strong method of lengthening a tie-beam; but the strength consists wholly in the bolts, and in the friction of the parts produced by screwing the pieces firmly together. Should the timber shrink to even a small degree, the strength would depend altogether on the bolts. It would be made much stronger by indenting the pieces together; as at the upper edge of the tie-beam in *Fig.* 263; or by placing keys in the joints, as at the lower edge in the same figure. This process, however, weakens the beam in proportion to the depth of the indents.

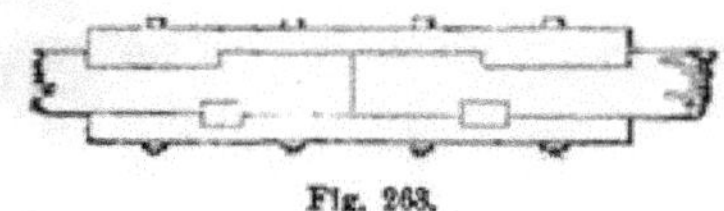

Fig. 263.

419.—*Fig.* 264 shows a method of scarfing, or splicing, a tie-beam without bolts. The keys are to be of well-seasoned,

Fig. 264.

hard wood, and, if possible, very cross-grained. The addition of bolts would make this a very strong splice, or even white-oak pins would add materially to its strength.

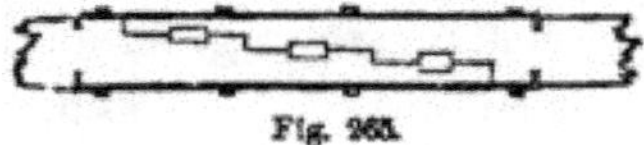

Fig. 265.

420.—*Fig.* 265 shows about as strong a splice, perhaps, as can well be made. It is to be recommended for its simplicity; as, on account of there being no oblique joints in it, it can be readily and accurately executed. A complicated joint is the worst that can be adopted; still, some have proposed joints that seem to have little else besides complication to recommend them.

421.—In proportioning the parts of these scarfs, the depths of all the indents taken together should be equal to one-third of the depth of the beam. In oak, ash or elm, the whole length of the scarf should be six times the depth, or thickness, of the beam, when there are no bolts; but, if bolts instead of indents are used, then three times the breadth; and, when both methods are combined, twice the depth of the beam. The length of the scarf in pine and similar soft woods, depending wholly on indents, should be about 12 times the thickness, or depth, of the beam; when depending wholly on bolts, 6 times the breadth; and, when both methods are combined, 4 times the depth.

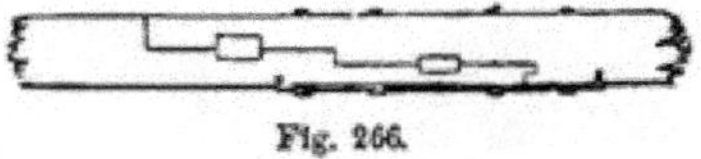

Fig. 266.

422.—Sometimes beams have to be pieced that are required to resist cross strains—such as a girder, or the tie-beam of a roof when supporting the ceiling. In such beams, the fibres of the wood in the upper part are compressed; and therefore a simple butt joint at that place, (as in *Fig.* 266.) is far preferable to any other. In such case, an oblique joint is the very worst. The under side of the beam being in a state of tension, it must be indented or bolted, or both; and an iron plate under the heads of the bolts, gives a great addition of strength.

Scarfing requires accuracy and care, as all the indents should bear equally; otherwise, one being strained more than another, there would be a tendency to splinter off the parts. Hence the simplest form that will attain the object, is by far the best. In all beams that are compressed endwise, abutting joints, formed at right angles to the direction of their length, are at once the simplest and the best. For a temporary purpose, *Fig.* 262 would do very well; it would be improved, however, by having a piece bolted on all four sides. *Fig.* 263, and indeed each of the others, since they have no oblique joints, would resist compression well.

423.—In framing one beam into another for bearing purposes, such as a floor-beam into a trimmer, the best place to make the mortice in the trimmer is in the neutral line, (*Arts* 317, 318,) which is in the middle of its depth. Some have thought that, as the fibres of the upper edge are compressed, a mortice might be made there, and the tenon driven in tight enough to make the parts as capable of resisting the compression, as they would be without it; and they have therefore concluded that plan to be the best. This could not be the case, even if the tenon would not shrink; for a joint between two pieces cannot possibly be made to resist compression, so well as a solid piece without joints. The proper place, therefore, for the mortice, is at the middle of the depth of the beam; but the best place for the tenon, in the floor-beam, is at its bottom edge. For the nearer this is placed to the upper edge, the greater is the liability for it to splinter off; if the joint is

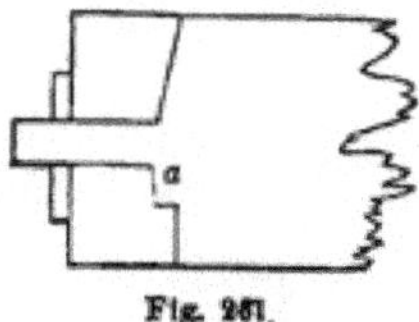

Fig. 267.

formed, therefore, as at *Fig.* 267, it will combine all the advantages that can be obtained. Double tenons are objectionable, because the piece framed into is needlessly weakened, and the tenons are seldom so accurately made as to bear equally. For this reason, unless the tusk at *a* in the figure fits exactly, so as to bear equally with the tenon, it had better be omitted. And in sawing the shoulders, care should be taken not to saw into the tenon in the least, as it would wound the beam in the place least able to bear it.

424.—Thus it will be seen that framing weakens both pieces, more or less. It should, therefore, be avoided as much as possible; and where it is practicable one piece should rest *upon* the other, rather than be framed into it. This remark applies to the bearing of floor-beams on a girder, to the purlins and jack-rafters of a roof, &c.

425.—In a framed truss for a roof, bridge, partition, &c., the joints should be so constructed as to direct the pressures through the axes of the several pieces, and also to avoid every tendency of the parts to slide. To attain this object, the abut-

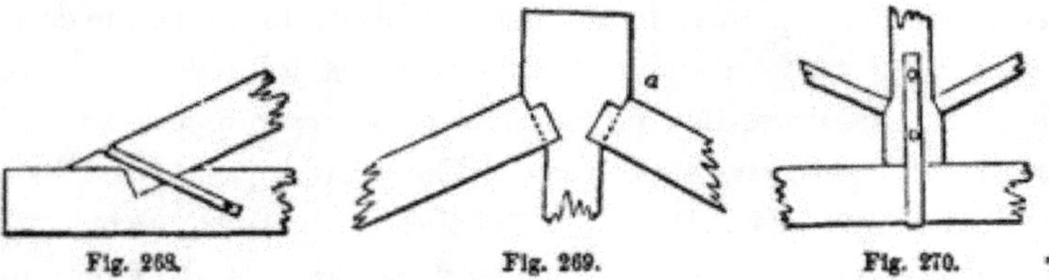

Fig. 268. Fig. 269. Fig. 270.

ting surface on the end of a strut should be at right angles to the direction of the pressure; as at the joint shown in *Fig.* 268 for the foot of a rafter, (see *Art.* 277,) in *Fig.* 269 for the head of a rafter, and in *Fig.* 270 for the foot of a strut or brace. The joint at *Fig.* 268 is not cut completely across the tie-beam, but a narrow lip is left standing in the middle, and a corresponding indent is made in the rafter, to prevent the parts from separating sideways. The abutting surface should be made as large as the attainment of other necessary objects will admit. The iron strap is added to prevent the rafter sliding out, should the end of the tie-beam, by decay or otherwise, splinter off. In making the joint shown at *Fig.* 269, it should be left a little open at *a*, so as to bring the parts to a fair bearing at the settling of the truss, which must necessarily take place from the shrinking of the king-post and other parts. If the joint is made fair at first, when the truss settles it will cause it to open at the under side of the rafter, thus throwing the whole pressure upon the sharp edge at *a*. This will cause an indentation in the king-post, by which the truss will be made to settle further; and this pressure not being in the axis of the rafter, it will be greatly increased, thereby rendering the rafter liable to split and break.

426.—If the rafters and struts were made to abut end to end, as in *Figs.* 271, 272 and 273, and the king or queen post notched on in halves and bolted, the ill effects of shrinking would be avoided. This method has been practised with success, in some of the most celebrated bridges and roofs in Eu-

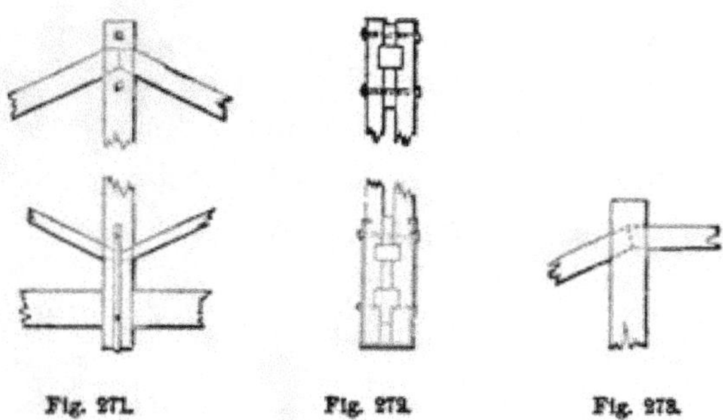

Fig. 271. Fig. 272. Fig. 273.

rope; and, were its use adopted in this country, the unseemly sight of a *hogged* ridge would seldom be met with. A plate of cast iron between the abutting surfaces will equalize the pressure.

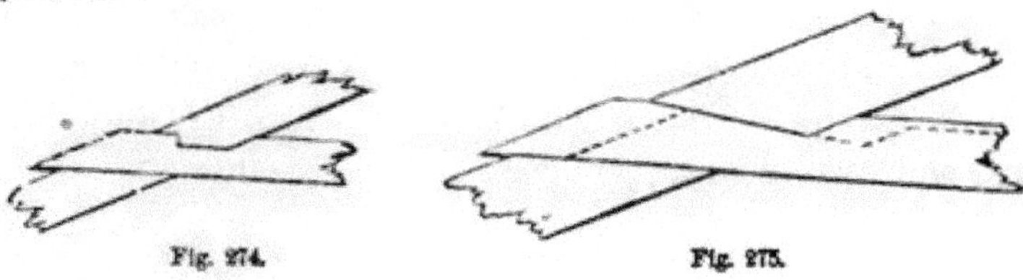

Fig. 274. Fig. 275.

427.—*Fig.* 274 is a proper joint for a collar-beam in a small roof: the principle shown here should characterize all tie-joints. The dovetail joint, although extensively practised in the above and similar cases, is the very worst that can be employed. The shrinking of the timber, if only to a small degree, permits the tie to withdraw—as is shown at *Fig.* 275. The dotted line shows the position of the tie after it has shrunk.

428.—Locust and white-oak pins are great additions to the strength of a joint. In many cases they would supply the place of iron bolts; and, on account of their small cost, they should be used in preference wherever the strength of iron is not requisite. In small framing, good cut nails are of great service at the joints; but they should not be trusted to bear any considerable pressure, as they are apt to be brittle. Iron straps are seldom necessary, as all the joinings in carpentry may be made without them. They can be used to advantage, however, at the foot of suspending-pieces, and for the rafter at the end of the tie-beam. In roofs for ordinary purposes, the iron straps for suspending-pieces may be as follows: When the longest unsupported part of the tie-beam is

10 feet, the strap may be 1 inch wide by $\frac{3}{16}$ thick.
15 " " " $1\frac{1}{2}$ " " $\frac{1}{4}$ "
20 " " " 2 " " $\frac{1}{4}$ "

In fastening a strap, its hold on the suspending-piece will be much increased by turning its ends into the wood. Iron straps should be protected from rust; for thin plates of iron decay very soon, especially when exposed to dampness. For this purpose, as soon as the strap is made, let it be heated to about a blue heat, and, while it is hot, pour over its entire surface raw linseed oil, or rub it with beeswax. Either of these will give it a coating which dampness will not penetrate.

IRON GIRDERS.

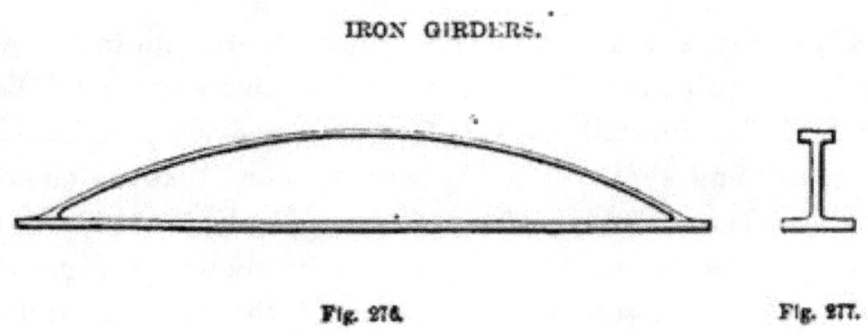

Fig. 276. Fig. 277.

429.—*Fig.* 276 represents the front view, and *Fig.* 277 the cross section at middle, of a cast iron girder of proper form for sustaining a weight equally diffused over its length. The curve is that of a parabola: generally an arc of a circle is used, and is near enough. Beams of this form are much used to sustain brick walls of buildings; the brickwork resting upon the bottom flange, and laid, not arching, but horizontal. In the cross section, the bottom flange is made to contain in area four times as much as the top flange. The strength will be in proportion to the area of the bottom flange, and to the height or depth. Hence, to obtain the greatest strength from a given amount of material, it is requisite to make the upright part, or the blade, rather thin; yet, in order to prevent injurious strains in the casting while it is cooling, the parts should be nearly equal in thickness. The thickness of the three parts—blade, top flange and bottom flange, may be made in proportion as 5, 6 and 8. For a beam of this form, the weight equally diffused over it equals

$$w = 9000 \frac{t a d}{l}. \qquad (199.)$$

The depth equals

$$d = \frac{l w}{9000 \, t a}. \qquad (200.)$$

The area of the bottom flange equals

$$a = \frac{l w}{9000 \, t d}. \qquad (201.)$$

where w equals the weight in pounds equally diffused over the length; d, the depth, or height in inches of the cross section at middle; a, the area of the bottom flange in inches; l, the length of the beam in feet, in the clear between the bearings; and t, a decimal in proportion to unity as the safe weight is to the breaking weight. This is usually from 0·2 to 0·3, or from one-fifth to one-third, at discretion.

430.—Beams of this form, laid in series, are much used in sustaining brick arches turned over vaults and other fire-proof rooms, forming a roof to the vault or room, and a floor above; the arches springing from the flanges, one on either side of the beam, as in *Fig.* 278.

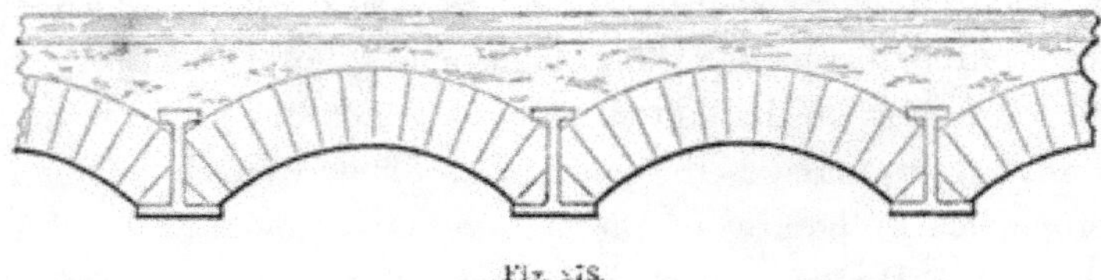

Fig. 278.

For this use the depth of cross section at middle equals

$$d = \frac{c f l^2}{9000\, t\, a}. \qquad (202.)$$

The area of the bottom flange equals

$$a = \frac{c f l^2}{9000\, t\, d}, \qquad (203.)$$

where the symbols signify as before, and c equals the distance apart from centres in feet at which the beams are placed, and f the weight per superficial foot, in pounds, including the weight of the material of which the floor is constructed.

Practical Rules and Examples.

431.—For a single girder the dimensions may be found by the following rule, ((200) and (201):)

Rule LXVIII.—Divide the weight in pounds equally diffused over the length of the girder by a decimal in proportion to unity as the safe weight is to the breaking weight, multiply the quotient by the length in feet, and divide the product by 9000. Then this quotient, divided by the depth of the beam at middle, will give the area of the bottom flange; or, if divided by the area of the bottom flange, will give the depth—the area and depth both in inches.

Example.—Let the weight equally diffused over a girder equal 60000 pounds; the decimal that is in proportion to unity as the safe weight is to be to the breaking weight, equal 0·3 the length in the clear of the bearings equal 20 feet. Then 60000 divided by 0·3 equals 200000, and this by 20 equals 4000000; this divided by 9000 equals 444$\frac{4}{9}$. Now if the *depth* is fixed, say at 20 inches, then 444$\frac{4}{9}$, divided by 20, equals 22$\frac{2}{9}$, equals the area of the bottom flange in inches. But if the *area* is given, say 24 inches, then to find the depth, divide 444$\frac{4}{9}$ by 24, and the quotient, 18·5, equals the depth in inches; and such a girder may be made with a bottom flange of 2 by 12 inches, top flange, (equal to ¼ of bottom flange,) 1½ by 4 inches, and the blade 1¼ inches thick.

432.—For a series of girders or iron beams, the dimensions may be found by the following rule: (202) and (203).

Rule LXIX.—Divide the weight per superficial foot, in pounds, by a decimal in proportion to unity as the safe weight is to the breaking weight, and multiply the quotient by the square of the length of the beams and by the distance apart at which the beams are placed from centres, both in feet, and divide the product by 9000. Then this quotient, divided by the depth of the beams at middle, will give the area of the bottom flange; or, if divided by the area of the bottom flange, will give the depth of the beam—the depth and area both in inches.

Example.—Let the weight per superficial foot resting upon an arched floor be 200 pounds, and the weight of the arches, concrete, &c., equal 100 pounds, total 300 pounds per superficial foot. Let the proportion of the breaking weight to be trusted on the beams equal 0·3, the length of the beams in the clear of the bearings equal 12 feet, and the distance apart from centres at which they are placed equal 4 feet. Then 300 divided by 0·3 equals 1000; this multiplied by 144 (the square of 12), equals 144000, and this by 4, equals 576000; this divided by 9000, equals 64. Now if the depth is fixed, and at 8 inches, then 64 divided by 8 equals 8, equals the area of the bottom flange. But if the area of the bottom flange is fixed, and at 6 inches, then 64, divided by 6, equals 10⅔, the depth required. Such a beam may be made with the bottom flange 1 by 6 inches, the top flange, (equal to one-quarter of the bottom flange,) $\frac{3}{4}$ by 2 inches, and the blade $\frac{5}{8}$ inch thick.

433.—The kind of girder shown at *Fig.* 280, (a cast iron arch with a wrought iron tie rod,) is extensively used as a support upon which to build brick walls where the space below is required to be free. The objections to its use are, the disproportion between the material and the strains, and the enhanced cost over the cast iron girder formed as in *Figs.* 276 and 277. The material in the cast arch, (*Fig.* 280,) is greatly in *excess* over the amount needed to resist effectually the compressive strains induced by the load through the axis of the arch, while the wrought metal in the tie is usually much *less* than is required to resist the horizontal thrust of the arch; absolute failure being prevented, partly by the weight of the walls resting on the haunches, and partly by the presence of adjoining buildings, their walls acting as buttresses to the arch. Some instances have occurred where the tie has parted.

Where this arched girder is used it is customary to lay the first courses of brick in the form of an arch. This brick arch of itself is quite sufficient to sustain the compressive strain, and, were there proper resistance to the horizontal thrust provided, the brick arch would entirely supersede the necessity for the girder. Indeed, the instances are not rare where constructions of this nature have proved quite satisfactory, the horizontal thrust of the arch being sustained by a tie rod secured to a pair of cast iron heel plates, as in *Fig.* 279. The

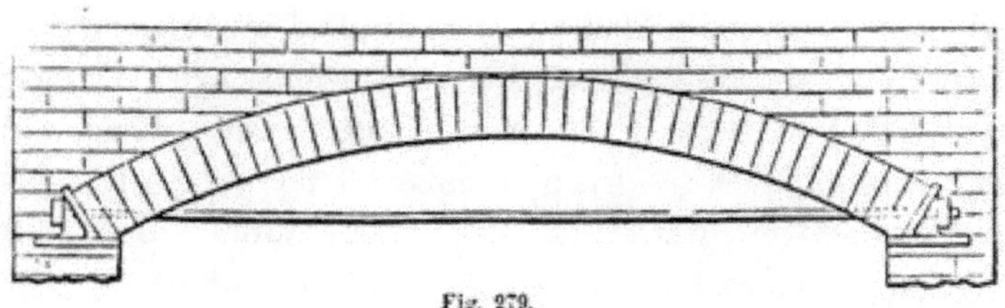

Fig. 279.

brick arch, in this case, being built upon a wooden centre, which was afterwards removed.

The diameter of the rod required for an arch of this kind is equal to

$$D = \sqrt{\frac{w\,s}{3000\,h}}, \qquad (204.)$$

where w equals the weight in pounds equally diffused over the arch; s, the length of the rod, clear of the heel plates, in feet; and h, the height at the middle, or rise, of the arc, in inches; D, the diameter, being also in inches.

When the diameter found by formula (204) is impracticably large, this difficulty may be overcome by dividing the metal into two rods. In the bow-string girder, (*Fig.* 280,) two rods cannot be used with advantage, because of the difficulty in adjusting their lengths so as to ensure to each an equal amount of the strain. But in the case of the brick arch, the two heel plates being disconnected, any discrepancy of length in the rods is adjusted simply by the pressure of the arch acting on the plates. When there are to be two rods, the diameter of each rod equals

$$D = \sqrt{\frac{w\,s}{6000\,h}}, \qquad (205.)$$

Practical Rule and Example.

434.—To obtain the diameter of wrought iron tie-rods for heel plates, as in *Fig.* 279, proceed by this rule.

Rule LXX.—Multiply the weight in pounds equally distributed over the arch by the length of the tie-rod in feet, clear of the heel plates, and divide the product by the height of the arc in inches, (that is, the height at the middle, from the axis of the tie-rod to the centre of the depth of the brick arch,) then, if there is to be but one tie-rod, divide the quotient by 3000;

but if two, then divide by 6000, and the square root of the quotient, in either case, will be the required diameter.

Example.—The weight to be supported on a brick arch, equally distributed, is 24000 pounds; the length of the tie-rod, clear of the heel plates, is 10 feet; and the height, or rise, of the arc is 10 inches. Now by the rule, 24000 × 10 = 240000. This divided by 10, equals 24000. Upon the presumption that one tie-rod only will be needed, divide by 3000, and the quotient is 8, the square root of which is 2·82 inches. This is rather large, therefore there had better be two rods. In this case the quotient, 24000, divided by 6000, equals 4, the square root of which is 2, the diameter required. The arch should, therefore, have two rods of 2 inches diameter. Two rods are preferable to one. The iron is stronger per inch in small rods than in large ones, and the rules require no more metal in the two rods than in the one.

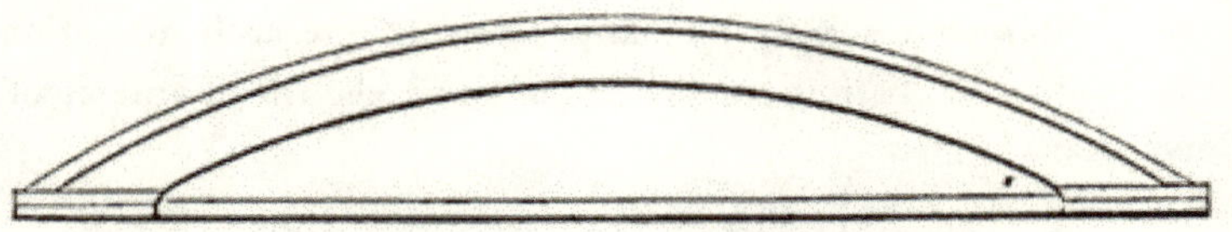

Fig. 280.

435.—The *Bow-string Girder*, as per *Fig.* 280, has little to recommend it, (see *Art.* 433,) yet because it has by some been much used, it is well to show the rules that govern its strength, if only for the benefit of those who are willing to be governed by reason rather than precedent. To resist the horizontal thrust of the cast arch, the diameter of the rod must equal (204)

$$D=\sqrt{\frac{w\,s}{3000\,h}},$$

But the cast iron arch has a certa'n amount of strength to re sist cross strains: this strength must be considered. Upon the presumption that the cross section of the cast arch at the middle is of the most favorable form, as in *Fig.* 277, or at least that it have a bottom flange, (although the most of those cast are without it), the strength of the cast arch to resist cross strains is shown by formula (199), when l, its length, is changed to s, its span. The weight in pounds equally diffused over the arch will then equal

$$w = \frac{9000\,t\,a\,d}{s}.$$

This is the weight borne by the cast arch acting simply as a beam. Deducting this weight from the whole weight, the remainder is the weight to be sustained by the rod. Calling the whole weight w, then

$$w - \frac{9000\,t\,a\,d}{s} = \frac{w\,s - 9000\,t\,a\,d}{s} = W$$

Therefore, from (204), the diameter equals

$$D = \sqrt{\frac{W\,s}{3000\,h}}$$

$$= \sqrt{\frac{\left(\frac{w\,s - 9000\,t\,a\,d}{s}\right)s}{3000\,h}}$$

$$= \sqrt{\frac{w\,s - 9000\,t\,a\,d}{3000\,h}}, \qquad (206.)$$

where D equals the diameter of the rod in inches; w, the weight in pounds equally diffused over the arch; s, the span of the arch in feet; h, the rise or height of the arc at middle, in inches; d, the height or depth of the cross section of the cast arch in inches; a, the area of the bottom flange of the cross section of the cast arch in inches; and t, a decimal in proportion to unity as the safe weight is to be to the breaking weight.

The rule in words at length, is

Rule LXXI.—Multiply the decimal in proportion to unity

as the safe weight is to be to the breaking weight, by 9000 times the depth of the cross section of the cast arch at middle, and by the area of the bottom flange of said section, both in inches, and deduct the product from the weight in pounds equally diffused over the arch multiplied by the span in feet, and divide the remainder by 3000 times the height of the arc in inches, measured from the axis of the tie-rod to the centre of the depth of the cast arch at middle, and the square root of the quotient will be the diameter of the rod in inches.

Example.—The rear wall of a building is of brick, and is 40 feet high, and 21 feet wide in the clear between the piers of the story below. Allowing for the voids for windows, this wall will weigh about 63000 pounds; and it is proposed to support it by a bow-string girder, of which the cross section at middle of the cast arch is 8 inches deep, and has a bottom flange containing 12 inches area. The rise of the curve or arc is 24 inches. What must be the diameter of the rod, the portion of the breaking weight of the cast arch, considered safe to trust, being three-tenths or 0·3? By the rule, $0{\cdot}3 \times 9000 \times 8 \times 12 = 259200$; then $\overline{63000 \times 21} - 259200 = 1063800$. This remainder divided by $(3000 \times 24 =)$ 72000, the quotient equals 14·775; the square root of which, 3·84, or nearly $3\frac{7}{8}$ inches, is the required diameter.

This size, though impracticably large, is as small as a due regard for safety will permit; yet it is not unusual to find the rods in girders intended for as heavy a load as in this example, only $2\frac{1}{4}$ and $2\frac{1}{2}$ inches! Were it possible to attach the rod so as not to injure its strength in the process of shrinking it in—putting it to its place hot, and depending on the contraction of the metal in cooling to bring it to a proper bearing —and were it possible to have the bearings so true as to induce the strain through the *axis* of the rod, and not along its *side*, (*Art.* 308,) then a less diameter than that given by the rule would suffice. But while these contingencies remain, the rule cannot safely be reduced, for, in the rule, the value of T, for wrought iron, (Table III., *Art.* 308,) is taken at nearly 6000 pounds, a point rather high in consideration of the size of the rod and the injuries, before stated, to which it is subjected. In cases where a girder wholly of cast iron (*Fig.* 276) is not preferred, it were better to build a brick arch resting on heel plates, (*Fig.* 279,) in which the metal required to resist the thrust may be divided into *two* rods instead of *one*, thus rendering the size more practical, and at the same time avoiding the injuries to which rods in arch girders are subjected. The heel-plate arch is also to be preferred to the cast arch on the score of economy; inasmuch as the brick which is substituted for the cast arch will cost less than iron. For example, suppose the cross section of the iron arch to be thus: the blade or upright part 8 by $1\frac{1}{2}$ inches, the top flange 12 by $1\frac{1}{4}$ inches, and the bottom flange 6 by $1\frac{3}{4}$ inches. At these dimensions, the area of the cross section will equal $12 + 15 + 10\frac{1}{2} = 37\frac{1}{2}$ inches. A bar of cast iron, one foot long and one inch square, will weigh 3·2 pounds; therefore, $37\frac{1}{2} \times 3{\cdot}2 = 120$ pounds, equals the weight of the cast arch per lineal foot. The price of castings per pound, as also the price of brickwork per cubic foot, of course will depend upon the locality and the state of the market at the time, but for a comparison they may be stated, the one at three and a half cents per pound, and the other at thirty cents per cubic foot. At these prices the cast arch will cost $120 \times 3\frac{1}{2} =$ \$4 20 per lineal foot; while the brick arch—12 inches high and 12 inches thick—will cost 30 cents per lineal foot. The difference is \$3 90. This amount is not all to be credited to the account of the brick arch Proper allowance is to be made for the cost of the heel plates, and of the wooden centre; also for the cost of a small addition to the size of the tie rods, which is required to sustain the strain otherwise borne by the cast arch in its resistance to a cross strain (*Art.* 435). Deducting the cost of these items, the difference in favor of the brick arch will be about $3 per foot. This, on a girder 25 feet long, amounts to $75. The difference in all cases will not equal this, but will be sufficiently great to be worth saving.

SECTION V.—DOORS, WINDOWS, &c.

DOORS.

436.—Among the several architectural arrangements of an edifice, the door is by no means the least in importance; and, if properly constructed, it is not only an article of use, but also of ornament, adding materially to the regularity and elegance of the apartments. The dimensions and style of finish of a door, should be in accordance with the size and style of the building, or the apartment for which it is designed. As regards the utility of doors, the principal door to a public building should be of sufficient width to admit of a free passage for a crowd of people; while that of a private apartment will be wide enough, if it permit one person to pass without being incommoded. Experience has determined that the least width allowable for this is 2 feet 8 inches; although doors leading to inferior and unimportant rooms may, if circumstances require it, be as narrow as 2 feet 6 inches; and doors for closets, where an entrance is seldom required, may be but 2 feet wide. The width of the principal door to a public building may be from 6 to 12 feet, according to the size of the building; and the width of doors for a dwelling may be from 2 feet 8 inches, to 3 feet 6 inches. If the importance of an apartment in a dwelling be such as to require a door of greater width than 3 feet 6 inches, the opening should be closed with two doors, or a door in two folds; generally, in such cases, where the opening is from 5 to 8 feet, folding or sliding doors are adopted. As to the height of a door, it should in no case be less than about 6 feet 3 inches; and generally not less than 6 feet 8 inches.

437.—The proportion between the width and height of single doors, for a dwelling, should be as 2 is to 5; and, for entrance-doors to public buildings, as 1 is to 2. If the width is given and the height required of a door for a dwelling, multiply the width by 5, and divide the product by 2; but, if the height is given and the width required, divide by 5, and multiply by 2. Where two or more doors of different widths show in the same room, it is well to proportion the dimensions of the more important by the above rule, and make the narrower doors of the same height as the wider ones; as all the doors in a suit of apartments, except the folding or sliding doors, have the best appearance when of one height. The proportions for folding or sliding doors should be such that the width may be equal to $\frac{1}{3}$ of the height; yet this rule needs some qualification: for, if the width of the opening be greater than one-half the width of the room, there will not be a sufficient space left for opening the doors; also, the height should be about one-tenth greater than that of the adjacent single doors.

438.—Where doors have but two panels in width, let the stiles and muntins be each $\frac{1}{7}$ of the width; or, whatever number of panels there may be, let the united widths of the stiles and the muntins, or the whole width of the solid, be equal to $\frac{3}{7}$ of the width of the door. Thus: in a door, 35 inches wide, containing two panels in width, the stiles should be 5 inches wide; and in a door, 3 feet 6 inches wide, the stiles should be 6 inches. If a door, 3 feet 6 inches wide, is to have 3 panels in width, the stiles and muntins should be each $4\frac{1}{2}$ inches wide, each panel being 8 inches. The bottom rail and the lock rail ought to be each equal in width to $\frac{1}{10}$ of the height of the door; and the top rail, and all others, of the same width as the stiles. The moulding on the panel should be equal in width to ¼ of the width of the stile

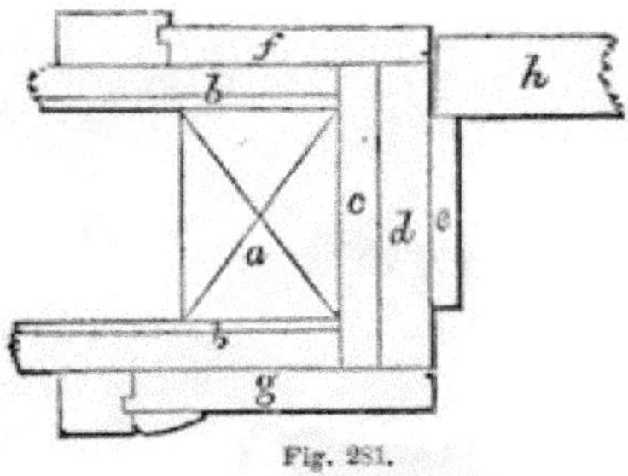

Fig. 281.

439.—*Fig.* 281 shows an approved method of trimming doors: *a* is the door stud; *b*, the lath and plaster; *c*, the *ground*; *d*, the jamb; *e*, the *stop*; *f* and *g*, architrave casings; and *h*, the door stile. It is customary in ordinary work to form the stop for the door by *rebating* the jamb. But, when the door is thick and heavy, a better plan is to nail on a piece as at *e* in the figure. This piece can be fitted to the door, and put on after the door is hung; so, should the door be a trifle *winding*, this will correct the evil, and the door be made to shut solid.

440.—*Fig.* 282 is an elevation of a door and trimmings suitable for the best rooms of a dwelling. (For trimmings generally, see Sect. III.) The number of panels into which a door should be divided, is adjusted at pleasure; yet the present style of finishing requires, that the number be as small as a proper regard for strength will admit. In some of our best dwellings, doors have been made having only two upright panels. A few years experience, however, has proved that the omission of the lock rail is at the expense of the strength and durability of the door; a four-panel door, therefore, is the best that can be made.

441.—The doors of a dwelling should all be hung so as to open into the principal rooms; and, in general, no door should be hung to open into the hall, or passage. As to the proper edge of the door on which to affix the hinges, no general rule can be assigned

Fig. 282.

WINDOWS.

442.—A window should be of such dimensions, and in such a position, as to admit a sufficiency of light to that part of the apartment for which it is designed. No definite rule for the size can well be given, that will answer in all cases; yet, as an approximation, the following has been used for general purposes. Multiply together the length and the breadth in feet of the apartment to be lighted, and the product by the height in feet; then the square-root of this product will show the required number of square feet of glass.

443.—To ascertain the dimensions of window frames, add 4½ inches to the width of the glass for their width, and 6½ inches to the height of the glass for their height. These give the dimensions, in the clear, of ordinary frames for 12-light windows; the height being taken at the inside edge of the sill. In a brick wall, the width of the opening is 8 inches more than the width of the glass—4½ for the stiles of the sash, and 3½ for hanging stiles—and the height between the stone sill and lintel is about 10½ inches more than the height of the glass, it being varied according to the thickness of the sill of the frame.

444.—In hanging inside shutters to fold into *boxes*, it is necessary to have the box shutter about one inch wider than the flap, in order that the flap may not interfere when both are folded into the box. The usual margin shown between the face of the shutter when folded into the box and the quirk of the stop bead, or edge of the casing, is half an inch; and, in the usual method of letting the *whole* of the thickness of the butt hinge into the edge of the box shutter, it is necessary to make allowance for the *throw* of the hinge. This may, in general, be estimated at ¼ of an inch at each hinging; which being added to the margin, the entire width of the shutters will be 1½ inches more than the width of the frame in the clear. Then, to ascertain the width of the box shutter, add 1½ inches to the width of the frame in the clear, between the pulley stiles; divide this product by 4, and add half an inch to the quotient; and the last product will be the required width. For example, suppose the window to have 3 lights in width, 11 inches each. Then, 3 times 11 is 33, and 4½ added for the wood of the sash, gives 37½——37½ and 1½ is 39 and 39, divided by 4, gives $9\frac{3}{4}$; to which add half an inch, and the result will be $10\frac{1}{4}$ inches, the width required for the box shutter.

445.—In disposing and proportioning windows for the walls of a building, the rules of architectural taste require that they be of different heights in different stories, but of the same width. The windows of the upper stories should all range perpendicularly over those of the first, or principal, story; and they should be disposed so as to exhibit a balance of parts throughout the front of the building. To aid in this, it is always proper to place the front door in the middle of the front of the building; and, where the size of the house will admit of it, this plan should be adopted. (See the latter part of *Art.* 224.) The proportion that the height should bear to the width, may be, in accordance with general usage, as follows:

The height	of	basement windows,		$1\frac{1}{3}$	of the width.
"	"	principal-story	"	$2\frac{1}{8}$	"
"	"	second-story	"	$1\frac{7}{8}$	"
"	"	third-story	"	$1\frac{3}{4}$	'
"	"	fourth-story	"	$1\frac{1}{2}$	"
"	"	attic-story	"	the same as the width.	

But, in determining the height of the windows for the several stories, it is necessary to take into consideration the height of the story in which the window is to be placed. For, in addition to the height from the floor, which is generally required to be from 28 to 30 inches, room is wanted above the head of the window for the window-trimming and the cornice of the room, besides some respectable space which there ought to be between these.

446.—Doors and windows are usually *square-headed*, or terminate in a horizontal line at top. These require no special directions for their trimmings. But circular-headed doors and windows are more difficult of execution, and require some attention. If the jambs of a door or window be placed at right angles to the face of the wall, the edges of the *soffit*, or surface of the head, would be straight, and its length be found by getting the stretch-out of the circle, (*Art.* 92;) but, when the jambs are placed obliquely to the face of the wall, occasioned by the demand for light in an oblique direction, the form of the soffit will be obtained by the following article: and, when the face of the wall is circular, as in the succeeding one.

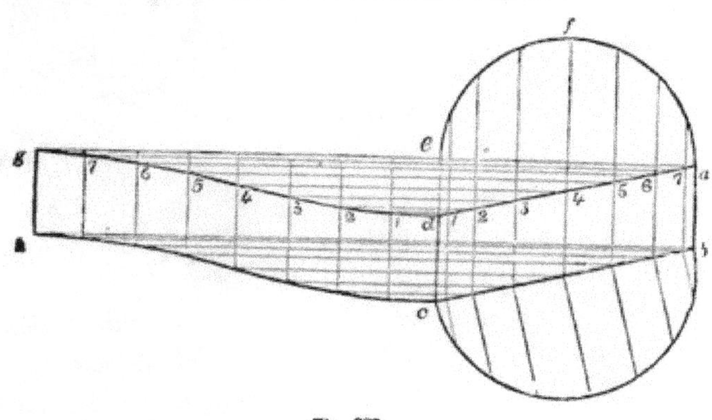

Fig. 283.

447.—*To find the form of the soffit for circular window heads, when the light is received in an oblique direction.* Let *a b c d*, (*Fig.* 283,) be the ground-plan of a given window, and *e f a*, a vertical section taken at right angles to the face of the jambs. From *a*, through *e*, draw *a g*, at right angles to *a b*; obtain the stretch-out of *e f a*, and make *e g* equal to it; divide *e g* and *e f a*, each into a like number of equal parts, and drop perpendiculars from the points of division in each; from the points of intersection, 1, 2, 3, &c., in the line, *a d*, draw horizontal lines to meet corresponding perpendiculars from *e g*; then those points of intersection will give the curve line, *d g*, which will be the one required for the edge of the soffit. The other edge, *c h*, is found in the same manner.

448.—*To find the form of the soffit for circular window-heads, when the face of the wall is curved.* Let *a b c d*, (*Fig.* 284,) be the ground-plan of a given window, and *e f a*, a vertical section of the head taken at right angles to the face of the jambs.

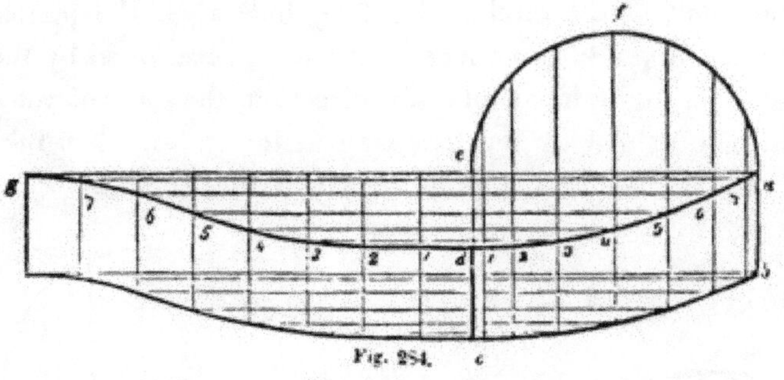

Fig. 284.

Proceed as in the foregoing article to obtain the line, *d g*; then that will be the curve required for the edge of the soffit; the other edge being found in the same manner.

If the given vertical section be taken in a line with the face of the wall, instead of at right angles to the face of the jambs, place it upon the line, *c b*, (*Fig.* 283;) and, having drawn ordinates at right angles to *c b*, transfer them to *e f a*; in this way, a section at right angles to the jambs can be obtained.

449.—The STAIRS is that mechanical arrangement in a building by which access is obtained from one story to another. Their position, form and finish, when determined with discriminating taste, add greatly to the comfort and elegance of a structure. As regards their position, the first object should be to have them near the middle of the building, in order that an equally easy access may be obtained from all the rooms and passages. Next in importance is light; to obtain which they would seem to be best situated near an outer wall, in which windows might be constructed for the purpose; yet a sky-light, or opening in the roof, would not only provide light, and so secure a central position for the stairs, but may be made, also, to assist materially as an ornament to the building, and, what is of more importance, afford an opportunity for better ventilation.

450.—It would seem that the length of the raking side of the *pitch-board*, or the distance from the top of one riser to the top of the next, should be about the same in all cases; for, whether stairs be intended for large buildings or for small, for public or for private, the accommodation of men of the same stature is to be consulted in every instance. But it is evident that, with the same effort, a longer step can be taken on level than on rising ground and that, although the tread and rise cannot be proportioned merely in accordance with the style and importance of the building, yet this may be done according to the angle at which the flight rises. If it is required to ascend gradually and easy, the length from the top of one rise to that of another, or the hypothenuse of the pitch-board, may be long; but, if the flight is steep the length must be shorter. Upon this data the following problem is constructed.

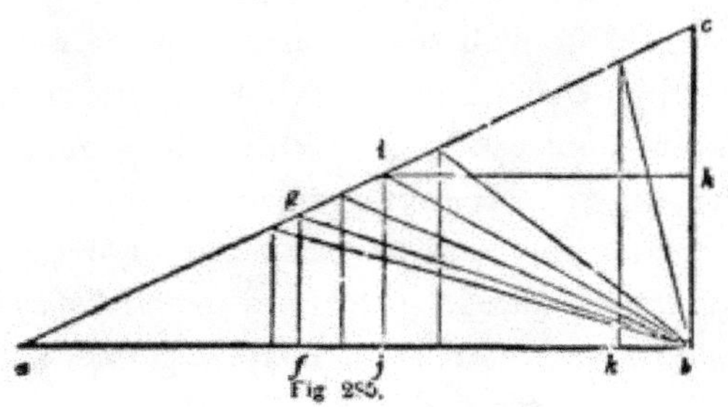

Fig 285.

451.—*To proportion the rise and tread to one another.* Make the line, *a b*, (*Fig.* 285,) equal to 21 inches; from *b*, erect *b c*, at right angles to *a b*, and make *b c* equal to 12 inches; join *a* and *c*, and the triangle, *a b c*, will form a scale upon which to graduate the sides of the pitch-board. For example, suppose a very easy stairs is required, and the tread is fixed at 14 inches. Place it from *b* to *f*, and from *f*, draw *f g*, at right angles to *a b*; then the length of *f g* will be found to be 5 inches, which is a proper rise for 14 inches tread, and the angle, *f b g*, will show the degree of inclination at which the flight will ascend. But, in a majority of instances, the height of a story is fixed, while the length of tread, or the space that the stairs occupy on the lower floor, is optional. The height of a story being determined, the height of each rise will of course depend upon the number into which the whole height is divided; the angle of ascent being more easy if the number be great, than if it be smaller. By dividing the whole height of a story into a certain number of rises, suppose the length of each is found to be 6 inches. Place this length from *b* to *h*, and draw *h i*, parallel to *a b*; then *h i*, or *b j* will be the proper tread for that rise, and *j b i* will show the angle of ascent. On the other hand, if the angle of ascent be given, as *a b l*, (*b l* being 10½ inches, the proper *length of run* for a step-ladder,) drop the perpendicular, *l k*, from *l* to *k*; then *l k b* will be the proper proportion for the sides of a pitch-board for that *run*.

452.—The angle of ascent will vary according to circumstances. The following treads will determine about the right inclination for the different classes of buildings specified.

In public edifices,	tread about	14	inches.
In first-class dwellings	"	12½	"
In second-class "	"	11	"
In third-class " and cottages	"	9	"

Step-ladders to ascend to scuttles, &c., should have from 10 to 11 inches *run* on the rake of the string. (See notes at *Art.* 103.)

453.—The length of the steps is regulated according to the extent and importance of the building in which they are placed, varying from 3 to 12 feet, and sometimes longer. Where two persons are expected to pass each other conveniently, the shortest length that will admit of it is 3 feet; still, in crowded cities where land is so valuable, the space allowed for passages being very small, they are frequently executed at 2½ feet.

454.—*To find the dimensions of the pitch-board.* The first thing in commencing to build a stairs, is to make the *pitch*-board; this is done in the following manner. Obtain very accurately, in feet and inches, the perpendicular height of the story in which the stairs are to be placed. This must be taken from the top of the floor in the lower story to the top of the floor in the upper story. Then, to obtain the number of rises, the height in inches thus obtained must be divided by 5, 6, 7, 8, or 9, according to the quality and style of the building in which the stairs are to be built. For instance, suppose the building to be a first-class dwelling, and the height ascertained is 13 feet 4 inches, or 160 inches. The proper rise for a stairs in a house of this class is about 6 inches. Then, 160 divided by 6, gives 26$\frac{2}{3}$ inches. This being nearer 27 than 26, the number of risers, should be 27. Then divide the height, 160 inches, by 27, and the quotient will give the height of one rise. On performing this operation, the quotient will be found to be 5 inches, $\frac{7}{8}$ and $\frac{1}{18}$ of an inch.

Then, if the space for the extension of the stairs is not limited, the tread can be found as at *Art.* 451. But, if the contrary is the case, the whole distance given for the treads must be divided by the number of treads required. On account of the upper floor forming a step for the last riser, the number of treads is always one less than the number of risers. Having obtained this rise and tread, the pitch-board may be made in the following manner. Upon a piece of well-seasoned board about $\frac{3}{4}$ of an inch thick, having one edge jointed straight and square, lay the corner of a carpenters'-square, as shown at *Fig.* 286. Make *a b*

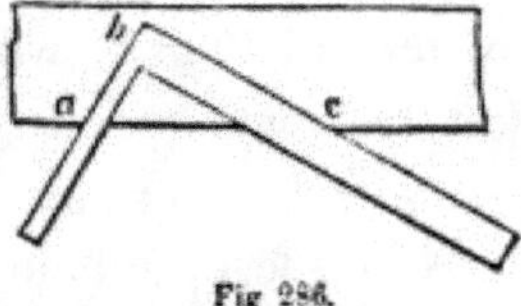

Fig 286.

equal to the rise, and *b c* equal to the tread; mark along those edges with a knife, and cut it out by the marks, making the edges perfectly square. The grain of the wood must run in the direction indicated in the figure, because, if it shrinks a trifle, the rise and the tread will be equally affected by it. When a pitch-board is first made, the dimensions of the rise and tread should be preserved in figures, in order that, should the first shrink, a second could be made.

455.—*To lay out the string.* The space required for timber

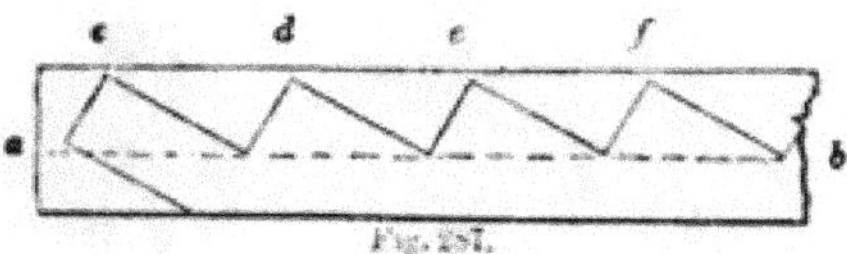

Fig. 287.

and plastering under the steps, is about 5 inches for ordinary stairs; set a gauge, therefore, at 5 inches, and run it on the lower edge of the plank, as *a b*, (*Fig.* 287.) Commencing at one end, lay the longest side of the pitch-board against the gauge-mark, *a b*, as at *c*, and draw by the edges the lines for the first rise and tread; then place it successively as at *d*, *e* and *f*, until the required number of risers shall be laid down.

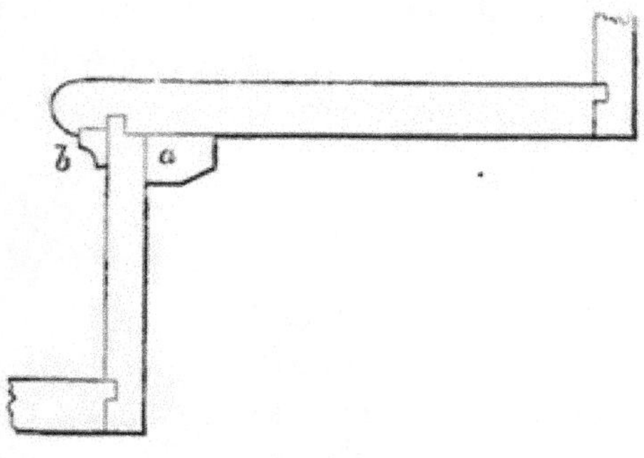

Fig. 288.

456.—*Fig.* 288 represents a section of a step and riser, joined after the most approved method. In this, *a* represents the end of a block about 2 inches long, two of which are glued in the corner in the length of the step. The cove at *b* is planed up square, glued in, and *stuck* after the glue is set.

PLATFORM STAIRS.

457.—A platform stairs ascends from one story to another in two or more flights, having platforms between for resting and to change their direction. This kind of stairs is the most easily constructed, and is therefore the most common. The cylin-

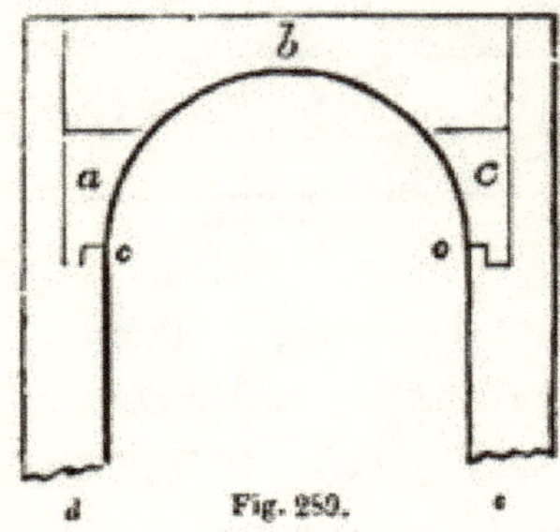

Fig. 289.

der is generally of small diameter, in most cases about 6 inches. It may be worked out of one solid piece, but a better way is to glue together three pieces, as in *Fig.* 289; in which the pieces, *a*, *b* and *c*, compose the cylinder, and *d* and *e* represent parts of the strings. The strings, after being glued to the cylinder, are secured with screws. The joining at *o* and *o* is the most proper for that kind of joint.

458.—*To obtain the form of the lower edge of the cylinder.* Find the stretch-out, *d e*, (*Fig.* 290,) of the face of the cylinder *a b c*, according to *Art.* 92; from *d* and *e*, draw *d f* and *e g*, at right angles to *d e*; draw *h g*, parallel to *d e*, and make *h f* and *g i*, each equal to one rise; from *i* and *f*, draw *i j* and *f k*, parallel to *h g*; place the tread of the pitch-board at these last lines, and draw by the lower edge the lines, *k h* and *i l*; parallel to these, draw *m n* and *o p*, at the requisite distance for the dimensions of the string; from *s*, the centre of the plan, draw *s q*, parallel to *d f*; divide *h q* and *q g*, each into 2 equal parts, as at *v* and *w*; from *v* and *w*, draw *v n* and *w o*, parallel to *f d*; join *n* and *o*, cutting *q s* in *r*; then the angles, *u n r* and *r o t*, being eased off according to *Art.* 89, will give the proper curve for the bottom edge of the cylinder. A centre may be found upon which to describe these curves thus: from *u*, draw *u x*, at right angles to *m n*; from *r*, draw *r x*, at right angles to *n o*; then *x* will be the centre for the curve, *u r*. The centre for the curve, *r t*, is found in the same manner.

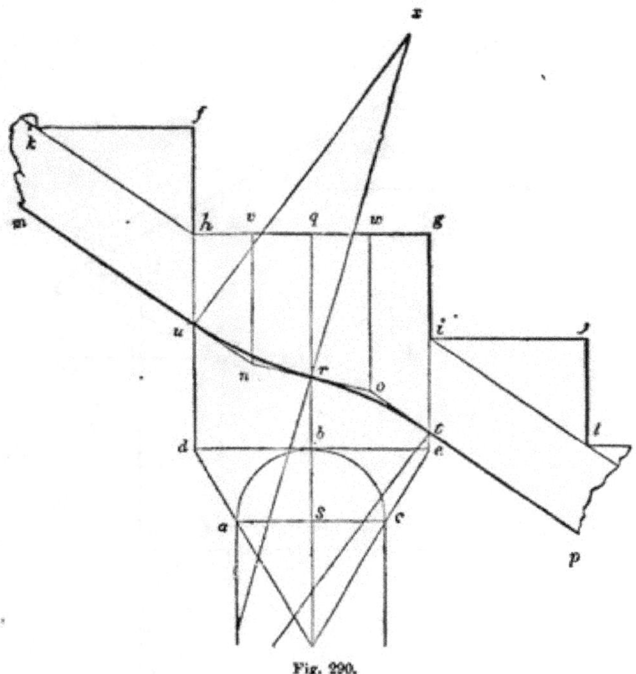

Fig. 290.

459.—*To find the position for the balusters.* Place the centre of the first baluster, (*b. Fig.* 291,) ½ its diameter from the face of the riser, *c d*, and ½ its diameter from the end of the step, *e d*; and place the centre of the other baluster, *a*, half the tread from the centre of the first. The centre of the rail must be placed over the centre of the balusters. Their usual length is 2 feet 5 inches, and 2 feet 9 inches, for the short and the long balusters respectively.

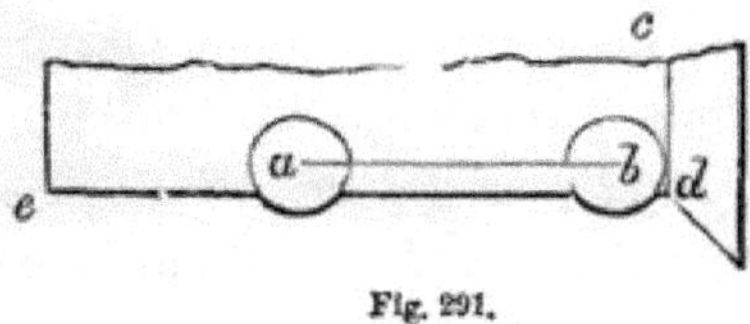

Fig. 291.

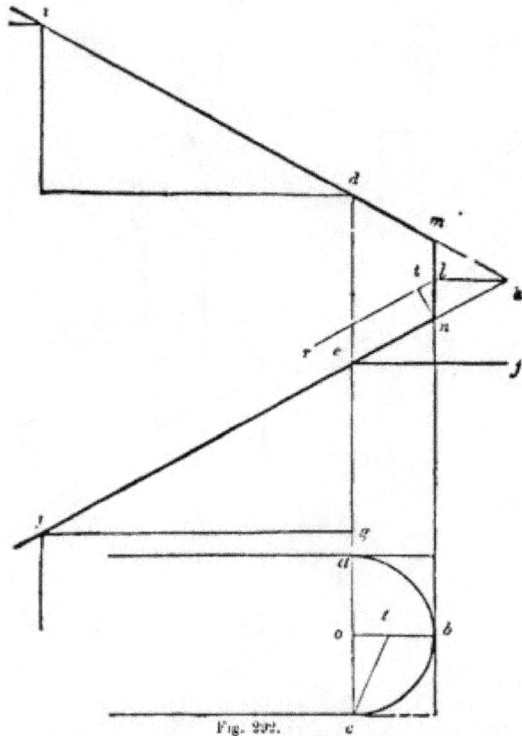

Fig. 292.

460.—*To find the face-mould for a round hand-rail to platform stairs.* Case 1.—*When the cylinder is small.* In *Fig.* 292, *j* and *e* represent a vertical section of the last two steps of the first flight, and *d* and *i* the first two steps of the second flight, of a platform stairs, the line, *e f*, being the platform; and *a b c* is the plan of a line passing through the centre of the rail around the cylinder. Through *i* and *d*, draw *i k*, and through *j* and *e*, draw *j k*; from *k*, draw *k l*, parallel to *f e*; from *b*, draw *b m*, parallel to *g d*; from *l*, draw *l r*, parallel to *k j*; from *n*, draw *n t*, at right angles to *j k*; on the line, *o b*, make *o t* equal to *n t*; join *c* and *t*: on the line, *j c*, (*Fig.* 293.) make *e c* equal to *e n* at *Fig.* 292; from *c*, draw *c t*, at right angles to *j c*, and make *c t*

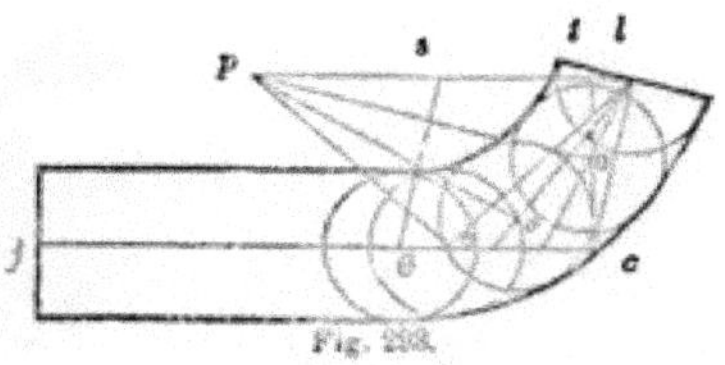

Fig. 293.

equal to *c t* at *Fig.* 292; through *t*, draw *p l*, parallel to *j c*, and make *t l* equal to *t l* at *Fig.* 292; join *l* and *c*, and complete the parallelogram, *e c l s*; find the points, *o*, *o*, *o*, according to *Art.* 118; upon *c*, *o*, *o*, *o*, and *l*, successively, with a radius equal to half the width of the rail, describe the circles shown in the figure; then a curve traced on both sides of these circles and just touching them, will give the proper form for the mould. The joint at *l* is drawn at right angles to *c l*.

461.—*Elucidation of the foregoing method.* This excellent plan for obtaining the face-moulds for the hand-rail of a platform stairs, has never before been published. It was communicated to me by an eminent stair-builder of this city: and having seen rails put up from it, I am enabled to give it my unqualified recommendation. In order to have it fully understood, I have introduced *Fig.* 294; in which the cylinder, for this purpose, is made rectangular instead of circular. The figure gives a perspective view of a part of the upper and of the lower flights, and a part of the platform about the cylinder. The heavy lines, *i m*, *m c* and *c j*, show the direction of the rail, and are supposed to pass through the centre of it. When the rake of the second flight is the same as that of the first, which is here and is generally the case, the face-mould for the lower twist will, when reversed, do for the upper flight: that part of the rail, therefore, which passes from *e* to *c* and from *c* to *l*, is all that will need explanation.

Suppose, then, that the parallelogram, *e a o c*, represent a plane lying perpendicularly over *e a b f*, being inclined in the direction, *e c*, and level in the direction, *c o*; suppose this plane, *e a o c*,

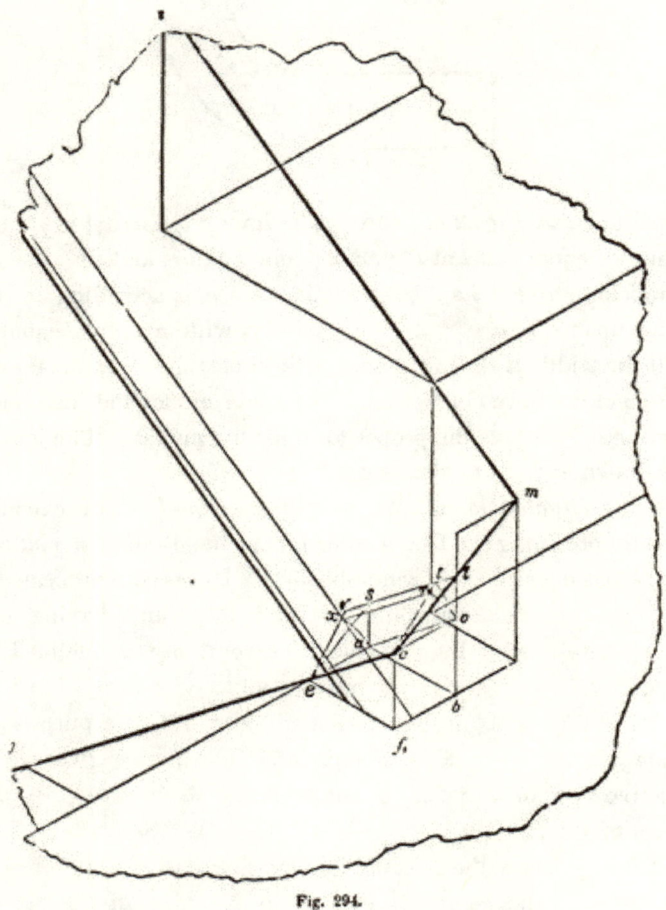

Fig. 294.

be revolved on *e c* as an axis, in the manner indicated by the arcs, *o n* and *a x*, until it coincides with the plane, *e r t c*; the line, *a o*, will then be represented by the line, *x n*; then add the parallelogram, *x r t n*, and the triangle, *c t l*, deducting the triangle, *e r s*; and the edges of the plane, *e s l c*, inclined in the direction, *e c*, and also in the direction, *c l*, will lie perpendicularly over the plane, *e a b f*. From this we gather that the line, *c o*, being at right angles to *e c*, must, in order to reach the point, *l*, be lengthened the distance, *n t*, and the right angle, *e c t*, be made obtuse by the addition to it of the angle, *t c l*. By reference to *Fig.* 292, it will be seen that this lengthening is performed by forming the right-angled triangle, *c o t*, corresponding to the triangle, *c o t*, in *Fig*. 294. The line, *c t*, is then transferred to *Fig*. 293, and placed at right angles to *e c;* this angle, *e c t*, being increased by adding the angle, *t c l*, corresponding to *t c l*, *Fig*. 294, the point, *l*, is reached, and the proper position and length of the lines, *e c* and *c l* obtained. To obtain the face-mould for a rail over a cylindrical well-hole, the same process is necessary to be followed until the the length and position of these lines are found; then, by forming the parallelogram, *e c l s*, and describing a quarter of an ellipse therein, the proper form will be given.

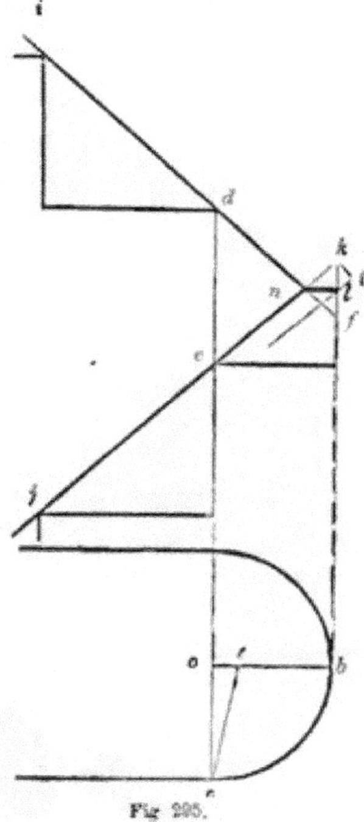

Fig. 295.

462.—CASE 2.—*When the cylinder is large.* *Fig.* 295 represents a plan and a vertical section of a line passing through the centre of the rail as before. From *b*, draw *b k*, parallel to *c d*; extend the lines, *i d* and *j e*, until they meet *k b* in *k* and *f*; from *n*, draw *n l*, parallel to *o b*; through *l*, draw *l t*, parallel to *j k*, from *k*, draw *k t*, at right angles to *j k*; on the line, *o b*, make *o t* equal to *k t*. Make *e c*, (*Fig*. 296.) equal to *e k* at *Fig*. 295; from *c*,

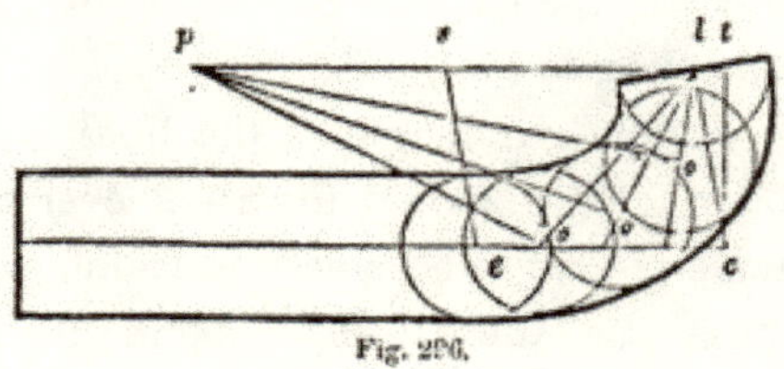

Fig. 296.

draw *c t*, at right angles to *e c*, and equal to *c t* at *Fig*. 295 · from *t*, draw *t p*, parallel to *c e*, and make *t l* equal to *t l* at *Fig*. 295; complete the parallelogram, *e c l s*, and find the points, *o*, *o*, *o*, as before; then describe the circles and complete the mould as in *Fig*. 293 The difference between this and Case 1 is, that the line, *c t*, instead of being raised and thrown out, is lowered and drawn in. (See note at page 381.)

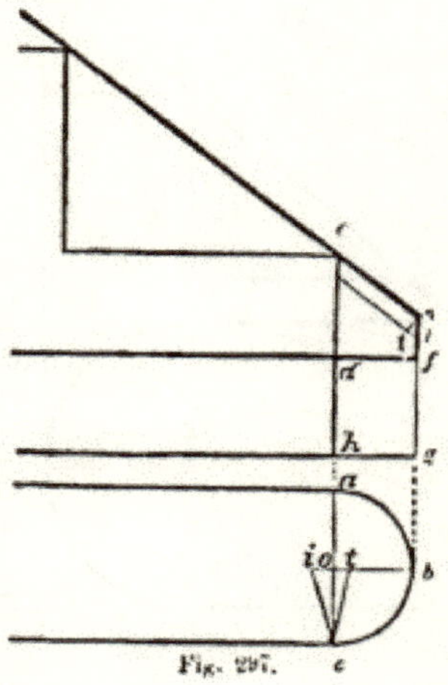

Fig. 297.

463.—Case 3.—*Where the rake meets the level.* In *Fig*

297, ***a b c*** is the plan of a line passing through the centre of the rail around the cylinder as before, and *j* and *e* is a vertical section of two steps starting from the floor, *h g*. Bisect *e h* in *d*, and through *d*, draw *d f*, parallel to *h g*; bisect *f n* in *l*, and from *l*, draw *l t*, parallel to *n j*; from *n*, draw *n t*, at right angles to *j n*. on the line, *o b*, make *o t* equal to *n t*. Then, to obtain a mould for the twist going up the flight, proceed as at *Fig.* 293; making *e c* in that figure equal to *e n* in *Fig.* 297, and the other lines of a length and position such as is indicated by the letters of reference in each figure. To obtain the mould for the level rail, extend *h o*, (*Fig.* 297,) to *i*; make *o i* equal to *f l*, and join *i* and *c*; make *c i*, (*Fig.* 298,) equal to *c i* at *Fig.* 297; through *c*, draw *c d*, at

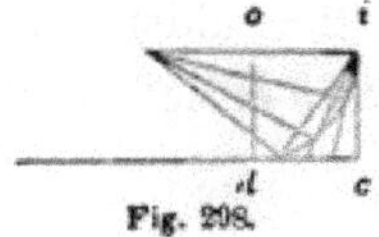

Fig. 298.

right angles to *c i*; make *d c* equal to *d f* at *Fig.* 297, and complete the parallelogram, *o d c i*; then proceed as in the previous cases to find the mould.

464.—All the moulds obtained by the preceding examples have been for round rails. For these, the mould may be applied to a plank of the same thickness as the rail is intended to be, and the plank sawed square through, the joints being cut square from the face of the plank. A twist thus cut and truly rounded will hang in a proper position over the plan, and present a perfect and graceful wreath.

465.—*To bore for the balusters of a round rail before rounding it.* Make the angle, *o c t*, (*Fig.* 299,) equal to the angle, *o c t*, at *Fig.* 292; upon *c*, describe a circle with a radius equal to half the thickness of the rail; draw the tangent, *b d*, parallel to *t c*, and complete the rectangle, *e b d f*, having sides tangical to the circle; from *c*, draw *c a*, at right angles to *o c*; then, *b d* being the bottom of the rail, set a gauge from *b* to *a*, and run it the whole length of the stuff; in boring, place the centre of the

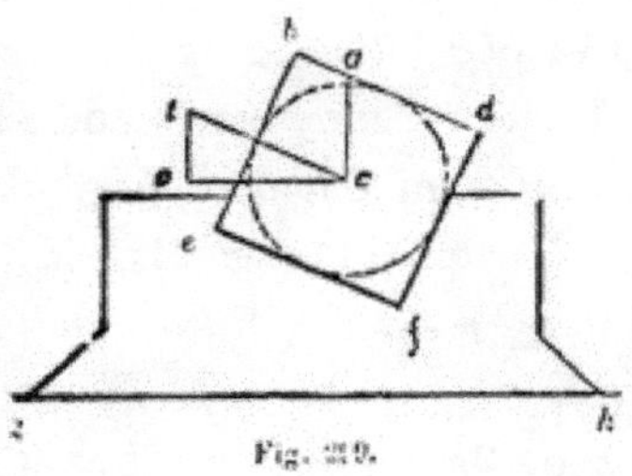

Fig. 229.

bit in the gauge-mark at *a*, and bore in the direction, *a c*. To do this easily, make *chucks* as represented in the figure, the bottom edge, *g h*, being parallel to *o c*, and having a place sawed out, as *e f*, to receive the rail. These being nailed to the bench, the rail will be held steadily in its proper place for boring vertically. The distance apart that the balusters require to be, on the under side of the rail, is one-half the length of the *rake-side* of the pitch-board.

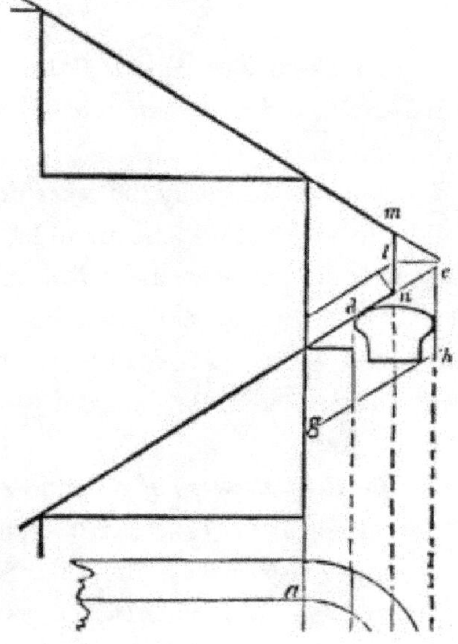

466.—*To obtain, by the foregoing principles, the face-mould for the twists of a moulded rail upon platform stairs* In *Fig.* 300, *a b c* is the plan of a line passing through the centre of the rail around the cylinder as before, and the lines above it are a vertical section of steps, risers and platform, with the lines for the rail obtained as in *Fig.* 292. Set half the width of the rail from *b* to *f* and from *b* to *r*, and from *f* and *r*, draw *f e* and *r d*, parallel to *c a* At *Fig.* 301, the centre lines of the

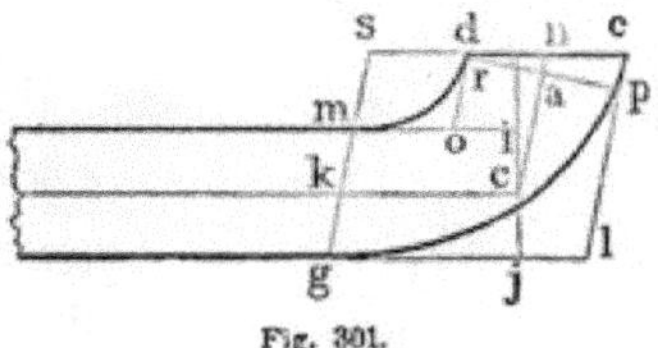

Fig. 301.

rai', *k c* and *c n*, are obtained as in the previous examples. Make *c i* and *c j*, each equal to *c i* at *Fig.* 300, and draw the lines, *i m* and *j g*, parallel to *c k*; make *n e* and *n d* equal to *n e* and *n d* at *Fig.* 300, and draw *d o* and *e l*, parallel to *n c*; also, through *k*, draw *s g*, parallel to *n c*; then, in the parallelograms, *m s d o* and *g s e l*, find the elliptic curves, *d m* and *e g*, according to *Art.* 118, and they will define the curves. The line, *d e*, being drawn through *n* parallel to *k c*, defines the joint, which is to be cut through the plank vertically. If the rail crosses the platform rather steep, a butt joint will be preferable, to obtain which see *Art.* 498.

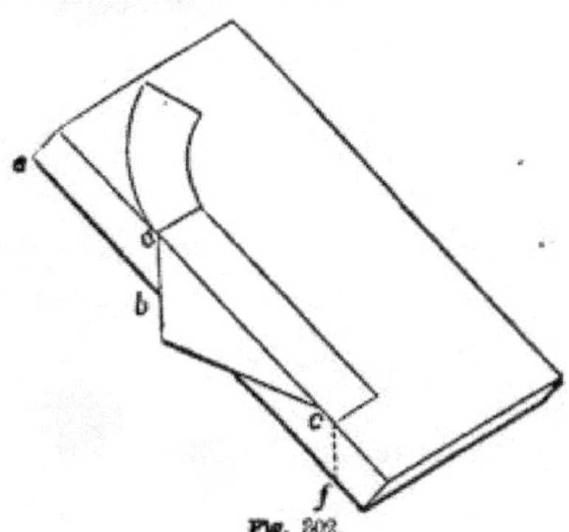

Fig. 302.

467.—*To apply the mould to the plank.* The mould obtained according to the last article must be applied to both sides of the plank, as shown at *Fig.* 302. Before applying the mould, the edge, *e f*, must be bevilled according to the angle, *c t x*, at *Fig* 300; if the rail is to be canted *up*, the edge must be bevilled at an *obtuse* angle with the upper face; but if it is to be canted *down*, the angle that the edge makes with the upper face must be *acute*. From the spring of the curve, *a*, and the end, *c*, draw vertical lines across the edge of the plank by applying the pitch-board, *a b c;* then, in applying the mould to the other side, place the points, *a* and *c*, at *b* and *f;* and, after marking around it, saw the rail out vertically. After the rail is sawed out, the bottom and the top surfaces must be squared from the sides.

468.—*To ascertain the thickness of stuff required for the twists.* The thickness of stuff required for the twists of a round rail, as before observed, is the same as that for the straight; but for a moulded rail, the stuff for the twists must be thicker than that for the straight. In *Fig.* 300, draw a section of the rail between the lines, *d r* and *e f*, and as close to the line, *d e*, as possible; at the lower corner of the section, draw *g h*, parallel to *d e;* then the distance that these lines are apart, will be the thickness required for the twists of a moulded rail.

The foregoing method of finding moulds for rails is applicable to all stairs which have continued rails around cylinders, and are without winders.

WINDING STAIRS.

469.—Winding stairs have steps tapering narrower at one end than at the other. In some stairs, there are steps of parallel width incorporated with tapering steps; the former are then called *flyers* and the latter *winders.*

470.—*To describe a regular geometrical winding stairs.* In *Fig.* 303, *a b c d* represents the inner surface of the wall enclosing the space allotted to the stairs, *a e* the length of the steps, and *e f g h* the cylinder, or face of the front string. The line,

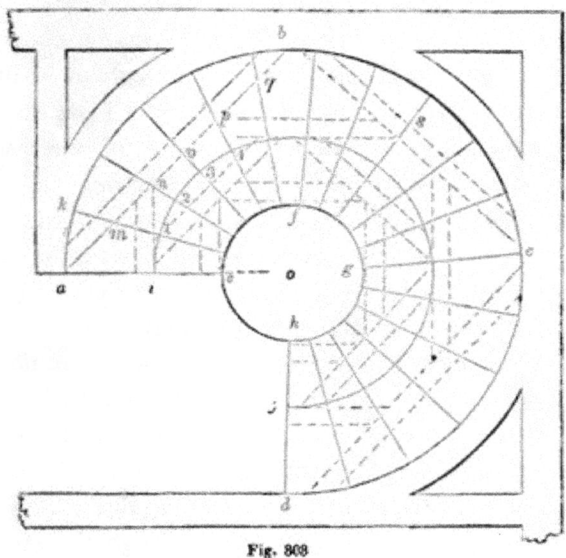

Fig. 308

a e, is given as the face of the first riser, and the point, *j*, for the limi: of the last. Make *e i* equal to 18 inches, and upon *o*, with *o i* for radius, describe the arc, *i j*; obtain the number of risers and of treads required to ascend to the floor at *j*, according to *Art.* 454, and divide the arc, *i j*, into the same number of equal parts as there are to be treads; through the points of division, 1, 2, 3, &c., and from the wall-string to the front-string, draw lines tending to the centre, *o*; then these lines will represent the face of each riser, and determine the form and width of the steps. Allow the necessary projection for the nosing beyond *a e*, which should be equal to the thickness of the step, and then *a e l k* will be the dimensions for each step. Make a pitch-board for the wall-string having *a k* for the tread, and the rise as previously ascertained; with this, lay out on a thicknessed plank the several risers and treads, as at *Fig.* 287, gauging from the upper edge of the string for the line at which to set the pitch-board.

Upon the back of the string, with a 1¼ inch dado plane, make a succession of grooves $1\frac{1}{4}$ inches apart, and parallel with the lines for the risers on the face. These grooves must be cut along the whole length of the plank, and deep enough to admit of the plank's bending around the curve, *a b c d*. Then construct a drum, or cylinder, of any common kind of stuff, and made to fit a curve having a radius the thickness of the string less than *o a* ; upon this the string must be bent, and the grooves filled with strips of wood, called *keys*, which must be very nicely fitted and glued in. After it has dried, a board thin enough to bend around on the outside of the string, must be glued on from one end to the other and nailed with clout nails. In doing this, be careful not to nail into any place where a riser or step is to enter on the face.

After the string has been on the drum a sufficient time for the glue to set, take it off, and cut the mortices for the steps and risers on the face at the lines previously made; which may be done by boring with a centre-bit half through the string, and nicely chiseling to the line. The drum need not be made so large as the whole space occupied by the stairs, but merely large enough to receive one piece of the wall-string at once—for it is evident that more than one will be required. The front string may be constructed in the same manner; taking *e l* instead of *a k* for the tread of the pitch-board, dadoing it with a smaller dado plane, and bending it on a drum of the proper size.

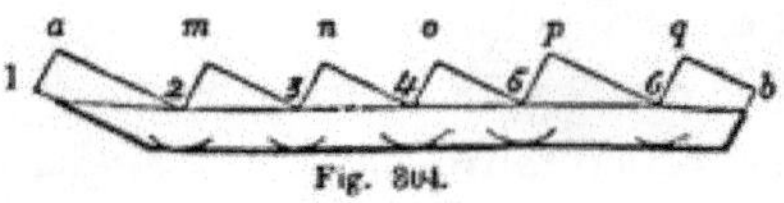

Fig. 304.

471.—*To find the shape and position of the timbers necessary to support a winding stairs.* The dotted lines in *Fig.* 303 show the proper position of the timbers as regards the plan: the shape of each is obtained as follows. In *Fig.* 304, the line, 1 *a*, is equal to a riser, less the thickness of the floor, and the lines, 2 *m*, 3 *n*, 4 *o*, 5 *p* and 6 *q*, are each equal to one riser. The line, *a* 2, is equal to *a m* in *Fig*. 303, the line, *m* 3 to *m n* in that figure, &c. In drawing this figure, commence at *a*, and make the lines, *a* 1 and *a* 2, of the length above specified, and draw them at right angles to each other; draw 2 *m*, at right angles to *a* 2, and *m* 3, at right angles to *m* 2, and make 2 *m* and *m* 3 of the lengths as above specified; and so proceed to the end. Then, through the points, 1, 2, 3, 4, 5 and 6, trace the line, 1 *b;* upon the points, 1, 2, 3, 4, &c., with the size of the timber for radius, describe arcs as shown in the figure, and by these the lower line may be traced parallel to the upper. This will give the proper shape for the timber, *a b*, in *Fig*. 303; and that of the others may be found in the same manner. In ordinary cases, the shape of one face of the timber will be sufficient, for a good workman can easily hew it to its proper level by that; but where great accuracy is desirable, a pattern for the other side may be found in the same manner as for the first.

472.—*To find the falling-mould for the rail of a winding stairs.* In *Fig*. 305, *a c b* represents the plan of a rail around half the cylinder, *A* the cap of the newel, and 1, 2, 3, &c., the face of the risers in the order they ascend. Find the stretch-out, *e f*, of *a c b*, according to *Art*. 92; from *o*, through the point of the mitre at the newel-cap, draw *o s;* obtain on the tangent, *e d*, the position of the points, *s* and h^2,* as at *t* and f^2; from *e t* f^2 and *f*, draw *e x*, *t u*, f^2 g^2 and *f h*, all at right angles to *e d;* make *e g* equal to one rise and f^2 g^2 equal to 12, as this line is drawn from the 12th riser; from *g*, through g^2, draw *g i;* make *g x* equal to about three-fourths of a rise, (the top of the newel, *x*, should be 3½ feet from the floor;) draw *x u*, at right angles to *e x*, and ease off the angle at *u;* at a distance equal to the thickness of

* In the above, the references, a^2, b^2, &c., are introduced for the first time. During the time taken to refer to the figure, the memory of the *form* of these may pass from the mind, while that of the sound alone remains; they may then be mistaken for *a* 2, *b* 2, &c. This can be avoided in reading by giving them a sound corresponding to their meaning, which is *second a second b*, &c. or *a second*, *b second*.

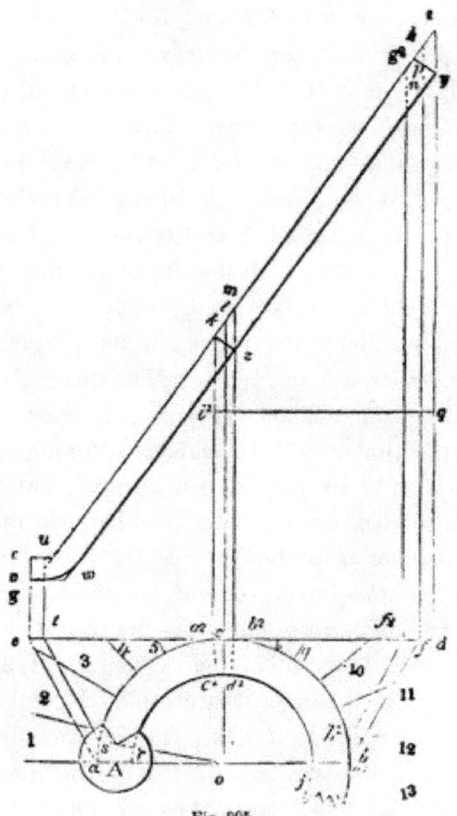

Fig. 305.

the rail, draw *v w y*, parallel to *x u i*; from the centre of the plan, *o*, draw *o l*, at right angles to *e d*; bisect *h n* in *p*, and through *p*, at right angles to *g i*, draw a line for the joint; in the same manner, draw the joint at *k*; then *x y* will be the falling-mould for that part of the rail which extends from *s* to *b* on the plan.

473.—*To find the face-mould for the rail of a winding-stairs.* From the extremities of the joints in the falling-mould, as *k*, *z* and *y*, (*Fig.* 305,) draw *k* a^2, *z* b^2 and *y d*, at right angles to *e d*; make *b* e^2 equal to *f d*. Then, to obtain the direction of the joint, a^2 c^2, or b^2 d', proceed as at *Fig* 306, at which the parts are

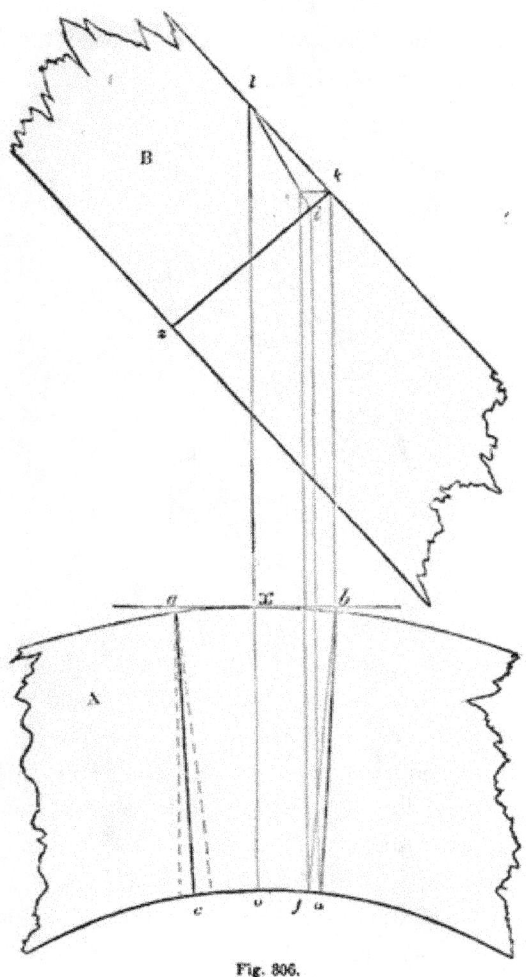

Fig. 306.

shown at half their full size. *A* is the plan of the rail, and *B* is the falling-mould; in which *k z* is the direction of the butt-joint. From *k*, draw *k b*, parallel to *l o*, and *k e*, at right angles to *k b*: from *b*, draw *b f*, tending to the centre of the plan, and from *f*, draw *f e*, parallel to *b k*; from *l*, through *e* draw *l i*, and from *i*, draw *i d*, parallel to *e f*; join *d* and *b*, and *d b* will be the proper direction for the joint on the plan. The direction of the joint on the other side, *a c*, can be found by transferring the distances, *x b* and *o d* to *x a* and *o c*. (See *Art.* 477.)

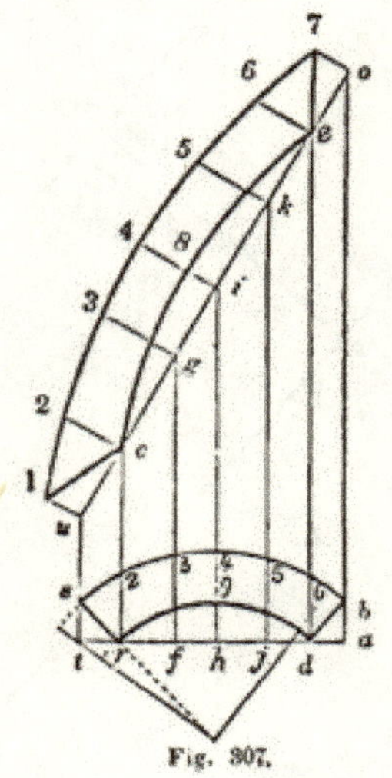

Fig. 307.

Having obtained the direction of the joint, make *s r d b*, (*Fig.* 307,) equal to *s r d* b^2 in *Fig.* 305; through *r* and *d*, draw *t a*, through *s* and from *d*, draw *t u* and *d e*, at right angles to *t a*, make *t u* and *d e* equal to *t u* and b^2 *m*, respectively, in *Fig.* 305; from *u*, through *e*, draw *u o*; through *b*, from *r*, and from as many other points in the line, *t a*, as is thought necessary, as *f*, *h* and *j* draw the ordinates, *r c*, *f g*, *h i*, *j k* and *a o*; from *u*, *c*, *g*, *i*, *k*, *e* and *o*, draw the ordinates, *u* 1, *c* 2, *g* 3, *i* 4, *k* 5, *e* 6 and *o* 7, at right angles to *u o*; make *u* 1 equal to *t s*, *c* 2 equal to *r* 2, *g* 3 equal to *f* 3, &c., and trace the curve, 1 7, through the points thus found; find the curve, *c e*, in the same manner, by transferring the distances between the line, *t a*, and the arc, *r d*; join 1 and *c*, also *e* and 7; then, 1 *c e* 7 will be the face-mould required for that part of the rail which is denoted by the letters, *s r* d^2 b^2, on the plan at *Fig.* 305.

To ascertain the mould for the next quarter, make *a c j e*, (*Fig*

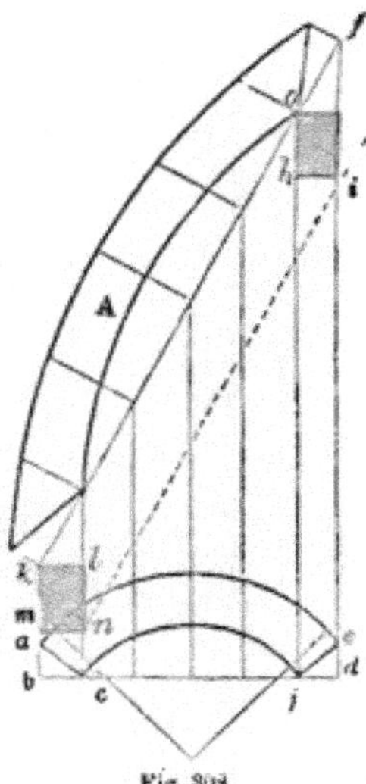

Fig. 308.

308,) equal to a^2 c^2 j e^2 at *Fig*. 305; at any convenient height on the line, *d i*, in that figure, draw *q* i^2, parallel to *e d;* through *c* and *j*, (*Fig*. 308,) draw *b d;* through *a*, and from *j*, draw *b k* and *j o*, at right angles to *b d;* make *b k* and *j o* equal to i^2 *k* and *q i*, respectively, in *Fig*. 305; from *k*, through *o*, draw *k f;* and proceed as in the last figure to obtain the face-mould, *A*.

474.—*To ascertain the requisite thickness of stuff.* CASE 1.—*When the falling-mould is straight.* Make *o h* and *k m*, (*Fig*. 308,) equal to *i y* at *Fig*. 305; draw *h i* and *m n*, parallel to *b d;* through the corner farthest from *k f*, as *n* or *i*, draw *n i*, parallel to *k f;* then the distance between *k f* and *n i* will give the thickness required.

475.—CASE 2.—*When the falling-mould is curved.* In *Fig*. 309, *s r d b* is equal to *s r* d^2 b^2 in *Fig*. 305. Make *a c* equal to the stretch-out of the arc, *s b*, according to *Art*. 92, and divide *a c* and *s b*, each into a like number of equal parts; from *a* and *c*, and from each point of division in the line, *a c*, draw *a k*, *e l*, &c., at right angles to *a c*, make *a k* equal to *t u* in *Fig*. 305, and *c j* equal to b^2 *m*

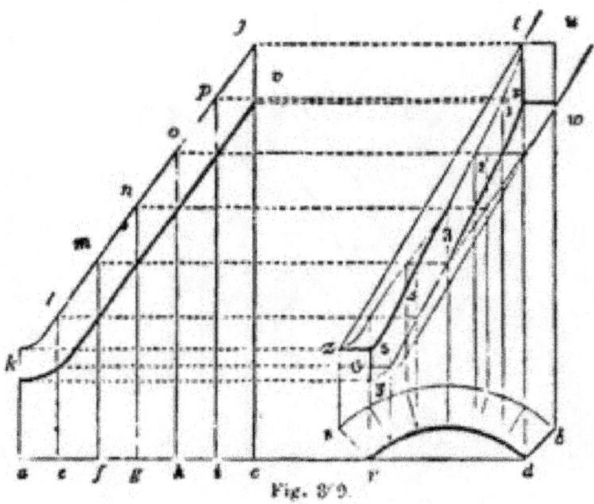

Fig. 309.

in that figure, and complete the falling-mould, *k j*, every way equal to *u m* in *Fig.* 305; from the points of division in the arc, *s b*, draw lines radiating towards the centre of the circle, dividing the arc. *r d*, in the same proportion as *s b* is divided; from *d* and *b*, draw *d t* and *b u*, at right angles to *a d*, and from *j* and *r*, draw *j u* and *v w*, at right angles to *j c;* then *x t u w* will be a vertical projection of the joint, *d b*. Supposing every radiating line across *s r d b*—corresponding to the vertical lines across *k j*—to represent a joint, find their vertical projection, as at 1, 2, 3, 4, 5 and 6; through the corners of those parallelograms, trace the curve lines shown in the figure; then 6 *u* will be a *helinet*, or vertical projection, of *s r d b*. To find the thickness of plank necessary to get out this part of the rail, draw the line, *z t*, touching the upper side of the helinet in two places: through the corner farthest projecting from that line, as *w*, draw *y w*, parallel to *z t;* then the distance between those lines will be the proper thickness of stuff for this part of the rail. The same process is necessary to find the thickness of stuff in all cases in which the falling-mould is in any way curved.

476.—*To apply the face-mould to the plank.* In *Fig.* 310, *A* represents the plank with its best side and edge in view, and *B* the same plank turned up so as to bring in view the other side

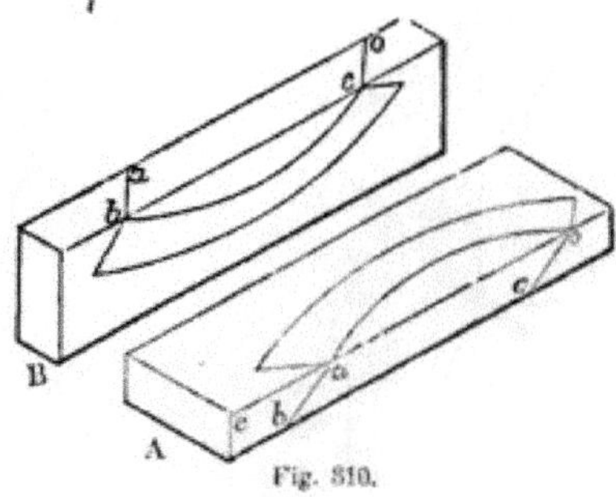

Fig. 310.

and the same edge, this being square from the face. Apply the tips of the mould at the edge of the plank, as at *a* and *o*, (*A*,) and mark out the shape of the twist; from *a* and *o*, draw the lines, *a b* and *o c*, across the edge of the plank, the angles, *e a b* and *e o c*, corresponding with *k f d* at *Fig.* 308; turning the plank up as at *B*, apply the tips of the mould at *b* and *c*, and mark it out as shown in the figure. In sawing out the twist, the saw must be be moved in the direction, *a b*; which direction will be perpendicular when the twist is held up in its proper position.

In sawing by the face-mould, the *sides* of the rail are obtained; the top and bottom, or the upper and the lower surfaces, are obtained by squaring from the sides, after having bent the falling-mould around the outer, or convex side, and marked by its edges. Marking across by the ends of the falling-mould will give the position of the butt-joint.

477.—*Elucidation of the process by which the direction of the butt-joint is obtained in Art.* 473. Mr. Nicholson, in his *Carpenter's Guide*, has given the joint a different direction to that here shown; he radiates it towards the centre of the cylinder. This is erroneous—as can be shown by the following operation:

In *Fig.* 311, *a r j i* is the plan of a part of the rail about the joint, *s u* is the stretch-out of *a i*, and *g p* is the helinet, or vertical projection of the plan, *a r j i*, obtained according to *Art*

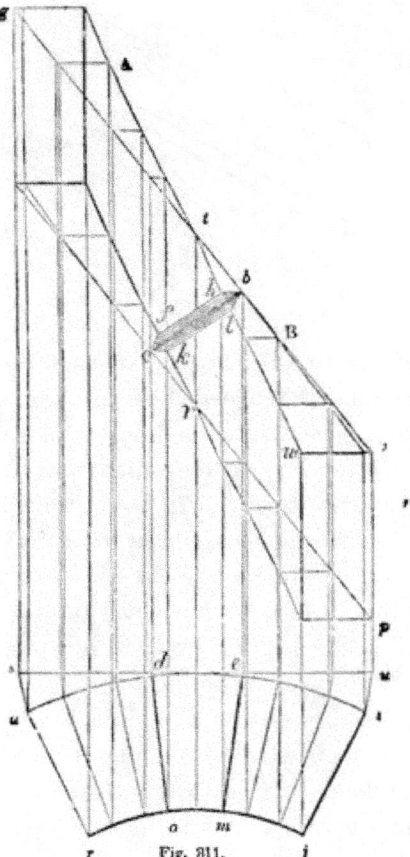

Fig. 311.

475. Bisect *r t*, part of an ordinate from the centre of the plan, and through the middle, draw *c b*, at right angles to *g v*; from *b* and *c*, draw *c d* and *b e*, at right angles to *s u*; from *d* and *e*, draw lines radiating towards the centre of the plan: then *d o* and *e m* will be the direction of the joint on the plan, according to Nicholson, and *c b* its direction on the falling-mould. It will be admitted that all the lines on the upper or the lower side of the rail which radiate towards the centre of the cylinder, as *d o*, *e m* or *i j*, are level; for instance, the level line, *w v*, on the top of the

rail in the helinet, is a true representation of the radiating line, *j i*, on the plan. The line, *b h*, therefore, on the top of the rail in the helinet, is a true representation of *e m* on the plan, and *k c* on the bottom of the rail truly represents *d o*. From *k*, draw *k l*, parallel to *c b*, and from *h*, draw *h f*, parallel to *b c*; join *l* and *b*, also *c* and *f*; then *c k l b* will be a true representation of the end of the lower piece, *B*, and *c f h b* of the end of the upper piece, *A*; and *f k* or *h l* will show how much the joint is open on the inner, or concave side of the rail.

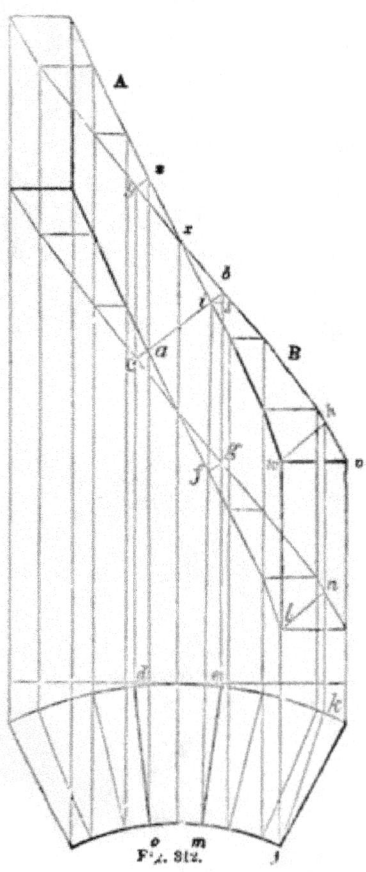

Fig. 312.

To show that the process followed in *Art.* 473 is correct, let *d o* and *e m*, (*Fig.* 312,) be the direction of the butt-joint found as at *Fig.* 306. Now, to project, on the top of the rail in the helinet, a line that does not radiate towards the centre of the cylinder, as *j k*, draw vertical lines from *j* and *k* to *w* and *h*, and join *w* and *h*; then it will be evident that *w h* is a true representation in the helinet of *j k* on the plan, it being in the same plane as *j k*, and also in the same winding surface as *w v*. The line, *l n*, also, is a true representation on the bottom of the helinet of the line, *j k*, in the plan. The line of the joint, *e m*, therefore, is projected in the same way and truly by *i b* on the top of the helinet; and the line, *d o*, by *c a* on the bottom. Join *a* and *i*, and then it will be seen that the lines, *c a*, *a i* and *i b*, exactly coincide with *c b*, the line of the joint on the convex side of the rail; thus proving the lower end of the upper piece, *A*, and the upper end of the lower piece, *B*, to be in one and the same plane, and that the direction of the joint on the plan is the true one. By reference to *Fig.* 306 it will be seen that the line, *l i*, corresponds to *x i* in *Fig.* 312; and that *e k* in that figure is a representation of *f b*, and *i k* of *d b*.

Fig. 318.

In getting out the twists, the joints, before the falling-mould is applied, are cut perpendicularly, the face-mould being long enough to include the overplus necessary for a butt-joint. The face-mould for *A*, therefore, would have to extend to the line, *i b*; and that for *B*, to the line, *y z*. Being sawed vertically at first, a section of the joint at the end of the face-mould for *A*, would be represented in the helinet by *b i f g*. To obtain the position of the line, *b i*, on the end of the twist, draw *i s*, (*Fig.* 313,) at right angles to *i f*, and make *i s* equal to *m e* at *Fig.* 312; through *s*. draw *s g*, parallel to *i f*, and make *s b* equal to *s b* at *Fig.* 312; join *b* and *i*; make *i f* equal to *i f* at *Fig.* 312, and from *f*, draw *f g*, parallel to *i b*; then *i b g f* will be a perpendicular section of the rail over the line, *e m*, on the plan at *Fig.* 312, corresponding to *i b g f* in the helinet at that figure; and when the rail is squared, the top, or back, must be trimmed off to the line, *i b*, and the bottom to the line, *f g*.

478.—*To grade the front string of a stairs, having winders in a quarter-circle at the top of the flight connected with flyers at the bottom.* In *Fig.* 314, *a b* represents the line of the facia along the floor of the upper story, *b e c* the face of the cylinder, and *c d* the face of the front string. Make *g b* equal to ⅓ of the diameter of the baluster, and draw the centre-line of the rail, *f g*, *g h i* and *i j*, parallel to *a b*, *b e c* and *c d*; make *g k* and *g l* each equal to half the width of the rail, and through *k* and *l*, draw lines for the convex and the concave sides of the rail, parallel to the centre-line; tangical to the convex side of the rail, and parallel to *k m*, draw *n o*; obtain the stretch-out, *q r*, of the semi-circle, *k p m*, according to *Art.* 92; extend *a b* to *t*, and *k m* to *s*; make *c s* equal to the length of the steps, and *i u* equal to 18 inches, and describe the arcs, *s t* and *u* 6, parallel to *m p*; from *t*, draw *t w*, tending to the centre of the cylinder; from 6, and on the line, 6 *u x*, run off the regular tread, as at 5, 4, 3, 2, 1 and *v*; make *u x* equal to half the arc, *u* 6, and make the point of division nearest to *x*, as *v*, the limit of the parallel steps, or flyers; make *r o* equal to *m z*; from *o*, draw *o* a^2, at right angles to *n o*, and equal to one rise;

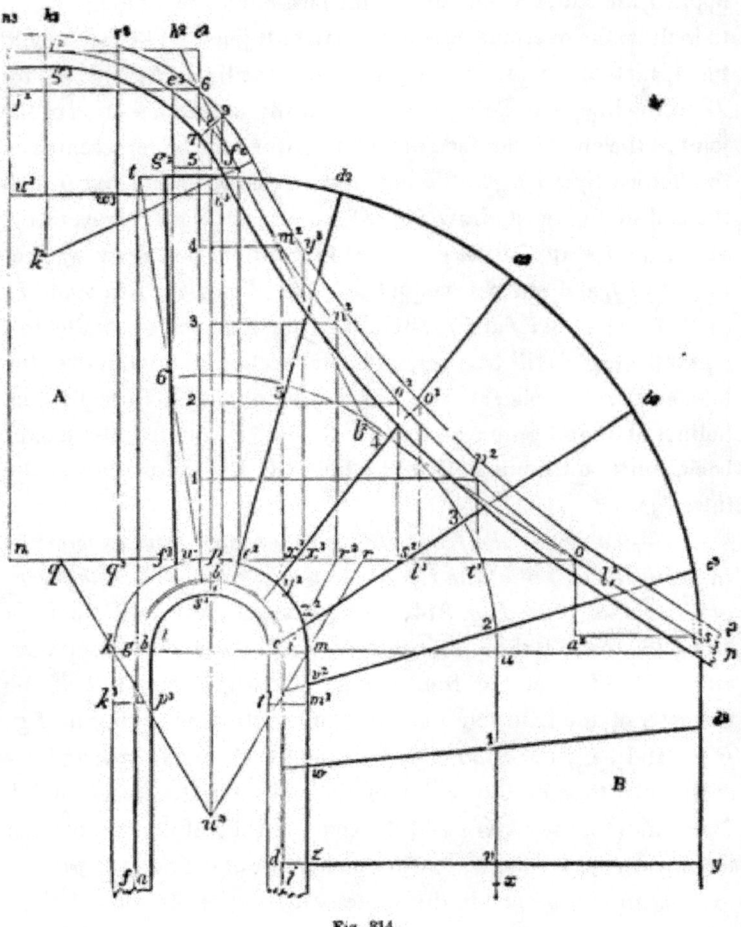

Fig. 814.

from a^2, draw $a^2\,s$, parallel to $n\,o$, and equal to one tread; from s through o, draw $s\,b^2$.

Then from w, draw $w\,c^2$, at right angles to $n\,o$, and set up, on the line, $w\,c^2$, the same number of risers that the floor, A, is above the first winder, B, as at 1, 2, 3, 4, 5 and 6; through 5, (on the arc, 6 u,) draw $d^2\,e^2$, tending to the centre of the cylinder; from e^2, draw $e^2\,f^2$, at right angles to $n\,o$, and through 5, (on the line,

$w\,c^2$,) draw $g^2 f^2$, parallel to $n\,o$; through 6, (on the line, $w\,c^2$,) and f^2, draw the line, $h^2\,b^2$; make 6 c^2 equal to half a rise, and from c^2 and 6, draw $c^2\,i^2$ and 6 j^2, parallel to $n\,o$; make $h^2\,i^2$ equal to $h^2 f^2$; from i^2, draw $i^2\,k^2$, at right angles to $i^2\,h^2$, and from f^2, draw $f^2\,k^2$, at right angles to $f^2\,h^2$; upon k^2, with $k^2 f^2$ for radius, describe the arc, $f^2\,i^2$; make $b^2\,l^2$ equal to $b^2 f^2$, and ease off the angle at b^2 by the curve, $f^2\,l^2$. In the figure, the curve is described from a centre, but in a full-size plan, this would be impracticable; the best way to ease the angle, therefore, would be with a tanged curve, according to *Art.* 89. Then from 1, 2, 3 and 4, (on the line, $w\,c^2$,) draw lines parallel to $n\,o$, meeting the curve in m^2, n^2, o^2 and p^2; from these points, draw lines at right angles to $n\,o$, and meeting it in x^2, r^2, s^2 and t^2; from x^2 and r^2, draw lines tending to u^2, and meeting the convex side of the rail in y^2 and z^2; make $m\,v^2$ equal to $r\,s^2$, and $m\,w^2$ equal to $r\,t^2$; from y^2, z^2, v^2, and w^2, through 4, 3, 2 and 1, draw lines meeting the line of the wall-string in a^3, b^3, c^3 and d^3; from e^3, where the centre-line of the rail crosses the line of the floor, draw $e^3 f^3$, at right angles to $n\,o$, and from f^3, through 6, draw $f^3\,g^3$; then the heavy lines, $f^3\,g^3$, $e^3\,d^3$, $y^2\,a^3$, $z^2\,b^3$, $v^2\,c^3$, $w^2\,d^3$, and $z\,y$, will be the lines for the risers, which, being extended to the line of the front string, $b\,e\,c\,d$, will give the dimensions of the winders, and the grading of the front string, as was required.

479.—*To obtain the falling-mould for the twists of the last-mentioned stairs.* Make $i^2\,g^3$ and $i^2\,h^3$, (*Fig.* 314,) each equal to half the thickness of the rail; through h^3 and g^3, draw $h^3\,i^3$ and $g^3\,j^3$, parallel to $i^2\,s$; assuming $k\,k^3$ and $m\,m^3$ on the plan as the amount of straight to be got out with the twists, make $n\,q$ equal to $k\,k^3$, and $r\,l^3$ equal to $m\,m^3$; from n and l^3, draw lines at right angles to $n\,o$, meeting the top of the falling-mould in n^3 and o^3; from o^3, draw a line crossing the falling-mould at right angles to a chord of the curve, $f^2\,l^2$; through the centre of the cylinder, draw u^2 8, at right angles to $n\,o$; through 8, draw 7 9, tending to k^3; then n^3 7 will be the falling-mould for the upper twist, and 7 o^3 the falling-mould for the lower twist.

480.—*To obtain the face-moulds.* The moulds for the twists of this stairs may be obtained as at *Art.* 473; but, as the falling-mould in its course departs considerably from a straight line, it would, according to that method, require a very thick plank for the rail, and consequently cause a great waste of stuff. In order, therefore, to economize the material, the following method is to be preferred—in which it will be seen that the heights are taken in three places instead of two only, as is done in the previous method.

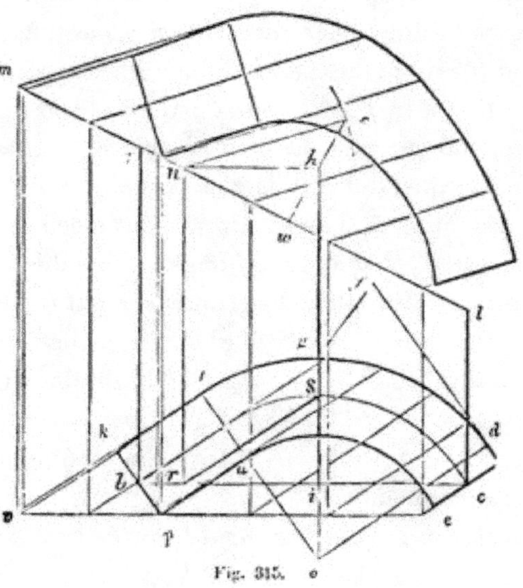

Fig. 315. o

CASE 1.—*When the middle height is above a line joining the other two.* Having found at *Fig.* 314 the direction of the joint, $w\ s^3$ and $p\ e$, according to *Art.* 473, make $k\ p\ e\ a$, (*Fig.* 315,) equal to $k^3\ p^3\ e\ p$ in *Fig.* 314; join b and c, and from o, draw $o\ h$, at right angles to $b\ c$; obtain the stretch-out of $d\ g$, as $d\ f$, and at *Fig.* 314, place it from the axis of the cylinder, p, to q^3; from q^3 in that figure, draw $q^3\ r^3$, at right angles to $n\ o$; also, at a convenient height on the line, $n\ n^3$, in that figure, and at right angles to that line, draw $u^3\ v^3$; from b and c, in *Fig.* 315, draw $b\,j$ and $c\,l$, at right angles to $b\,c$; make $b\,j$ equal to $u^3\,n^3$ in *Fig*. 314, $i\,h$ equal to $w^3\,r^3$ in that figure, and $c\,l$ equal to $v^3\,9$; from l, through j, draw $l\,m$; from h, draw $h\,n$, parallel to $c\,b$; from n, draw $n\,r$, at right angles to $b\,c$, and join r and s; through the lowest corner of the plan, as p, draw $v\,e$, parallel to $b\,c$; from a, e, u, p, k, t, and from as many other points as is thought necessary, draw ordinates to the base-line, $v\,e$, parallel to $r\,s$; through h, draw $w\,x$, at right angles to $m\,l$; upon n, with $r\,s$ for radius, describe an intersecting arc at x, and join n and x; from the points at which the ordinates from the plan meet the base-line, $v\,e$, draw ordinates to meet the line, $m\,l$, at right angles to $v\,e$; and from the points of intersection on $m\,l$, draw corresponding ordinates, parallel to $n\,x$; make the ordinates which are parallel to $n\,x$ of a length corresponding to those which are parallel to $r\,s$, and through the points thus found, trace the face-mould as required.

CASE 2.—*When the middle height is below a line joining the other two.* The lower twist in *Fig*. 314 is of this nature. The face-mould for this is found at *Fig*. 316 in a manner similar to that at *Fig*. 315. The heights are all taken from the top of the falling-mould at *Fig*. 314; $b\,j$ being equal to w 6 in *Fig*. 314, $i\,h$ equal to $x^3\,y^3$ in that figure, and $c\,l$ to $l^3\,o^3$. Draw a line through j and l, and from h, draw $h\,n$, parallel to $b\,c$; from n, draw $n\,r$, at right angles to $b\,c$, and join r and s; then $r\,s$ will be the bevil for the lower ordinates. From h, draw $h\,x$, at right angles to $j\,l$; upon n, with $r\,s$ for radius, describe an intersecting arc at x, and join n and x; then $n\,x$ will be the bevil for the upper ordinates, upon which the face-mould is found as in Case 1.

481.—*Elucidation of the foregoing method.*—This method of finding the face-moulds for the handrailing of winding stairs, being founded on principles which govern cylindric sections, may be illustrated by the following figures. *Fig*. 317 and 318 represent solid blocks, or prisms, standing upright on a level base, $b\,d$; the upper surface, $j\,a$ forming oblique angles with the face, $b\,l$—

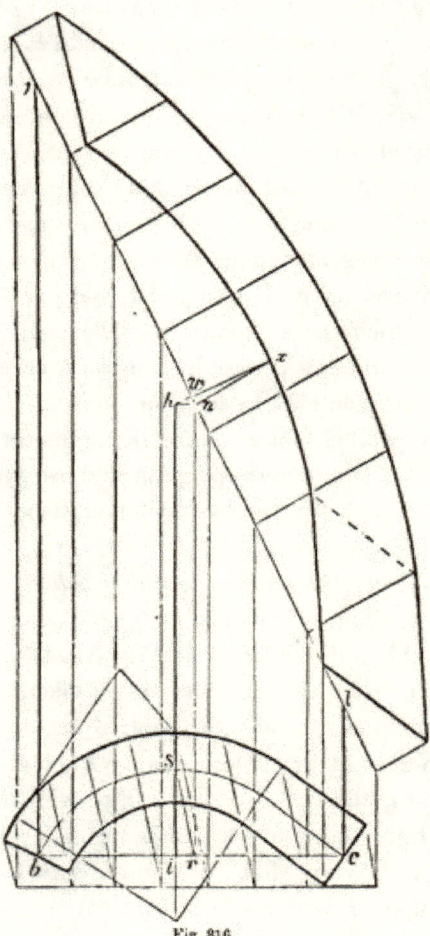

Fig. 316.

in *Fig.* 317 obtuse, and in *Fig.* 318 acute. Upon the base, describe the semi-circle, *b s c*; from the centre, *i*, draw *i s*, at right angles to *b c*; from *s*, draw *s x*, at right angles to *c d*, and from *i* draw *i h*, at right angles to *b c*: make *i h* equal to *s x*, and join *h* and *x*; then, *h* and *x* being of the same height, the line, *h x*, joining them, is a level line. From *h*, draw *h n*, parallel to *b c*, and from *n*, draw *n r*, at right angles to *b c*; join *r* and *s*, also *n*

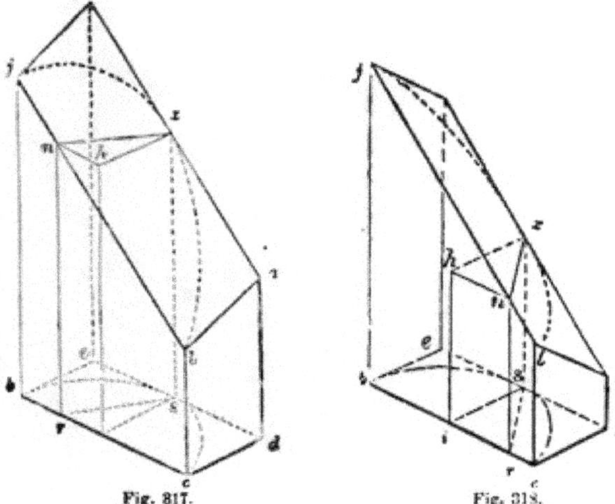
Fig. 317. Fig. 318.

and x; then, n and x being of the same height, $n\,x$ is a level line; and this line lying perpendicularly over $r\,s$, $n\,x$ and $r\,s$ must be of the same length. So, all lines on the top, drawn parallel to $n\,x$, and perpendicularly over corresponding lines drawn parallel to $r\,s$ on the base, must be equal to those lines on the base; and by drawing a number of these on the semi-circle at the base and others of the same length at the top, it is evident that a curve, $j\,x\,l$, may be traced through the ends of those on the top, which shall lie perpendicularly over the semi-circle at the base.

It is upon this principle that the process at *Fig.* 315 and 316 is founded. The plan of the rail at the bottom of those figures is supposed to lie perpendicularly under the face-mould at the top; and each ordinate at the top over a corresponding one at the base. The ordinates, $n\,x$ and $r\,s$, in those figures, correspond to $n\,x$ and $r\,s$ in *Fig.* 317 and 318.

In *Fig.* 319, the top, $e\,a$, forms a right angle with the face, $d\,c$; all that is necessary, therefore, in this figure, is to find a line corresponding to $h\,x$ in the last two figures, and that will lie level and in the upper surface; so that all ordinates at right angles to $d\,r$ on the base, will correspond to those that are at right angles

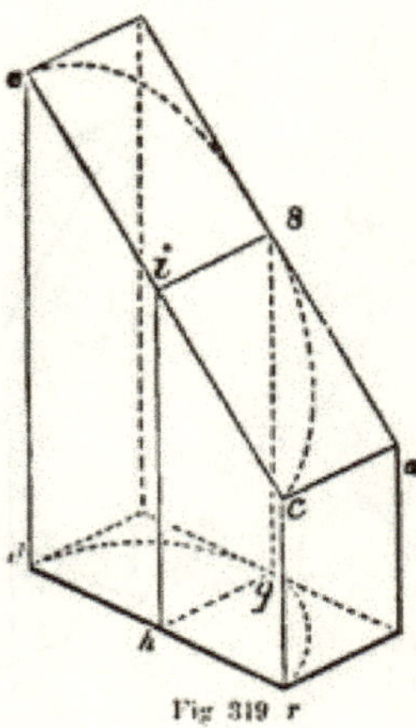

Fig. 319

to *e c* on the top. This elucidates *Fig.* 307; at which the lines, *h* 9 and *i* 8, correspond to *h* 9 and *i* 8 in this figure.

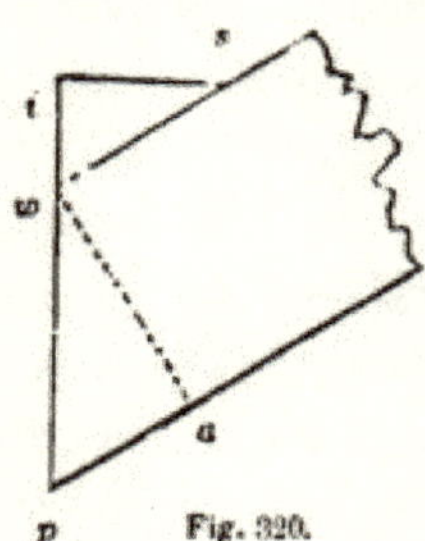

Fig. 320.

482.—*To find the bevil for the edge of the plank.* The plank, before the face-mould is applied, must be bevilled according to the angle which the top of the imaginary block, or prism, in the previous figures, makes with the face. This angle is determined in the following manner: draw *w i*, (*Fig.* 320,) at right angles to *i s*, and equal to *w h* at *Fig.* 315; make *i s* equal to *i s* in that figure, and join *w* and *s*; then *s w p* will be the bevil required in order to apply the face-mould at *Fig.* 315. In *Fig.* 316, the middle height being below the line joining the other two, the bevil is therefore acute. To determine this, draw *i s*, (*Fig.* 321,) at

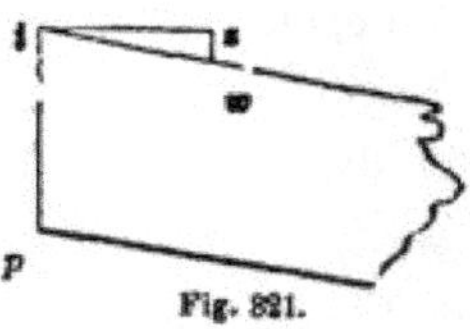

Fig. 321.

right angles to *i p*, and equal to *i s* in *Fig*. 316; make *s w* equal to *h w* in *Fig*. 316, and join *w* and *i*; then *w i p* will be the bevil required in order to apply the face-mould at *Fig*. 316. Although the falling-mould in these cases is curved, yet, as the plank is *sprung*, or bevilled on its edge, the thickness necessary to get out the twist may be ascertained according to *Art*. 474—taking the vertical distance across the falling-mould at the joints, and placing it down from the two outside heights in *Fig*. 315 or 316. After bevilling the plank, the moulds are applied as at *Art*. 476—applying the pitch-board on the bevilled instead of a square edge, and placing the tips of the mould so that they will bear the same relation to the edge of the plank, as they do to the line, *j l*, in *Fig*. 315 or 316.

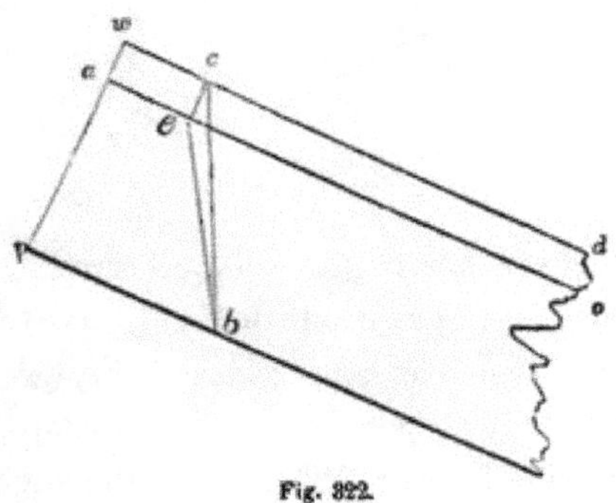

Fig. 322.

483.—*To apply the moulds without bevilling the plank.* Make *w p*, (*Fig*. 322,) equal to *w p* at *Fig*. 320, and the angle, *b c d*, equal to *b j l* in *Fig*. 315; make *p a* equal to the thickness of the plank, as *w a* in *Fig*. 320, and from *a* draw *a o*, parallel to *w d*; from c, draw *c e*, at right angles to *w d*, and join *e* and *b*; then the angle, *b e o*, on a square edge of the plank, having a line on the upper face at the distance, *p a*, in *Fig.* 320, at which to apply the tips of the mould—will answer the same purpose as bevilling the edge.

If the bevilled edge of the plank, which reaches from *p* to *w*, is supposed to be in the plane of the paper, and the point, *a*, to be above the plane of the paper as much as *a*, in *Fig.* 320, is distant from the line, *w p*; and the plank to be revolved on *p b* as an axis until the line, *p w*, falls below the plane of the paper, and the line, *p a*, arrives in it; then, it is evident that the point, *c*, will fall, in the line, *c e*, until it lies directly behind the point, *e*, and the line, *b c*, will lie directly behind *b e*.

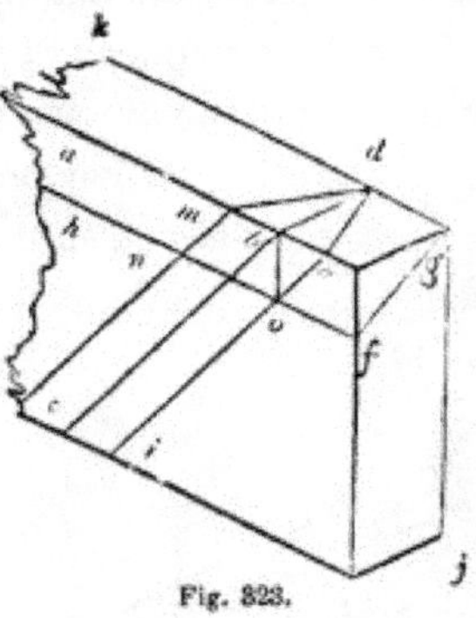

Fig. 323.

484.—*To find the bevils for splayed work.* The principle employed in the last figure is one that will serve to find the bevils for *splayed* work—such as hoppers, bread-trays, &c.—and a way of applying it to that purpose had better, perhaps, be introduced in this connection. In *Fig.* 323, *a b c* is the angle at which the work is splayed, and *b d*, on the upper edge of the board, is at right angles to *a b*; make the angle, *f g j*, equal to *a b c*, and from *f*, draw *f h*, parallel to *e a*; from *b*, draw *b o*, at right angles to *a b*; through *o*, draw *i e*, parallel to *c b*, and join *e* and *d*; then the angle, *a e d*, will be the proper bevil for the ends from the inside, or *k d e* from the outside. If a mitre-joint is required, set *f g*, the thickness of the stuff on the level, from *e* to *m*, and join *m* and *d*; then *k d m* will be the proper bevil for a mitre-joint.

If the upper edges of the splayed work is to be bevilled, so as to be horizontal when the work is placed in its proper position, *f g j*, being the same as *a b c*, will be the proper bevil for that purpose. Suppose, therefore, that a piece indicated by the lines, *k g*, *g f* and *f h*, were taken off; then a line drawn upon the bevilled surface from *d*, at right angles to *k d*, would show the true position of the joint, because it would be in the direction of the board for the other side; but a line so drawn would pass through the point, *o*,—thus proving the principle correct. So, if a line were drawn upon the bevilled surface from *d*, at an angle of 45 degrees to *k d*, it would pass through the point, *n*.

485.—*Another method for face-moulds.* It will be seen by reference to *Art.* 481, that the principal object had in view in the preparatory process of finding a face-mould, is to ascertain upon it the direction of a horizontal line. This can be found by a method different from any previously proposed; and as it requires fewer lines, and admits of less complication, it is probably to be preferred. It can be best introduced, perhaps, by the following explanation.

In *Fig.* 324, *j d* represents a prism standing upon a level base, *b d*, its upper surface forming an acute angle with the face, *b l*, as at *Fig.* 318. Extend the base line, *b c*, and the raking line, *j l*, to meet at *f*; also, extend *e d* and *g a*, to meet at *k*; from *f*, through *k*, draw *f m*. If we suppose the prism to stand upon a level floor, *o f m*, and the plane, *j g a l*, to be extended to meet that floor, then it will be obvious that the intersection between that plane and the plane of the floor would be in the line, *f k*; and the line, *f k*, being in the plane of the floor, and also in the inclined plane, *j g k f*, any line made in the plane, *j g k f*, parallel to *f k*, must be a level line. By finding the position of a perpendicular plane, at right angles to the raking plane, *j f k g*, we shall greatly shorten the process for obtaining ordinates.

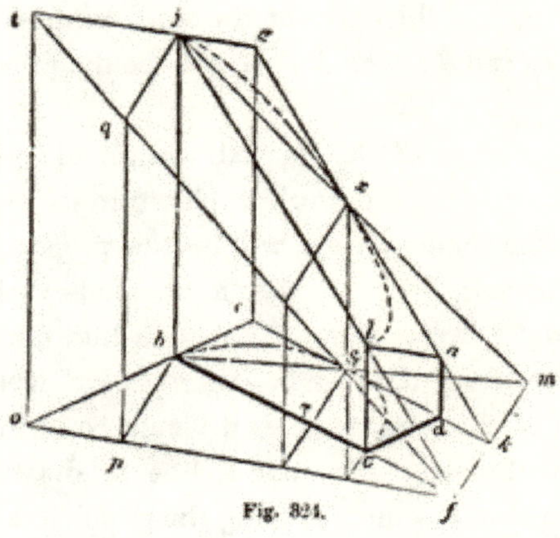

Fig. 324.

This may be done thus: from *f*, draw *f o*, at right angles to *f m*; extend *e b* to *o*, and *g j*, to *t*; from *o*, draw *o t*, at right angles to *o f*, and join *t* and *f*; then *t o f* will be a perpendicular plane, at right angles to the inclined plane, *t g k f*; because the base of the former, *o f*, is at right angles to the base of the latter, *f k*, both these lines being in the same plane. From *b*, draw *b p*, at right angles to *o f*, or parallel to *f m*; from *p*, draw *p q*, at right angles to *o f*, and from *q*, draw a line on the upper plane, parallel to *f m*, or at right angles to *t f*; then this line will obviously be drawn to the point, *j*, and the line, *q j*, be equal to *p b*. Proceed, in the same way, from the points, *s* and *c*, to find *x* and *l*.

Now, to apply the principle here explained, let the curve, *b s c*, (*Fig.* 325,) be the base of a cylindric segment, and let it be required to find the shape of a section of this segment, cut by a plane passing through three given points in its curved surface: one perpendicularly over *b*, at the height, *b j*; one perpendicularly over *s*, at the height, *s x*; and the other over *c*, at the height, *c l*—these lines being drawn at right angles to the chord of the base, *b c*. From *j*, through *l*, draw a line to meet the chord line extended to *f*; from *s*, draw *s k*, parallel to *b f*, and from *x*, draw *x k*, parallel to *j f*; from *f*, through *k*, draw *f m*; then *f m* will be the intersecting line of the plane of the section with the

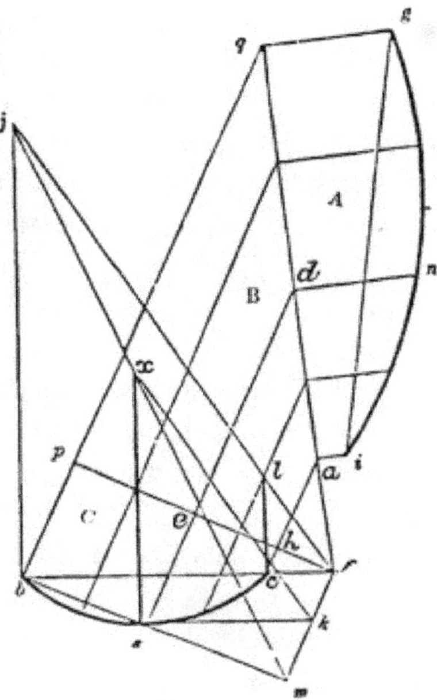

Fig. 325.

plane of the base. This line can be proved to be the intersection of these planes in another way; from *b*, through *s*, and from *j*, through *x*, draw lines meeting at *m*; then the point, *m*, will be in the intersecting line, as is shown in the figure, and also at *Fig*. 324.

From *f*, draw *f p*, at right angles to *f m*; from *b* and *c*, and from as many other points as is thought necessary, draw ordinates, parallel to *f m*; make *p q* equal to *b j*, and join *q* and *f*; from the points at which the ordinates meet the line, *q f*, draw others at right angles to *q f*; make each ordinate at *A* equal to its corresponding ordinate at *C*, and trace the curve, *g n i*, through the points thus found.

Now it may be observed that *A* is the plane of the section, *B* the plane of the segment, corresponding to the plane, *q p f*, of *Fig*. 324, and *C* is the plane of the base. To give these planes their proper position, let *A* be turned on *q f* as an axis until it stands perpendicularly over the line, *q f*, and at right angles to the plane, *B*; then, while *A* and *B* are fixed at right angles, let *B* be turned on the line, *p f*, as an axis until it stands perpendicularly over *p f*, and at right angles to the plane, *C*; then the plane, *A*, will lie over the plane, *C*, with the several lines on one corresponding to those on the other; the point, *i*, resting at *l*, the point, *n*, at *x*, and *g* at *j*; and the curve, *g n i*, lying perpendicularly over *b s c*—as was required. If we suppose the cylinder to be cut by a level plane passing through the point, *l*, (as is done in finding a face-mould,) it will be obvious that lines corresponding to *q f* and *p f* would meet in *l*; and the plane of the section, *A*, the plane of the segment, *B*, and the plane of the base, *C*, would all meet in that point.

486.—*To find the face-mould for a hand-rail according to the principles explained in the previous article.* In *Fig.* 326, *a e c f* is the plan of a hand-rail over a quarter of a cylinder; and in *Fig.* 327, *a b c d* is the falling-mould; *f e* being equal to the stretch-out of *a d f* in *Fig.* 326. From *c*, draw *c h*, parallel to *e f*; bisect *h c* in *i*, and find a point, as *b*, in the arc, *d f*, (*Fig.* 326,) corresponding to *i* in the line, *h c*; from *i*, (*Fig.* 327,) to the top of the falling-mould, draw *i j*, at right angles to *h c*; at *Fig.* 326, from *c*, through *b*, draw *c g*, and from *b* and *c*, draw *b j* and *c k*, at right angles to *g c*; make *c k* equal to *h g* at *Fig.* 327, and *b j* equal to *i j* at that figure; from *k*, through *j*, draw *k g*, and from *g*, through *a*, draw *g p*; then *g p* will be the intersecting line, corresponding to *f m* in *Fig.* 324 and 325; through *e*, draw *p* 6, at right angles to *g p*, and from *c*, draw *c q*, parallel to *g p*; make *r q* equal to *h g* at *Fig.* 327; join *p* and *q*, and proceed as in the previous examples to find the face-mould, *A*. The joint of the face-mould, *u v*, will be more accurately determined by finding the projection of the centre of the plan, *o*, as at *w*; joining *s* and *w*, and drawing *u v*, parallel to *s w*.

It may be noticed that *c k* and *b j* are not of a length corresponding to the above directions: they are but ½ the length given.

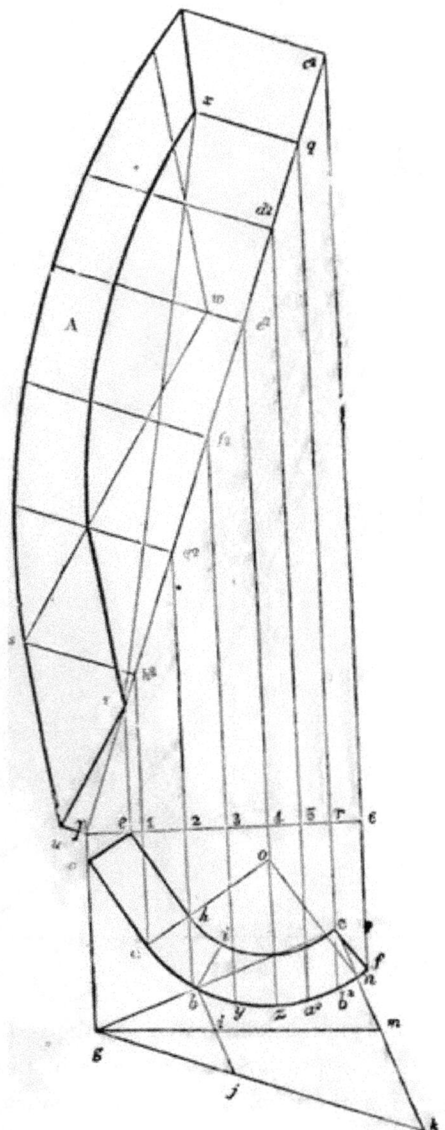

Fig. 826.

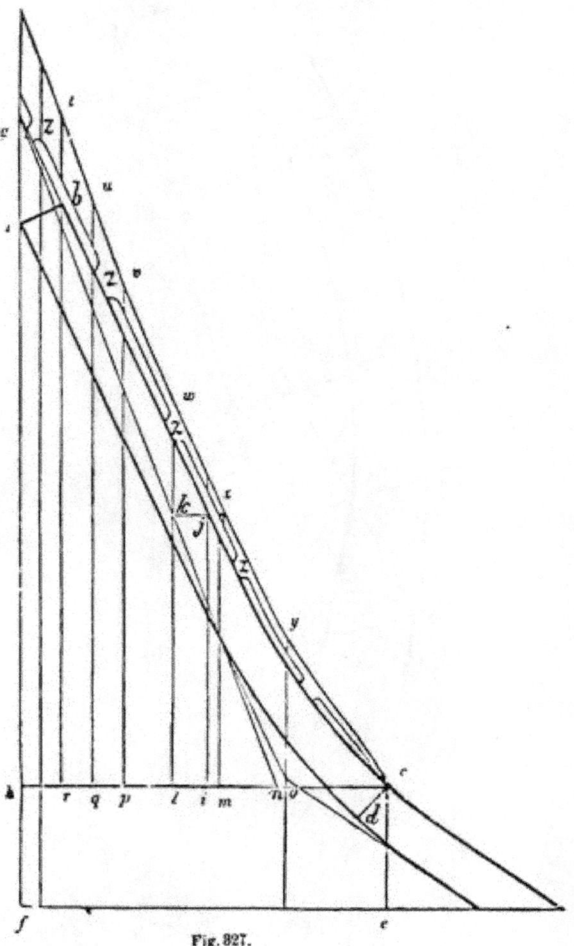

Fig. 327.

The object of drawing these lines is to find the point, *g*, and that can be done by taking any proportional parts of the lines given, as well as by taking the whole lines. For instance, supposing *c k* and *b j* to be the full length of the given lines, bisect one in *i* and the other in *m*; then a line drawn from *m*, through *i*, will give the point, *g*, as was required. The point, *g*, may also be obtained thus: at *Fig.* 327, make *h l* equal to *c b* in *Fig.* 326 from *l*, draw *l k*, at right angles to *h c*; from *j*, draw *j k*, parallel to *h c*; from *g*, through *k*, draw *g n*; at *Fig.* 326, make *b g* equal to *l n* in *Fig.* 327; then *g* will be the point required.

The reason why the points, *a*, *b* and *c*, in the plan of the rail at *Fig.* 326, are taken for resting points instead of *e*, *i* and *f*, is this: the top of the rail being level, it is evident that the points, *a* and *e*, in the section *a e*, are of the same height; also that the point, *i*, is of the same height as *b*, and *c* as *f*. Now, if *a* is taken for a point in the inclined plane rising from the line *g p*, *e* must be below that plane; if *b* is taken for a point in that plane, *i* must be below it; and if *c* is in the plane, *f* must be below it. The rule, then, for taking these points, is to take in each section the one that is nearest to the line, *g p*. Sometimes the line of intersection, *g p*, happens to come almost in the direction of the line, *e r*: in such case, after finding the line, see if the points from which the heights were taken agree with the above rule; if the heights were taken at the wrong points, take them according to the rule above, and then find the true line of intersection, which will not vary much from the one already found.

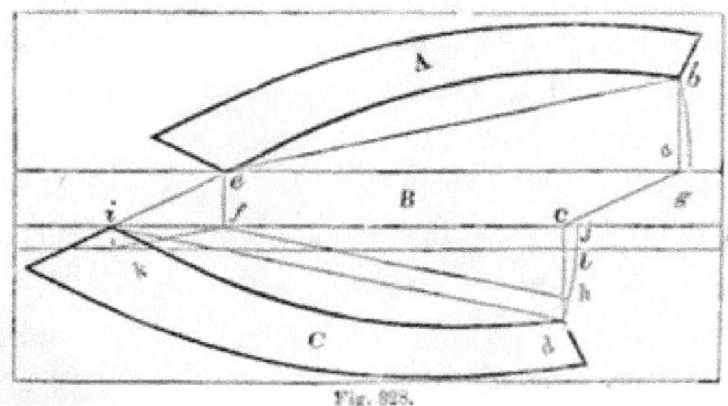

Fig. 328.

487.—*To apply the face-mould thus found to the plank.* The face-mould, when obtained by this method, is to be applied to a square-edged plank, as directed at *Art.* 476, with this difference: instead of applying both tips of the mould to the edge of the plank, one of them is to be set as far from the edge of the plank, as x, in *Fig.* 326, is from the chord of the section $p\ q$—as is shown at *Fig.* 328. A, in this figure, is the mould applied on the upper side of the plank, B, the edge of the plank, and C, the mould applied on the under side; $a\ b$ and $c\ d$ being made equal to $q\ x$ in *Fig.* 326, and the angle, $e\ a\ c$, on the edge, equal to the angle, $p\ q\ r$, at *Fig.* 326. In order to avoid a waste of stuff, it would be advisable to apply the tips of the mould, e and b, immediately at the edge of the plank. To do this, suppose the moulds to be applied as shown in the figure; then let A be revolved upon e until the point, b, arrives at g, causing the line, $e\ b$, to coincide with $e\ g$: the mould upon the under side of the plank must now be revolved upon a point that is perpendicularly beneath e, as f; from f, draw $f\ h$, parallel to $i\ d$, and from d, draw $d\ h$, at right angles to $i\ d$; then revolve the mould, C, upon f, until the point, h, arrives at j, causing the line, $f\ h$, to coincide with $f\ j$, and the line, $i\ d$, to coincide with $k\ l$; then the tips of the mould will be at k and l.

The rule for doing this, then, will be as follows: make the angle, $i\ f\ k$, equal to the angle $q\ v\ x$, at *Fig.* 326; make $f\ k$ equal to $f\ i$, and through k, draw $k\ l$, parallel to $i\ j$; then apply the corner of the mould, i, at k, and the other corner d, at the line, $k\ l$.

The thickness of stuff is found as at *Art.* 474.

488.—*To regulate the application of the falling-mould.* Obtain, on the line, $h\ c$, (*Fig.* 327,) the several points, r, q, p, l and m, corresponding to the points, b^7, a^2, z, y, &c., at *Fig.* 326, from $r\ q\ p$, &c., draw the lines, $r\ t$, $q\ u$, $p\ v$, &c., at right angles to $h\ c$; make $h\ s$, $r\ t$, $q\ u$, &c., respectively equal to $6\ c^2$, $r\ q$, $5\ d^1$, &c., at *Fig.* 326; through the points thus found, trace the curve, $s\ w\ c$. Then get out the piece, $g\ s\ c$, attached to the falling-mould at several places along its length, as at z, z, z, &c. In applying the falling-mould with this strip thus attached, the edge, $s\ w\ c$, will coincide with the upper surface of the rail piece before it is squared; and thus show the proper position of the falling-mould along its whole length. (See *Art.* 496.)

SCROLLS FOR HAND-RAILS.

489.—*General rule for finding the size and position of the regulating square.* The breadth which the scroll is to occupy, the number of its revolutions, and the relative size of the regulating square to the eye of the scroll, being given, multiply the number of revolutions by 4, and to the product add the number of times a side of the square* is contained in the diameter of the eye, and the sum will be the number of equal parts into which the breadth is to be divided. Make a side of the regulating square equal to one of these parts. To the breadth of the scroll add one of the parts thus found, and half the sum will be the length of the longest ordinate.

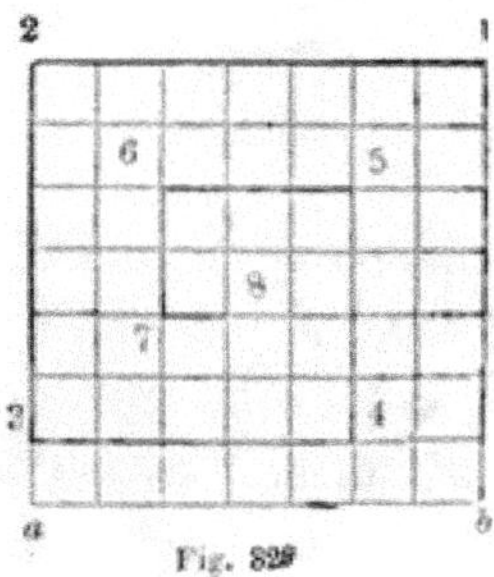

Fig. 329

490.—*To find the proper centres in the regulating square.* Let *a* 2 1 *b*, (*Fig.* 329,) be the size of a regulating square, found according to the previous rule, the required number of revolutions being $1\frac{3}{4}$. Divide two adjacent sides, as *a* 2 and 2 1, into as many equal parts as there are quarters in the number of revolutions, as seven; from those points of division, draw lines across the square, at right angles to the lines divided; then, 1 being the first centre, 2, 3, 4, 5, 6 and 7, are the centres for the other quarters, and 8 is the centre for the eye; the heavy lines that determine these centres being each one part less in length than its preceding line.

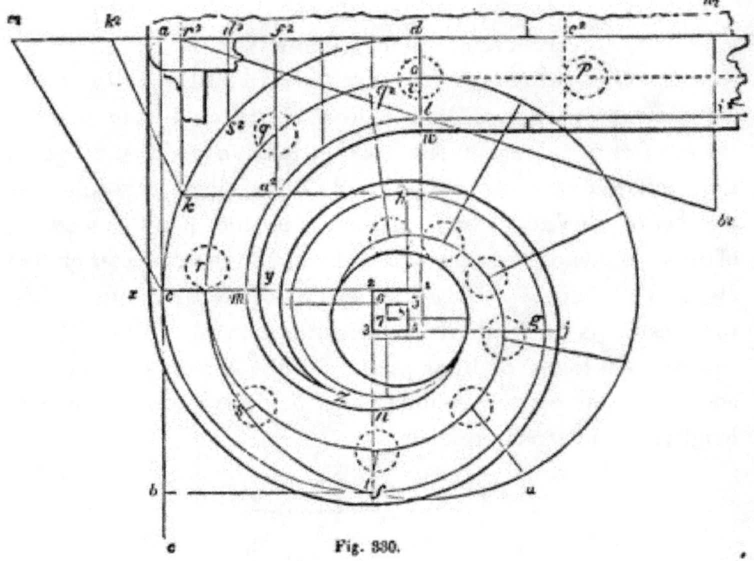

Fig. 330.

491.—*To describe the scroll for a hand-rail over a curtail step.* Let *a b*, (*Fig.* 330,) be the given breadth, 1$\frac{3}{4}$ the given number of revolutions, and let the relative size of the regulating square to the eye be $\frac{1}{3}$ of the diameter of the eye. Then, by the rule, 1$\frac{3}{4}$ multiplied by 4 gives 7, and 3, the number of times a side of the square is contained in the eye, being added, the sum is 10. Divide *a b*, therefore, into 10 equal parts, and set one from *b* to *c*; bisect *a c* in *e*; then *a e* will be the length of the longest ordinate, (1 *d* or 1 *e*.) From *a*, draw *a d*, from *e*, draw *e* 1, and from *b*, draw *b f*, all at right angles to *a b*; make *e* 1 equal to *e a*, and through 1, draw 1 *d*, parallel to *a b*; set *b c* from 1 to 2, and upon 1 2, complete the regulating square; divide this square as at *Fig.* 329; then describe the arcs that compose the scroll, as follows: upon 1, describe *d e*; upon 2, describe *e f*; upon 3, describe *f g*; upon 4, describe *g h*, &c.; make *d l* equal to the width of the rail, and upon 1, describe *l m*; upon 2, describe *m n*, &c.; describe the eye upon 8, and the scroll is completed.

492.—*To describe the scroll for a curtail step.* Bisect *d l*, (*Fig.* 330,) in *o*, and make *o v* equal to ⅛ of the diameter of a baluster; make *v w* equal to the projection of the nosing, and *e x* equal to *w l*; upon 1, describe *w y*, and upon 2, describe *y z*; also upon 2, describe *x i*; upon 3, describe *i j*, and so around to *z*; and the scroll for the step will be completed.

493.—*To determine the position of the balusters under the scroll.* Bisect *d l*, (*Fig.* 330,) in *o*, and upon 1, with 1 *o* for radius, describe the circle, *o r u*; set the baluster at *p* fair with the face of the second riser, c^2, and from *p*, with half the tread in the dividers, space off as at *o*, *q*, *r*, *s*, *t*, *u*, &c., as far as q^2; upon 2, 3, 4 and 5, describe the centre-line of the rail around to the eye of the scroll; from the points of division in the circle, *o r u*, draw lines to the centre-line of the rail, tending to the centre of the eye, 8; then, the intersection of these radiating lines with the centre-line of the rail, will determine the position of the balusters, as shown in the figure.

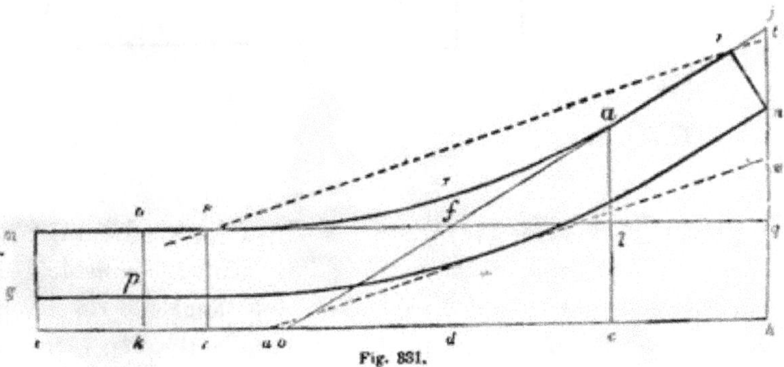

Fig. 331.

494.—*To obtain the falling-mould for the raking part of the scroll.* Tangical to the rail at *h*, (*Fig.* 330,) draw *h k*, parallel to *d a*; then *k* a^2 will be the joint between the twist and the other part of the scroll. Make *d* e^2 equal to the stretch-out of *d e*, and upon *d*

e^2, find the position of the point, *k*, as at k^2; at *Fig.* 331, make *e d* equal to e^2 *d* in *Fig.* 330, and *d c* equal to *d* c^2 in that figure; from *c*, draw *c a*, at right angles to *e c*, and equal to one rise; make *c b* equal to one tread, and from *b*, through *a*, draw *b j*, bisect *a c* in *l*, and through *l*, draw *m q*, parallel to *e h*; *m q* is the height of the level part of a scroll, which should always be about 3½ feet from the floor; ease off the angle, *m f j*, according to *Art.* 89, and draw *g w n*, parallel to *m x j*, and at a distance equal to the thickness of the rail; at a convenient place for the joint, as *i*, draw *i n*, at right angles to *b j*; through *n*, draw *j h*, at right angles to *e h*; make *d k* equal to *d* k^2 in *Fig.* 330, and from *k*, draw *k o*, at right angles to *e h*; at *Fig.* 330, make *d* h^2 equal to *d h* in *Fig.* 331, and draw h^2 b^2, at right angles to *d* h^2; then *k* a^2 and h^2 i^2 will be the position of the joints on the plan, and at *Fig.* 331, *o p* and *i n*, their position on the falling-mould; and *p o i n*, (*Fig.* 331,) will be the falling-mould required.

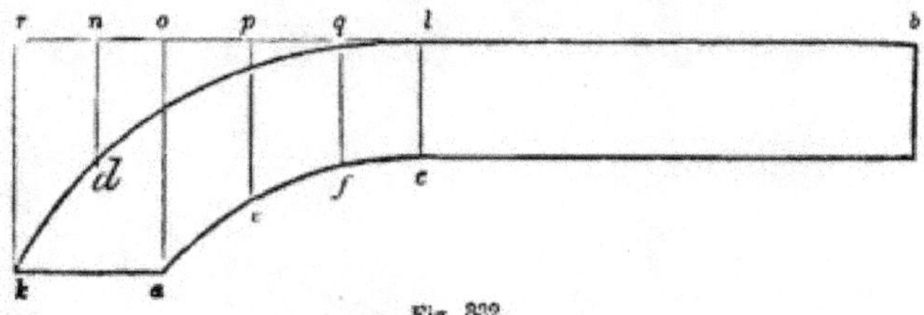

Fig. 332.

495.—*To describe the face-mould.* At *Fig.* 330, from *k*, draw *k* r^2, at right angles to r^2 *d*; at *Fig.* 331, make *h r* equal to h^2 r^2 in *Fig.* 330, and from *r*, draw *r s*, at right angles to *r h*; from the intersection of *r s* with the level line, *m q*, through *i*, draw *s t*; at *Fig.* 330, make h^2 b^2 equal to *q t* in *Fig.* 331, and join b^2 and r^2; from a^2, and from as many other points in the arcs, a^2 *l* and *k d*, as is thought necessary, draw ordinates to r^2 *d*, at right angles to the latter; make *r b*, (*Fig.* 332,) equal in its length and in its divisions to the line, r^2 b^2, in *Fig.* 330; from *r*, *n*, *o*, *p*, *q* and *l*, draw the lines, *r k*, *n d*, *o a*, *p e*, *q f* and *l c*, at right angles to *r b*, and equal to $r^2 k$, $d^2 s^2$, $f^2 a^2$, &c., in *Fig.* 330; through the points thus found, trace the curves, *k l* and *a c*, and complete the face-mould, as shown in the figure. This mould is to be applied to a square-edged plank, with the edge, *l b*, parallel to the edge of the plank. The rake lines upon the edge of the plank are to be made to correspond to the angle, *s t h*, in *Fig.* 331. The thickness of stuff required for this mould is shown at *Fig.* 331, between the lines *s t* and *u v*—*u v* being drawn parallel to *s t*.

496.—All the previous examples given for finding face-moulds over winders, are intended for *moulded* rails. For *round* rails, the same process is to be followed with this difference: instead of working from the sides of the rail, work from a centre-line. After finding the projection of that line upon the upper plane, describe circles upon it, as at *Fig.* 293, and trace the sides of the moulds by the points so found. The thickness of stuff for the twists of a round rail, is the same as for the straight; and the twists are to be sawed square through.

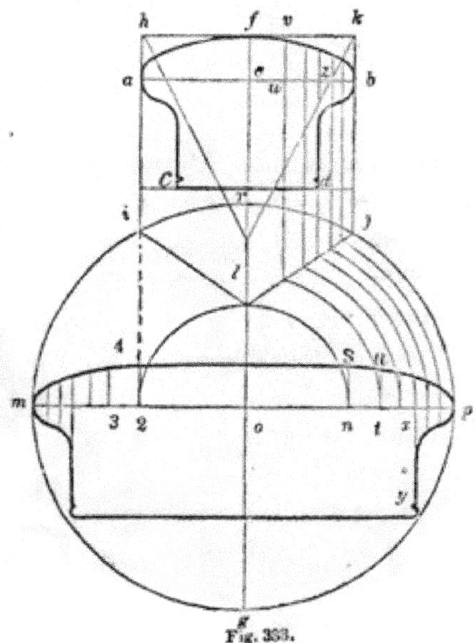

Fig. 333.

497.—*To ascertain the form of the newel-cap from a section of the rail.* Draw *a b*, (*Fig.* 333,) through the widest part of the given section, and parallel to *c d*; bisect *a b* in *e*, and through *a*, *e* and *b*, draw *h i*, *f g*, and *k j*, at right angles to *a b*; at a convenient place on the line, *f g*, as *o*, with a radius equal to half the width of the cap, describe the circle, *i j g*; make *r l* equal to *e b* or *e a*; join *l* and *j*, also *l* and *i*; from the curve, *f b*, to the line, *l j*, draw as many ordinates as is thought necessary, parallel to *f g*; from the points at which these ordinates meet the line, *l j*, and upon the centre, *o*, describe arcs in continuation to meet *o p*; from *n*, *t*, *x*, &c., draw *n s*, *t u*, &c., parallel to *f g*; make *n s*, *t u*, &c., equal to *e f*, *w v*, &c.; make *x y*, &c., equal to *z d*, &c.; make *o* 2, *o* 3, &c., equal to *o n*, *o t*, &c.; make 2 4 equal to *n s*, and in this way find the length of the lines crossing *o m*; through the points thus found, describe the section of the newel-cap, as shown in the figure.

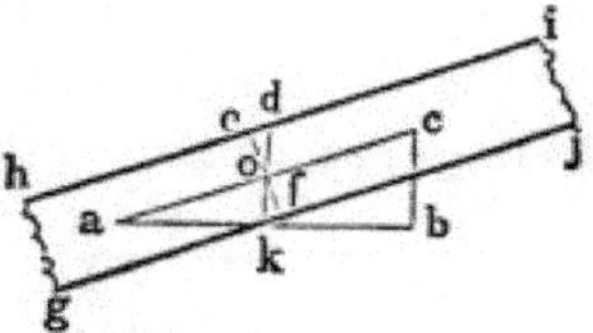

Fig 334.

498.—*To find the true position of a butt joint for the twists of a moulded rail over platform stairs.* Obtain the shape of the mould according to *Art.* 466, and make the line *a b*, *Fig.* 334, equal to *a c*, *Fig.* 300; from *b*, draw *b c*, at right angles to *a b*, and equal in length to *n m*, *Fig.* 300: join *a* and *c*, and bisect *a c* in *o*; through *o* draw *e f*, at right angles to *a c*, and *d k*, parallel to *c b*; make *o d* and *o k* each equal to half *e h* at *Fig.* 300; through *e* and *f*, draw *h i* and *g j*, parallel to *a c*. At *Fig.* 301, make *n a* equal to *e d*, *Fig.* 334, and through *a*, draw *r p*, at right angles to *n c*; then *r p* will be the true position on the face-mould for a butt joint, as was required. The sides must be sawn verti cally as described at *Art.* 467, but the joint is to be sawn square through the plank. The moulds obtained for round rails, (*Art.* 464,) give the line for the joint, when applied to either side of the plank; but here, for moulded rails, the line for the joint can be obtained from only one side. When the rail is canted up, the joint is taken from the mould laid on the upper side of the lower twist, and on the under side of the upper twist; but when it is canted down, a course just the reverse of this is to be pursued. When the rail is not canted, either up or down, the vertical joint, obtained as at *Art.* 466, will be a butt joint, and therefore, in such a case, the process described in this article will be unnecessary.

NOTE TO ARTICLE 462.

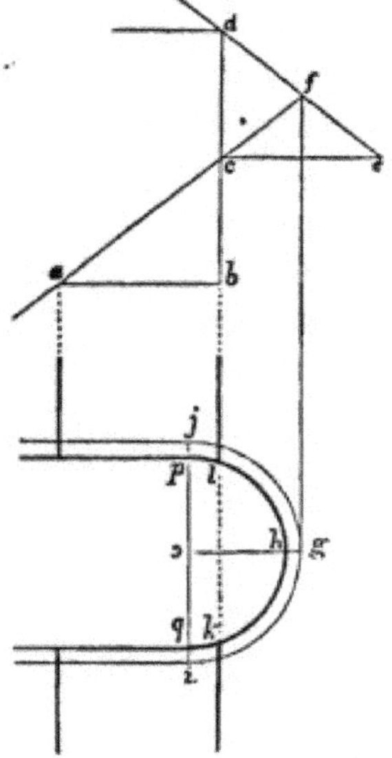

Platform stairs with a large cylinder. Instead of placing the platform-risers at the spring of the cylinder, a more easy and graceful appearance may be given to the rail, and the necessity of canting either of the twists entirely obviated, by fixing the place of the above risers at a certain distance within the cylinder, as shown in the annexed cut—the lines indicating the face of the risers cutting the cylinder at *k* and *l*, instead of at *p* and *q*, the spring of the cylinder. To ascertain the position of the risers, let *a b c* be the pitch-board of the lower flight, and *c d e* that of the upper flight, these being placed so that *b c* and *c d* shall form a right line. Extend *a c* to cut *d e* in *f*; draw *f g* parallel to *d b*, and of indefinite length: draw *g o* at right angles to *f g*, and equal in length to the radius of the circle formed by the centre of the rail in passing around the cylinder; on *o* as centre describe the semicircle *j g i*; make *o h* equal to the radius of the cylinder, and describe on *o* the face of the cylinder *p h q*; then extend *d b* across the cylinder, cutting it in *l* and *k*—giving the position of the face of the risers, as required. To find the face-mould for the twists is simple and obvious: it being merely a quarter of an ellipse, having *o j* for semi-minor axis, and the distance on the rake corresponding to *o g*, on the plan, for the semi-major axis, found thus,—extend *i j* to meet *a f*, then from this point of meeting to *f* is the semi-major axis.

499.—The art of drawing consists in representing solids upon a plane surface: so that a curious and nice adjustment of lines is made to present the same appearance to the eye, as does the human figure, a tree, or a house. It is by the effects of light, in its reflection, shade, and shadow, that the presence of an object is made known to us; so, upon paper, it is necessary, in order that the delineation may appear real, to represent fully all the shades and shadows that would be seen upon the object itself. In this section I propose to illustrate, by a few plain examples, the simple elementary principles upon which shading, in architectural subjects, is based. The necessary knowledge of drawing, preliminary to this subject, is treated of in the Introduction, from *Art.* 1 to 14.

500.—*The inclination of the line of shadow.* This is always, in architectural drawing, 45 degrees, both on the elevation and the plan; and the sun is supposed to be behind the spectator, and over his left shoulder. This can be illustrated by reference to *Fig.* 335, in which *A* represents a horizontal plane, and *B* and *C* two vertical planes placed at right angles to each other. *A* represents the plan, *C* the elevation, and *B* a vertical projection from the elevation. In finding the shadow of the plane, *B*, the

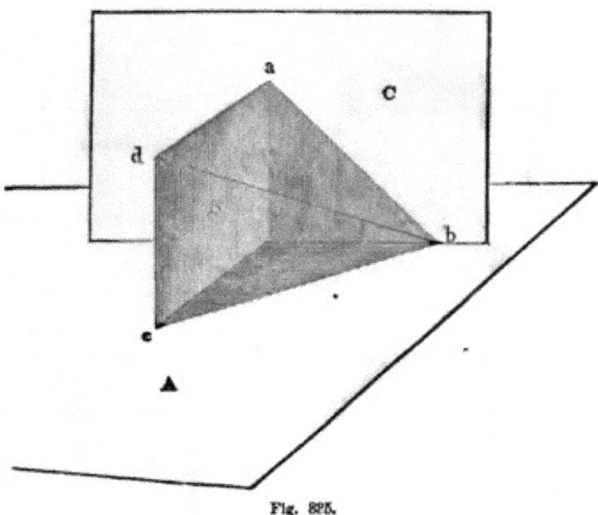

Fig. 825.

line, *a b*, is drawn at an angle of 45 degrees with the horizon, and the line, *c b*, at the same angle with the vertical plane, *B*. The plane, *B*, being a rectangle, this makes the true direction of the sun's rays to be in a course parallel to *d b*; which direction has been proved to be at an angle of 35 degrees and 16 minutes with the horizon. It is convenient, in shading, to have a set-square with the two sides that contain the right angle of equal length; this will make the two acute angles each 45 degrees; and will give the requisite bevil when worked upon the edge of the T-square. One reason why this angle is chosen in preference to another, is, that when shadows are properly made upon the drawing by it, the depth of every recess is more readily known, since the breadth of shadow and the depth of the recess will be equal.

To distinguish between the terms *shade* and *shadow*, it will be understood that all such parts of a body as are not exposed to the direct action of the sun's rays, are in *shade;* while those parts which are deprived of light by the interposition of other bodies, are in *shadow*.

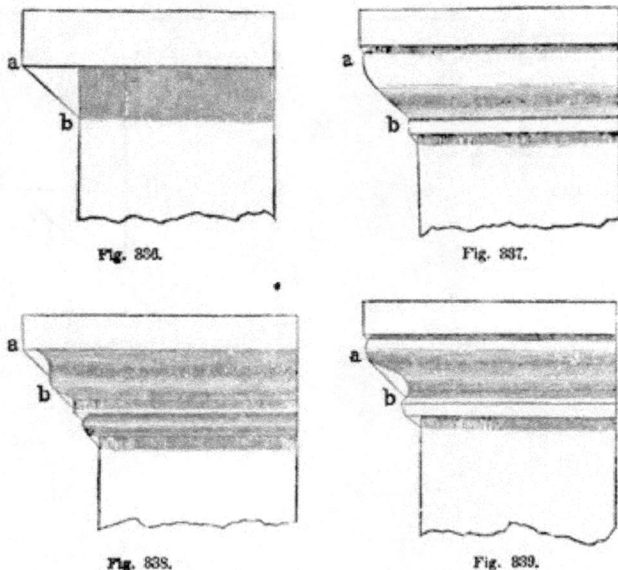

Fig. 336. Fig. 337.

Fig. 338. Fig. 339.

501.—*To find the line of shadow on mouldings and other horizontally straight projections.* *Fig.* 336, 337, 338, and 339, represent various mouldings in elevation, returned at the left, in the usual manner of mitreing around a projection. A mere inspection of the figures is sufficient to see how the line of shadow is obtained; bearing in mind that the ray, *a b*, is drawn from the projections at an angle of 45 degrees. Where there is no return at the end, it is necessary to draw a section, at any place in the length of the mouldings, and find the line of shadow from that.

502.—*To find the line of shadow cast by a shelf.* In *Fig.* 340, *A* is the plan, and *B* is the elevation of a shelf attached to a wall. From *a* and *c*, draw *a b* and *c d*, according to the angle previously directed; from *b*, erect a perpendicular intersecting *c d* at *d*; from *d*, draw *d e*, parallel to the shelf; then the lines, *c d* and *d e*, will define the shadow cast by the shelf. There is another method of finding the shadow, without the plan, *A*. Extend the lower line of the shelf to *f*, and make *c f* equal to the projection of the shelf

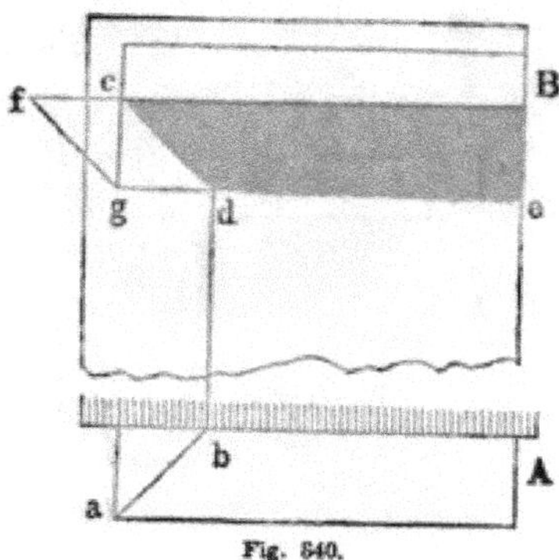

Fig. 340.

from the wall; from *f*, draw *f g*, at the customary angle, and from *c*, drop the vertical line, *c g*, intersecting *f g* at *g*; from *g*, draw *g e*, parallel to the shelf, and from *c*, draw *c d*, at the usual angle; then the lines, *c d* and *d e*, will determine the extent of the shadow as before.

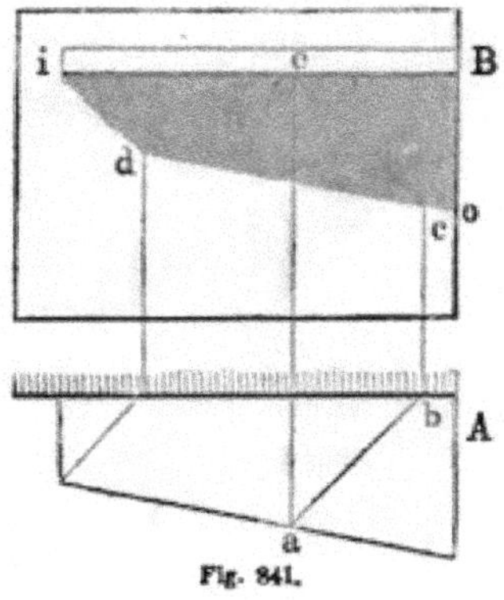

Fig. 341.

503.—*To find the shadow cast by a shelf, which is wider at one end than at the other.* In *Fig.* 341, *A* is the plan, and *B* the elevation. Find the point, *d*, as in the previous example, and from any other point in the front of the shelf, as *a*, erect the perpendicular, *a e*; from *a* and *e*, draw *a b* and *e c*, at the proper angle, and from *b*, erect the perpendicular, *b c*, intersecting *e c* in *c*:

from *d*, through *c*, draw *d o*; then the lines, *i d* and *d o*, will give the limit of the shadow cast by the shelf.

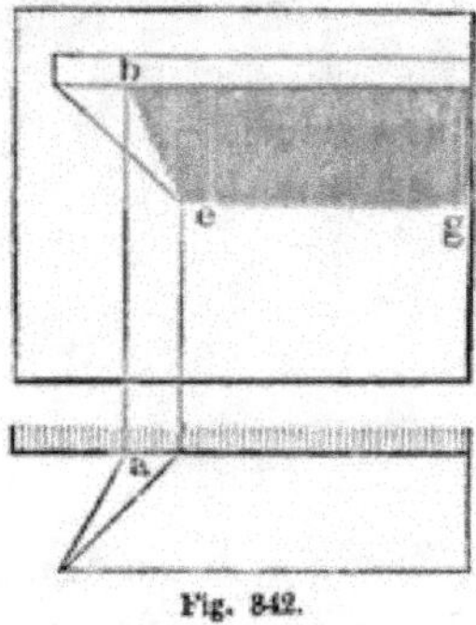

Fig. 342.

504.—*To find the shadow of a shelf having one end acute or obtuse angled.* *Fig.* 342 shows the plan and elevation of an acute-angled shelf. Find the line, *e g*, as before; from *a*, erect the perpendicular, *a b*; join *b* and *e*; then *b e* and *e g* will define the boundary of shadow.

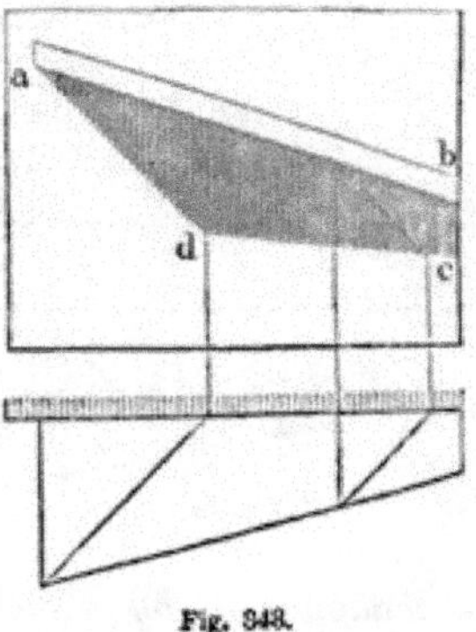

Fig. 343.

505.—*To find the shadow cast by an inclined shelf.* In *Fig.* 343, the plan and elevation of such a shelf is shown, having also one end wider than the other. Proceed as directed for finding the shadows of *Fig.* 341, and find the points, *d* and *c*; then *a d* and *d c* will be the shadow required. If the shelf had been parallel in width on the plan, then the line, *d c*, would have bee parallel with the shelf, *a b*.

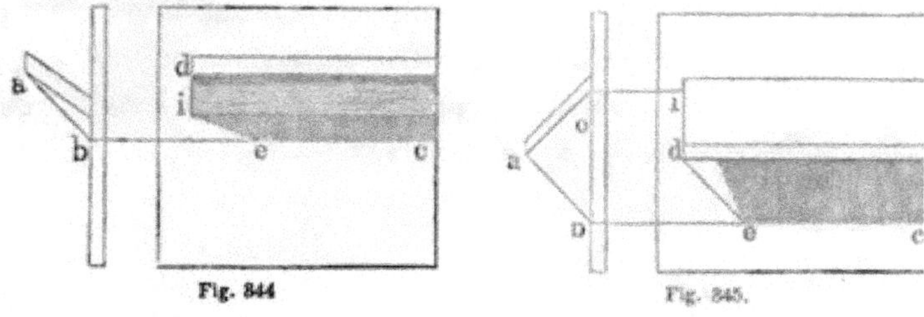

Fig. 844 Fig. 845.

506.—*To find the shadow cast by a shelf inclined in its ve tical section either upward or downward.* From *a*, (*Fig.* 34 and 345,) draw *a b*, at the usual angle, and from *b*, draw *b* parallel with the shelf; obtain the point, *e*, by drawing a lin from *d*, at the usual angle. In *Fig.* 344, join *e* and *i*; then *i* and *e c* will define the shadow. In *Fig.* 345, from *o*, draw *o* parallel with the shelf; join *i* and *e*; then *i e* and *e c* will be th shadow required.

The projections in these several examples are bounded b straight lines; but the shadows of curved lines may be found i the same manner, by projecting shadows from several points i the curved line, and tracing the curve of shadow through thes points. Thus—

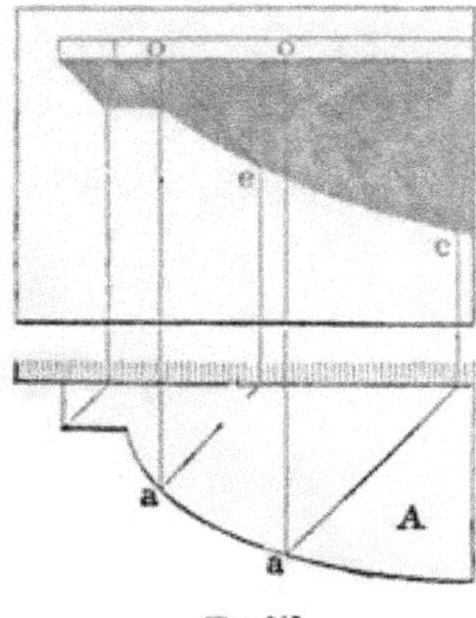

Fig. 846.

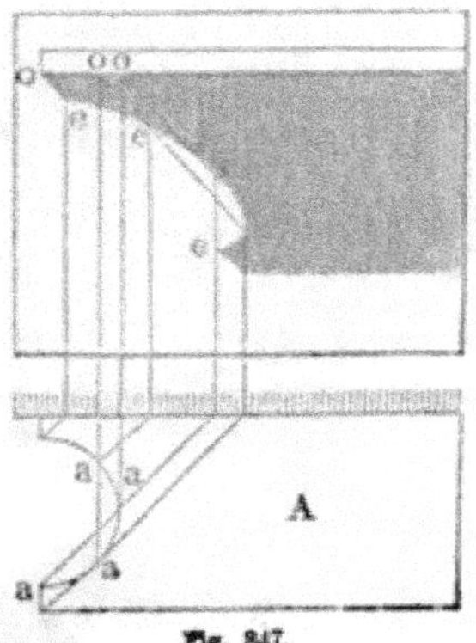

Fig. 847.

507.—*To find the shadow of a shelf having its front edge, or end, curved on the plan.* In *Fig.* 346 and 347, *A* and *A* show an example of each kind. From several points, as *a*, *a*, in the plan, and from the corresponding points, *o*, *o*, in the elevation, draw rays and perpendiculars intersecting at *e*, *e*, &c.; through these points of intersection trace the curve, and it will define the shadow.

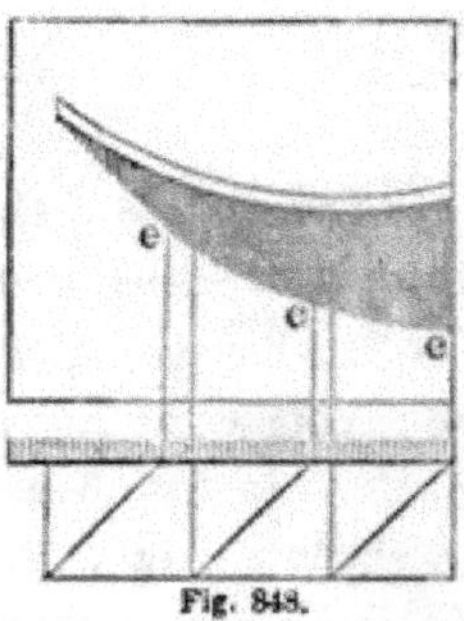

Fig. 348.

508.—*To find the shadow of a shelf curved in the elevation.* In *Fig.* 348, find the points of intersection, *e*, *e* and *e*, as in the last examples, and a curve traced through them will define the shadow.

The preceding examples show how to find shadows when cast upon a *vertical plane*; shadows thrown upon *curved surfaces* are ascertained in a similar manner. Thus—

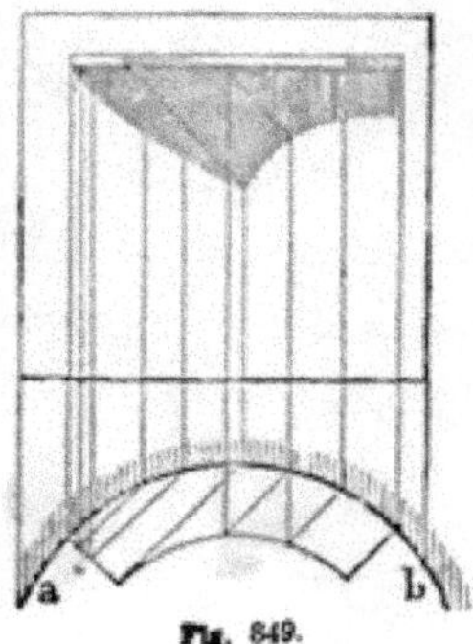

Fig. 349.

509.—*To find the shadow cast upon a cylindrical wall by a projection of any kind.* By an inspection of *Fig.* 349, it will be seen that the only difference between this and the last examples is, that the rays in the plan die against the circle, *a b*, instead of a straight line.

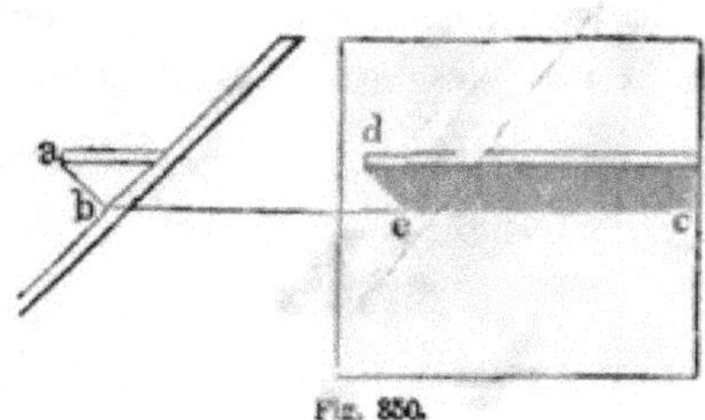

Fig. 350.

510.—*To find the shadow cast by a shelf upon an inclined wall.* Cast the ray, *a b*, (*Fig.* 350,) from the end of the shelf to the face of the wall, and from *b*, draw *b c*, parallel to the shelf; cast the ray, *d e*, from the end of the shelf; then the lines, *d e* and *e c*, will define the shadow.

These examples might be multiplied, but enough has been given to illustrate the general principle, by which shadows in all instances are found. Let us attend now to the application of this principle to such familiar objects as are likely to occur in practice.

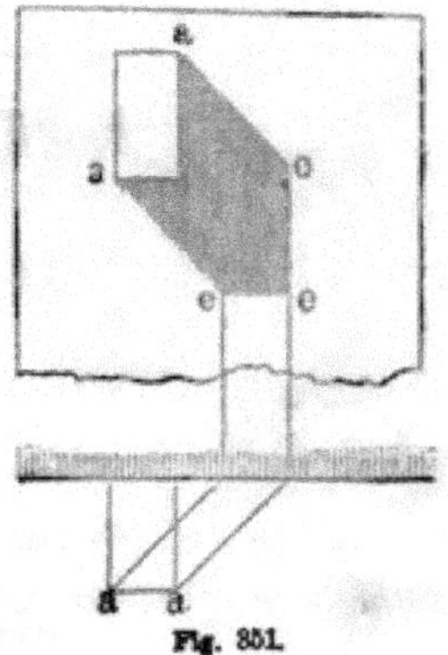

Fig. 351.

511.—*To find the shadow of a projecting horizontal beam* From the points, *a*, *a*, &c., (*Fig.* 351,) cast rays upon the wall; the intersections, *e*, *e*, *e*, of those rays with the perpendiculars drawn from the plan, will define the shadow. If the beam be inclined, either on the plan or elevation, at any angle other than a right angle, the difference in the manner of proceeding can be seen by reference to the preceding examples of inclined shelves &c.

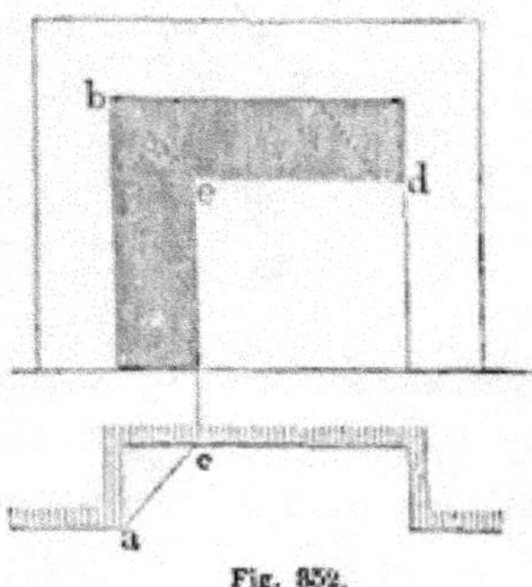

Fig. 352.

512.—*To find the shadow in a recess.* From the point, *a*, (*Fig.* 352,) in the plan, and *b* in the elevation, draw the rays, *a c* and *b e*; from *c*, erect the perpendicular, *c e*, and from *e*, draw the horizontal line, *e d*; then the lines, *c e* and *e d*, will show the extent of the shadow. This applies only where the back of the recess is parallel with the face of the wall.

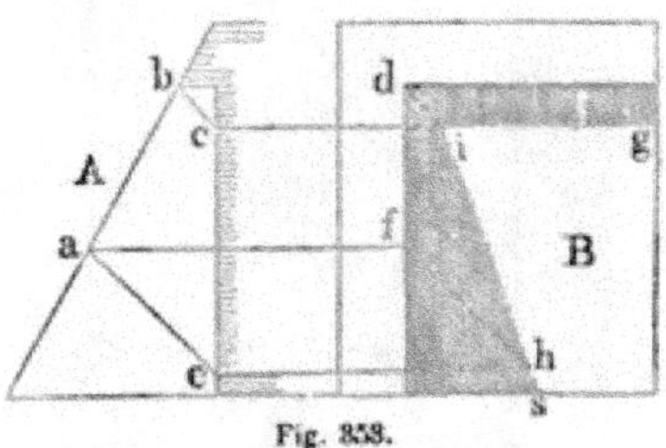

Fig. 353.

513.—*To find the shadow in a recess, when the face of the wall is inclined, and the back of the recess is vertical.* In *Fig.* 353, *A* shows the section and *B* the elevation of a recess of this kind. From *b*, and from any other point in the line, *b a* as *a* draw the rays, *b c* and *a e*; from *c*, *a*, and *e*, draw the horizontal lines, *c g*, *a f*, and *e h*; from *d* and *f*, cast the rays, *d i* and *f h*: from *i*, through *h*, draw *i s*; then *s i* and *i g* will define the shadow.

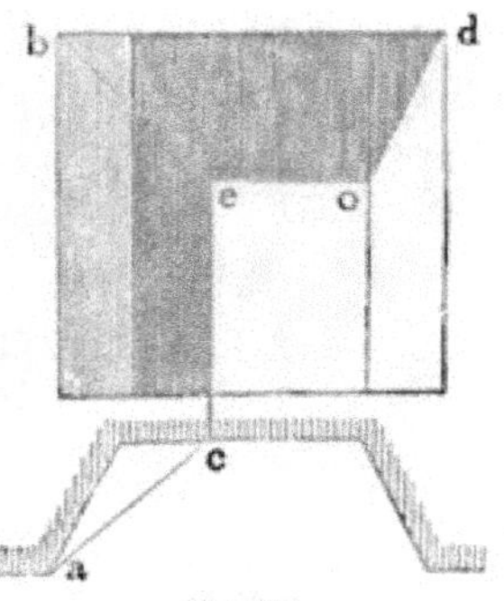

Fig. 354.

514.—*To find the shadow in a fireplace.* From *a* and *b*, (*Fig.* 354,) cast the rays, *a c* and *b e*, and from *c*, erect the perpendicular, *c e*; from *e*, draw the horizontal line, *e o*, and join *o* and *d*; then *c e*, *e o*, and *o d*, will give the extent of the shadow.

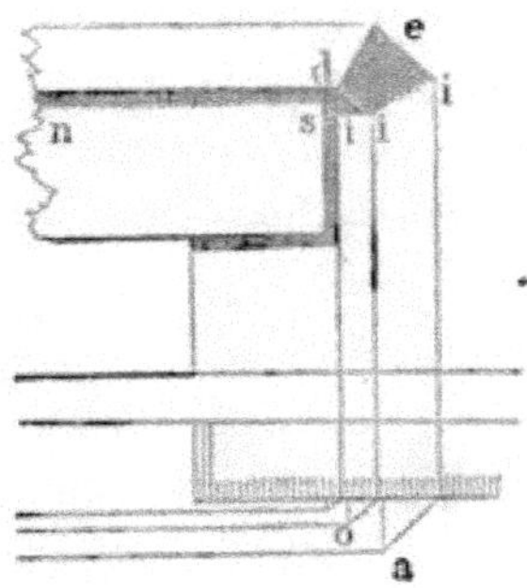

Fig. 355.

515.—*To find the shadow of a moulded window-lintel.* Cast rays from the projections, *a*, *c*, &c., in the plan, (*Fig.* 355,) and *d*, *e*, &c., in the elevation, and draw the usual perpendiculars intersecting the rays at *i*, *i*, and *i*; these intersections connected and horizontal lines drawn from them, will define the shadow The shadow on the face of the lintel is found by casting a ray back from *i* to *s*, and drawing the horizontal line, *s n*.

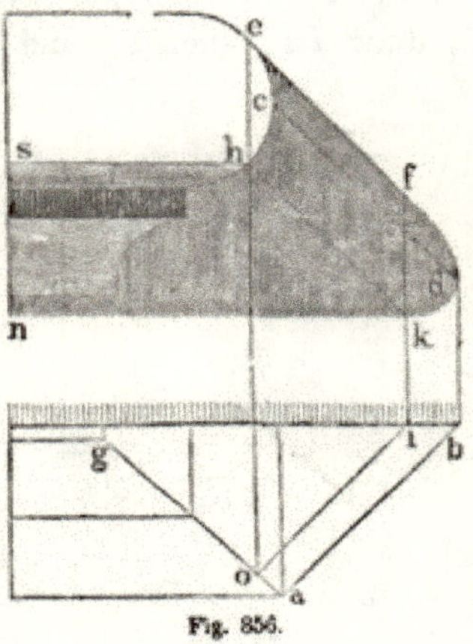

Fig. 356.

516.—*To find the shadow cast by the nosing of a step.* From *a*, (*Fig.* 356,) and its corresponding point, *c*, cast the rays, *a b* and *c d*, and from *b*, erect the perpendicular, *b d*; tangical to the curve at *e*, cast the ray, *e f*, and from *e*, drop the perpendicular, *e o*, meeting the mitre-line, *a g*, in *o*; cast a ray from *o* to *i*, and from *i*, erect the perpendicular, *i f*; from *h*, draw the ray, *h k*; from *f* to *d* and from *d* to *k*, trace the curve as shown in the figure; from *k* and *h*, draw the horizontal lines, *k n* and *h s*; then the limit of the shadow will be completed.

517.—*To find the shadow thrown by a pedestal upon steps.* From *a*, (*Fig.* 357,) in the plan, and from *c* in the elevation, draw the rays, *a b* and *c e*; then *a o* will show the extent of the shadow on the first riser, as at *A*; *f g* will determine the shadow on the second riser, as at *B*; *c d* gives the amount of shadow on the first tread, as at *C*, and *h i* that on the second tread, as at *D*; which completes the shadow of the left-hand pedestal, both on the plan and elevation. A mere inspection of the figure will be suf-

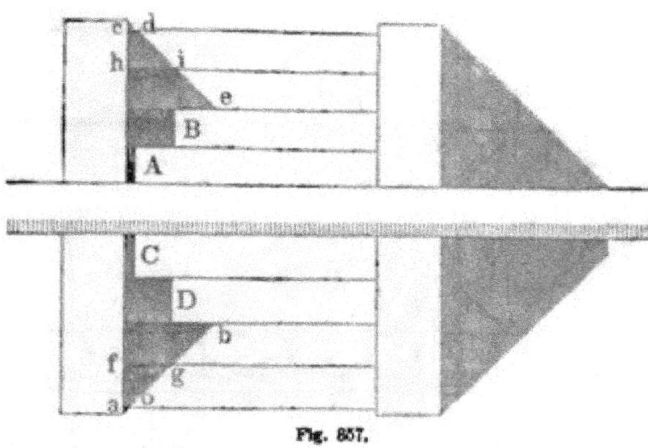

Fig. 357.

ficient to show how the shadow of the right-hand pedestal is obtained.

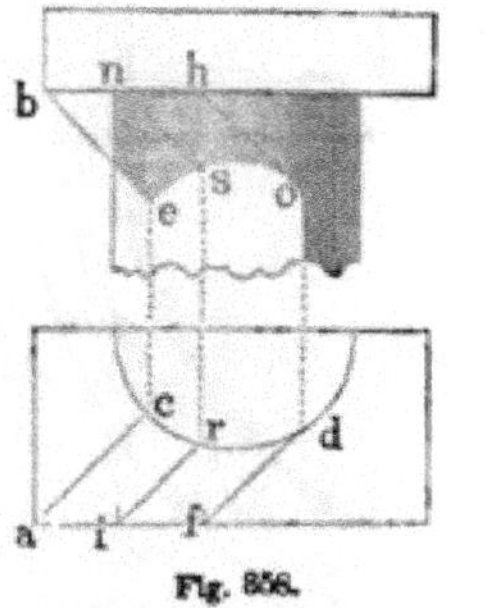

Fig. 358.

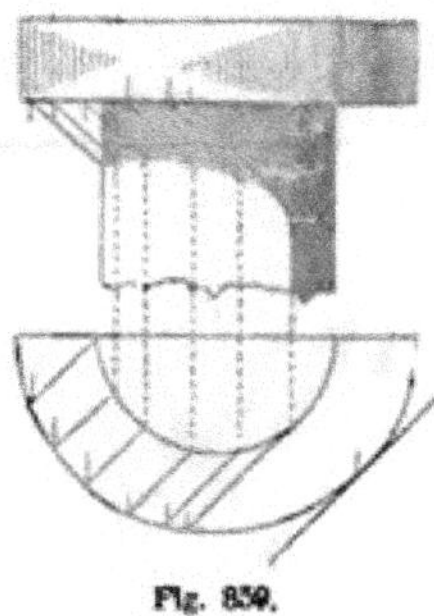

Fig. 359.

518.—*To find the shadow thrown on a column by a square abacus.* From *a* and *b*, (*Fig.* 358,) draw the rays, *a c* and *b e*, and from c, erect the perpendicular, *c e*; tangical to the curve at *d*, draw the ray, *d f*, and from *h*, corresponding to *f* in the plan, draw the ray, *h o*; take any point between *a* and *f*, as *i*, and from this, as also from a corresponding point, *n*, draw the rays, *i r* and *n s*; from *r*, and from *d*, erect the perpendiculars, *r s* and *d o*; through the points, *e*, *s*, and *o*, trace the curve as shown in the figure ; then the extent of the shadow will be defined.

519.—*To find the shadow thrown on a column by a circular abacus.* This is so near like the last example, that no explanation will be necessary farther than a reference to the preceding article

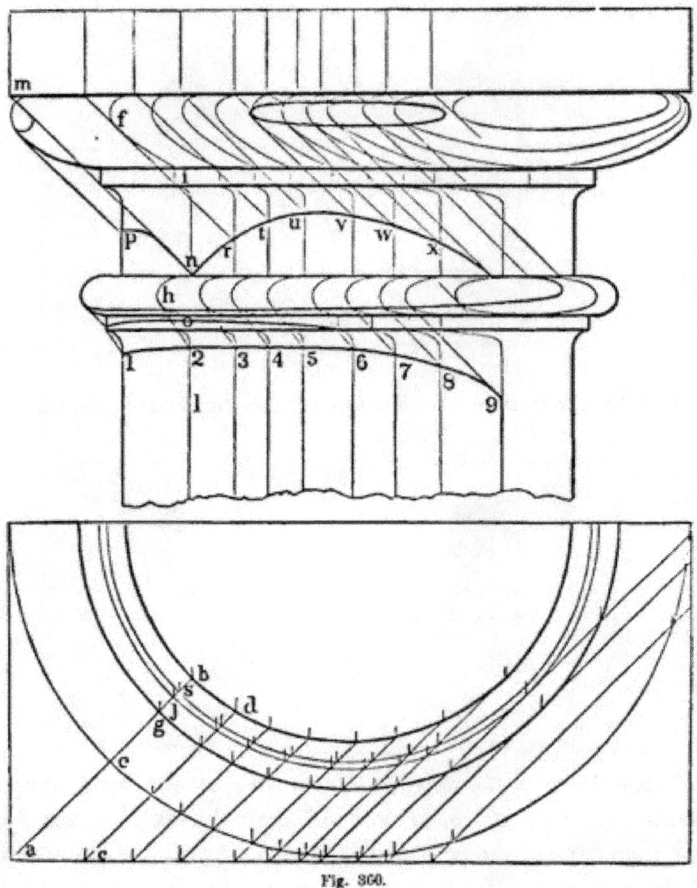

Fig. 360.

520.—*To find the shadows on the capital of a column.* This may be done according to the principles explained in the examples already given; a quicker way of doing it, however, is as follows. If we take into consideration one ray of light in connection with all those perpendicularly under and over it, it is evident that these several rays would form a vertical plane, standing at an angle of 45 degrees with the face of the elevation. Now, we may suppose the column to be *sliced,* so to speak, with planes of this

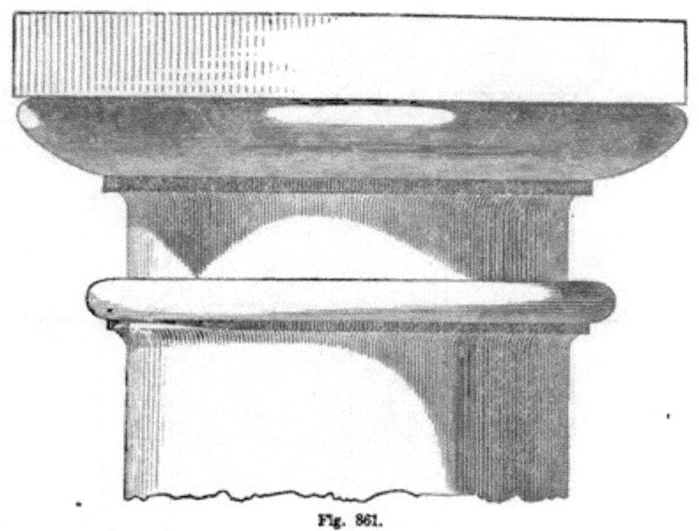

Fig. 361.

nature—cutting it in the lines, *a b*, *c d*, &c., (*Fig.* 360,) and, in the elevation, find, by squaring up from the plan, the *lines of section* which these planes would make thereupon. For instance: in finding upon the elevation the line of section, *a b*, the plane cuts the ovolo at *e*, and therefore *f* will be the corresponding point upon the elevation; *h* corresponds with *g*, *i* with *j*, *o* with *s*, and *l* with *b*. Now, to find the shadows upon this line of section, cast from *m*, the ray, *m n*, from *h*, the ray, *h o*, &c.; then that part of the section indicated by the letters, *m f i n*, and that part also between *h* and *o*, will be under shadow. By an inspection of the figure, it will be seen that the same process is applied to each line of section, and in that way the points, *p*, *r*, *t*, *u*, *v*, *w*, *x*, as also 1, 2, 3, &c., are successively found, and the lines of shadow traced through them.

Fig. 361 is an example of the same capital with all the shadows finished in accordance with the lines obtained on *Fig.* 360.

521.—*To find the shadow thrown on a vertical wall by a column and entablature standing in advance of said wall.* Cast

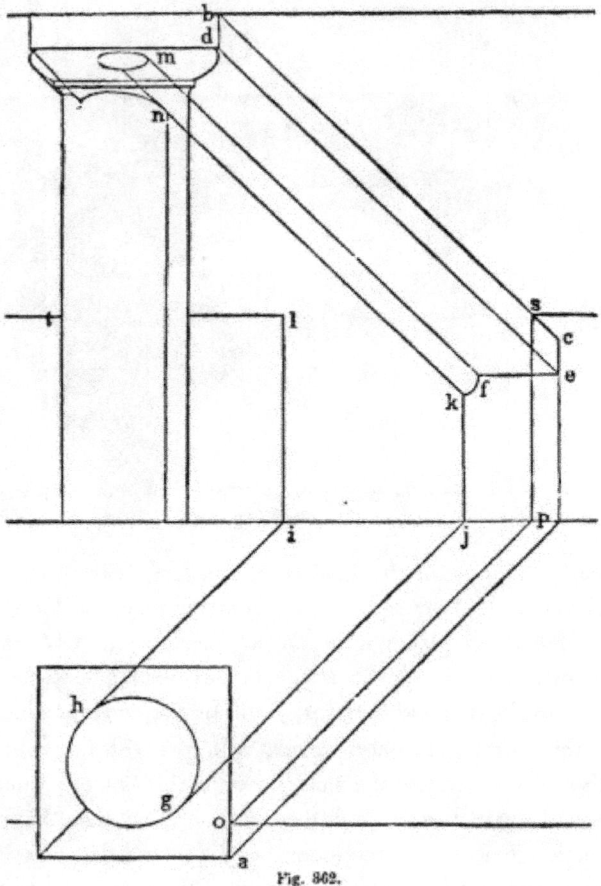

Fig. 362.

rays from *a* and *b*, (*Fig.* 362,) and find the point, *c*, as in the previous examples; from *d*, draw the ray, *d e*, and from *e*, the horizontal line, *e f;* tangical to the curve at *g* and *h*, draw the rays, *g j* and *h i*, and from *i* and *j*, erect the perpendiculars, *i l* and *j k;* from *m* and *n*, draw the rays, *m f* and *n k*, and trace the curve between *k* and *f;* cast a ray from *o* to *p*, a vertical line from *p* to *s*, and through *s*, draw the horizontal line, *s t;* the shadow as required will then be completed.

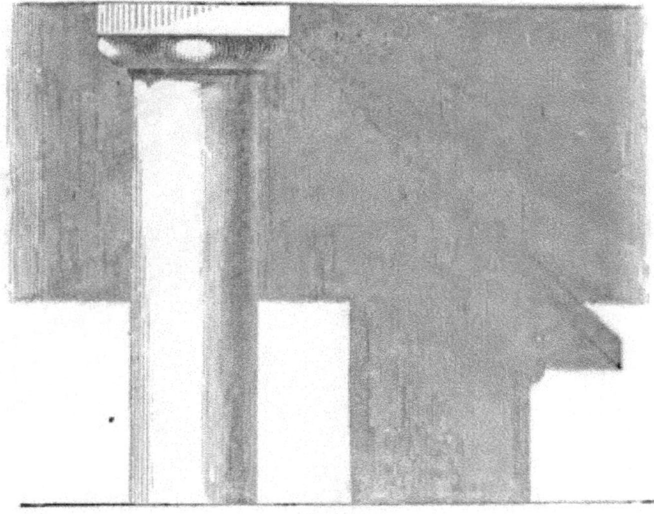

Fig. 363.

Fig. 363 is an example of the same kind as the last, with all the shadows filled in, according to the lines obtained in the preceding figure.

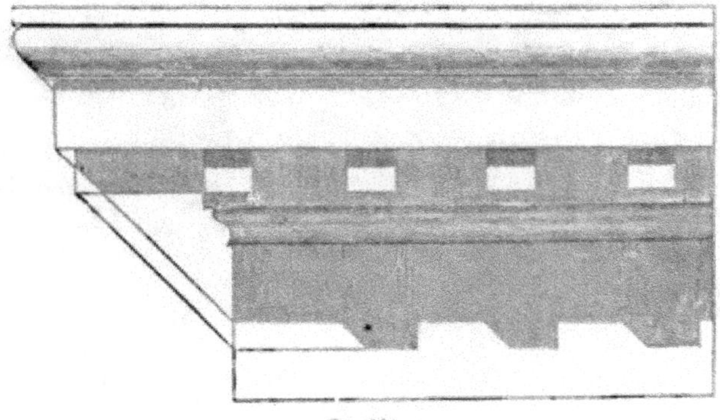

Fig. 364.

522.—*Fig.* 364 and 365 are examples of the Tuscan cornice. The manner of obtaining the shadows is evident.

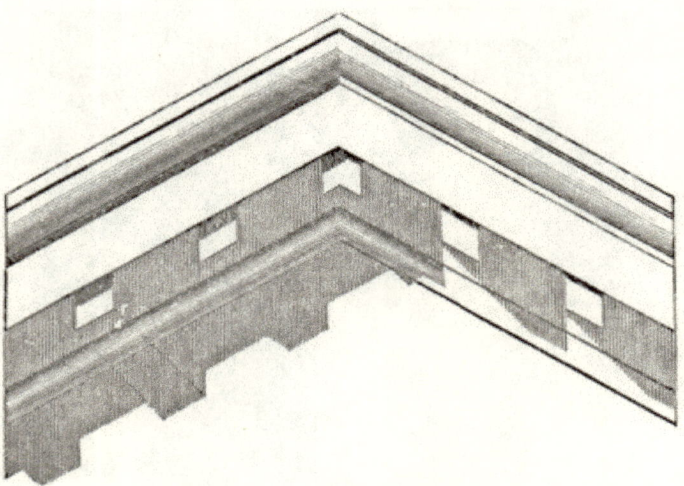

Fig. 365.

523.—*Reflected light.* In shading, the finish and life of an object depend much on reflected light This is seen to advantage in *Fig.* 361 and on the column in *Fig.* 363. Reflected rays are thrown in a direction exactly the reverse of direct rays; therefore, on that part of an object which is subject to reflected light, the shadows are reversed. The fillet of the ovolo in *Fig.* 361 is an example of this. On the right-hand side of the column, the face of the fillet is much darker than the cove directly under it. The reason of this is, the face of the fillet is deprived both of direct and reflected light, whereas the cove is subject to the latter. Other instances of the effect of reflected light will be seen in the other examples.

APPENDIX.

+, *plus*, signifies addition, and that the two quantities between which it stands are to be added together; as $a + b$, read a added to b.

−, *minus*, signifies subtraction, or that of the two quantities between which it occurs, the latter is to be subtracted from the former; as $a - b$, read a minus b.

×, *multiplied by*, or the sign of multiplication. It denotes that the two quantities between which it occurs are to be multiplied together; as $a \times b$, read a multiplied by b, or a times b. This sign is usually omitted between symbols or letters, and is then understood, as ab. This has the same meaning as $a \times b$. It is never omitted between arithmetical numbers; as 9×5, read nine times five.

÷, *divided by*, or the sign of division, and denotes that of the two quantities between which it occurs, the former is to be divided by the latter; as $a \div b$, read a divided by b. Division is also represented thus: $\frac{a}{b}$, in the form of a fraction. This signifies that a is to be divided by b. When more than one symbol occurs above or below the line, or both, as $\frac{a\,n\,r}{c\,m}$, it denotes that the product of the symbols above the line is to be divided by the product of those below the line.

=, *is equal to*, or sign of equality, and denotes that the quantity or quantities on its left are equal to those on its right; as $a - b = c$, read a minus b is equal to c, or equals c; or, $9 - 5 = 4$, read nine minus five equals four. This sign, together with the symbols on each side of it, when spoken of as a whole, is called an *equation*.

a^2 denotes a squared, or a multiplied by a, or the second power of a, and

a^3 denotes a cubed, or a multiplied by a and again multiplied by a, or the third power of a. The small figure, 2, 3, or 4, &c., is termed the index or exponent of the power. It indicates how many times the symbol is to be taken. Thus, $a^2 = a\,a$, $a^3 = a\,a\,a$, $a^4 = a\,a\,a\,a$.

√ is the *radical* sign, and denotes that the *square* root of the quantity following it is to be extracted, and

$\sqrt[3]{}$ denotes that the *cube* root of the quantity following it is to be extracted. Thus, $\sqrt{9} = 3$, and $\sqrt[3]{27} = 3$. The extraction of roots is also denoted by a fractional index or exponent, thus

$a^{1/2}$ denotes the square root of a,

$a^{1/3}$ denotes the cube root of a,

$a^{2/3}$ denotes the cube root of the square of a, &c.

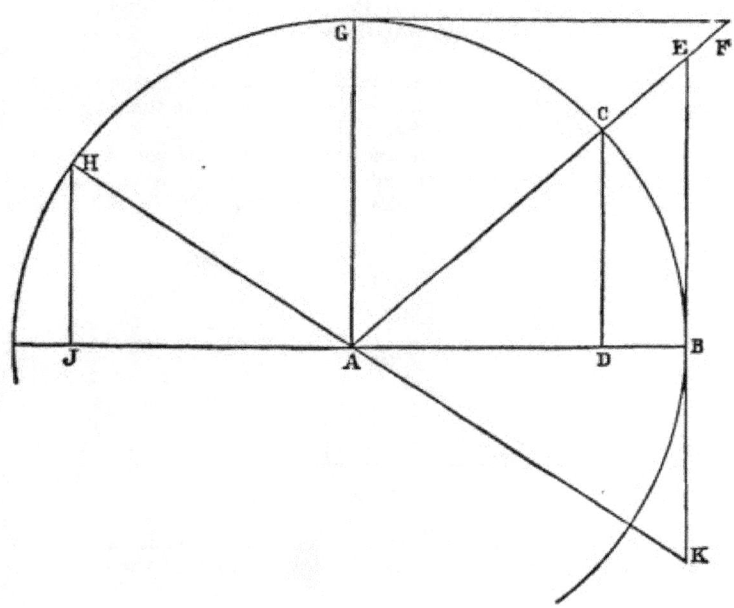

Fig. 366.

In *Fig.* 366, where *A B* is the radius of the circle *B C H*, draw a line *A F*, from *A*, through any point, *C*, of the arc *B G*. From *C* draw *C D* perpendicular to *A B*; from *B* draw *B E* perpendicular to *A B*; and from *G* draw *G F* perpendicular to *A G*.

Then, for the angle *F A B*, when the radius *A C* equals unity, *C D* is the *sine*; *A D* the *cosine*; *D B* the *versed sine*; *B E* the *tangent*, *G F* the *cotangent*; *A E* the *secant*; and *A F* the *cosecant*.

But if the angle be larger than one right angle, yet less than two right angles, as $B\ A\ H$, extend $H\ A$ to K and $E\ B$ to K, and from H draw $H\ J$ perpendicular to $A\ J$.

Then, for the angle BAH, when the radius $A\ H$ equals unity, HJ is the *sine*; $A\ J$ the *cosine*; $B\ J$ the *versed sine*; $B\ K$ the *tangent*; and $A\ K$ the *secant*.

When the number of degrees contained in a given angle is known, then the value of the *sine, cosine, &c.*, corresponding to that angle, may be found in a table of Natural Sines, Cosines, &c.

In the absence of such a table, and when the degrees contained in the given angle are unknown, the values of the sine, cosine, &c., may be found by computation, as follows:—Let $B\ A\ C$, (*Fig.* 367,) be the

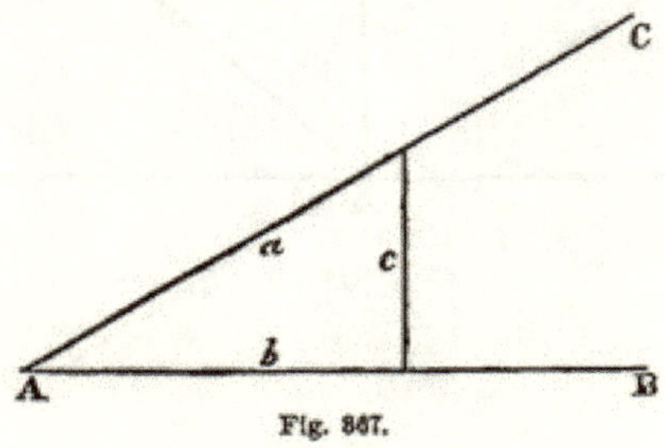

Fig. 367.

given angle. At any distance from A, draw c perpendicular to $A\ B$. By any scale of equal parts obtain the length of each of the three lines a, b, c. Then for the angle at A we have, by proportion,

$$a : c :: 1{\cdot}0 : \text{sin.} = \frac{c}{a}.$$

$$a : b :: 1{\cdot}0 : \text{cos.} = \frac{b}{a}.$$

$$b : c :: 1{\cdot}0 : \text{tan.} = \frac{c}{b}.$$

$$c : b :: 1{\cdot}0 : \text{cot.} = \frac{b}{c}.$$

$$b : a :: 1{\cdot}0 : \text{sec.} = \frac{a}{b}.$$

$$c : a :: 1{\cdot}0 : \text{cosec.} = \frac{a}{c}.$$

Or, in any right angled triangle, for the angle contained between the base and hypothenuse—

When perp. divided by hyp., the quotient equals the *sine*.
" base " " hyp., " " " " *cosine*.
" perp. " " base, " " " " *tangent*.
" base " " perp., " " " " *cotangent*.
" hyp. " " base, " " " " *secant*.
" hyp. " " perp., " " " " *cosecant*.

GLOSSARY.

Terms not found here can be found in the lists of definitions in other parts of this book, or in common dictionaries.

Abacus.—The uppermost member of a capital.

Abbatoir.—A slaughter-house.

Abbey.—The residence of an abbot or abbess.

Abutment.—That part of a pier from which the arch springs.

Acanthus.—A plant called in English, *bear's-breech.* Its leaves are employed for decorating the Corinthian and the Composite capitals.

Acropolis.—The highest part of a city; generally the citadel.

Acroteria.—The small pedestals placed on the extremities and apex of a pediment, originally intended as a base for sculpture.

Aisle.—Passage to and from the pews of a church. In Gothic architecture, the lean-to wings on the sides of the *nave.*

Alcove.—Part of a chamber separated by an *estrade*, or partition of columns. Recess with seats, &c., in gardens.

Altar.—A pedestal whereon sacrifice was offered. In modern churches, the area within the railing in front of the pulpit.

Alto-relievo.—High relief; sculpture projecting from a surface so as to appear nearly isolated.

Amphitheatre.—A double theatre, employed by the ancients for the exhibition of gladiatorial fights and other shows.

Ancones.—Trusses employed as an apparent support to a cornice upon the flanks of the architrave.

Annulet.—A small square moulding used to separate others; the fillets in the Doric capital under the ovolo, and those which separate the flutings of columns, are known by this term.

Antæ.—A pilaster attached to a wall.

Apiary.—A place for keeping beehives.

Arabesque.—A building after the Arabian style.

Areostyle.—An intercolumniation of from four to five diameters.

Arcade—A series of arches.

Arch.—An arrangement of stones or other material in a curvilinear form, so as to perform the office of a lintel and carry superincumbent weights.

Architrave.—That part of the entablature which rests upon the capital of a column, and is beneath the frieze. The casing and mouldings about a door or window.

Archivolt.—The ceiling of a vault: the under surface of an arch.

Area.—Superficial measurement. An open space, below the level of the ground, in front of basement windows.

Arsenal.—A public establishment for the deposition of arms and warlike stores.

Astragal.—A small moulding consisting of a half-round with a fillet on each side.

Attic.—A low story erected over an order of architecture. A low additional story immediately under the roof of a building.

Aviary.—A place for keeping and breeding birds.

Balcony.—An open gallery projecting from the front of a building.

Baluster.—A small pillar or pilaster supporting a rail.

Balustrade.—A series of balusters connected by a rail.

Barge-course.—That part of the covering which projects over the gable of a building.

Base.—The lowest part of a wall, column, &c.

Basement-story.—That which is immediately under the principal story, and included within the foundation of the building.

Basso-relievo.—Low relief; sculptured figures projecting from a surface one-half their thickness or less. See *Alto-relievo.*

Battering.—See *Talus.*

Battlement.—Indentations on the top of a wall or parapet.

Bay-window.—A window projecting in two or more planes, and not forming the segment of a circle.

Bazaar.—A species of mart or exchange for the sale of various articles of merchandise.

Bead.—A circular moulding.

Bed-mouldings.—Those mouldings which are between the corona and the frieze.

Belfry.—That part of a steeple in which the bells are hung: anciently called *campanile.*

Belvedere.—An ornamental turret or observatory commanding a pleasant prospect.

Bow-window.—A window projecting in curved lines.

Bressummer.—A beam or iron tie supporting a wall over a gateway or other opening.

Brick-nogging.—The brickwork between studs of partitions.

Buttress.—A projection from a wall to give additional strength.

Cable.—A cylindrical moulding placed in flutes at the lower part of the column.

Camber.—To give a convexity to the upper surface of a beam.

Campanile.—A tower for the reception of bells, usually, in Italy, separated from the church.

Canopy.—An ornamental covering over a seat of state.

Cantalivers.—The ends of rafters under a projecting roof. Pieces of wood or stone supporting the eaves.

Capital.—The uppermost part of a column included between the shaft and the architrave.

Caravansera.—In the East, a large public building for the reception of travellers by caravans in the desert.

Carpentry.—(From the Latin, *carpentum*, carved wood.) That department of science and art which treats of the disposition, the construction and the relative strength of timber. The first is called descriptive, the second constructive, and the last mechanical carpentry.

Caryatides.—Figures of women used instead of columns to support an entablature.

Casino.—A small country-house.

Castellated.—Built with battlements and turrets in imitation of ancient castles.

Castle.—A building fortified for military defence. A house with towers, usually encompassed with walls and moats, and having a donjon, or keep, in the centre.

Catacombs.—Subterraneous places for burying the dead.

Cathedral.—The principal church of a province or diocese, wherein the throne of the archbishop or bishop is placed.

Cavetto.—A concave moulding comprising the quadrant of a circle.

Cemetery.—An edifice or area where the dead are interred.

Cenotaph.—A monument erected to the memory of a person buried in another place.

Centring.—The temporary woodwork, or framing, whereon any vaulted work is constructed.

Cesspool.—A well under a drain or pavement to receive the waste-water and sediment.

Chamfer.—The bevilled edge of any thing originally right-angled.

Chancel.—That part of a Gothic church in which the altar is placed.

Chantry.—A little chapel in ancient churches, with an endowment for one or more priests to say mass for the relief of souls out of purgatory.

Chapel.—A building for religious worship, erected separately from a church, and served by a chaplain.

Chaplet.—A moulding carved into beads, olives, &c.

Cincture.—The ring, listel, or fillet, at the top and bottom of a column, which divides the shaft of the column from its capital and base.

Circus.—A straight, long, narrow building used by the Romans for the exhibition of public spectacles and chariot races. At the present day, a building enclosing an arena for the exhibition of feats of horsemanship.

Clere-story.—The upper part of the nave of a church above the roofs of the aisles.

Cloister.—The square space attached to a regular monastery or large church, having a peristyle or ambulatory around it, covered with a range of buildings.

Coffer-dam.—A case of piling, water-tight, fixed in the bed of a river, for the purpose of excluding the water while any work, such as a wharf, wall, or the pier of a bridge, is carried up.

Collar-beam.—A horizontal beam framed between two principal rafters above the tie-beam.

Collonade.—A range of columns.

Columbarium.—A pigeon-house.

Column.—A vertical, cylindrical support under the entablature of an order.

Common-rafters.—The same as *jack-rafters*, which see

Conduit.—A long, narrow, walled passage underground, for secret communication between different apartments. A canal or pipe for the conveyance of water.

Conservatory.—A building for preserving curious and rare exotic plants.

Consoles.—The same as *ancones*, which see.

Contour.—The external lines which bound and terminate a figure.

Convent.—A building for the reception of a society of religious persons.

Coping.—Stones laid on the top of a wall to defend it from the weather.

Corbels.—Stones or timbers fixed in a wall to sustain the timbers of a floor or roof.

Cornice.—Any moulded projection which crowns or finishes the part to which it is affixed.

Corona.—That part of a cornice which is between the crown-moulding and the bed-mouldings.

Cornucopia.—The horn of plenty.

Corridor.—An open gallery or communication to the different apartments of a house.

Cove.—A concave moulding.

Cripple-rafters.—The short rafters which are spiked to the hip-rafter of a roof.

Crockets.—In Gothic architecture, the ornaments placed along the angles of pediments, pinnacles, &c.

Crosettes.—The same as *ancones*, which see.

Crypt.—The under or hidden part of a building

Culvert.—An arched channel of masonry or brickwork, built beneath the bed of a canal for the purpose of conducting water under it. Any arched channel for water underground.

Cupola.—A small building on the top of a dome.

Curtail-step.—A step with a spiral end, usually the first of the flight.

Cusps.—The pendents of a pointed arch.

Cyma.—An ogee. There are two kinds; the *cyma-recta*, having the upper part concave and the lower convex, and the *cyma-reversa*, with the upper part convex and the lower concave.

Dado.—The die, or part between the base and cornice of a pedestal.

Dairy.—An apartment or building for the preservation of milk, and the manufacture of it into butter, cheese, &c.

Dead-shoar.—A piece of timber or stone stood vertically in brickwork, to support a superincumbent weight until the brickwork which is to carry it has set or become hard.

Decastyle.—A building having ten columns in front.

Dentils.—(From the Latin, *dentes*, teeth.) Small rectangular blocks used in the bed-mouldings of some of the orders.

Diastyle.—An intercolumniation of three, or, as some say, four diameters.

Die.—That part of a pedestal included between the base and the cornice; it is also called a *dado.*

Dodecastyle.—A building having twelve columns in front.

Donjon.—A massive tower within ancient castles to which the garrison might retreat in case of necessity.

Dooks.—A Scotch term given to wooden bricks.

Dormer.—A window placed on the roof of a house, the frame being placed vertically on the rafters.

Dormitory.—A sleeping-room.

Dovecote.—A building for keeping tame pigeons. A columbarium.

Echinus.—The Grecian ovolo.

Elevation.—A geometrical projection drawn on a plane at right angles to the horizon.

Entablature.—That part of an order which is supported by the columns; consisting of the architrave, frieze, and cornice.

Eustyle.—An intercolumniation of two and a quarter diameters.

Exchange.—A building in which merchants and brokers meet to transact business.

Extrados.—The exterior curve of an arch.

Façade.—The principal front of any building.

Face-mould—The pattern for marking the plank, out of which hand-railing is to be cut for stairs, &c.

Facia, or *Fascia.*—A flat member like a band or broad fillet.

Falling-mould.—The mould applied to the convex, vertical surface of the rail-piece, in order to form the back and under surface of the rail, and finish the squaring.

Festoon.—An ornament representing a wreath of flowers and leaves.

Fillet.—A narrow flat band, listel, or annulet, used for the separation of one moulding from another, and to give breadth and firmness to the edges of mouldings.

Flutes.—Upright channels on the shafts of columns.

Flyers.—Steps in a flight of stairs that are parallel to each other.

Forum.—In ancient architecture, a public market; also, a place where the common courts were held, and law pleadings carried on.

Foundry.—A building in which various metals are cast into moulds or shapes.

Frieze.—That part of an entablature included between the architrave and the cornice.

Gable.—The vertical, triangular piece of wall at the end of a roof, from the level of the eaves to the summit.

Gain.—A recess made to receive a tenon or tusk.

Gallery.—A common passage to several rooms in an upper story. A long room for the reception of pictures. A platform raised on columns, pilasters, or piers.

Girder.—The principal beam in a floor for supporting the binding and other joists, whereby the bearing or length is lessened.

Glyph.—A vertical, sunken channel. From their number, those in the Doric order are called *triglyphs.*

Granary.—A building for storing grain, especially that intended to be kept for a considerable time.

Groin.—The line formed by the intersection of two arches, which cross each other at any angle.

Guttæ.—The small cylindrical pendent ornaments, otherwise called *drops*, used in the Doric order under the triglyphs, and also pendent from the mutuli of the cornice.

Gymnasium.—Orig nally, a space measured out and covered with sand for the exercise of athletic games: afterwards, spacious buildings devoted to the mental as well as corporeal instruction of youth.

Hall.—The first large apartment on entering a house. The public room of a corporate body. A manor-house.

Ham.—A house or dwelling-place. A street or village: hence Notting*ham*, Bucking*ham*, &c. *Hamlet*, the diminutive of *ham*, is a small street or village.

Helix.—The small volute, or twist, under the abacus in the Corinthian capital.

Hem.—The projecting spiral fillet of the Ionic capital.

Hexastyle.—A building having six columns in front.

Hip-rafter.—A piece of timber placed at the angle made by two adjacent inclined roofs.

Homestall.—A mansion-house, or seat in the country.

Hotel, or *Hostel.*—A large inn or place of public entertainment. A large house or palace.

Hot-house.—A glass building used in gardening.

Hovel.—An open shed.

Hut.—A small cottage or hovel generally constructed of earthy materials, as strong loamy clay, &c.

Impost.—The capital of a pier or pilaster which supports an arch.

Intaglio.—Sculpture in which the subject is hollowed out, so that the impression from it presents the appearance of a bas-relief.

Intercolumniation.—The distance between two columns.

Intrados.—The interior and lower curve of an arch.

Jack-rafters.—Rafters that fill in between the principal rafters of a roof; called also *common-rafters*.

Jail.—A place of legal confinement.

Jambs.—The vertical sides of an aperture.

Joggle-piece.—A post to receive struts.

Joists.—The timbers to which the boards of a floor or the laths of a ceiling are nailed.

Keep.—The same as *donjon*, which see.

Key-stone.—The highest central stone of an arch.

Kiln.—A building for the accumulation and retention of heat, in order to dry or burn certain materials deposited within it.

King-post.—The centre-post in a trussed roof.

Knee.—A convex bend in the back of a hand-rail. See *Ramp*.

Lactarium.—The same as *dairy*, which see.

Lantern.—A cupola having windows in the sides for lighting an apartment beneath.

Larmier.—The same as *corona*, which see.

Lattice.—A reticulated window for the admission of air, rather than light, as in dairies and cellars.

Lever-boards.—Blind-slats: a set of boards so fastened that they may be turned at any angle to admit more or less light, or to lap upon each other so as to exclude all air or light through apertures.

Lintel.—A piece of timber or stone placed horizontally over a door, window, or other opening.

Listel.—The same as *fillet*, which see.

Lobby.—An enclosed space, or passage, communicating with the principal room or rooms of a house.

Lodge.—A small house near and subordinate to the mansion. A cottage placed at the gate of the road leading to a mansion.

Loop.—A small narrow window. *Loophole* is a term applied to the vertical series of doors in a warehouse, through which goods are delivered by means of a crane.

Luffer-boarding.—The same as *lever-boqrds*, which see.

Luthern.—The same as *dormer*, which see.

Mausoleum.—A sepulchral building—so called from a very celebrated one erected to the memory of Mausolus, king of Caria, by his wife Artemisia.

Metopa.—The square space in the frieze between the triglyphs of the Doric order.

Mezzanine.—A story of small height introduced between two of greater height.

Minaret.—A slender, lofty turret having projecting balconies, common in Mohammedan countries.

Minster.—A church to which an ecclesiastical fraternity has been or is attached.

Moat.—An excavated reservoir of water, surrounding a house, castle or town.

Modillion.—A projection under the corona of the richer orders, resembling a bracket.

Module.—The semi-diameter of a column, used by the architect as a measure by which to proportion the parts of an order.

Monastery.—A building or buildings appropriated to the reception of monks.

Monopteron.—A circular collonade supporting a dome without an enclosing wall.

Mosaic.—A mode of representing objects by the inlaying of small cubes of glass, stone, marble, shells, &c.

Mosque.—A Mohammedan temple, or place of worship.

Mullions.—The upright posts or bars, which divide the lights in a Gothic window.

Muniment-house.—A strong, fire-proof apartment for the keeping and preservation of evidences, charters, seals, &c., called muniments.

Museum.—A repository of natural, scientific and literary, curiosities or of works of art.

Mutule.—A projecting ornament of the Doric cornice supposed to represent the ends of rafters.

Nave.—The main body of a Gothic church.

Newel.—A post at the starting or landing of a flight of stairs.

Niche.—A cavity or hollow place in a wall for the reception of a statue, vase, &c.

Nogs.—Wooden bricks.

Nosing.—The rounded and projecting edge of a step in stairs.

Nunnery.—A building or buildings appropriated for the reception of nuns.

Obelisk.—A lofty pillar of a rectangular form.

Octastyle.—A building with eight columns in front.

Odeum.—Among the Greeks, a species of theatre wherein the poets and musicians rehearsed their compositions previous to the public production of them.

Ogee.—See *Cyma*.

Orangery.—A gallery or building in a garden or parterre fronting the south.

Oriel-window.—A large bay or recessed window in a hall, chapel, or other apartment.

Ovolo.—A convex projecting moulding whose profile is the quadrant of a circle.

Pagoda.—A temple or place of worship in India.

Palisade.—A fence of pales or stakes driven into the ground.

Parapet.—A small wall of any material for protection on the sides of bridges, quays, or high buildings.

Pavilion.—A turret or small building generally insulated and comprised under a single roof.

Pedestal.—A square foundation used to elevate and sustain a column, statue, &c.

Pediment.—The triangular crowning part of a portico or aperture which terminates vertically the sloping parts of the roof: this, in Gothic architecture, is called a *gable*.

Penitentiary.—A prison for the confinement of criminals whose crimes are not of a very heinous nature.

Piazza.—A square, open space surrounded by buildings. This term is often improperly used to denote a *portico*.

Pier.—A rectangular pillar without any regular base or capital. The upright, narrow portions of walls between doors and windows are known by this term.

Pilaster.—A square pillar, sometimes insulated, but more commonly engaged in a wall, and projecting only a part of its thickness.

Piles.—Large timbers driven into the ground to make a secure foundation in marshy places, or in the bed of a river.

Pillar.—A column of irregular form, always disengaged, and always deviating from the proportions of the orders; whence the distinction between a pillar and a column.

Pinnacle.—A small spire used to ornament Gothic buildings.

Planceer.—The same as *soffit*, which see.

Plinth.—The lower square member of the base of a column, pedestal, or wall.

Porch.—An exterior appendage to a building, forming a covered approach to one of its principal doorways.

Portal.—The arch over a door or gate; the framework of the gate; the lesser gate, when there are two of different dimensions at one entrance.

Portcullis.—A strong timber gate to old castles, made to slide up and down vertically.

Portico.—A colonnade supporting a shelter over a walk, or ambulatory.

Priory.—A building similar in its constitution to a monastery or abbey, the head whereof was called a prior or prioress.

Prism.—A solid bounded on the sides by parallelograms, and on the ends by polygonal figures in parallel planes.

Prostyle.—A building with columns in front only.

Purlines.—Those pieces of timber which lie under and at right angles to the rafters to prevent them from sinking.

Pycnostyle.—An intercolumniation of one and a half diameters.

Pyramid.—A solid body standing on a square, triangular or polygonal basis, and terminating in a point at the top.

Quarry.—A place whence stones and slates are procured.

Quay.—(Pronounced, *key.*) A bank formed towards the sea or on the side of a river for free passage, or for the purpose of unloading merchandise.

Quoin.—An external angle. See *Rustic quoins.*

Rabbet, or *Rebate.*—A groove or channel in the edge of a board.

Ramp.—A concave bend in the back of a hand-rail.

Rampant arch.—One having abutments of different heights.

Regula.—The band below the tænia in the Doric order.

Riser.—In stairs, the vertical board forming the front of a step.

Rostrum.—An elevated platform from which a speaker addresses an audience.

Rotunda.—A circular building.

Rubble-wall.—A wall built of unhewn stone.

Rudenture.—The same as *cable*, which see.

Rustic quoins.—The stones placed on the external angle of a building, projecting beyond the face of the wall, and having their edges bevilled.

Rustic-work.—A mode of building masonry wherein the faces of the stones are left rough, the sides only being wrought smooth where the union of the stones takes place.

Salon, or *Saloon*.—A lofty and spacious apartment comprehending the height of two stories with two tiers of windows.

Sarcophagus.—A tomb or coffin made of one stone.

Scantling.—The measure to which a piece of timber is to be or has been cut.

Scarfing.—The joining of two pieces of timber by bolting or nailing transversely together, so that the two appear but one.

Scotia.—The hollow moulding in the base of a column, between the fillets of the tori.

Scroll.—A carved curvilinear ornament, somewhat resembling in profile the turnings of a ram's horn.

Sepulchre.—A grave, tomb, or place of interment.

Sewer.—A drain or conduit for carrying off soil or water from any place.

Shaft.—The cylindrical part between the base and the capital of a column.

Shoar.—A piece of timber placed in an oblique direction to support a building or wall.

Sill.—The horizontal piece of timber at the bottom of framing; the timber or stone at the bottom of doors and windows.

Soffit—The underside of an architrave, corona, &c. The underside of the heads of doors, windows, &c.

Summer.—The lintel of a door or window; a beam tenoned into a girder to support the ends of joists on both sides of it.

Systyle.—An intercolumniation of two diameters.

Tænia.—The fillet which separates the Doric frieze from the architrave.

Talus.—The slope or inclination of a wall, among workmen called *battering*.

Terrace.—An area raised before a building, above the level of the ground, to serve as a walk.

Tesselated pavement.—A curious pavement of Mosaic work, composed of small square stones.

Tetrastyle.—A building having four columns in front.

Thatch.—A covering of straw or reeds used on the roofs of cottages, barns, &c.

Theatre.—A building appropriated to the representation of dramatic spectacles.

Tile.—A thin piece or plate of baked clay or other material used for the external covering of a roof.

Tomb.—A grave, or place for the interment of a human body, including also any commemorative monument raised over such a place.

Torus.—A moulding of semi-circular profile used in the bases of columns.

Tower.—A lofty building of several stories, round or polygonal.

Transept.—The transverse portion of a cruciform church.

Transom.—The beam across a double-lighted window; if the window have no transom, it is called a *clere-story* window.

Tread.—That part of a step which is included between the face of its riser and that of the riser above.

Trellis.—A reticulated framing made of thin bars of wood for screens, windows, &c.

Triglyph.—The vertical tablets in the Doric frieze, chamfered on the two vertical edges, and having two channels in the middle.

Tripod.—A table or seat with three legs.

Trochilus.—The same as *scotia*, which see.

Truss.—An arrangement of timbers for increasing the resistance to cross-strains, consisting of a tie, two struts and a suspending-piece.

Turret.—A small tower, often crowning the angle of a wall, &c.

Tusk—A short projection under a tenon to increase its strength.

Tympanum.—The naked face of a pediment, included between the level and the raking mouldings.

Underpinning.—The wall under the ground-sills of a building.

University.—An assemblage of colleges under the supervision of a senate, &c.

Vault.—A concave arched ceiling resting upon two opposite parallel walls.

Venetian-door.—A door having side-lights.

Venetian-window.—A window having three separate apertures.

Veranda.—An awning. An open portico under the extended roof of a building.

Vestibule.—An apartment which serves as the medium of communication to another room or series of rooms.

Vestry.—An apartment in a church, or attached to it, for the preservation of the sacred vestments and utensils.

Villa.—A country-house for the residence of an opulent person.

Vinery.—A house for the cultivation of vines.

Volute.—A spiral scroll, which forms the principal feature of the Ionic and the Composite capitals.

Voussoirs.—Arch-stones

Wainscoting.—Wooden lining of walls, generally in panels.

Water-table.—The stone covering to the projecting foundation or other walls of a building.

Well.—The space occupied by a flight of stairs. The space left beyond the ends of the steps is called the *well-hole*.

Wicket.—A small door made in a gate.

Winders.—In stairs, steps not parallel to each other.

Zophorus.—The same as *frieze*, which see.

Zystos.—Among the ancients, a portico of unusual length, commonly appropriated to gymnastic exercises.

TABLE OF SQUARES, CUBES, AND ROOTS.

(From Hutton's Mathematics.)

No.	Square.	Cube.	Sq. Root.	CubeRoot.	No.	Square.	Cube.	Sq. Root.	CubeRoot.
1	1	1	1·0000000	1·000000	68	4624	314432	8·2462113	4·081655
2	4	8	1·4142136	1·259921	69	4761	328509	8·3066239	4·101566
3	9	27	1·7320508	1·442250	70	4900	343000	8·3666003	4·121285
4	16	64	2·0000000	1·587401	71	5041	357911	8·4261498	4·140818
5	25	125	2·2360680	1·709976	72	5184	373248	8·4852814	4·160168
6	36	216	2·4494897	1·817121	73	5329	389017	8·5440037	4·179339
7	49	343	2·6457513	1·912931	74	5476	405224	8·6023253	4·198336
8	64	512	2·8284271	2·000000	75	5625	421875	8·6602540	4·217163
9	81	729	3·0000000	2·080084	76	5776	438976	8·7177979	4·235824
10	100	1000	3·1622777	2·154435	77	5929	456533	8·7749644	4·254321
11	121	1331	3·3166248	2·223980	78	6084	474552	8·8317609	4·272659
12	144	1728	3·4641016	2·289429	79	6241	493039	8·8881944	4·290840
13	169	2197	3·6055513	2·351335	80	6400	512000	8·9442719	4·308869
14	196	2744	3·7416574	2·410142	81	6561	531441	9·0000000	4·326749
15	225	3375	3·8729833	2·466212	82	6724	551368	9·0553851	4·344481
16	256	4096	4·0000000	2·519842	83	6889	571787	9·1104336	4·362071
17	289	4913	4·1231056	2·571282	84	7056	592704	9·1651514	4·379519
18	324	5832	4·2426407	2·620741	85	7225	614125	9·2195445	4·396830
19	361	6859	4·3588989	2·668402	86	7396	636056	9·2736185	4·414005
20	400	8000	4·4721360	2·714418	87	7569	658503	9·3273791	4·431048
21	441	9261	4·5825757	2·758924	88	7744	681472	9·3808315	4·447960
22	484	10648	4·6904158	2·802039	89	7921	704969	9·4339811	4·464745
23	529	12167	4·7958315	2·843867	90	8100	729000	9·4868330	4·481405
24	576	13824	4·8989795	2·884499	91	8281	753571	9·5393920	4·497941
25	625	15625	5·0000000	2·924018	92	8464	778688	9·5916630	4·514357
26	676	17576	5·0990195	2·962496	93	8649	804357	9·6436508	4·530655
27	729	19683	5·1961524	3·000000	94	8836	830584	9·6953597	4·546836
28	784	21952	5·2915026	3·036589	95	9025	857375	9·7467943	4·562903
29	841	24389	5·3851648	3·072317	96	9216	884736	9·7979590	4·578857
30	900	27000	5·4772256	3·107233	97	9409	912673	9·8488578	4·594701
31	961	29791	5·5677644	3·141381	98	9604	941192	9·8994949	4·610436
32	1024	32768	5·6568542	3·174802	99	9801	970299	9·9498744	4·626065
33	1089	35937	5·7445626	3·207534	100	10000	1000000	10·0000000	4·641589
34	1156	39304	5·8309519	3·239612	101	10201	1030301	10·0498756	4·657010
35	1225	42875	5·9160798	3·271066	102	10404	1061208	10·0995049	4·672329
36	1296	46656	6·0000000	3·301927	103	10609	1092727	10·1488916	4·687548
37	1369	50653	6·0827625	3·332222	104	10816	1124864	10·1980390	4·702669
38	1444	54872	6·1644140	3·361975	105	11025	1157625	10·2469508	4·717694
39	1521	59319	6·2449980	3·391211	106	11236	1191016	10·2956301	4·732623
40	1600	64000	6·3245553	3·419952	107	11449	1225043	10·3440804	4·747459
41	1681	68921	6·4031242	3·448217	108	11664	1259712	10·3923048	4·762203
42	1764	74088	6·4807407	3·476027	109	11881	1295029	10·4403065	4·776856
43	1849	79507	6·5574385	3·503398	110	12100	1331000	10·4880885	4·791420
44	1936	85184	6·6332496	3·530348	111	12321	1367631	10·5356538	4·805896
45	2025	91125	6·7082039	3·556893	112	12544	1404928	10·5830052	4·820284
46	2116	97336	6·7823300	3·583048	113	12769	1442897	10·6301458	4·834588
47	2209	103823	6·8556546	3·608826	114	12996	1481544	10·6770783	4·848808
48	2304	110592	6·9282032	3·634241	115	13225	1520875	10·7238053	4·862944
49	2401	117649	7·0000000	3·659306	116	13456	1560896	10·7703296	4·876999
50	2500	125000	7·0710678	3·684031	117	13689	1601613	10·8166538	4·890973
51	2601	132651	7·1414284	3·708430	118	13924	1643032	10·8627805	4·904868
52	2704	140608	7·2111026	3·732511	119	14161	1685159	10·9087121	4·918685
53	2809	148877	7·2801099	3·756286	120	14400	1728000	10·9544512	4·932424
54	2916	157464	7·3484692	3·779763	121	14641	1771561	11·0000000	4·946087
55	3025	166375	7·4161985	3·802952	122	14884	1815848	11·0453610	4·959676
56	3136	175616	7·4833148	3·825862	123	15129	1860867	11·0905365	4·973190
57	3249	185193	7·5498344	3·848501	124	15376	1906624	11·1355287	4·986631
58	3364	195112	7·6157731	3·870877	125	15625	1953125	11·1803399	5·000000
59	3481	205379	7·6811457	3·892996	126	15876	2000376	11·2249722	5·013298
60	3600	216000	7·7459667	3·914868	127	16129	2048383	11·2694277	5·026526
61	3721	226981	7·8102497	3·936497	128	16384	2097152	11·3137085	5·039684
62	3844	238328	7·8740079	3·957892	129	16641	2146689	11·3578167	5·052774
63	3969	250047	7·9372539	3·979057	130	16900	2197000	11·4017543	5·065797
64	4096	262144	8·0000000	4·000000	131	17161	2248091	11·4455231	5·078753
65	4225	274625	8·0622577	4·020726	132	17424	2299968	11·4891253	5·091643
66	4356	287496	8·1240384	4·041240	133	17689	2352637	11·5325626	5·104469
67	4489	300763	8·1853528	4·061548	134	17956	2406104	11·5758369	5·117230

No.	Square.	Cube.	Sq. Root.	Cube Root.	No.	Square.	Cube.	Sq. Root.	Cube Root.
135	18225	2460375	11·6189500	5·129928	202	40804	8242408	14·2126704	5·867464
136	18496	2515456	11·6619038	5·142563	203	41209	8365427	14·2478068	5·877131
137	18769	2571353	11·7046999	5·155137	204	41616	8489664	14·2828569	5·886765
138	19044	2628072	11·7473401	5·167649	205	42025	8615125	14·3178211	5·896368
139	19321	2685619	11·7898261	5·180101	206	42436	8741816	14·3527001	5·905941
140	19600	2744000	11·8321596	5·192494	207	42849	8869743	14·3874946	5·915482
141	19881	2803221	11·8743422	5·204828	208	43264	8998912	14·4222051	5·924992
142	20164	2863288	11·9163753	5·217103	209	43681	9129329	14·4568323	5·934473
143	20449	2924207	11·9582607	5·229321	210	44100	9261000	14·4913767	5·943922
144	20736	2985984	12·0000000	5·241483	211	44521	9393931	14·5258390	5·953342
145	21025	3048625	12·0415946	5·253588	212	44944	9528128	14·5602198	5·962732
146	21316	3112136	12·0830460	5·265637	213	45369	9663597	14·5945195	5·972093
147	21609	3176523	12·1243557	5·277632	214	45796	9800344	14·6287388	5·981424
148	21904	3241792	12·1655251	5·289572	215	46225	9938375	14·6628783	5·990726
149	22201	3307949	12·2065556	5·301459	216	46656	10077696	14·6969385	6·000000
150	22500	3375000	12·2474487	5·313293	217	47089	10218313	14·7309199	6·009245
151	22801	3442951	12·2882057	5·325074	218	47524	10360232	14·7648231	6·018462
152	23104	3511808	12·3288280	5·336803	219	47961	10503459	14·7986486	6·027650
153	23409	3581577	12·3693169	5·348481	220	48400	10648000	14·8323970	6·036811
154	23716	3652264	12·4096736	5·360108	221	48841	10793861	14·8660687	6·045943
155	24025	3723875	12·4498996	5·371685	222	49284	10941048	14·8996644	6·055049
156	24336	3796416	12·4899960	5·383213	223	49729	11089567	14·9331845	6·064127
157	24649	3869893	12·5299641	5·394691	224	50176	11239424	14·9666295	6·073178
158	24964	3944312	12·5698051	5·406120	225	50625	11390625	15·0000000	6·082202
159	25281	4019679	12·6095202	5·417501	226	51076	11543176	15·0332964	6·091199
160	25600	4096000	12·6491106	5·428835	227	51529	11697083	15·0665192	6·100170
161	25921	4173281	12·6885775	5·440122	228	51984	11852352	15·0996689	6·109115
162	26244	4251528	12·7279221	5·451362	229	52441	12008989	15·1327460	6·118033
163	26569	4330747	12·7671453	5·462556	230	52900	12167000	15·1657509	6·126926
164	26896	4410944	12·8062485	5·473704	231	53361	12326391	15·1986842	6·135792
165	27225	4492125	12·8452326	5·484807	232	53824	12487168	15·2315462	6·144634
166	27556	4574296	12·8840987	5·495865	233	54289	12649337	15·2643375	6·153449
167	27889	4657463	12·9228480	5·506878	234	54756	12812904	15·2970585	6·162240
168	28224	4741632	12·9614814	5·517848	235	55225	12977875	15·3297097	6·171006
169	28561	4826809	13·0000000	5·528775	236	55696	13144256	15·3622915	6·179747
170	28900	4913000	13·0384048	5·539658	237	56169	13312053	15·3948043	6·188463
171	29241	5000211	13·0766968	5·550499	238	56644	13481272	15·4272486	6·197154
172	29584	5088448	13·1148770	5·561298	239	57121	13651919	15·4596248	6·205822
173	29929	5177717	13·1529464	5·572055	240	57600	13824000	15·4919334	6·214465
174	30276	5268024	13·1909060	5·582770	241	58081	13997521	15·5241747	6·223084
175	30625	5359375	13·2287566	5·593445	242	58564	14172488	15·5563492	6·231680
176	30976	5451776	13·2664992	5·604079	243	59049	14348907	15·5884573	6·240251
177	31329	5545233	13·3041347	5·614672	244	59536	14526784	15·6204994	6·248800
178	31684	5639752	13·3416641	5·625226	245	60025	14706125	15·6524758	6·257325
179	32041	5735339	13·3790882	5·635741	246	60516	14886936	15·6843871	6·265827
180	32400	5832000	13·4164079	5·646216	247	61009	15069223	15·7162336	6·274305
181	32761	5929741	13·4536240	5·656653	248	61504	15252992	15·7480157	6·282761
182	33124	6028568	13·4907376	5·667051	249	62001	15438249	15·7797338	6·291195
183	33489	6128487	13·5277493	5·677411	250	62500	15625000	15·8113883	6·299605
184	33856	6229504	13·5646600	5·687734	251	63001	15813251	15·8429795	6·307994
185	34225	6331625	13·6014705	5·698019	252	63504	16003008	15·8745079	6·316360
186	34596	6434856	13·6381817	5·708267	253	64009	16194277	15·9059737	6·324704
187	34969	6539203	13·6747943	5·718479	254	64516	16387064	15·9373775	6·333026
188	35344	6644672	13·7113092	5·728654	255	65025	16581375	15·9687194	6·341326
189	35721	6751269	13·7477271	5·738794	256	65536	16777216	16·0000000	6·349604
190	36100	6859000	13·7840488	5·748897	257	66049	16974593	16·0312195	6·357861
191	36481	6967871	13·8202750	5·758965	258	66564	17173512	16·0623784	6·366097
192	36864	7077888	13·8564065	5·768998	259	67081	17373979	16·0934769	6·374311
193	37249	7189057	13·8924440	5·778996	260	67600	17576000	16·1245155	6·382504
194	37636	7301384	13·9283883	5·788960	261	68121	17779581	16·1554944	6·390677
195	38025	7414875	13·9642400	5·798890	262	68644	17984728	16·1864141	6·398828
196	38416	7529536	14·0000000	5·808786	263	69169	18191447	16·2172747	6·406959
197	38809	7645373	14·0356688	5·818648	264	69696	18399744	16·2480768	6·415069
198	39204	7762392	14·0712473	5·828477	265	70225	18609625	16·2788206	6·423158
199	39601	7880599	14·1067360	5·838272	266	70756	18821096	16·3095064	6·431228
200	40000	8000000	14·1421356	5·848035	267	71289	19034163	16·3401346	6·439277
201	40401	8120601	14·1774469	5·857766	268	71824	19248832	16·3707055	6·447306

No.	Square.	Cube.	Sq. Root.	CubeRoot.
269	72361	19465109	16·4012195	6·455315
270	72900	19683000	16·4316767	6·463304
271	73441	19902511	16·4620776	6·471274
272	73984	20123648	16·4924225	6·479224
273	74529	20346417	16·5227116	6·487154
274	75076	20570824	16·5529454	6·495065
275	75625	20796875	16·5831240	6·502957
276	76176	21024576	16·6132477	6·510830
277	76729	21253933	16·6433170	6·518684
278	77284	21484952	16·6733320	6·526519
279	77841	21717639	16·7032931	6·534335
280	78400	21952000	16·7332005	6·542133
281	78961	22188041	16·7630546	6·549912
282	79524	22425768	16·7928556	6·557672
283	80089	22665187	16·8226038	6·565414
284	80656	22906304	16·8522995	6·573139
285	81225	23149125	16·8819430	6·580844
286	81796	23393656	16·9115345	6·588532
287	82369	23639903	16·9410743	6·596202
288	82944	23887872	16·9705627	6·603854
289	83521	24137569	17·0000000	6·611489
290	84100	24389000	17·0293864	6·619106
291	84681	24642171	17·0587221	6·626705
292	85264	24897088	17·0880075	6·634287
293	85849	25153757	17·1172428	6·641852
294	86436	25412184	17·1464282	6·649400
295	87025	25672375	17·1755640	6·656930
296	87616	25934336	17·2046505	6·664444
297	88209	26198073	17·2336879	6·671940
298	88804	26463592	17·2626765	6·679420
299	89401	26730899	17·2916165	6·686883
300	90000	27000000	17·3205081	6·694330
301	90601	27270901	17·3493516	6·701759
302	91204	27543608	17·3781472	6·709173
303	91809	27818127	17·4068952	6·716570
304	92416	28094464	17·4355958	6·723951
305	93025	28372625	17·4642492	6·731316
306	93636	28652616	17·4928557	6·738664
307	94249	28934443	17·5214155	6·745997
308	94864	29218112	17·5499288	6·753313
309	95481	29503629	17·5783958	6·760614
310	96100	29791000	17·6068169	6·767899
311	96721	30080231	17·6351921	6·775169
312	97344	30371328	17·6635217	6·782423
313	97969	30664297	17·6918060	6·789661
314	98596	30959144	17·7200451	6·796884
315	99225	31255875	17·7482393	6·804092
316	99856	31554496	17·7763888	6·811285
317	100489	31855013	17·8044938	6·818462
318	101124	32157432	17·8325545	6·825624
319	101761	32461759	17·8605711	6·832771
320	102400	32768000	17·8885438	6·839904
321	103041	33076161	17·9164729	6·847021
322	103684	33386248	17·9443584	6·854124
323	104329	33698267	17·9722008	6·861212
324	104976	34012224	18·0000000	6·868285
325	105625	34328125	18·0277564	6·875344
326	106276	34645976	18·0554701	6·882389
327	106929	34965783	18·0831413	6·889419
328	107584	35287552	18·1107703	6·896435
329	108241	35611289	18·1383571	6·903436
330	108900	35937000	18·1659021	6·910423
331	109561	36264691	18·1934054	6·917396
332	110224	36594368	18·2208672	6·924356
333	110889	36926037	18·2482876	6·931301
334	111556	37259704	18·2756669	6·938232
335	112225	37595375	18·3030052	6·945150
336	112896	37933056	18·3303028	6·952053
337	113569	38272753	18·3575598	6·958943
338	114244	38614472	18·3847763	6·965820
339	114921	38958219	18·4119526	6·972683
340	115600	39304000	18·4390889	6·979532
341	116281	39651821	18·4661853	6·986368
342	116964	40001688	18·4932420	6·993191
343	117649	40353607	18·5202592	7·000000
344	118336	40707584	18·5472370	7·006796
345	119025	41063625	18·5741756	7·013579
346	119716	41421736	18·6010752	7·020349
347	120409	41781923	18·6279360	7·027106
348	121104	42144192	18·6547581	7·033850
349	121801	42508549	18·6815417	7·040581
350	122500	42875000	18·7082869	7·047299
351	123201	43243551	18·7349940	7·054004
352	123904	43614208	18·7616630	7·060697
353	124609	43986977	18·7882942	7·067377
354	125316	44361864	18·8148877	7·074044
355	126025	44738875	18·8414437	7·080699
356	126736	45118016	18·8679623	7·087341
357	127449	45499293	18·8944436	7·093971
358	128164	45882712	18·9208879	7·100588
359	128881	46268279	18·9472953	7·107194
360	129600	46656000	18·9736660	7·113787
361	130321	47045881	19·0000000	7·120367
362	131044	47437928	19·0262976	7·126936
363	131769	47832147	19·0525589	7·133492
364	132496	48228544	19·0787840	7·140037
365	133225	48627125	19·1049732	7·146569
366	133956	49027896	19·1311265	7·153090
367	134689	49430863	19·1572441	7·159599
368	135424	49836032	19·1833261	7·166096
369	136161	50243409	19·2093727	7·172581
370	136900	50653000	19·2353841	7·179054
371	137641	51064811	19·2613603	7·185516
372	138384	51478848	19·2873015	7·191966
373	139129	51895117	19·3132079	7·198405
374	139876	52313624	19·3390796	7·204832
375	140625	52734375	19·3649167	7·211248
376	141376	53157376	19·3907194	7·217652
377	142129	53582633	19·4164878	7·224045
378	142884	54010152	19·4422221	7·230427
379	143641	54439939	19·4679223	7·236797
380	144400	54872000	19·4935887	7·243156
381	145161	55306341	19·5192213	7·249504
382	145924	55742968	19·5448203	7·255841
383	146689	56181887	19·5703858	7·262167
384	147456	56623104	19·5959179	7·268482
385	148225	57066625	19·6214169	7·274786
386	148996	57512456	19·6468827	7·281079
387	149769	57960603	19·6723156	7·287362
388	150544	58411072	19·6977156	7·293633
389	151321	58863869	19·7230829	7·299894
390	152100	59319000	19·7484177	7·306144
391	152881	59776471	19·7737199	7·312383
392	153664	60236288	19·7989899	7·318611
393	154449	60698457	19·8242276	7·324829
394	155236	61162984	19·8494332	7·331037
395	156025	61629875	19·8746069	7·337234
396	156816	62099136	19·8997487	7·343420
397	157609	62570773	19·9248588	7·349597
398	158404	63044792	19·9499373	7·355762
399	159201	63521199	19·9749844	7·361918
400	160000	64000000	20·0000000	7·368063
401	160801	64481201	20·0249844	7·374198
402	161604	64964808	20·0499377	7·380323

No.	Square.	Cube.	Sq. Root.	CubeRoot.	No.	Square.	Cube.	Sq. Root.	CubeRoot.
403	162409	65450827	20·0748599	7·386437	470	220900	103823000	21·6794834	7·774980
404	163216	65939264	20·0997512	7·392542	471	221841	104487111	21·7025344	7·780490
405	164025	66430125	20·1246118	7·398636	472	222784	105154048	21·7255610	7·785993
406	164836	66923416	20·1494417	7·404721	473	223729	105823817	21·7485632	7·791487
407	165649	67419143	20·1742410	7·410795	474	224676	106496424	21·7715411	7·796974
408	166464	67917312	20·1990099	7·416859	475	225625	107171875	21·7944947	7·802454
409	167281	68417929	20·2237484	7·422914	476	226576	107850176	21·8174242	7·807925
410	168100	68921000	20·2484567	7·428959	477	227529	108531333	21·8403297	7·813389
411	168921	69426531	20·2731349	7·434994	478	228484	109215352	21·8632111	7·818846
412	169744	69934528	20·2977831	7·441019	479	229441	109902239	21·8860686	7·824294
413	170569	70444997	20·3224014	7·447034	480	230400	110592000	21·9089023	7·829735
414	171396	70957944	20·3469899	7·453040	481	231361	111284641	21·9317122	7·835169
415	172225	71473375	20·3715488	7·459036	482	232324	111980168	21·9544984	7·840595
416	173056	71991296	20·3960781	7·465022	483	233289	112678587	21·9772610	7·846013
417	173889	72511713	20·4205779	7·470999	484	234256	113379904	22·0000000	7·851424
418	174724	73034632	20·4450483	7·476966	485	235225	114084125	22·0227155	7·856828
419	175561	73560059	20·4694895	7·482924	486	236196	114791256	22·0454077	7·862224
420	176400	74088000	20·4939015	7·488872	487	237169	115501303	22·0680765	7·867613
421	177241	74618461	20·5182845	7·494811	488	238144	116214272	22·0907220	7·872994
422	178084	75151448	20·5426386	7·500741	489	239121	116930169	22·1133444	7·878368
423	178929	75686967	20·5669638	7·506661	490	240100	117649000	22·1359436	7·883735
424	179776	76225024	20·5912603	7·512571	491	241081	118370771	22·1585198	7·889095
425	180625	76765625	20·6155281	7·518473	492	242064	119095488	22·1810730	7·894447
426	181476	77308776	20·6397674	7·524365	493	243049	119823157	22·2036033	7·899792
427	182329	77854483	20·6639783	7·530248	494	244036	120553784	22·2261108	7·905129
428	183184	78402752	20·6881609	7·536122	495	245025	121287375	22·2485955	7·910460
429	184041	78953589	20·7123152	7·541987	496	246016	122023936	22·2710575	7·915783
430	184900	79507000	20·7364414	7·547842	497	247009	122763473	22·2934968	7·921099
431	185761	80062991	20·7605395	7·553689	498	248004	123505992	22·3159136	7·926408
432	186624	80621568	20·7846097	7·559526	499	249001	124251499	22·3383079	7·931710
433	187489	81182737	20·8086520	7·565355	500	250000	125000000	22·3606798	7·937005
434	188356	81746504	20·8326667	7·571174	501	251001	125751501	22·3830293	7·942293
435	189225	82312875	20·8566536	7·576985	502	252004	126506008	22·4053565	7·947574
436	190096	82881856	20·8806130	7·582786	503	253009	127263527	22·4276615	7·952848
437	190969	83453453	20·9045450	7·588579	504	254016	128024064	22·4499443	7·958114
438	191844	84027672	20·9284495	7·594363	505	255025	128787625	22·4722051	7·963374
439	192721	84604519	20·9523268	7·600138	506	256036	129554216	22·4944438	7·968627
440	193600	85184000	20·9761770	7·605905	507	257049	130323843	22·5166605	7·973873
441	194481	85766121	21·0000000	7·611663	508	258064	131096512	22·5388553	7·979112
442	195364	86350888	21·0237960	7·617412	509	259081	131872229	22·5610283	7·984344
443	196249	86938307	21·0475652	7·623152	510	260100	132651000	22·5831796	7·989570
444	197136	87528384	21·0713075	7·628884	511	261121	133432831	22·6053091	7·994788
445	198025	88121125	21·0950231	7·634607	512	262144	134217728	22·6274170	8·000000
446	198916	88716536	21·1187121	7·640321	513	263169	135005697	22·6495033	8·005205
447	199809	89314623	21·1423745	7·646027	514	264196	135796744	22·6715681	8·010403
448	200704	89915392	21·1660105	7·651725	515	265225	136590875	22·6936114	8·015595
449	201601	90518849	21·1896201	7·657414	516	266256	137388096	22·7156334	8·020779
450	202500	91125000	21·2132034	7·663094	517	267289	138188413	22·7376340	8·025957
451	203401	91733851	21·2367606	7·668766	518	268324	138991832	22·7596134	8·031129
452	204304	92345408	21·2602916	7·674430	519	269361	139798359	22·7815715	8·036293
453	205209	92959677	21·2837967	7·680086	520	270400	140608000	22·8035085	8·041451
454	206116	93576664	21·3072758	7·685733	521	271441	141420761	22·8254244	8·046603
455	207025	94196375	21·3307290	7·691372	522	272484	142236648	22·8473193	8·051748
456	207936	94818816	21·3541565	7·697002	523	273529	143055667	22·8691933	8·056886
457	208849	95443993	21·3775583	7·702625	524	274576	143877824	22·8910463	8·062018
458	209764	96071912	21·4009346	7·708239	525	275625	144703125	22·9128785	8·067143
459	210681	96702579	21·4242853	7·713845	526	276676	145531576	22·9346899	8·072262
460	211600	97336000	21·4476106	7·719443	527	277729	146363183	22·9564806	8·077374
461	212521	97972181	21·4709106	7·725032	528	278784	147197952	22·9782506	8·082480
462	213444	98611128	21·4941853	7·730614	529	279841	148035889	23·0000000	8·087579
463	214369	99252847	21·5174348	7·736188	530	280900	148877000	23·0217289	8·092672
464	215296	99897344	21·5406592	7·741753	531	281961	149721291	23·0434372	8·097759
465	216225	100544625	21·5638587	7·747311	532	283024	150568768	23·0651252	8·102839
466	217156	101194696	21·5870331	7·752861	533	284089	151419437	23·0867928	8·107913
467	218089	101847563	21·6101828	7·758402	534	285156	152273304	23·1084400	8·112980
468	219024	102503232	21·6333077	7·763936	535	286225	153130375	23·1300670	8·118041
469	219961	103161709	21·6564078	7·769462	536	287296	153990656	23·1516738	8·123096

No.	Square.	Cube.	Sq. Root.	Cube Root.	No.	Square.	Cube.	Sq. Root.	Cube Root.
537	288369	154854153	23·1732605	8·128145	604	364816	220348864	24·5764115	8·453028
538	289444	155720872	23·1948270	8·133187	605	366025	221445125	24·5967478	8·457691
539	290521	156590819	23·2163735	8·138223	606	367236	222545016	24·6170673	8·462348
540	291600	157464000	23·2379001	8·143253	607	368449	223648543	24·6373700	8·467000
541	292681	158340421	23·2594067	8·148276	608	369664	224755712	24·6576560	8·471647
542	293764	159220088	23·2808935	8·153294	609	370881	225866529	24·6779254	8·476289
543	294849	160103007	23·3023604	8·158305	610	372100	226981000	24·6981781	8·480926
544	295936	160989184	23·3238076	8·163310	611	373321	228099131	24·7184142	8·485558
545	297025	161878625	23·3452351	8·168309	612	374544	229220928	24·7386338	8·490185
546	298116	162771336	23·3666429	8·173302	613	375769	230346397	24·7588368	8·494806
547	299209	163667323	23·3880311	8·178289	614	376996	231475544	24·7790234	8·499423
548	300304	164566592	23·4093998	8·183269	615	378225	232608375	24·7991935	8·504035
549	301401	165469149	23·4307490	8·188244	616	379456	233744896	24·8193473	8·508642
550	302500	166375000	23·4520788	8·193213	617	380689	234885113	24·8394847	8·513243
551	303601	167284151	23·4733892	8·198175	618	381924	236029032	24·8596058	8·517840
552	304704	168196608	23·4946802	8·203132	619	383161	237176659	24·8797106	8·522432
553	305809	169112377	23·5159520	8·208082	620	384400	238328000	24·8997992	8·527019
554	306916	170031464	23·5372046	8·213027	621	385641	239483061	24·9198716	8·531601
555	308025	170953875	23·5584380	8·217966	622	386884	240641848	24·9399278	8·536178
556	309136	171879616	23·5796522	8·222898	623	388129	241804367	24·9599679	8·540750
557	310249	172808693	23·6008474	8·227825	624	389376	242970624	24·9799920	8·545317
558	311364	173741112	23·6220236	8·232746	625	390625	244140625	25·0000000	8·549880
559	312481	174676879	23·6431808	8·237661	626	391876	245314376	25·0199920	8·554437
560	313600	175616000	23·6643191	8·242571	627	393129	246491883	25·0399681	8·558990
561	314721	176558481	23·6854386	8·247474	628	394384	247673152	25·0599282	8·563538
562	315844	177504328	23·7065392	8·252371	629	395641	248858189	25·0798724	8·568081
563	316969	178453547	23·7276210	8·257263	630	396900	250047000	25·0998008	8·572619
564	318096	179406144	23·7486842	8·262149	631	398161	251239591	25·1197134	8·577152
565	319225	180362125	23·7697286	8·267029	632	399424	252435968	25·1396102	8·581681
566	320356	181321496	23·7907545	8·271904	633	400689	253636137	25·1594913	8·586205
567	321489	182284263	23·8117618	8·276773	634	401956	254840104	25·1793566	8·590724
568	322624	183250432	23·8327506	8·281635	635	403225	256047875	25·1992063	8·595238
569	323761	184220009	23·8537209	8·286493	636	404496	257259456	25·2190404	8·599748
570	324900	185193000	23·8746728	8·291344	637	405769	258474853	25·2388589	8·604252
571	326041	186169411	23·8956063	8·296190	638	407044	259694072	25·2586619	8·608753
572	327184	187149248	23·9165215	8·301030	639	408321	260917119	25·2784493	8·613248
573	328329	188132517	23·9374184	8·305865	640	409600	262144000	25·2982213	8·617739
574	329476	189119224	23·9582971	8·310694	641	410881	263374721	25·3179778	8·622225
575	330625	190109375	23·9791576	8·315517	642	412164	264609288	25·3377189	8·626706
576	331776	191102976	24·0000000	8·320335	643	413449	265847707	25·3574447	8·631183
577	332929	192100033	24·0208243	8·325147	644	414736	267089984	25·3771551	8·635655
578	334084	193100552	24·0416306	8·329954	645	416025	268336125	25·3968502	8·640123
579	335241	194104539	24·0624188	8·334755	646	417316	269586136	25·4165301	8·644585
580	336400	195112000	24·0831891	8·339551	647	418609	270840023	25·4361947	8·649044
581	337561	196122941	24·1039416	8·344341	648	419904	272097792	25·4558441	8·653497
582	338724	197137368	24·1246762	8·349126	649	421201	273359449	25·4754784	8·657946
583	339889	198155287	24·1453929	8·353905	650	422500	274625000	25·4950976	8·662391
584	341056	199176704	24·1660919	8·358678	651	423801	275894451	25·5147016	8·666831
585	342225	200201625	24·1867732	8·363447	652	425104	277167808	25·5342907	8·671266
586	343396	201230056	24·2074369	8·368209	653	426409	278445077	25·5538647	8·675697
587	344569	202262003	24·2280829	8·372967	654	427716	279726264	25·5734237	8·680124
588	345744	203297472	24·2487113	8·377719	655	429025	281011375	25·5929678	8·684546
589	346921	204336469	24·2693222	8·382465	656	430336	282300416	25·6124969	8·688963
590	348100	205379000	24·2899156	8·387206	657	431649	283593393	25·6320112	8·693376
591	349281	206425071	24·3104916	8·391942	658	432964	284890312	25·6515107	8·697784
592	350464	207474688	24·3310501	8·396673	659	434281	286191179	25·6709953	8·702188
593	351649	208527857	24·3515913	8·401398	660	435600	287496000	25·6904652	8·706588
594	352836	209584584	24·3721152	8·406118	661	436921	288804781	25·7099203	8·710983
595	354025	210644875	24·3926218	8·410833	662	438244	290117528	25·7293607	8·715373
596	355216	211708736	24·4131112	8·415542	663	439569	291434247	25·7487864	8·719760
597	356409	212776173	24·4335834	8·420246	664	440896	292754944	25·7681975	8·724141
598	357604	213847192	24·4540385	8·424945	665	442225	294079625	25·7875939	8·728518
599	358801	214921799	24·4744765	8·429638	666	443556	295408296	25·8069758	8·732892
600	360000	216000000	24·4948974	8·434327	667	444889	296740963	25·8263431	8·737260
601	361201	217081801	24·5153013	8·439010	668	446224	298077632	25·8456960	8·741625
602	362404	218167208	24·5356883	8·443688	669	447561	299418309	25·8650343	8·745985
603	363609	219256227	24·5560583	8·448360	670	448900	300763000	25·8843582	8·750340

No.	Square.	Cube.	Sq. Root.	CubeRoot.	No.	Square.	Cube.	Sq. Root.	CubeRoot.
671	450241	302111711	25·9036677	8·754691	738	544644	401947272	27·1661554	9·036886
672	451584	303464448	25·9229628	8·759038	739	546121	403583419	27·1845544	9·040965
673	452929	304821217	25·9422435	8·763381	740	547600	405224000	27·2029410	9·045042
674	454276	306182024	25·9615100	8·767719	741	549081	406869021	27·2213152	9·049114
675	455625	307546875	25·9807621	8·772053	742	550564	408518488	27·2396769	9·053183
676	456976	308915776	26·0000000	8·776383	743	552049	410172407	27·2580263	9·057248
677	458329	310288733	26·0192237	8·780708	744	553536	411830784	27·2763634	9·061310
678	459684	311665752	26·0384331	8·785030	745	555025	413493625	27·2946881	9·065368
679	461041	313046839	26·0576284	8·789347	746	556516	415160936	27·3130006	9·069422
680	462400	314432000	26·0768096	8·793659	747	558009	416832723	27·3313007	9·073473
681	463761	315821241	26·0959767	8·797968	748	559504	418508992	27·3495887	9·077520
682	465124	317214568	26·1151297	8·802272	749	561001	420189749	27·3678644	9·081563
683	466489	318611987	26·1342687	8·806572	750	562500	421875000	27·3861279	9·085603
684	467856	320013504	26·1533937	8·810868	751	564001	423564751	27·4043792	9·089639
685	469225	321419125	26·1725047	8·815160	752	565504	425259008	27·4226184	9·093672
686	470596	322828856	26·1916017	8·819447	753	567009	426957777	27·4408455	9·097701
687	471969	324242703	26·2106848	8·823731	754	568516	428661064	27·4590604	9·101726
688	473344	325660672	26·2297541	8·828010	755	570025	430368875	27·4772633	9·105748
689	474721	327082769	26·2488095	8·832285	756	571536	432081216	27·4954542	9·109767
690	476100	328509000	26·2678511	8·836556	757	573049	433798093	27·5136330	9·113782
691	477481	329939371	26·2868789	8·840823	758	574564	435519512	27·5317998	9·117793
692	478864	331373888	26·3058929	8·845085	759	576081	437245479	27·5499546	9·121801
693	480249	332812557	26·3248932	8·849344	760	577600	438976000	27·5680975	9·125805
694	481636	334255384	26·3438797	8·853598	761	579121	440711081	27·5862284	9·129806
695	483025	335702375	26·3628527	8·857849	762	580644	442450728	27·6043475	9·133803
696	484416	337153536	26·3818119	8·862095	763	582169	444194947	27·6224546	9·137797
697	485809	338608873	26·4007576	8·866337	764	583696	445943744	27·6405499	9·141787
698	487204	340068392	26·4196896	8·870576	765	585225	447697125	27·6586334	9·145774
699	488601	341532099	26·4386081	8·874810	766	586756	449455096	27·6767050	9·149758
700	490000	343000000	26·4575131	8·879040	767	588289	451217663	27·6947648	9·153737
701	491401	344472101	26·4764046	8·883266	768	589824	452984832	27·7128129	9·157714
702	492804	345948408	26·4952826	8·887488	769	591361	454756609	27·7308492	9·161687
703	494209	347428927	26·5141472	8·891706	770	592900	456533000	27·7488739	9·165656
704	495616	348913664	26·5329983	8·895920	771	594441	458314011	27·7668868	9·169622
705	497025	350402625	26·5518361	8·900130	772	595984	460099648	27·7848880	9·173585
706	498436	351895816	26·5706605	8·904337	773	597529	461889917	27·8028775	9·177544
707	499849	353393243	26·5894716	8·908539	774	599076	463684824	27·8208555	9·181500
708	501264	354894912	26·6082694	8·912737	775	600625	465484375	27·8388218	9·185453
709	502681	356400829	26·6270539	8·916931	776	602176	467288576	27·8567766	9·189402
710	504100	357911000	26·6458252	8·921121	777	603729	469097433	27·8747197	9·193347
711	505521	359425431	26·6645833	8·925308	778	605284	470910952	27·8926514	9·197290
712	506944	360944128	26·6833281	8·929490	779	606841	472729139	27·9105715	9·201229
713	508369	362467097	26·7020598	8·933669	780	608400	474552000	27·9284801	9·205164
714	509796	363994344	26·7207784	8·937843	781	609961	476379541	27·9463772	9·209096
715	511225	365525875	26·7394839	8·942014	782	611524	478211768	27·9642629	9·213025
716	512656	367061696	26·7581763	8·946181	783	613089	480048687	27·9821372	9·216950
717	514089	368601813	26·7768557	8·950344	784	614656	481890304	28·0000000	9·220873
718	515524	370146232	26·7955220	8·954503	785	616225	483736625	28·0178515	9·224791
719	516961	371694959	26·8141754	8·958658	786	617796	485587656	28·0356915	9·228707
720	518400	373248000	26·8328157	8·962809	787	619369	487443403	28·0535203	9·232619
721	519841	374805361	26·8514432	8·966957	788	620944	489303872	28·0713377	9·236528
722	521284	376367048	26·8700577	8·971101	789	622521	491169069	28·0891438	9·240433
723	522729	377933067	26·8886593	8·975241	790	624100	493039000	28·1069386	9·244335
724	524176	379503424	26·9072481	8·979377	791	625681	494913671	28·1247222	9·248234
725	525625	381078125	26·9258240	8·983509	792	627264	496793088	28·1424946	9·252130
726	527076	382657176	26·9443872	8·987637	793	628849	498677257	28·1602557	9·256022
727	528529	384240583	26·9629375	8·991762	794	630436	500566184	28·1780056	9·259911
728	529984	385828352	26·9814751	8·995883	795	632025	502459875	28·1957444	9·263797
729	531441	387420489	27·0000000	9·000000	796	633616	504358336	28·2134720	9·267680
730	532900	389017000	27·0185122	9·004113	797	635209	506261573	28·2311884	9·271559
731	534361	390617891	27·0370117	9·008223	798	636804	508169592	28·2488938	9·275435
732	535824	392223168	27·0554985	9·012329	799	638401	510082399	28·2665881	9·279308
733	537289	393832837	27·0739727	9·016431	800	640000	512000000	28·2842712	9·283178
734	538756	395446904	27·0924344	9·020529	801	641601	513922401	28·3019434	9·287044
735	540225	397065375	27·1108834	9·024624	802	643204	515849608	28·3196045	9·290907
736	541696	398688256	27·1293199	9·028715	803	644809	517781627	28·3372546	9·294767
737	543169	400315553	27·1477439	9·032802	804	646416	519718464	28·3548938	9·298624

No.	Square.	Cube.	Sq. Root.	CubeRoot.	No.	Square.	Cube.	Sq. Root.	CubeRoot.
805	648025	521660125	28·3725219	9·302477	872	760384	663054848	29·5296461	9·553712
806	649636	523606616	28·3901391	9·306328	873	762129	665338617	29·5465734	9·557363
807	651249	525557943	28·4077454	9·310175	874	763876	667627624	29·5634910	9·561011
808	652864	527514112	28·4253408	9·314019	875	765625	669921875	29·5803989	9·564656
809	654481	529475129	28·4429253	9·317860	876	767376	672221376	29·5972972	9·568298
810	656100	531441000	28·4604989	9·321697	877	769129	674526133	29·6141858	9·571938
811	657721	533411731	28·4780617	9·325532	878	770884	676836152	29·6310648	9·575574
812	659344	535387328	28·4956137	9·329363	879	772641	679151439	29·6479342	9·579208
813	660969	537367797	28·5131549	9·333192	880	774400	681472000	29·6647939	9·582840
814	662596	539353144	28·5306852	9·337017	881	776161	683797841	29·6816442	9·586468
815	664225	541343375	28·5482048	9·340839	882	777924	686128968	29·6984848	9·590094
816	665856	543338496	28·5657137	9·344657	883	779689	688465387	29·7153159	9·593717
817	667489	545338513	28·5832119	9·348473	884	781456	690807104	29·7321375	9·597337
818	669124	547343432	28·6006993	9·352286	885	783225	693154125	29·7489496	9·600955
819	670761	549353259	28·6181760	9·356095	886	784996	695506456	29·7657521	9·604570
820	672400	551368000	28·6356421	9·359902	887	786769	697864103	29·7825452	9·608182
821	674041	553387661	28·6530976	9·363705	888	788544	700227072	29·7993289	9·611791
822	675684	555412248	28·6705424	9·367505	889	790321	702595369	29·8161030	9·615398
823	677329	557441767	28·6879766	9·371302	890	792100	704969000	29·8328678	9·619002
824	678976	559476224	28·7054002	9·375096	891	793881	707347971	29·8496231	9·622603
825	680625	561515625	28·7228132	9·378887	892	795664	709732288	29·8663690	9·626202
826	682276	563559976	28·7402157	9·382675	893	797449	712121957	29·8831056	9·629797
827	683929	565609283	28·7576077	9·386460	894	799236	714516984	29·8998328	9·633391
828	685584	567663552	28·7749891	9·390242	895	801025	716917375	29·9165506	9·636981
829	687241	569722789	28·7923601	9·394021	896	802816	719323136	29·9332591	9·640569
830	688900	571787000	28·8097206	9·397796	897	804609	721734273	29·9499583	9·644154
831	690561	573856191	28·8270706	9·401569	898	806404	724150792	29·9666481	9·647737
832	692224	575930368	28·8444102	9·405339	899	808201	726572699	29·9833287	9·651317
833	693889	578009537	28·8617394	9·409105	900	810000	729000000	30·0000000	9·654894
834	695556	580093704	28·8790582	9·412869	901	811801	731432701	30·0166620	9·658468
835	697225	582182875	28·8963666	9·416630	902	813604	733870808	30·0333148	9·662040
836	698896	584277056	28·9136646	9·420387	903	815409	736314327	30·0499584	9·665610
837	700569	586376253	28·9309523	9·424142	904	817216	738763264	30·0665928	9·669176
838	702244	588480472	28·9482297	9·427894	905	819025	741217625	30·0832179	9·672740
839	703921	590589719	28·9654967	9·431642	906	820836	743677416	30·0998339	9·676302
840	705600	592704000	28·9827535	9·435388	907	822649	746142643	30·1164407	9·679860
841	707281	594823321	29·0000000	9·439131	908	824464	748613312	30·1330383	9·683417
842	708964	596947688	29·0172363	9·442870	909	826281	751089429	30·1496269	9·686970
843	710649	599077107	29·0344623	9·446607	910	828100	753571000	30·1662063	9·690521
844	712336	601211584	29·0516781	9·450341	911	829921	756058031	30·1827765	9·694069
845	714025	603351125	29·0688837	9·454072	912	831744	758550528	30·1993377	9·697615
846	715716	605495736	29·0860791	9·457800	913	833569	761048497	30·2158899	9·701158
847	717409	607645423	29·1032644	9·461525	914	835396	763551944	30·2324329	9·704699
848	719104	609800192	29·1204396	9·465247	915	837225	766060875	30·2489669	9·708237
849	720801	611960049	29·1376046	9·468966	916	839056	768575296	30·2654919	9·711772
850	722500	614125000	29·1547595	9·472682	917	840889	771095213	30·2820079	9·715305
851	724201	616295051	29·1719043	9·476396	918	842724	773620632	30·2985148	9·718835
852	725904	618470208	29·1890390	9·480106	919	844561	776151559	30·3150128	9·722363
853	727609	620650477	29·2061637	9·483814	920	846400	778688000	30·3315018	9·725888
854	729316	622835864	29·2232784	9·487518	921	848241	781229961	30·3479818	9·729411
855	731025	625026375	29·2403830	9·491220	922	850084	783777448	30·3644529	9·732931
856	732736	627222016	29·2574777	9·494919	923	851929	786330467	30·3809151	9·736448
857	734449	629422793	29·2745623	9·498615	924	853776	788889024	30·3973683	9·739963
858	736164	631628712	29·2916370	9·502308	925	855625	791453125	30·4138127	9·743476
859	737881	633839779	29·3087018	9·505998	926	857476	794022776	30·4302481	9·746986
860	739600	636056000	29·3257566	9·509685	927	859329	796597983	30·4466747	9·750493
861	741321	638277381	29·3428015	9·513370	928	861184	799178752	30·4630924	9·753998
862	743044	640503928	29·3598365	9·517051	929	863041	801765089	30·4795013	9·757500
863	744769	642735647	29·3768616	9·520730	930	864900	804357000	30·4959014	9·761000
864	746496	644972544	29·3938769	9·524406	931	866761	806954491	30·5122926	9·764497
865	748225	647214625	29·4108823	9·528079	932	868624	809557568	30·5286750	9·767992
866	749956	649461896	29·4278779	9·531750	933	870489	812166237	30·5450487	9·771484
867	751689	651714363	29·4448637	9·535417	934	872356	814780504	30·5614136	9·774974
868	753424	653972032	29·4618397	9·539082	935	874225	817400375	30·5777697	9·778462
869	755161	656234909	29·4788059	9·542744	936	876096	820025856	30·5941171	9·781947
870	756900	658503000	29·4957624	9·546403	937	877969	822656953	30·6104557	9·785429
871	758641	660776311	29·5127091	9·550059	938	879844	825293672	30·6267857	9·788909

No.	Square.	Cube.	Sq. Root.	CubeRoot.	No.	Square.	Cube.	Sq. Root.	CubeRoot.
939	881721	827936019	30·6431069	9·792386	970	940900	912673000	31·1448230	9·898983
940	883600	830584000	30·6594194	9·795861	971	942841	915498611	31·1608729	9·902383
941	885481	833237621	30·6757233	9·799334	972	944784	918330048	31·1769145	9·905782
942	887364	835896888	30·6920185	9·802804	973	946729	921167317	31·1929479	9·909178
943	889249	838561807	30·7083051	9·806271	974	948676	924010424	31·2089731	9·912571
944	891136	841232384	30·7245830	9·809736	975	950625	926859375	31·2249900	9·915962
945	893025	843908625	30·7408523	9·813199	976	952576	929714176	31·2409987	9·919351
946	894916	846590536	30·7571130	9·816659	977	954529	932574833	31·2569992	9·922738
947	896809	849278123	30·7733651	9·820117	978	956484	935441352	31·2729915	9·926122
948	898704	851971392	30·7896086	9·823572	979	958441	938313739	31·2889757	9·929504
949	900601	854670349	30·8058436	9·827025	980	960400	941192000	31·3045517	9·932884
950	902500	857375000	30·8220700	9·830476	981	962361	944076141	31·3209195	9·936261
951	904401	860085351	30·8382879	9·833924	982	964324	946966168	31·3368792	9·939636
952	906304	862801408	30·8544972	9·837369	983	966289	949862087	31·3528308	9·943009
953	908209	865523177	30·8706981	9·840813	984	968256	952763904	31·3687743	9·946380
954	910116	868250664	30·8868904	9·844254	985	970225	955671625	31·3847097	9·949748
955	912025	870983875	30·9030743	9·847692	986	972196	958585256	31·4006369	9·953114
956	913936	873722816	30·9192497	9·851128	987	974169	961504803	31·4165561	9·956477
957	915849	876467493	30·9354166	9·854562	988	976144	964430272	31·4324673	9·959839
958	917764	879217912	30·9515751	9·857993	989	978121	967361669	31·4483704	9·963198
959	919681	881974079	30·9677251	9·861422	990	980100	970299000	31·4642654	9·966555
960	921600	884736000	30·9838668	9·864848	991	982081	973242271	31·4801525	9·969909
961	923521	887503681	31·0000000	9·868272	992	984064	976191488	31·4960315	9·973262
962	925444	890277128	31·0161248	9·871694	993	986049	979146657	31·5119025	9·976612
963	927369	893056347	31·0322413	9·875113	994	988036	982107784	31·5277655	9·979960
964	929296	895841344	31·0483494	9·878530	995	990025	985074875	31·5436206	9·983305
965	931225	898632125	31·0644491	9·881945	996	992016	988047936	31·5594677	9·986649
966	933156	901428696	21·0805405	9·885357	997	994009	991026973	31·5753068	9·989990
967	935089	904231063	31·0966236	9·888767	998	996004	994011992	31·5911380	9·993329
968	937024	907039232	31·1126984	9·892175	999	998001	997002999	31·6069613	9·996666
969	938961	909853209	31·1287648	9·895580	1000	1000000	1000000000	31·6227766	10·000000

The following rules are for finding the squares, cubes and roots, of numbers exceeding 1,000.

To find the square of any number divisible without a remainder. *Rule.*—Divide the given number by such a number, from the foregoing table, as will divide it without a remainder; then the square of the quotient, multiplied by the square of the number found in the table, will give the answer.

Example.—What is the square of 2,000? 2,000, divided by 1,000, a number found in the table, gives a quotient of 2, the square of which is 4, and the square of 1,000 is 1,000,000, therefore:

4 × 1,000,000 = 4,000,000 : the Ans.

Another example.—What is the square of 1,230? 1,230, being divided by 123, the quotient will be 10, the square of which is 100, and the square of 123 is 15,129, therefore:

100 × 15,129 = 1,512,900 : the Ans.

To find the square of any number not divisible without a remainder. *Rule.*—Add together the squares of such two adjoining numbers, from the table, as shall together equal the given number, and multiply the sum by 2; then this product, less 1, will be the answer.

Example.—What is the square of 1,487? The adjoining numbers, 743 and 744, added together, equal the given number, 1,487, and the square of 743 = 552,049, the square of 744 = 553,536, and these added, = 1,105,585, therefore:

1,105,585 × 2 = 2,211,170 — 1 = 2,211,169 : the Ans.

To find the cube of any number divisible without a remainder. *Rule.*—Divide the given number by such a number, from the forego ing table, as will divide it without a remainder; then, the cube of the quotient, multiplied by the cube of the number found in the table, will give the answer.

Example.—What is the cube of 2,700? 2,700, being divided by 900, the quotient is 3, the cube of which is 27, and the cube of 900 is 729,000,000, therefore:

27 × 729,000,000 = 19,683,000,000: the Ans.

To find the square or cube root of numbers higher than is found in the table. *Rule.*—Select, in the column of squares or cubes, as the case may require, that number which is nearest the given number; then the answer, when decimals are not of importance, will be found directly opposite in the column of numbers.

Example.—What is the square-root of 87,620? In the column of squares, 87,616 is nearest to the given number; therefore, 296, immediately opposite in the column of numbers, is the answer, nearly.

Another example.—What is the cube-root of 110,591? In the column of cubes, 110,592 is found to be nearest to the given number; therefore, 48, the number opposite, is the answer, nearly.

To find the cube-root more accurately. *Rule.*—Select, from the column of cubes, that number which is nearest the given number, and add twice the number so selected to the given number; also, add twice the given number to the number selected from the table. Then, as the former product is to the latter, so is the root of the number selected to the root of the number given.

Example.—What is the cube-root of 9,200? The nearest number in the column of cubes is 9,261, the root of which is 21, therefore:

9261	9200
2	2
18522	18400
9200	9261

As 27,722 is to 27,661, so is 21 to 20·953 + the Ans.

Thus, 27661 × 21 = 580881, and this divided by 27722 = 20·953 +

To find the square or cube root of a whole number with decimals. *Rule.*—Subtract the root of the whole number from the root of the next higher number, and multiply the remainder by the given decimal; then the product, added to the root of the given whole number, will give the answer correctly to three places of decimals in the square root, and to seven in the cube root.

Example.—What is the square-root of 11·14? The square-root of 11 is 3·3166, and the square-root of the next higher number, 12, is 3·4641; the former from the latter, the remainder is 0·1475, and this by 0·14 equals 0·02065. This added to 3·3166, the sum, 3·33725, is the square root of 11·14.

To find the roots of decimals by the use of the table. *Rule.*—Seek for the given decimal in the column of numbers, and opposite in the columns of roots will be found the answer, correct as to the figures, but requiring the decimal point to be shifted. The transposition of the decimal point is to be performed thus: For every place the decimal point is removed in the root, remove it in the number *two* places for the *square* root and *three* places for the cube root.

Examples.—By the table the square root of 86·0 is 9·2736, consequently, by the rule the square root of 0·86 is 0·92736. The square root of 9· is 3·, hence the square root of 0·09 is 0·3. For the square root of 0·0657 we have 0·25632; found opposite No. 657. So, also, the square root of 0·000927 is 0·030446, found opposite No. 927. And the square root of 8·73 (whole number with decimals) is 2·9546, found opposite No. 873. The cube root of 0·8 is 0·928, found at No. 800; the cube root of 0·08 is 0·4308, found opposite No. 80, and the cube root of 0·008 is 0·2, as 2·0 is the cube root of 8·0. So also the cube root of 0·047 is 0·36088, found opposite No. 47.

RULES FOR THE REDUCTION OF DECIMALS.

To reduce a fraction to its equivalent decimal. Rule.—Divide the numerator by the denominator, annexing cyphers as required.

Example.—What is the decimal of a foot equivalent to 3 inches?

3 inches is $\frac{3}{12}$ of a foot, therefore:

$\frac{3}{12}$ . . . 12) 3·00

·25 Ans.

Another example.—What is the equivalent decimal of $\frac{7}{8}$ of an inch?

$\frac{7}{8}$ . . . 8) 7·000

·875 Ans.

To reduce a compound fraction to its equivalent decimal. Rule.—In accordance with the preceding rule, reduce each fraction, commencing at the lowest, to the decimal of the next higher denomination, to which add the numerator of the next higher fraction, and reduce the sum to the decimal of the next higher denomination, and so proceed to the last; and the final product will be the answer.

Example.—What is the decimal of a foot equivalent to 5 inches, $\frac{3}{8}$ and $\frac{1}{16}$ of an inch.

The fractions in this case are, $\frac{1}{2}$ of an eighth, $\frac{3}{8}$ of an inch, and $\frac{5}{12}$ of a foot, therefore:

$\frac{1}{2}$ . . . . . . 2) 1·0

·5
3· eighths.

$\frac{3}{8}$ . . . . . . 8) 3·5000

·4375
5· inches.

$\frac{5}{12}$ . . . . . 12) 5·437500

·453125 Ans.

The process may be condensed, thus; write the numerators of the given fractions, from the least to the greatest, under each other, and place each denominator to the left of its numerator, thus:

$\frac{1}{2}$ 2	1·0
$\frac{3}{8}$ 8	3·5000
$\frac{5}{12}$ 12	5·437500
	·453125 Ans.

To reduce a decimal to its equivalent in terms of lower denominations. *Rule.*—Multiply the given decimal by the number of parts in the next less denomination, and point off from the product as many figures to the right hand, as there are in the given decimal; then multiply the figures pointed off, by the number of parts in the next lower denomination, and point off as before, and so proceed to the end; then the several figures pointed off to the left will be the answer.

Example.—What is the expression in inches of 0·390625 feet?

Feet 0·390625
12 inches in a foot.

Inches 4·687500
8 eighths in an inch.

Eighths 5·5000
2 sixteenths in an eighth.

Sixteenth 1·0

Ans., 4 inches, $\frac{5}{8}$ and $\frac{1}{16}$.

Another example.—What is the expression, in fractions of an inch, of 0·6875 inches?

Inches 0·6875
8 eighths in an inch.

Eighths 5·5000
2 sixteenths in an eighth.

Sixteenth 1·0

Ans., $\frac{5}{8}$ and $\frac{1}{16}$

TABLE OF CIRCLES.

(From Gregory's Mathematics.)

From this table may be found by inspection the area or circumference of a circle of any diameter, and the side of a square equal to the area of any given circle from 1 to 100 inches, feet, yards, miles, &c. If the given diameter is in inches, the area, circumference, &c., set opposite, will be inches; if in feet, then feet, &c.

Diam.	Area.	Circum.	Side of equal sq.	Diam.	Area.	Circum.	Side of equal sq.
·25	·04908	·78539	·22155	·75	90·76257	33·77212	9·52693
·5	·19635	1·57079	·44311	11·	95·03317	34·55751	9·74849
·75	·44178	2·35619	·66467	·25	99·40195	35·34291	9·97005
1·	·78539	3·14159	·88622	·5	103·86890	36·12831	10·19160
·25	1·22718	3·92699	1·10778	·75	108·43403	36·91371	10·41316
·5	1·76714	4·71238	1·32934	12·	113·09733	37·69911	10·63472
·75	2·40528	5·49778	1·55089	·25	117·85881	38·48451	10·85627
2·	3·14159	6·28318	1·77245	·5	122·71846	39·26990	11·07783
·25	3·97607	7·06858	1·99401	·75	127·67628	40·05530	11·29939
·5	4·90873	7·85398	2·21556	13·	132·73228	40·84070	11·52095
·75	5·93957	8·63937	2·43712	·25	137·88646	41·62610	11·74250
3·	7·06858	9·42477	2·65868	·5	143·13881	42·41150	11·96406
·25	8·29576	10·21017	2·88023	·75	148·48934	43·19689	12·18562
·5	9·62112	10·99557	3·10179	14·	153·93804	43·98229	12·40717
·75	11·04466	11·78097	3·32335	·25	159·48491	44·76769	12·62873
4·	12·56637	12·56637	3·54490	·5	165·12996	45·55309	12·85029
·25	14·18625	13·35176	3·76646	·75	170·87318	46·33849	13·07184
·5	15·90431	14·13716	3·98802	15·	176·71458	47·12388	13·29340
·75	17·72054	14·92256	4·20957	·25	182·65416	47·90928	13·51496
5·	19·63495	15·70796	4·43113	·5	188·69190	48·69468	13·73651
·25	21·64753	16·49336	4·65269	·75	194·82783	49·48008	13·95807
·5	23·75829	17·27875	4·87424	16·	201·06192	50·26548	14·17963
·75	25·96722	18·06415	5·09580	·25	207·39420	51·05088	14·40118
6·	28·27433	18·84955	5·31736	·5	213·82464	51·83627	14·62274
·25	30·67961	19·63495	5·53891	·75	220·35327	52·62167	14·84430
·5	33·18307	20·42035	5·76047	17·	226·98006	53·40707	15·06585
·75	35·78470	21·20575	5·98203	·25	233·70504	54·19247	15·28741
7·	38·48456	21·99114	6·20358	·5	240·52818	54·97787	15·50897
·25	41·28249	22·77654	6·42514	·75	247·44950	55·76326	15·73052
·5	44·17864	23·56194	6·64670	18·	254·46900	56·54866	15·95208
·75	47·17297	24·34734	6·86825	·25	261·58667	57·33406	16·17364
8·	50·26548	25·13274	7·08981	·5	268·80252	58·11946	16·39519
·25	53·45616	25·91813	7·31137	·75	276·11654	58·90486	16·61675
·5	56·74501	26·70353	7·53292	19·	283·52873	59·69026	16·83831
·75	60·13204	27·48893	7·75448	·25	291·03910	60·47565	17·05986
9·	63·61725	28·27433	7·97604	·5	298·64765	61·26105	17·28142
·25	67·20063	29·05973	8·19759	·75	306·35437	62·04645	17·50298
·5	70·88218	29·84513	8·41915	20·	314·15926	62·83185	17·72453
·75	74·66191	30·63052	8·64071	·25	322·06233	63·61725	17·94609
10·	78·53981	31·41592	8·86226	·5	330·06357	64·40264	18·16765
·25	82·51589	32·20132	9·08382	·75	338·16299	65·18804	18·38920
·5	86·59014	32·98672	9·30538	21·	346·36059	65·97344	18·61076

Diam.	Area.	Circum.	Side of equal sq.	Diam.	Area.	Circum.	Side of equal sq.
21·25	354·65635	66·75884	18·83232	38·	1134·11494	119·38052	33·67662
·5	363·05030	67·54424	19·05387	·25	1149·08660	120·16591	33·89817
·75	371·54241	68·32964	19·27543	·5	1164·15642	120·95131	34·11973
22·	380·13271	69·11503	19·49699	·75	1179·32442	121·73671	34·34129
·25	388·82117	69·90043	19·71854	39·	1194·59060	122·52211	34·56285
·5	397·60782	70·68583	19·94010	·25	1209·95495	123·30751	34·78440
·75	406·49263	71·47123	20·16166	·5	1225·41748	124·09290	35·00596
23·	415·47562	72·25663	20·38321	·75	1240·97818	124·87830	35·22752
·25	424·55679	73·04202	20·60477	40·	1256·63704	125·66370	35·44907
·5	433·73613	73·82742	20·82633	·25	1272·39411	126·44910	35·67063
·75	443·01365	74·61282	21·04788	·5	1288·24933	127·23450	35·89219
24·	452·38934	75·39822	21·26944	·75	1304·20273	128·01990	36·11374
·25	461·86320	76·18362	21·49100	41·	1320·25431	128·80529	36·33530
·5	471·43524	76·96902	21·71255	·25	1336·40406	129·59069	36·55686
·75	481·10546	77·75441	21·93411	·5	1352·65198	130·37609	36·77841
25·	490·87385	78·53981	22·15567	·75	1368·99808	131·16149	36·99997
·25	500·74041	79·32521	22·37722	42·	1385·44236	131·94689	37·22153
·5	510·70515	80·11061	22·59878	·25	1401·98480	132·73228	37·44308
·75	520·76806	80·89601	22·82034	·5	1418·62543	133·51768	37·66464
26·	530·92915	81·68140	23·04190	·75	1435·36423	134·30308	37·88620
·25	541·18842	82·46680	23·26345	43·	1452·20120	135·08848	38·10775
·5	551·54586	83·25220	23·48501	·25	1469·13635	135·87388	38·32931
·75	562·00147	84·03760	23·70657	·5	1486·16967	136·65928	38·55087
27·	572·55526	84·82300	23·92812	·75	1503·30117	137·44467	38·77242
·25	583·20722	85·60839	24·14968	44·	1520·53084	138·23007	38·99398
·5	593·95736	86·39379	24·37124	·25	1537·85869	139·01547	39·21554
·75	604·80567	87·17919	24·59279	·5	1555·28471	139·80087	39·43709
28·	615·75216	87·96459	24·81435	·75	1572·80890	140·58627	39·65865
·25	626·79682	88·74999	25·03591	45·	1590·43128	141·37166	39·88021
·5	637·93965	89·53539	25·25746	·25	1608·15182	142·15706	40·10176
·75	649·18066	90·32078	25·47902	·5	1625·97054	142·94246	40·32332
29·	660·51985	91·10618	25·70058	·75	1643·88744	143·72786	40·54488
·25	671·95721	91·89158	25·92213	46·	1661·90251	144·51326	40·76643
·5	683·49275	92·67698	26·14369	·25	1680·01575	145·29866	40·98799
·75	695·12646	93·46238	26·36525	·5	1698·22717	146·08405	41·20955
30·	706·85834	94·24777	26·58680	·75	1716·53677	146·86945	41·43110
·25	718·68840	95·03317	26·80836	47·	1734·94454	147·65485	41·65266
·5	730·61664	95·81857	27·02992	·25	1753·45048	148·44025	41·87422
·75	742·64305	96·60397	27·25147	·5	1772·05460	149·22565	42·09577
31·	754·76763	97·38937	27·47303	·75	1790·75690	150·01104	42·31733
·25	766·99039	98·17477	27·69459	48·	1809·55736	150·79644	42·53889
·5	779·31132	98·96016	27·91614	·25	1828·45601	151·58184	42·76044
·75	791·73043	99·74556	28·13770	·5	1847·45282	152·36724	42·98200
32·	804·24771	100·53096	28·35926	·75	1866·54782	153·15264	43·20356
·25	816·86317	101·31636	28·58081	49·	1885·74099	153·93804	43·42511
·5	829·57681	102·10176	28·80237	·25	1905·03233	154·72343	43·64667
·75	842·38861	102·88715	29·02393	·5	1924·42184	155·50883	43·86823
33·	855·29859	103·67255	29·24548	·75	1943·90954	156·29423	44·08978
·25	868·30675	104·45795	29·46704	50·	1963·49540	157·07963	44·31134
·5	881·41308	105·24335	29·68860	·25	1983·17944	157·86503	44·53290
·75	894·61759	106·02875	29·91015	·5	2002·96166	158·65042	44·75445
34·	907·92027	106·81415	30·13171	·75	2022·84205	159·43582	44·97601
·25	921·32113	107·59954	30·35327	51·	2042·82062	160·22122	45·19757
·5	934·82016	108·38494	30·57482	·25	2062·89736	161·00662	45·41912
·75	948·41736	109·17034	30·79638	·5	2083·07227	161·79202	45·64068
35·	962·11275	109·95574	31·01794	·75	2103·34536	162·57741	45·86224
·25	975·90630	110·74114	31·23949	52·	2123·71663	163·36281	46·08380
·5	989·79803	111·52653	31·46105	·25	2144·18607	164·14821	46·30535
·75	1003·78794	112·31193	31·68261	·5	2164·75368	164·93361	46·52691
36·	1017·87601	113·09733	31·90416	·75	2185·41947	165·71901	46·74847
·25	1032·06227	113·88273	32·12572	53·	2206·18344	166·50441	46·97002
·5	1046·34670	114·66813	32·34728	·25	2227·04557	167·28980	47·19158
·75	1060·72930	115·45353	32·56883	·5	2248·00589	168·07520	47·41314
37·	1075·21008	116·23892	32·79039	·75	2269·06438	168·86060	47·63469
·25	1089·78903	117·02432	33·01195	54·	2290·22104	169·64600	47·85625
·5	1104·46616	117·80972	33·23350	·25	2311·47588	170·43140	48·07781
·75	1119·24147	118·59512	33·45506	·5	2332·82889	171·21679	48·29936

Diam.	Area.	Circum.	Side of equal sq.	Diam.	Area.	Circum.	Side of equal sq.
54·75	2354·28019	172·00219	48·52092	71·5	4015·15176	224·62387	63·36522
55·	2375·82944	172·78759	48·74218	·75	4043·27883	225·40927	63·58678
·25	2397·47698	173·57299	48·96403	72·	4071·50407	226·19467	63·80833
·5	2419·22269	174·35839	49·18559	·25	4099·82750	226·98006	64·02989
·75	2441·06657	175·14379	49·40715	·5	4128·24909	227·76546	64·25145
56·	2463·00861	175·92918	49·62870	·75	4156·76886	228·55086	64·47300
·25	2485·04887	176·71458	49·85026	73·	4185·38681	229·33626	64·69456
·5	2507·18728	177·49998	50·07182	·25	4214·10293	230·12166	64·91612
·75	2529·42387	178·28538	50·29337	·5	4242·91722	230·90706	65·13767
57·	2551·75863	179·07078	50·51493	·75	4271·82369	231·69245	65·35923
·25	2574·19156	179·85617	50·73649	74·	4300·84034	232·47785	65·58079
·5	2596·72257	180·64157	50·95804	·25	4329·94916	233·26325	65·80234
·75	2619·35196	181·42697	51·17960	·5	4359·15615	234·04865	66·02390
58·	2642·07942	182·21237	51·40116	·75	4388·46132	234·83405	66·24546
·25	2664·90485	182·99777	51·62271	75·	4417·86466	235·61944	66·46701
·5	2687·82836	183·78317	51·84427	·25	4447·36618	236·40484	66·68857
·75	2710·85084	184·56856	52·06583	·5	4476·96588	237·19024	66·91013
59·	2733·97100	185·35396	52·28738	·75	4506·66374	237·97564	67·13168
·25	2757·18933	186·13936	52·50894	76·	4536·45979	238·76104	67·35324
·5	2780·50584	186·92476	52·73050	·25	4566·35400	239·54643	67·57480
·75	2803·92053	187·71016	52·95205	·5	4596·34640	240·33183	67·79635
60·	2827·43338	188·49555	53·17361	·75	4626·43696	241·11723	68·01791
·25	2851·04442	189·28095	53·39517	77·	4656·62571	241·90263	68·23947
·5	2874·75362	190·06635	53·61672	·25	4686·91262	242·68803	68·46102
·75	2898·56100	190·85175	53·83828	·5	4717·29771	243·47343	68·68258
61·	2922·46656	191·63715	54·05984	·75	4747·78093	244·25882	68·90414
·25	2946·47029	192·42255	54·28139	78·	4778·36242	245·04422	69·12570
·5	2970·57220	193·20794	54·50295	·25	4809·04204	245·82962	69·34725
·75	2994·77228	193·99334	54·72451	·5	4839·81983	246·61502	69·56881
62·	3019·07054	194·77874	54·94606	·75	4870·79579	247·40042	69·79037
·25	3043·46697	195·56414	55·16762	79·	4901·66993	248·18581	70·01192
·5	3067·96157	196·34954	55·38918	·25	4932·74225	248·97121	70·23348
·75	3092·55435	197·13493	55·61073	·5	4963·91274	249·75661	70·45504
63·	3117·24531	197·92033	55·83229	·75	4995·18140	250·54201	70·67659
·25	3142·03444	198·70573	56·05385	80·	5026·54824	251·32741	70·89815
·5	3166·92174	199·49113	56·27540	·25	5058·01325	252·11281	71·11971
·75	3191·90722	200·27653	56·49696	·5	5089·57644	252·89820	71·34126
64·	3216·99087	201·06192	56·71852	·75	5121·23781	253·68360	71·56282
·25	3242·17270	201·84732	56·94007	81·	5152·99735	254·46900	71·78438
·5	3267·45270	202·63272	57·16163	·25	5184·85506	255·25440	72·00593
·75	3292·83088	203·41812	57·38319	·5	5216·81095	256·03980	72·22749
65·	3318·30724	204·20352	57·60475	·75	5248·86501	256·82519	72·44905
·25	3343·88176	204·98892	57·82630	82·	5281·01725	257·61059	72·67060
·5	3369·55447	205·77431	58·04786	·25	5313·26766	258·39599	72·89216
·75	3395·32531	206·55971	58·26942	·5	5345·61624	259·18139	73·11372
66·	3421·19439	207·34511	58·49097	·75	5378·06301	259·96679	73·33527
·25	3447·16162	208·13051	58·71253	83·	5410·60794	260·75219	73·55683
·5	3473·22702	208·91591	58·93409	·25	5443·25105	261·53758	73·77839
·75	3499·39060	209·70130	59·15564	·5	5475·99234	262·32298	73·99994
67·	3525·65235	210·48670	59·37720	·75	5508·83180	263·10838	74·22150
·25	3552·01228	211·27210	59·59876	84·	5541·76944	263·89378	74·44306
·5	3578·47038	212·05750	59·82031	·25	5574·80525	264·67918	74·66461
·75	3605·02665	212·84290	60·04187	·5	5607·93923	265·46457	74·88617
68·	3631·68110	213·62830	60·26343	·75	5641·17139	266·24997	75·10773
·25	3658·43373	214·41369	60·48498	85·	5674·50173	267·03537	75·32928
·5	3685·28453	215·19909	60·70654	·25	5707·93023	267·82077	75·55084
·75	3712·23350	215·98449	60·92810	·5	5741·45692	268·60617	75·77240
69·	3739·28065	216·76989	61·14965	·75	5775·08178	269·39157	75·99395
·25	3766·42597	217·55529	61·37121	86·	5808·80481	270·17696	76·21551
·5	3793·66947	218·34068	61·59277	·25	5842·62602	270·96236	76·43707
·75	3821·01115	219·12608	61·81432	·5	5876·54540	271·74776	76·65862
70	3848·45100	219·91148	62·03588	·75	5910·56296	272·53316	76·88018
·25	3875·98902	220·69688	62·25744	87·	5944·67869	273·31856	77·10174
·5	3903·62522	221·48228	62·47899	·25	5978·89260	274·10395	77·32329
·75	3931·35959	222·26768	62·70055	·5	6013·20468	274·88935	77·54485
71	3959·19214	223·05307	62·92211	·75	6047·61494	275·67475	77·76641
·25	3987·12286	223·83347	63·14366	88·	6082·12337	276·46015	77·98796

Diam.	Area.	Circum.	Side of equal sq.	Diam.	Area.	Circum.	Side of equal sq.
88·25	6116·72994	277·24555	78·20952	94·25	6976·74097	296·09510	83·52688
·5	6151·43476	278·03094	78·43103	·5	7013·80194	296·88050	83·74844
·75	6186·23772	278·81634	78·65263	·75	7050·96109	297·66590	83·97000
89·	6221·13885	279·60174	78·87419	95·	7088·21842	298·45130	84·19155
·25	6256·13815	280·38714	79·09575	·25	7125·57992	299·23670	84·41311
·5	6291·23563	281·17254	79·31730	·5	7163·02759	300·02209	84·63467
·75	6326·43129	281·95794	79·53886	·75	7200·57941	300·80749	84·85622
90·	6361·72512	282·74333	79·76042	96·	7238·22947	301·59289	85·07778
·25	6397·11712	283·52873	79·98198	·25	7275·97767	302·37829	85·29934
·5	6432·60730	284·31413	80·20353	·5	7313·82404	303·16369	85·52089
·75	6468·19536	285·09953	80·42509	·75	7351·76859	303·94908	85·74245
91·	6503·88219	285·88493	80·64669	97·	7389·81131	304·73448	85·96401
·25	6539·66689	286·67032	80·86820	·25	7427·95221	305·51988	86·18556
·5	6575·51577	287·45572	81·08976	·5	7466·19129	306·30528	86·40712
·75	6611·53082	288·24112	81·31132	·75	7504·52853	307·09068	86·62868
92·	6647·61005	289·02652	81·53287	98·	7542·96396	307·87608	86·85023
·25	6683·78745	289·81192	81·75443	·25	7581·49755	308·66147	87·07179
·5	6720·06303	290·59732	81·97599	·5	7620·12933	309·44687	87·29335
·75	6756·43678	291·38271	82·19754	·75	7658·85927	310·23227	87·51490
93·	6792·90871	292·16811	82·41910	99·	7697·68739	311·01767	87·73646
·25	6829·47881	292·95351	82·64066	·25	7736·61369	311·80307	87·95802
·5	6866·14709	293·73891	82·86221	·5	7775·63816	312·58846	88·17957
·75	6902·91354	294·52431	83·08377	·75	7814·76081	313·37386	88·40113
94·	6939·77817	295·30970	83·30533	100·	7853·98163	314·15926	88·62269

The following rules are for extending the use of the above table.

To find the area, circumference, or side of equal square, of a circle having a diameter of more than 100 *inches, feet, &c.* *Rule.*—Divide the given diameter by a number that will give a quotient equal to some one of the diameters in the table; then the circumference or side of equal square, opposite that diameter, multiplied by that divisor, or, the area opposite that diameter, multiplied by the square of the aforesaid divisor, will give the answer.

Example.—What is the circumference of a circle whose diameter is 228 feet? 228, divided by 3, gives 76, a diameter of the table, the circumference of which is 238·761, therefore:

238·761
3

716·283 feet. Ans.

Another example.—What is the area of a circle having a diameter of 150 inches? 150, divided by 10, gives 15, one of the diameters in the table, the area of which is 176·71458, therefore:

176·71458
100 = 10 × 10

17,671·45800 inches. Ans.

To find the area, circumference, or side of equal square, of a circle having an intermediate diameter to those in the table. *Rule.*—Multiply the given diameter by a number that will give a product equal to some one of the diameters in the table; then the circumference or side of equal square opposite that diameter, divided by that multiplier, or, the area opposite that diameter divided by the square of the aforesaid multiplier, will give the answer.

Example.—What is the circumference of a circle whose diameter is 6⅛, or 6·125 inches? 6·125, multiplied by 2, gives 12·25, one of the diameters of the table, whose circumference is 38·484, therefore:

$$\frac{2)38\cdot484}{19\cdot242}$$ inches. Ans.

Another example.—What is the area of a circle, the diameter of which is 3·2 feet? 3·2, multiplied by 5, gives 16, and the area of 16 is 201·0619, therefore:

```
5 × 5 = 25)201·0619(8·0424 + feet. Ans.
           200
           ---
             106
             100
             ---
               61
               50
               ---
               119
               100
               ---
                19
```

Note.—The diameter of a circle, multiplied by 3·14159, will give its circumference; the square of the diameter, multiplied by ·78539, will give its area; and the diameter, multiplied by ·88622, will give the side of a square equal to the area of the circle.

TABLE SHOWING THE CAPACITY OF WELLS, CISTERNS, &C.

The gallon of the State of New York, by an act passed April 11, 1851, is required to conform to the standard gallon of the United States government. This standard gallon contains 231 cubic inches. In conformity with this standard the following table has been computed.

One foot in depth of a cistern of

Diameter				Gallons	
3	feet diameter will contain		. . . . .	52·872	gallons.
3½	"	"	. . . .	71·965	"
4	"	"	. . . .	93·995	"
4½	"	"	. . . .	118·963	"
5	"	"	. . . .	146·868	"
5½	"	"	. . . .	177·710	"
6	"	"	. . . .	211·490	"
6½	"	"	. . . .	248·207	"
7	"	"	. . . .	287·861	"
8	"	"	. . . .	375·982	"
9	"	"	. . . .	475·852	"
10	"	"	. . . .	587·472	"
12	"	"	. . . .	845·959	"

Note—To reduce cubic feet to gallons, multiply by 7·48.

TABLE OF POLYGONS.

(From Gregory's Mathematics.)

No. of sides.	Names.	Multipliers for areas.	Radius of circum. circle.	Factors for sides.
3	Trigon - -	0·4330127	0·5773503	1·732051
4	Tetragon, or Square	1·0000000	0·7071068	1·414214
5	Pentagon - -	1·7204774	0·8506508	1·175570
6	Hexagon - -	2·5980762	1·0000000	1·000000
7	Heptagon - -	3·6339124	1·1523824	0·867767
8	Octagon - -	4·8284271	1·3065628	0·765367
9	Nonagon - -	6·1818242	1·4619022	0·684040
10	Decagon - -	7·6942088	1·6180340	0·618034
11	Undecagon -	9·3656399	1·7747324	0·563465
12	Dodecagon - -	11·1961524	1·9318517	0·517638

To find the area of any regular polygon, whose sides do not exceed twelve. Rule.—Multiply the square of a side of the given polygon by the number in the column termed *Multipliers for areas,* standing opposite the name of the given polygon, and the product will be the answer. *Example.*—What is the area of a regular heptagon, whose sides measure each 2 feet?

3·6339124
4 = 2 × 2

14·5356496 : Ans.

To find the radius of a circle which will circumscribe any regular polygon given, whose sides do not exceed twelve. Rule.—Multiply a side of the given polygon by the number in the column termed *Radius of circumscribing circle,* standing opposite the name of the given polygon, and the product will give the answer. *Example.*—What is the radius of a circle which will circumscribe a regular pentagon, whose sides measure each 10 feet?

·8506508
10

8·5065080 : Ans.

To find the side of any regular polygon that may be inscribed within a given circle. Rule.—Multiply the radius of the given circle by the number in the column termed *Factors for sides,* standing opposite the name of the given polygon, and the product will be the answer. *Example.*—What is the side of a regular octagon that may be inscribed within a circle, whose radius is 5 feet?

·765367
5

3·826835 : Ans.

WEIGHT OF MATERIALS

Woods.	lbs. in a cubic foot.
Apple,	49
Ash,	45
Beach,	40
Birch,	45
Box,	60
Cedar,	28
Virginian red cedar,	40
Cherry,	38
Sweet chestnut,	36
Horse-chestnut,	34
Cork,	15
Cypress,	28
Ebony,	83
Elder,	43
Elm,	34
Fir, (white spruce,)	29
Hickory,	52
Lance-wood,	59
Larch,	31
Larch, (whitewood,)	22
Lignum-vitæ,	83
Logwood,	57
St. Domingo mahogany,	45
Honduras, or bay mahogany,	35
Maple,	47
White oak,	43 to 53
Canadian oak,	54
Red oak,	47
Live oak,	76
White pine,	23 to 30
Yellow pine,	34 to 44
Pitch pine,	46 to 58
Poplar,	25
Sycamore,	36
Walnut,	40

Metals.	lbs. in a cubic foot.
Wire-drawn brass,	534
Cast brass,	506
Sheet-copper,	549
Pure cast gold,	1210
Bar-iron,	475 to 487
Cast iron,	450 to 475
Milled lead,	713
Cast lead,	709
Pewter,	453
Pure platina,	1345
Pure cast silver,	654
Steel,	486 to 490
Tin,	456
Zinc,	439
Stone, Earths, &c.	
Brick, Phila. stretchers,	105
North river common hard brick,	107
Do. salmon brick,	100
Brickwork, about	95
Cast Roman cement,	100
Do. and sand in equal parts,	113
Chalk,	144 to 166
Clay,	115
Potter's clay,	112 to 130
Common earth,	95 to 124
Flint,	163
Plate-glass,	172
Crown-glass,	157
Granite,	158 to 187
Quincy granite,	166
Gravel,	109
Grindstone,	134
Gypsum, (Plaster-stone,)	142
Unslaked lime,	52

	lbs. in a cubic foot.		lbs. in a cubic foot.
Limestone, - -	118 to 198	Common blue stone, -	160
Marble, - -	161 to 177	Silver-gray flagging, -	185
New mortar, - - -	107	Stonework, about, -	120
Dry mortar, - -	90	Common plain tiles, -	115
Mortar with hair, (Plastering,) - - - -	105	*Sundries.*	
Do. dry, - -	86	Atmospheric air, -	0·075
Do. do. including lath and nails, from 7 to 11 lbs. per superficial foot.		Yellow beeswax, - -	60
		Birch-charcoal, - -	34
		Oak-charcoal, - -	21
		Pine-charcoal, - -	17
Crystallized quartz, -	165	Solid gunpowder, - -	109
Pure quartz-sand, -	171	Shaken gunpowder, -	58
Clean and coarse sand,	100	Honey, - - - -	90
Welsh slate, - - -	180	Milk, - - - -	64
Paving stone, - -	151	Pitch, - - - - -	71
Pumice stone, - -	56	Sea-water, - -	64
Nyack brown stone, -	148	Rain-water, - - -	62·5
Connecticut brown stone,	170	Snow, - - -	8
Tarrytown blue stone, -	171	Wood-ashes, - -	58

THE END.

A TREATISE UPON THE THEORY OF THE CONSTRUCTION OF BRIDGES AND ROOFS. Illustrated by 124 Wood Engravings.

One vol. 8vo, cloth, $3.00.

This work is intended to be a logical development of the theory of the strains in trussed girders. It begins with the simplest structures, and passes, step by step, to the more complex, including in its consideration an analysis of all the important trussed bridges which are now in common use.

The chapter on Roofs is intended to be sufficiently comprehensive to enable a person who is acquainted with ordinary mathematics to proportion a structure correctly.

The last chapter reverses the process of the preceding chapters, and deduces all the consequences from the fundamental equations of statics.

The work is supposed to be especially adapted to the wants of those who desire to establish themselves on correct principles in regard to the elements of Bridge Construction.

JOHN WILEY & SON'S

LIST OF PUBLICATIONS,

15 ASTOR PLACE,

Under the Mercantile Library and Trade Salesrooms.

AGRICULTURE.

DOWNING. **FRUITS AND FRUIT-TREES OF AMERICA**; or the Culture, Propagation, and Management in the Garden and Orchard, of Fruit-trees generally, with descriptions of all the finest varieties of Fruit, Native and Foreign, cultivated in this country. By A. J. Downing. Second revision and correction, with large additions. By Chas. Downing. 1 vol. 8vo, over 1100 pages, with several hundred outline engravings. Price, with Supplement for 1872..........................$5 00

"As a work of reference it has no equal in this country, and deserves a place in the Library of every Pomologist in America."—*Marshall P. Wilder.*

" **ENCYCLOPEDIA OF FRUITS**; or, Fruits and Fruit-Trees of America. Part 1.—APPLES. With an Appendix containing many new varieties, and brought down to 1872. By Chas. Downing. With numerous outline engravings. 8vo, full cloth..........................$2 50

" **ENCYCLOPEDIA OF FRUITS**; or, Fruits and Fruit-Trees of America. Part 2.—CHERRIES, GRAPES, PEACHES, PEARS, &c. With an Appendix containing many new varieties, and brought down to 1872. By Chas. Downing. With numerous outline engravings. 8vo, full cloth.........$2 50

" **FRUITS AND FRUIT-TREES OF AMERICA.** By A. J. Downing. First revised edition. By Chas. Downing. 12mo, cloth..........................$2 00

" **SELECTED FRUITS.** From Downing's Fruits and Fruit-Trees of America. With some new varieties, including their Culture, Propagation, and Management in the Garden and Orchard, with a Guide to the selection of Fruits, with reference to the Time of Ripening. By Chas. Downing. Illustrated with upwards of four hundred outlines of Apples, Cherries, Grapes, Plums, Pears, &c. 1 vol., 12mo....$2 50

" **LOUDON'S GARDENING FOR LADIES, AND COMPANION TO THE FLOWER-GARDEN.** Second American from third London edition. Edited by A. J. Downing. 1 vol., 12mo $2 00

DOWNING & LINDLEY. **THE THEORY OF HORTICULTURE.** By J. Lindley. With additions by A. J. Downing. 12mo, cloth.......$2 00

DOWNING. **COTTAGE RESIDENCES.** A Series of Designs for Rural Cottages and Cottage Villas, with Garden Grounds. By A. J. Downing. Containing a revised List of Trees, Shrubs, and Plants, and the most recent and best selected Fruit, with some account of the newer style of Gardens. By Henry Winthrop Sargent and Charles Downing. With many new designs in Rural Architecture. By George E. Harney, Architect. 1 vol. 4to..........................$6 00

DOWNING & WIGHTWICK. **HINTS TO PERSONS ABOUT BUILDING IN THE COUNTRY.** By A. J. Downing. And **HINTS TO YOUNG ARCHITECTS,** calculated to facilitate their practical operations. By George Wightwick, Architect. Wood engravings. 8vo, cloth.......................$2 00

KEMP. **LANDSCAPE GARDENING; or, How to Lay Out a Garden.** Intended as a general guide in choosing, forming, or improving an estate (from a quarter of an acre to a hundred acres in extent), with reference to both design and execution. With numerous fine wood engravings. By Edward Kemp. 1 vol. 12mo, cloth.........................$2 50

LIEBIG. **CHEMISTRY IN ITS APPLICATION TO AGRICULTURE, &c.** By Justus Von Liebig. 12mo, cloth....$1 00

" **LETTERS ON MODERN AGRICULTURE.** By Baron Von Liebig. Edited by John Blyth, M.D. With addenda by a practical Agriculturist, embracing valuable suggestions, adapted to the wants of American Farmers. 1 vol. 12mo. cloth...$1 00

" **PRINCIPLES OF AGRICULTURAL CHEMISTRY,** with special reference to the late researches made in England. By Justus Von Liebig. 1 vol. 12mo ....75 cents.

PARSONS. **HISTORY AND CULTURE OF THE ROSE.** By S. B. Parsons. 1 vol. 12mo $1 25

ARCHITECTURE.

DOWNING. **COTTAGE RESIDENCES; or, a Series of Designs for Rural Cottages and Cottage Villas and their Gardens and Grounds,** adapted to North America. By A. J. Downing. Containing a revised List of Trees, Shrubs, Plants, and the most recent and best selected Fruits. With some account of the newer style of Gardens, by Henry Wentworth Sargent and Charles Downing. With many new designs in Rural Architecture by George E. Harney, Architect...........................$6 00

DOWNING & WIGHTWICK. **HINTS TO PERSONS ABOUT BUILDING IN THE COUNTRY.** By A. J. Downing. And **HINTS TO YOUNG ARCHITECTS,** calculated to facilitate their practical operations. By George Wightwick, Architect. With many wood-cuts. 8vo, cloth..................$2 00

HATFIELD. **THE AMERICAN HOUSE CARPENTER.** A Treatise upon Architecture, Cornices, and Mouldings, Framing, Doors, Windows, and Stairs; together with the most important principles of Practical Geometry. New, thoroughly revised, and improved edition, with about 150 additional pages, and numerous additional plates. By R. G. Hatfield. 1 vol. 8vo...$3 50

NOTICES OF THE WORK.

"The clearest and most thoroughly practical work on the subject."

"This work is a most excellent one, very comprehensive, and lucidly arranged."

"This work commends itself by its practical excellence."

"It is a valuable addition to the library of the architect, and almost indispensable to every scientific master-mechanic."—*R. R. Journal.*

HOLLY. **CARPENTERS' AND JOINERS' HAND-BOOK,** containing a Treatise on Framing, Roofs, etc., and useful Rules and Tables. By H. W. Holly. 1 vol. 18mo, cloth........$0 75

" **THE ART OF SAW-FILING SCIENTIFICALLY TREATED AND EXPLAINED.** With Directions f putting in order all kinds of Saws. By H. W. Holly. 18m cloth...$0 7.

RUSKIN **SEVEN LAMPS OF ARCHITECTURE.** 1 vol. 12mo, cloth, plates..$1 75

RUSKIN. **LECTURES ON ARCHITECTURE AND PAINTING.** 1 vol. 12mo, cloth, plates........................$1 50

" **LECTURE BEFORE SOCIETY OF ARCHITECTS.** 0 15

WOOD. **A TREATISE ON THE RESISTANCE OF MATERIALS,** and an Appendix on the Preservation of Timber. By De Volson Wood, Prof. of Engineering, University of Michigan. 1 vol. 8vo, cloth........................$2 50

This work is used as a Text-Book in Iowa University, Iowa Agricultural College, Illinois Industrial University, Sheffield Scientific School, New Haven, Cooper Institute, New York, Polytechnic College, Brooklyn, University of Michigan, and other institutions.

" **A TREATISE ON BRIDGES.** Designed as a Text-book and for Practical Use. By De Volson Wood. 1 vol. 8vo, numerous illustrations, cloth (shortly)..................$3 00

ASSAYING—ASTRONOMY.

BODEMANN. **A TREATISE ON THE ASSAYING OF LEAD, SILVER, COPPER, GOLD, AND MERCURY.** By Bodemann and Kerl. Translated by W. A. Goodyear. 1 vol. 12mo, cloth..$2 50

MITCHELL. **A MANUAL OF PRACTICAL ASSAYING.** By John Mitchell. Third edition, edited by William Crookes. 1 vol. thick 8vo, cloth..................................$10 00

NORTON. **A TREATISE ON ASTRONOMY, SPHERICAL AND PHYSICAL,** with Astronomical Problems and Solar, Lunar, and other Astronomical Tables for the use of Colleges and Scientific Schools. By William A. Norton. Fourth edition, revised, remodelled, and enlarged. Numerous plates. 8vo, cloth..$3 50

BIBLES, &c.

BAGSTER. **THE COMMENTARY WHOLLY BIBLICAL.** Contents: —The Commentary: an Exposition of the Old and New Testaments in the very words of Scripture. 2264 pp. II. An outline of the Geography and History of the Nations mentioned in Scripture. III. Tables of Measures, Weights, and Coins. IV. An Itinerary of the Children of Israel from Egypt to the Promised Land. V. A Chronological comparative Table of the Kings and Prophets of Israel and Judah. VI. A Chart of the World's History from Adam to the Third Century, A. D. VII. A complete Series of Illustrative Maps. IX. A Chronological Arrangement of the Old and New Testaments. X. An Index to Doctrines and Subjects, with numerous Selected Passages, quoted in full. XI. An Index to the Names of Persons mentioned in Scripture. XII. An Index to the Names of Places found in Scripture. XIII. The Names, Titles, and Characters of Jesus Christ our Lord, as revealed in the Scriptures, methodically arranged.

2 volumes 4to, cloth.................................$19 50
2 volumes 4to, half morocco, gilt edges.............. 26 00
2 volumes 4to, morocco, gilt edges................... 35 00
3 volumes 4to, cloth................................. 20 00
3 volumes 4to, half morocco, gilt edges.............. 33 00
3 volumes 4to, morocco, gilt edges................... 40 00

BLANK-PAGED BIBLE. **THE HOLY SCRIPTURES OF THE OLD AND NEW TESTAMENTS;** with copious references to parallel and illustrative passages, and the alternate pages ruled for MS. notes.

This edition of the Scriptures contains the Authorized Version, illustrated by the references of "Bagster's Polyglot Bible," and enriched with accurate maps, useful tables, and an Index of Subjects.

1 vol. 8vo, half morocco..............................$10 00
1 vol. 8vo, morocco.................................. 15 00

THE TREASURY BIBLE. Containing the authorized English version of the Holy Scriptures, interleaved with a Treasury of more than 500,000 Parallel Passages from Canne, Brown, Blayney, Scott, and others. With numerous illustrative notes.

1 vol., half bound....................................$7 50
1 vol., morocco..10 00

COMMON PRAYER, 48mo Size.

(Done in London expressly for us.)

COMMON PRAYER.
No. 1. Gilt and red edges, imitation morocco...........$0 62½
No. 2. Gilt and red edges, rims.......................87½
No. 3. Gilt and red edges, best morocco and calf....... 1 25
No. 4. Gilt and red edges, best morocco and calf, rims.. 1 50

BOOK-KEEPING.

JONES. **BOOKKEEPING AND ACCOUNTANTSHIP.** Elementary and Practical. In two parts, with a Key for Teachers. By Thomas Jones, Accountant and Teacher. 1 volume 8vo, cloth....................................$2 50

" **BOOKKEEPING AND ACCOUNTANTSHIP.** School Edition. By Thomas Jones. 1 vol. 8vo, half roan.......$1 50

" **BOOKKEEPING AND ACCOUNTANTSHIP.** Set of Blanks. In 6 parts. By Thomas Jones.............$1 50

" **BOOKKEEPING AND ACCOUNTANTSHIP.** Double Entry; Results obtained from Single Entry; Equation of Payments, etc. By Thomas Jones. 1 vol. thin 8vo...$0 75

CHEMISTRY.

CRAFTS. **A SHORT COURSE IN QUALITATIVE ANALYSIS;** with the new notation. By Prof. J. M. Crafts. Second edition. 1 vol. 12mo, cloth..........................$1 50

JOHNSON'S FRESENIUS. **A MANUAL OF QUALITATIVE CHEMICAL ANALYSIS.** By C. R. Fresenius. Edited by S. W. Johnson, Professor in Sheffield Scientific School, Yale College. With Chemical Notation and Nomenclature, old and new. 1 vol. 8vo, cloth....................................$4 50

" **A SYSTEM OF INSTRUCTION IN QUANTITATIVE CHEMICAL ANALYSIS.** By C. R. Fresenius. From latest editions, edited, with additions, by Prof. S. W. Johnson. With Chemical Notation and Nomenclature, old and new....................................$6 00

KIRKWOOD **COLLECTION OF REPORTS (CONDENSED) AND OPINIONS OF CHEMISTS IN REGARD TO THE USE OF LEAD PIPE FOR SERVICE PIPE,** in the Distribution of Water for the Supply of Cities. By Jas. P. Kirkwood. 8vo, cloth..........................$1 50

MILLER. **ELEMENTS OF CHEMISTRY, THEORETICAL AND PRACTICAL.** By Wm. Allen Miller. 3 vols. 8vo..$18 00

" Part I.—**CHEMICAL PHYSICS.** 1 vol. 8vo.......... $4 00

" Part II.—**INORGANIC CHEMISTRY.** 1 vol. 8vo..... 6 00

" Part III.—**ORGANIC CHEMISTRY.** 1 vol. 8vo.......10 00

"Dr. Miller's Chemistry is a work of which the author has every reason to feel proud. It is now by far the largest and most accurately written Treatise on Chemistry in the English language," etc.—*Dublin Med. Journal.*

" **MAGNETISM AND ELECTRICITY.** By Wm. Allen Miller. 1 vol. 8vo....................................$2 50

MUSPRATT. **CHEMISTRY—THEORETICAL, PRACTICAL, AND ANALYTICAL**—as applied and relating to the Arts and Manufactures. By Dr. Sheridan Muspratt. 2 vols. 8vo, cloth, $18.00; half russia........................$24 00

NOAD. **A MANUAL OF QUALITATIVE AND QUANTITATIVE CHEMICAL ANALYSIS.** For the use of Students. By H. M. Noad, author of "Manual of Electricity." 1 vol. 12mo. (London.) Complete........................$6 00

" **QUANTITATIVE ANALYSIS.** 1 vol. cloth.......... 4 00

PERKINS. **AN ELEMENTARY MANUAL OF QUALITATIVE CHEMICAL ANALYSIS.** By Maurice Perkins. 12mo, cloth........................$1 00

DRAWING AND PAINTING.

BOUVIER AND OTHERS. **HANDBOOK ON OIL PAINTING.** Handbook of Young Artists and Amateurs in Oil Painting; being chiefly a condensed compilation from the celebrated Manual of Bouvier, with additional matter selected from the labors of Merriwell, De Montalbert, and other distinguished Continental writers on the art. In 7 parts. Adapted for a Text-Book in Academies of both sexes, as well as for self-instruction. Appended, a new Explanatory and Critical Vocabulary. By an American Artist. 12mo, cloth........................$2 00

COE. **NEW SERIES OF DRAWING CARDS,** by Benj. H. Coe, viz.:

" **DRAWING FOR LITTLE FOLKS;** or, First Lessons for the Nursery. 30 cards. Neat case........................$0 30

" **FIRST STUDIES IN DRAWING.** Containing Elementary Exercises, Drawings from Objects, Animals, and Rustic Figures. Complete in *three numbers* of 18 cards each, in neat case. Each........................$0 30

" **COTTAGES.** An Introduction to Landscape Drawing. *Containing 72 Studies.* Complete in four numbers of 18 cards each, in neat case. Each........................$0 30

" **EASY LESSONS IN LANDSCAPE.** Complete in four numbers of 40 Studies each. In neat 8vo case. Each, $0 30

" **HEADS, ANIMALS, AND FIGURES.** Adapted to Pencil Drawing. Complete in three numbers of 10 Studies each. In neat 8vo case. Each........................$0 30

" **COPY BOOK, WITH INSTRUCTIONS**............$0 37½

RUSKIN. **THE ELEMENTS OF DRAWING.** In Three Letters to Beginners. By John Ruskin. 1 vol. 12mo............$1 00

" **THE ELEMENTS OF PERSPECTIVE.** Arranged for the use of Schools. By John Ruskin........................$1 00

SMITH. **A MANUAL OF TOPOGRAPHICAL DRAWING.** By Prof. R. S. Smith. Second edition. 1 vol. 8vo, cloth, plates........................$2 00

" **MANUAL OF LINEAR PERSPECTIVE.** Form, Shade, Shadow, and Reflection. By Prof. R. S. Smith. 1 vol. 8vo, plates, cloth........................$2 00

WARREN. **CONSTRUCTIVE GEOMETRY AND INDUSTRIAL DRAWING.** By S. Edward Warren, Professor in the Massachusetts Institute of Technology, Boston:—

I. ELEMENTARY WORKS.

1. ELEMENTARY FREE-HAND GEOMETRICAL DRAWING. A series of progressive exercises on regular lines and forms including systematic instruction in lettering; a training of the eye and hand for all who are learning to draw. 12mo, cloth, many cuts........................ 75 cts.
Vols. 1 and 3, bound in 1 vol........................$1 75

ELEMENTARY WORKS.—Continued.

WARREN

2. PLANE PROBLEMS IN ELEMENTARY GEOMETRY. With numerous wood-cuts. 12mo, cloth..................$1 25

3. DRAFTING INSTRUMENTS AND OPERATIONS. Containing full information about all the instruments and materials used by the draftsmen, with full directions for their use. With plates and wood-cuts. One vol. 12mo, cloth, $1 25

4. ELEMENTARY PROJECTION DRAWING. Revised and enlarged edition. In five divisions. This and the last volume are favorite text-books, especially valuable to all Mechanical Artisans, and are particularly recommended for the use of all higher public and private schools. New revised and enlarged edition, with numerous wood-cuts and plates. (1872.) 12mo, cloth..$1 50

5. ELEMENTARY LINEAR PERSPECTIVE OF FORMS AND SHADOWS. Part I.—Primitive Methods, with an Introduction. Part II.—Derivative Methods, with Notes on Aerial Perspective, and many Practical Examples. Numerous wood-cuts. 1 vol. 12mo, cloth..........................$1 00

II. HIGHER WORKS.

These are designed principally for Schools of Engineering and Architecture, and for the members generally of those professions; and the first three are also designed for use in those colleges which provide courses of study adapted to the preliminary general training of candidates for the scientific professions, as well as for those technical schools which undertake that training themselves.

1. GENERAL PROBLEMS OF ORTHOGRAPHIC PROJECTIONS. The foundation course for the subsequent theoretical and practical works. A new edition of this work will soon appear.

2. GENERAL PROBLEMS OF SHADES AND SHADOWS. A wider range of problems than can elsewhere be found in English, and the principles of shading. 1 vol. 8vo, with numerous plates. Cloth.......................... $3 50

3. HIGHER LINEAR PERSPECTIVE. Distinguished by its concise summary of various methods of perspective construction; a full set of standard problems, and a careful discussion of special higher ones. With numerous large plates 8vo, cloth..$4 00

4. ELEMENTS OF MACHINE CONSTRUCTION AND DRAWING; or, Machine Drawings. With some elements of descriptive and rational cinematics. A Text-Book for Schools of Civil and Mechanical Engineering, and for the use of Mechanical Establishments, Artisans, and Inventors. Containing the principles of gearings, screw propellers, valve motions, and governors, and many standard and novel examples, mostly from present American practice. By S. Edward Warren. 2 vols. 8vo. 1 vol. text and cuts, and 1 vol. large plates......$7 50

A FEW FROM MANY TESTIMONIALS.

"It seems to me that your Works only need a thorough examination to be introduced and permanently used in all the Scientific and Engineering Schools." —Prof. J. G. FOX. *Collegiate and Engineering Institute, New York City.*

"I have used several of your Elementary Works, and believe them to be better adapted to the purposes of instruction than any others with which I am acquainted."—H. F. WALLING, *Prof. of Civil and Topographical Engineering, Lafayette College, Easton, Pa.*

"Your Works appear to me to fill a very important gap in the literature of the subjects treated. Any effort to draw Artisans, etc., away from the 'rule of thumb,' and give them an insight into principles, is in the right direction, and meets my heartiest approval. This is the distinguishing feature of your Elementary Works."—Prof. H. L. EUSTIS, *Lawrence Scientific School, Cambridge, Mass.*

"The author has happily divided the subjects into two great portions; the former embracing those processes and problems proper to be taught to all students in Institutions of Elementary Instruction; the latter, those suited to advanced students preparing for technical purposes. The Elementary Books ought to be used in all High Schools and Academies; the Higher ones in Schools of Technology."—WM. W. FOLWELL, *President of University of Minnesota.*

DYEING, &c.

MACFARLANE. **A PRACTICAL TREATISE ON DYEING AND CALICO-PRINTING.** Including the latest Inventions and Improvements. With an Appendix, comprising definitions of chemical terms, with tables of Weights, Measures, &c. By an experienced Dyer. With a supplement, containing the most recent discoveries in color chemistry. By Robert Macfarlane. 1 vol. 8vo...$5 00

REIMANN. **A TREATISE ON THE MANUFACTURE OF ANILINE AND ANILINE COLORS.** By M. Reimann. To which is added the Report on the Coloring Matters derived from Coal Tar, as shown at the French Exhibition, 1867. By Dr. Hofmann. Edited by Wm. Crookes. 1 vol. 8vo, cloth, $2 50

"Dr. Reimann's portion of the Treatise, written in concise language, is profoundly practical, giving the minutest details of the processes for obtaining all the more important colors, with woodcuts of apparatus. Taken in conjunction with Hofmann's Report, we have now a complete history of Coal Tar Dyes, both theoretical and practical."—*Chemist and Druggist.*

ENGINEERING.

AUSTIN. **A PRACTICAL TREATISE ON THE PREPARATION, COMBINATION, AND APPLICATION OF CALCAREOUS AND HYDRAULIC LIMES AND CEMENTS.** To which is added many useful recipes for various scientific, mercantile, and domestic purposes. By James G. Austin. 1 vol. 12mo...$2 00

COLBURN. **LOCOMOTIVE ENGINEERING AND THE MECHANISM OF RAILWAYS.** A Treatise on the Principles and Construction of the Locomotive Engine, Railway Carriages, and Railway Plant, with examples. Illustrated by Sixty-four large engravings and two hundred and forty woodcuts. By Zerah Colburn. Complete, 20 parts, or 2 vols. cloth..$20 00
Or, half morocco, gilt top.................... .... 25 00

KNIGHT. **THE MECHANICIAN AND CONSTRUCTOR FOR ENGINEERS.** Comprising Forging, Planing, Lining, Slotting, Shaping, Turning, Screw-cutting, &c. Illustrated with ninety-six plates. By Cameron Knight. 1 vol. 4to, half morocco...$15 00

MAHAN. **AN ELEMENTARY COURSE OF CIVIL ENGINEERING,** for the use of the Cadets of the U. S. Military Academy. By H. P. Mahan. 1 vol. 8vo, with numerous woodcuts. New edition, with large Addenda, &c. Full cloth.....$4 00

" **DESCRIPTIVE GEOMETRY,** as applied to the Drawing of Fortifications and Stone-Cutting. For the use of the Cadets of the U. S. Military Academy. By Prof. D. H. Mahan. 1 vol. 8vo. Plates...................................$1 50

" **INDUSTRIAL DRAWING.** Comprising the Description and Uses of Drawing Instruments, the Construction of Plane Figures, the Projections and Sections of Geometrical Solids, Architectural Elements, Mechanism, and Topographical Drawing. With remarks on the method of Teaching the subject. For the use of Academies and Common Schools. By Prof. D. H. Mahan. 1 vol. 8vo. Twenty steel plates. Full cloth...$3 00

" **A TREATISE ON FIELD FORTIFICATIONS.** Containing instructions on the Methods of Laying Out, Constructing, Defending, and Attacking Entrenchments. With the General Outlines, also, of the Arrangement, the Attack, and Defence of Permanent Fortifications. By Prof. D. H. Mahan. New edition, revised and enlarged. 1 vol. 8vo, full cloth, with plates...$3 50

" **ELEMENTS OF PERMANENT FORTIFICATIONS.** By Prof. D. H. Mahan. 1 vol. 8vo, with numerous large plates. Cloth...$6 50

MAHAN. **ADVANCED GUARD, OUT-POST, and Detachment Service** of Troops, with the Essential Principles of Strategy and Grand Tactics. For the use of Officers of the Militia and Volunteers. By Prof. D. H. Mahan. New edition, with large additions and 12 plates. 1 vol. 18mo, cloth......$1 50

MAHAN & MOSELY. **MECHANICAL PRINCIPLES OF ENGINEERING AND ARCHITECTURE.** By Henry Mosely, M.A., F.R.S. From last London edition, with considerable additions, by Prof. D. H. Mahan, LL.D., of the U. S. Military Academy. 1 vol. 8vo, 700 pages. With numerous cuts. Cloth...$5 00

MAHAN & BRESSE. **HYDRAULIC MOTORS.** Translated from the French Cours de Mecanique, appliquée par M. Bresse. By Lieut. F. A. Mahan, and revised by Prof. D. H. Mahan. 1 vol. 8vo, plates..$2 50

WOOD. **A TREATISE ON THE RESISTANCE OF MATERIALS,** and an Appendix on the Preservation of Timber. By De Volson Wood, Professor of Engineering, University of Michigan. 1 vol. 8vo, cloth...........................$2 50

A TREATISE ON BRIDGES. Designed as a Text-book and for Practical Use. By De Volson Wood. 1 vol. 8vo, numerous illustrations, cloth (shortly)......................$3 00

GREEK.

BAGSTER. **GREEK TESTAMENT, ETC.** The Critical Greek and English New Testament in Parallel Columns, consisting of the Greek Text of Scholz, readings of Griesbach, etc., etc. 1 vol. 18mo, half morocco..........................$3 00

" —— do. Full morocco, gilt edges.................. 4 50

" —— With Lexicon, by T. S. Green. Half-bound........ 4 50

" —— do. Full morocco, gilt edges.................. 6 00

" —— do. With Concordance and Lexicon. Half mor., 6 00

" —— do. Limp morocco.............................. 7 50

" **THE ANALYTICAL GREEK LEXICON TO THE NEW TESTAMENT.** In which, by an alphabetical arrangement, is found every word in the Greek text *in every form in which it appears*—that is to say, every occurrent person, number, tense or mood of verbs, every case and number of nouns, pronouns, &c., is placed in its alphabetical order, fully explained by a careful grammatical analysis and referred to its root, so that no uncertainty as to the grammatical structure of any word can perplex the beginner, but, assured of the precise grammatical force of any word he may desire to interpret, he is able immediately to apply his knowledge of the English meaning of the root with accuracy and satisfaction. 1 vol. small 4to, half bound.................................$6 50

" **GREEK-ENGLISH LEXICON TO TESTAMENT.** By T. S. Green. Half morocco $1 50

HEBREW.

GREEN. **A GRAMMAR OF THE HEBREW LANGUAGE.** With copious Appendixes. By W. H. Green, D.D., Professor in Princeton Theological Seminary. 1 vol. 8vo, cloth....$3 50

" **AN ELEMENTARY HEBREW GRAMMAR.** With Tables, Reading Exercises, and Vocabulary. By Prof. W. H. Green, D.D. 1 vol. 12mo, cloth...........................$1 50

" **HEBREW CHRESTOMATHY;** or, Lessons in Reading and Writing Hebrew. By Prof. W. H. Green, D.D. 1 vol. 8vo, cloth...$2 00

LETTERIS. **A NEW AND BEAUTIFUL EDITION OF THE HEBREW BIBLE.** Revised and carefully examined by Myer Levi Letteris. 1 vol. 8vo, with key, marble edges.....$2 50

"This edition has a large and much more legible type than the known one volume editions, and the print is excellent, while the name of LETTERIS is a sufficient guarantee for correctness." —*Rev. Dr. J. M.* WISE, *Editor of the* ISRAELITE.

BAGSTER'S GESENIUS. **BAGSTER'S COMPLETE EDITION OF GESENIUS' HEBREW AND CHALDEE LEXICON.** In large, clear, and perfect type. Translated and edited with additions and corrections, by S. P. Tregelles, LL.D.

In this edition great care has been taken to guard the student from Neologian tendencies by suitable remarks whenever needed.

"The careful revisal to which the Lexicon has been subjected by a faithful and Orthodox translator exceedingly enhances the practical value of this edition." —*Edinburgh Ecclesiastical Journal.*

Small 4to, half bound........................... $7 50

BAGSTER'S **NEW POCKET HEBREW AND ENGLISH LEXICON.** The arrangement of this Manual Lexicon combines two things—the etymological order of roots and the alphabetical order of words. This arrangement tends to lead the learner onward; for, as he becomes more at home with roots and derivatives, he learns to turn at once to the root, without first searching for the particular word in its alphabetic order. 1 vol. 18mo, cloth........................... $2 00

"This is the most beautiful, and at the same time the most correct and perfect Manual Hebrew Lexicon we have ever used."—*Eclectic Review.*

IRON, METALLURGY, &c.

BODEMANN. **A TREATISE ON THE ASSAYING OF LEAD, SILVER, COPPER, GOLD, AND MERCURY.** By Bodemann & Kerl. Translated by W. A. Goodyear. 1 vol. 12mo, $2 50

CROOKES. **A PRACTICAL TREATISE ON METALLURGY.** Adapted from the last German edition of Prof. Kerl's Metallurgy. By William Crookes and Ernst Rohrig. In three vols. thick 8vo. Price........................... $30 00

Separately. Vol. 1. Lead, Silver, Zinc, Cadmium, Tin, Mercury, Bismuth, Antimony, Nickel, Arsenic, Gold, Platinum, and Sulphur........................... $10 00

Vol. 2. Copper and Iron........................... 10 00

Vol. 3. Steel, Fuel, and Supplement........................... 10 00

FAIRBAIRN. **CAST AND WROUGHT IRON FOR BUILDING.** By Wm. Fairbairn. 8vo, cloth........................... $2 00

FRENCH. **HISTORY OF IRON TRADE, FROM 1621 TO 1857.** By B. F. French. 8vo, cloth........................... $2 00

KIRKWOOD. **COLLECTION OF REPORTS (CONDENSED) AND OPINIONS OF CHEMISTS IN REGARD TO THE USE OF LEAD PIPE FOR SERVICE PIPE,** in the Distribution of Water for the Supply of Cities. By I. P. Kirkwood, C.E. 8vo, cloth........................... $1 50

LESLEY. **THE IRON MANUFACTURER'S GUIDE TO THE FURNACES, FORGES, AND ROLLING-MILLS OF THE UNITED STATES.** By J. P. Lesley. With maps and plates. 1 vol. 8vo, cloth........................... $8 00

MACHINISTS—MECHANICS.

FITZGERALD. **THE BOSTON MACHINIST.** A complete School for the Apprentice and Advanced Machinist. By W. Fitzgerald. 1 vol. 18mo, cloth........................... $0 75

HOLLY. **SAW FILING.** The Art of Saw Filing Scientifically Treated and Explained. With Directions for putting in order all kinds of Saws, from a Jeweller's Saw to a Steam Saw-mill. Illustrated by forty-four engravings. Third edition. By H. W. Holly. 1 vol. 18mo, cloth........................... $0 75

KNIGHT. **THE MECHANISM AND ENGINEER INSTRUCTOR.** Comprising Forging, Planing, Lining, Slotting, Shaping, Turning, Screw-Cutting, etc., etc. By Cameron Knight. 1 vol. 4to, half morocco........................... $15 00

TURNING, &c. **LATHE, THE, AND ITS USES, ETC.**; or, Instruction in the Art of Turning Wood and Metal. Including a description of the most modern appliances for the ornamentation of plane and curved surfaces, with a description also of an entirely novel form of *Lathe* for Eccentric and Rose Engine Turning, a Lathe and Turning Machine combined, and other valuable matter relating to the Art. 1 vol. 8vo, copiously illustrated. Including Supplement. 8vo, cloth......$7 00

"The most complete work on the subject ever published."—*American Artisan.*

"Here is an invaluable book to the practical workman and amateur."—*London Weekly Times.*

TURNING, &c. **SUPPLEMENT AND INDEX TO LATHE AND ITS USES.** Large type. Paper. 8vo....................$0 90

WILLIS. **PRINCIPLES OF MECHANISM.** Designed for the use of Students in the Universities and for Engineering Students generally. By Robert Willis, M.D., F.R.S., President of the British Association for the Advancement of Science, &c., &c. Second edition, enlarged. 1 vol. 8vo, cloth...........$7 50

*** It ought to be in every large Machine Workshop Office, in every School of Mechanical Engineering at least, and in the hands of every Professor of Mechanics, &c.—Prof. S. EDWARD WARREN.

MANUFACTURES.

BOOTH. **NEW AND COMPLETE CLOCK AND WATCH MAKERS' MANUAL.** Comprising descriptions of the various gearings, escapements, and Compensations now in use in French, Swiss, and English clocks and watches, Patents, Tools, etc., with directions for cleaning and repairing. With numerous engravings. Compiled from the French, with an Appendix containing a History of Clock and Watch Making in America. By Mary L. Booth. With numerous plates. 1 vol. 12mo, cloth....................................$2 00

GELDARD. **HANDBOOK ON COTTON MANUFACTURE**; or, A Guide to Machine-Building, Spinning, and Weaving. With practical examples, all needful calculations, and many useful and important tables. The whole intended to be a complete yet compact authority for the manufacture of cotton. By James Geldard. With steel engravings. 1 vol. 12mo, cloth.....................................$2 50

MEDICAL, &c.

BULL. **HINTS TO MOTHERS FOR THE MANAGEMENT OF HEALTH DURING THE PERIOD OF PREGNANCY, AND IN THE LYING-IN ROOM.** With an exposure of popular errors in connection with those subjects. By Thomas Bull, M.D. 1 vol. 12mo, cloth...........$1 00

FRANCKE. **OUTLINES OF A NEW THEORY OF DISEASE**, applied to Hydropathy, showing that water is the only true remedy. With observations on the errors committed in the practice of Hydropathy, notes on the cure of cholera by cold water, and a critique on Priessnitz's mode of treatment. Intended for popular use. By the late H. Francke. Translated from the German by Robert Blakie, M.D. 1 vol. 12mo, cloth...$1 50

GREEN. **A TREATISE ON DISEASES OF THE AIR PASSAGES.** Comprising an inquiry into the History, Pathology, Causes, and Treatment of those Affections of the Throat called Bronchitis, Chronic Laryngitis, Clergyman's Sore Throat, etc., etc. By Horace Green, M.D. Fourth edition, revised and enlarged. 1 vol. 8vo, cloth..$3 00

" **A PRACTICAL TREATISE ON PULMONARY TUBERCULOSIS**, embracing its History, Pathology, and Treatment. By Horace Green, M.D. Colored plates. 1 vol. 8vo, cloth..$5 00

GREEN. OBSERVATIONS ON THE PATHOLOGY OF CROUP. With Remarks on its Treatment by Topical Medications. By Horace Green, M.D. 1 vol. 8vo, cloth $1 25

" ON THE SURGICAL TREATMENT OF POLYPI OF THE LARYNX, AND ŒDEMA OF THE GLOTTIS. By Horace Green, M.D. 1 vol. 8vo $1 25

" FAVORITE PRESCRIPTIONS OF LIVING PRACTITIONERS. With a Toxicological Table, exhibiting the Symptoms of Poisoning, the Antidotes for each Poison, and the Test proper for their detection. By Horace Green. 1 vol. 8vo, cloth $2 50

TILT. ON THE PRESERVATION OF THE HEALTH OF WOMEN AT THE CRITICAL PERIODS OF LIFE. By E. G. Tilt, M.D. 1 vol. 18mo, cloth $0 50

VON DUBEN. GUSTAF VON DUBEN'S TREATISE ON MICROSCOPICAL DIAGNOSIS. With 71 engravings. Translated, with additions, by Prof. Louis Bauer, M.D. 1 vol. 8vo, cloth $1 00

MINERALOGY.

BRUSH. ON BLOW-PIPE ANALYSIS. By Prof. Geo. J. Brush. (In preparation.)

DANA. DESCRIPTIVE MINERALOGY. Comprising the most recent Discoveries. Fifth edition. Almost entirely re-written and greatly enlarged. Containing nearly 900 pages 8vo, and upwards of 600 wood engravings. By Prof. J. Dana. Cloth $10 00

"We have used a good many works on Mineralogy, but have met with none that begin to compare with this in fulness of plan, detail, and execution."—*American Journal of Mining.*

DANA & BRUSH. APPENDIX TO DANA'S MINERALOGY, bringing the work down to 1872. By Prof. G. J. Brush. 8vo $0 50

DANA. DETERMINATIVE MINERALOGY. 1 vol. (In preparation.)

" A TEXT-BOOK OF MINERALOGY. 1 vol. (In preparation.)

MISCELLANEOUS.

BAILEY. THE NEW TALE OF A TUB. An adventure in verse. By F. W. N. Bailey. With illustrations. 1 vol. 8vo $0 75

CARLYLE. ON HEROES, HERO-WORSHIP, AND THE HEROIC IN HISTORY. Six Lectures. Reported, with emendations and additions. By Thomas Carlyle. 1 vol. 12mo, cloth ... $0 75

CATLIN. THE BREATH OF LIFE; or, Mal-Respiration and its Effects upon the Enjoyments and Life of Man. By Geo. Catlin. With numerous wood engravings. 1 vol. 8vo, $0 75

CHEEVER. CAPITAL PUNISHMENT. A Defence of. By Rev. George B. Cheever, D.D. Cloth $0 50

" HILL DIFFICULTY, and other Miscellanies. By Rev. George B. Cheever, D.D. 1 vol. 12mo, cloth $1 00

" JOURNAL OF THE PILGRIMS AT PLYMOUTH ROCK. By Geo. B. Cheever, D.D. 1 vol. 12mo, cloth $1 00

" WANDERINGS OF A PILGRIM IN THE ALPS. By George B. Cheever, D.D. 1 vol. 12mo, cloth $1 00

" WANDERINGS OF THE RIVER OF THE WATER OF LIFE. By Rev. Dr. George B. Cheever. 1 vol. 12mo, cloth $1 00

CONYBEARE. ON INFIDELITY. 12mo, cloth 1 00

CHILD'S BOOK OF FAVORITE STORIES. Large colored plates. 4to, cloth $1 50

EDWARDS.	**FREE TOWN LIBRARIES.** The Formation, Management, and History in Britain, France, Germany, and America. Together with brief notices of book-collectors, and of the respective places of deposit of their surviving collections. By Edward Edwards. 1 vol. thick 8vo................$8 00
GREEN.	**THE PENTATEUCH VINDICATED FROM THE ASPERSIONS OF BISHOP COLENSO.** By Wm. Henry Green, Prof. Theological Seminary, Princeton, N. J. 1 vol. 12mo, cloth....................................$1 25
GOURAUD.	**PHRENO-MNEMOTECHNY; or, The Art of Memory.** The series of Lectures explanatory of the principles of the system. By Francis Fauvel-Gouraud. 1 vol. 8vo, cloth, $2 00
"	**PHRENO-MNEMOTECHNIC DICTIONARY.** Being a Philosophical Classification of all the Homophonic Words of the English Language. To be used in the application of the Phreno-Mnemotechnic Principles. By Francis Fauvel-Gouraud. 1 vol. 8vo, cloth.............................$2 00
HEIGHWAY.	**LEILA ADA.** 12mo, cloth............................ 1 00
"	**LEILA ADA'S RELATIVES.** 12mo, cloth........... 1 00
KELLY.	**CATALOGUE OF AMERICAN BOOKS.** The American Catalogue of Books, from January, 1861, to January, 1866. Compiled by James Kelly. 1 vol. 8vo, net cash.......$5 00
"	**CATALOGUE OF AMERICAN BOOKS.** The American Catalogue of Books from January, 1866, to January, 1871. Compiled by James Kelly. 1 vol. 8vo, net...........$7 50
MAVER'S	**COLLECTION OF GENUINE SCOTTISH MELODIES.** For the Piano-Forte or Harmonium, in keys suitable for the voice. Harmonized by C. H. Merine. Edited by Geo. Alexander. 1 vol. 4to, half calf.........................$10 00
NOTLEY.	**A COMPARATIVE GRAMMAR OF THE FRENCH, ITALIAN, SPANISH, AND PORTUGUESE LANGUAGES.** By Edwin A. Notley. 1 vol., cloth......$5 00
PARKER.	**POLAR MAGNETISM.** First and Second Lectures. By John A. Parker. Each....................................$0 25
"	**NON-EXISTENCE OF PROJECTILE FORCES IN NATURE.** By John A. Parker.....................$0 25
STORY OF	**A POCKET BIBLE.** Illustrated. 12mo, cloth........$1 00
TUPPER.	**PROVERBIAL PHILOSOPHY.** 12mo................ 1 00
WALTON & COTTON.	**THE COMPLETE ANGLER; or, The Contemplative Man's Recreation, by Isaac Walton, and Instructions how to Angle for a Trout or Grayling in a Clear Stream, by Charles Cotton, with copious notes, for the most part original. A bibliographical preface, giving an account of fishing and Fishing Books, from the earliest antiquity to the time of Walton, and a notice of Cotton and his writings, by Rev. Dr. Bethune. To which is added an appendix, including the most complete catalogue of books in angling ever printed, &c. Also a general index to the whole work. 1 vol. 12mo, cloth...$3 00**
WARREN.	**NOTES ON POLYTECHNIC OR SCIENTIFIC SCHOOLS IN THE UNITED STATES.** Their Nature, Position, Aims, and Wants. By S. Edward Warren. Paper....$0 40
WILLIAMS.	**THE MIDDLE KINGDOM.** A Survey of the Geography, Government, Education, Social Life, Arts, Religion, &c., of the Chinese Empire and its Inhabitants. With a new map of the Empire. By S. Wells Williams. Fourth edition, in 2 vols..$4 00

RUSKIN'S WORKS.

Uniform in size and style.

RUSKIN. **MODERN PAINTERS.** 5 vols. tinted paper, bevelled boards, plates, in box..$18 00

" **MODERN PAINTERS.** 5 vols. half calf............ 27 00

" " " " without plates....... 12 00

" " " " " half calf, 20 00

Vol. 1.—Part 1. General Principles. Part 2. Truth.
Vol. 2.—Part 3. Of Ideas of Beauty.
Vol. 3.—Part 4. Of Many Things.
Vol. 4.—Part 5. Of Mountain Beauty.
Vol. 5.—Part 6. Leaf Beauty. Part 7. Of Cloud Beauty. Part 8. Ideas of Relation of Invention, Formal. Part 9. Ideas of Relation of Invention, Spiritual.

" **STONES OF VENICE.** 3 vols., on tinted paper, bevelled boards, in box..$7 00

" **STONES OF VENICE.** 3 vols., on tinted paper, half calf..$12 00

" **STONES OF VENICE.** 3 vols., cloth.............. 6 00

Vol. 1.—The Foundations.
Vol. 2.—The Sea Stories.
Vol. 3.—The Fall.

" **SEVEN LAMPS OF ARCHITECTURE.** With illustrations, drawn and etched by the authors. 1 vol. 12mo, cloth, $1 75

" **LECTURES ON ARCHITECTURE AND PAINTING.** With illustrations drawn by the author. 1 vol. 12mo, cloth..$1 50

" **THE TWO PATHS.** Being Lectures on Art, and its Application to Decoration and Manufacture. With plates and cuts. 1 vol. 12mo, cloth..........................$1 25

" **THE ELEMENTS OF DRAWING.** In Three Letters to Beginners. With illustrations drawn by the author. 1 vol. 12mo, cloth..$1 00

" **THE ELEMENTS OF PERSPECTIVE.** Arranged for the use of Schools. 1 vol. 12mo, cloth...................$1 00

" **THE POLITICAL ECONOMY OF ART.** 1 vol. 12mo, cloth..$1 00

" **PRE-RAPHAELITISM.**
NOTES ON THE CONSTRUCTION OF SHEEPFOLDS.
KING OF THE GOLDEN RIVER; or, The Black Brothers. A Legend of Stiria.
} 1 vol. 12mo, cloth, $1 00

RUSKIN. **SESAME AND LILIES.** Three Lectures on Books, Women, &c. 1. Of Kings' Treasuries. 2. Of Queens' Gardens. 3. Of the Mystery of Life. 1 vol. 12mo, cloth..........$1 50

" **AN INQUIRY INTO SOME OF THE CONDITIONS AT PRESENT AFFECTING "THE STUDY OF ARCHITECTURE" IN OUR SCHOOLS.** 1 vol. 12mo, paper..$0 15

" **THE ETHICS OF THE DUST.** Ten Lectures to Little Housewives, on the Elements of Crystallization. 1 vol. 12mo, cloth..$1 25

" **"UNTO THIS LAST."** Four Essays on the First Principles of Political Economy. 1 vol. 12mo, cloth...............$1 00

RUSKIN. **THE CROWN OF WILD OLIVE.** Three Lectures on Work, Traffic, and War. 1 vol. 12mo. cloth.................$1 00

" **TIME AND TIDE BY WEARE AND TYNE.** Twenty-five Letters to a Workingman on the Laws of Work. 1 vol. 12mo, cloth..$1 00

" **THE QUEEN OF THE AIR.** Being a Study of the Greek Myths of Cloud and Storm. 1 vol. 12mo, cloth $1 00

" **LECTURES ON ART.** 1 vol. 12mo, cloth............ 1 00

" **FORS CLAVIGERA.** Letters to the Workmen and Labourers of Great Britain. Part 1. 1 vol. 12mo, cloth, plates, $1 00

" **FORS CLAVIGERA.** Letters to the Workmen and Labourers of Great Britain. Part 2. 1 vol. 12mo, cloth, plates, $1 00

" **MUNERA PULVERIS.** Six Essays on the Elements of Political Economy. 1 vol. 12mo, cloth................$1 00

" **ARATRA PENTELICI.** Six Lectures on the Elements of Sculpture, given before the University of Oxford. By John Ruskin. 12mo, cloth, $1 50, or with plates.........$3 00

" **THE EAGLE'S NEST.** Ten Lectures on the relation of Natural Science to Art. 1 vol. 12mo................$1 50

" **THE POETRY OF ARCHITECTURE:** Villa and Cottage. With numerous plates. 1 vol. 12mo, cloth (shortly)....$2 00

" **FORS CLAVIGERA.** Letters to the Workmen and Laborers of Great Britain. Part 3. 1 vol. 12mo, cloth (shortly).$1 50

BEAUTIFUL PRESENTATION VOLUMES.

Printed on tinted paper, and elegantly bound in crape cloth extra, bevelled boards, gilt head.

RUSKIN. **THE TRUE AND THE BEAUTIFUL IN NATURE, ART, MORALS, AND RELIGION.** Selected from the Works of John Ruskin, A.M. With a notice of the author by Mrs. L. C. Tuthill. Portrait. 1 vol. 12mo, cloth extra, gilt head.....................................$2 50

" **THE TRUE AND THE BEAUTIFUL IN NATURE, ART, MORALS, AND RELIGION.** Selected from the Works of John Ruskin, A.M. With a notice of the author by Mrs. L. C. Tuthill. Portrait. 1 vol. 12mo, plain cloth, $2 00

" **PRECIOUS THOUGHTS:** Moral and Religious. Gathered from the Works of John Ruskin, A.M. By Mrs. C. L. Tuthill. 1 vol. 12mo, cloth extra, gilt head...................$2 00

" **PRECIOUS THOUGHTS:** Moral and Religious. Gathered from the Works of John Ruskin, A.M. By Mrs. L. C. Tuthill. 1 vol. 12mo, plain cloth...........................$1 50

" **SELECTIONS FROM THE WRITINGS OF JOHN RUSKIN.** 1 vol. 12mo, cloth extra, gilt head........$2 50

" **SELECTIONS FROM THE WRITINGS OF JOHN RUSKIN.** 1 vol. 12mo, plain cloth...................$2 00

" **SESAME AND LILIES.** 1 vol. 12mo................$1 75

" **ETHICS OF THE DUST.** 12mo...................... 1 75

" **CROWN OF WILD OLIVE.** 12mo..................... 1 50

RUSKIN'S BEAUTIES.

" **THE TRUE AND BEAUTIFUL**
PRECIOUS THOUGHTS.
CHOICE SELECTIONS.
} 3 vols., in box, cloth extra, gilt head..........$6 00
do., half calf...10 00

RUSKIN'S POPULAR VOLUMES.

CROWN OF WILD OLIVE. SESAME AND LILIES. QUEEN OF THE AIR. ETHICS OF THE DUST. 4 vols. in box, cloth extra, gilt head.................$6 00

RUSKIN'S WORKS.

Revised edition.

RUSKIN. Vol. 1.—**SESAME AND LILIES.** Three Lectures. By John Ruskin, LL.D. 1. Of Kings' Treasuries. 2. Of Queens' Gardens. 3. Of the Mystery of Life. 1 vol. 8vo, cloth, $2 00. Large paper.................................$2 50

" Vol. 2.—**MUNERA PULVERIS.** Six Essays on the Elements of Political Economy. By John Ruskin. 1 volume 8vo, cloth...$2 00
Large paper... 2 50

" Vol. 3.—**ARATRA PENTELICI.** Six Lectures on the Elements of Sculpture, given before the University of Oxford. By John Ruskin. 1 vol. 8vo.........................$4 00
Large paper... 4 50

RUSKIN—COMPLETE WORKS.

THE COMPLETE WORKS OF JOHN RUSKIN. 27 vols., extra cloth, in a box..$40 00
Ditto 27 vols., extra cloth. *Plates*... 48 00
Ditto Bound in 17 vols., half calf. do... 70 00

SEAMANSHIP.

ALSTON. **SEAMANSHIP.** New edition, revised and enlarged, by Commander R. H. Harris, R. N. With a Treatise on Nautical Surveying by Staff-Commander May. Also, Instructions for Officers of the Merchant Service by W. H. Rosser. With two hundred illustrations. 1 vol. 12mo, cloth, price.......$5 00

SHIP-BUILDING, &c.

BOURNE. **A TREATISE ON THE SCREW PROPELLER, SCREW VESSELS, AND SCREW ENGINES,** as adapted for Purposes of Peace and War. Illustrated by numerous woodcuts and engravings. By John Bourne. New edition. 1867. 1 vol. 4to, cloth, $18.00; half russia.................$24 00

WATTS. **RANKINE (W. J. M.) AND OTHERS.** Ship-Building, Theoretical and Practical, consisting of the Hydraulics of Ship-Building, or Buoyancy, Stability, Speed and Design—The Geometry of Ship-Building, or Modelling, Drawing, and Laying Off—Strength of Materials as applied to Ship-Building —Practical Ship-Building—Masts, Sails, and Rigging—Marine Steam Engineering—Ship-Building for Purposes of War. By Isaac Watts, C.B., W. J. M. Rankine, C.B., Frederick K. Barnes, James Robert Napier, etc. Illustrated with numerous fine engravings and woodcuts. Complete in 30 numbers, boards, $35.00; 1 vol. folio, cloth, $37.50; half russia, $40 00

WILSON (T. D.) **SHIP-BUILDING, THEORETICAL AND PRACTICAL.** In Five Divisions.—Division I. Naval Architecture. II. Laying Down and Taking off Ships. III. Ship-Building. IV. Masts and Spar Making. V. Vocabulary of Terms used—intended as a Text-Book and for Practical Use in Public and Private Ship-Yards. By Theo. D. Wilson, Assistant Naval Constructor, U. S. Navy; Instructor of Naval Construction, U. S. Naval Academy; Member of the Institution of Naval Architects, England. With numerous plates, lithographic and wood. 1 vol. 8vo. Shortly.......................

SOAP.

MORFIT. **A PRACTICAL TREATISE ON THE MANUFACTURE OF SOAPS.** With numerous wood-cuts and elaborate working drawings. By Campbell Morfit, M.D., F.C.S. 1 vol. 8vo...$20 00

STEAM ENGINE.

TROWBRIDGE. **TABLES, WITH EXPLANATIONS, OF THE NON-CONDENSING STATIONERY STEAM ENGINE,** and of High-Pressure Steam Boilers. By Prof. W. P. Trowbridge, of Yale College Scientific School. 1 vol. 4to, plates...$3 50

THE LATHE, **AND ITS USES, ETC.** On Instructions in the Art of Turning Wood and Metal. Including a description of the most modern appliances for the ornamentation of plane and curved surfaces. With a description, also, of an entirely novel form of Lathe for Eccentric and Rose Engine Turning, a Lathe and Turning Machine combined, and other valuable matter relating to the Art. 1 vol. 8vo, copiously illustrated, cloth..........$7 00

" **SUPPLEMENT AND INDEX TO SAME.** Paper...$0 90

VENTILATION.

LEEDS (L. W.). **A TREATISE ON VENTILATION.** Comprising Seven Lectures delivered before the Franklin Institute, showing the great want of improved methods of Ventilation in our buildings, giving the chemical and physiological process of respiration, comparing the effects of the various methods of heating and lighting upon the ventilation, &c. Illustrated by many plans of all classes of public and private buildings, showing their present defects, and the best means of improving them. By Lewis W. Leeds. 1 vol. 8vo, with numerous wood-cuts and colored plates. Cloth.........$2 50

"It ought to be in the hands of every family in the country."—*Technologist.*

"Nothing could be clearer than the author's exposition of the principles of the principles and practice of both good and bad ventilation."—*Van Nostrand's Engineering Magazine.*

"The work is every way worthy of the widest circulation."—*Scientific American.*

REID. **VENTILATION IN AMERICAN DWELLINGS.** With a series of diagrams presenting examples in different classes of habitations. By David Boswell Reid, M.D. To which is added an introductory outline of the progress of improvement in ventilation. By Elisha Harris, M.D. 1 vol. 12mo, $1 50

WEIGHTS, MEASURES, AND COINS.

TABLES OF WEIGHTS, MEASURES, COINS, &c., OF THE UNITED STATES AND ENGLAND, with their Equivalents in the French Decimal System. Arranged by T. T. Egleston, Professor of Mineralogy, School of Mines, Columbia College. 1 vol. 18mo....................$0 75

"It is a most useful work for all chemists and others who have occasion to make the conversions from one system to another."—*American Chemist.*

"Every mechanic should have these tables at hand."—*American Horological Journal.*

J. W. & SON are Agents for and keep in stock

SAMUEL BAGSTER & SONS' PUBLICATIONS,

LONDON TRACT SOCIETY PUBLICATIONS,

COLLINS' SONS & CO.'S BIBLES,

MURRAY'S TRAVELLER'S GUIDES,

WEALE'S SCIENTIFIC SERIES.

Full Catalogues gratis on application.

J. W. & SON import to order, for the TRADE AND PUBLIC,

BOOKS, PERIODICALS, &c.,

FROM

ENGLAND, FRANCE, AND GERMANY.

*** JOHN WILEY & SON'S Complete Classified Catalogue of the most valuable and latest scientific publications supplied gratis to order.

www.ingramcontent.com/pod-product-compliance
Lightning Source LLC
Chambersburg PA
CBHW020906210326
41598CB00018B/1791

* 9 7 8 3 3 3 7 0 5 7 3 4 3 *